Stochastic and Chaotic Oscillations

Mathematics and Its Applications (*Soviet Series*)

Managing Editor:

M. HAZEWINKEL
Centre for Mathematics and Computer Science, Amsterdam, The Netherlands

Editorial Board:

A. A. KIRILLOV, *MGU, Moscow, U.S.S.R.*
Yu. I. MANIN, *Steklov Institute of Mathematics, Moscow, U.S.S.R.*
N. N. MOISEEV, *Computing Centre, Academy of Sciences, Moscow, U.S.S.R.*
S. P. NOVIKOV, *Landau Institute of Theoretical Physics, Moscow, U.S.S.R.*
Yu. A. ROZANOV, *Steklov Institute of Mathematics, Moscow, U.S.S.R.*

Volume 77

Stochastic and Chaotic Oscillations

by

Yu. I. Neimark
*Faculty of Computational Mathematics,
Gorky University, Gorky, Russia*

and

P. S. Landa
*Faculty of Physics,
Moscow University, Moscow, Russia*

SPRINGER SCIENCE+BUSINESS MEDIA, B.V.

ISBN 978-94-010-5146-0   ISBN 978-94-011-2596-3 (eBook)
DOI 10.1007/978-94-011-2596-3

Printed on acid-free paper

This is the translation of the original work
Stochastic and Chaotic Oscillations
Published by Nauka, Moscow, © 1987

All Rights Reserved
© 1992 Springer Science+Business Media Dordrecht
Originally published by Kluwer Academic Publishers in 1992
Softcover reprint of the hardcover 1st edition 1992
No part of the material protected by this copyright notice may be reproduced or
utilized in any form or by any means, electronic or mechanical,
including photocopying, recording or by any information storage and
retrieval system, without written permission from the copyright owner.

SERIES EDITOR'S PREFACE

'Et moi, ..., si j'avait su comment en revenir, je
n'y serais point allé.'
      Jules Verne

The series is divergent; therefore we may be
able to do something with it.
      O. Heaviside

One service mathematics has rendered the human race. It has put common sense back where it belongs, on the topmost shelf next to the dusty canister labelled 'discarded nonsense'.
      Eric T. Bell

Mathematics is a tool for thought. A highly necessary tool in a world where both feedback and nonlinearities abound. Similarly, all kinds of parts of mathematics serve as tools for other parts and for other sciences.

Applying a simple rewriting rule to the quote on the right above one finds such statements as: 'One service topology has rendered mathematical physics ...'; 'One service logic has rendered computer science ...'; 'One service category theory has rendered mathematics ...'. All arguably true. And all statements obtainable this way form part of the raison d'être of this series.

This series, *Mathematics and Its Applications*, started in 1977. Now that over one hundred volumes have appeared it seems opportune to reexamine its scope. At the time I wrote

> "Growing specialization and diversification have brought a host of monographs and textbooks on increasingly specialized topics. However, the 'tree' of knowledge of mathematics and related fields does not grow only by putting forth new branches. It also happens, quite often in fact, that branches which were thought to be completely disparate are suddenly seen to be related. Further, the kind and level of sophistication of mathematics applied in various sciences has changed drastically in recent years: measure theory is used (non-trivially) in regional and theoretical economics; algebraic geometry interacts with physics; the Minkowsky lemma, coding theory and the structure of water meet one another in packing and covering theory; quantum fields, crystal defects and mathematical programming profit from homotopy theory; Lie algebras are relevant to filtering; and prediction and electrical engineering can use Stein spaces. And in addition to this there are such new emerging subdisciplines as 'experimental mathematics', 'CFD', 'completely integrable systems', 'chaos, synergetics and large-scale order', which are almost impossible to fit into the existing classification schemes. They draw upon widely different sections of mathematics."

By and large, all this still applies today. It is still true that at first sight mathematics seems rather fragmented and that to find, see, and exploit the deeper underlying interrelations more effort is needed and so are books that can help mathematicians and scientists do so. Accordingly MIA will continue to try to make such books available.

If anything, the description I gave in 1977 is now an understatement. To the examples of interaction areas one should add string theory where Riemann surfaces, algebraic geometry, modular functions, knots, quantum field theory, Kac-Moody algebras, monstrous moonshine (and more) all come together. And to the examples of things which can be usefully applied let me add the topic 'finite geometry'; a combination of words which sounds like it might not even exist, let alone be applicable. And yet it is being applied: to statistics via designs, to radar/sonar detection arrays (via finite projective planes), and to bus connections of VLSI chips (via difference sets). There seems to be no part of (so-called pure) mathematics that is not in immediate danger of being applied. And, accordingly, the applied mathematician needs to be aware of much more. Besides analysis and numerics, the traditional workhorses, he may need all kinds of combinatorics, algebra, probability, and so on.

In addition, the applied scientist needs to cope increasingly with the nonlinear world and the extra mathematical sophistication that this requires. For that is where the rewards are. Linear models are honest and a bit sad and depressing: proportional efforts and results. It is in the nonlinear world that infinitesimal inputs may result in macroscopic outputs (or vice versa). To appreciate what I am hinting at: if electronics were linear we would have no fun with transistors and computers; we would have no TV; in fact you would not be reading these lines.

There is also no safety in ignoring such outlandish things as nonstandard analysis, superspace and anticommuting integration, p-adic and ultrametric space. All three have applications in both electrical engineering and physics. Once, complex numbers were equally outlandish, but they frequently proved the shortest path between 'real' results. Similarly, the first two topics named have already provided a number of 'wormhole' paths. There is no telling where all this is leading - fortunately.

Thus the original scope of the series, which for various (sound) reasons now comprises five subseries: white (Japan), yellow (China), red (USSR), blue (Eastern Europe), and green (everything else), still applies. It has been enlarged a bit to include books treating of the tools from one subdiscipline which are used in others. Thus the series still aims at books dealing with:

- a central concept which plays an important role in several different mathematical and/or scientific specialization areas;
- new applications of the results and ideas from one area of scientific endeavour into another;
- influences which the results, problems and concepts of one field of enquiry have, and have had, on the development of another.

The discovery of stochastic and chaotic motion in (quite simple) but nonlinear dynamical systems has been one of the greatest discoveries in science in recent years. Even more important is that there is order and systematics to be found in this chaos. It is a large field by now and next to a great many surveys and proceedings also a number of monographs have already appeared.

The present volume is a most welcome addition to the monographic literature on irregular oscillations. Its main aim is to present engineers, physicists, and mathematicians with an array of 'real life' dynamical systems which exhibit chaotic behaviour. It does this systematically and mathematically and concentrates on dissipative systems.

Both authors have contributed significantly and they are well aware (witness a bibliography of some 700 items) of what is going on, notably in the (former) Soviet Union.

All in all this is a volume that will be valuable and accessible to many scientists, ranging from mathematicians to biologists and it is a pleasure to welcome it in this series.

The shortest path between two truths in the real domain passes through the complex domain.
J. Hadamard

La physique ne nous donne pas seulement l'occasion de résoudre des problèmes ... elle nous fait pressentir la solution.
H. Poincaré

Never lend books, for no one ever returns them; the only books I have in my library are books that other folk have lent me.
Anatole France

The function of an expert is not to be more right than other people, but to be wrong for more sophisticated reasons.
David Butler

Bussum, 9 February 1992

Michiel Hazewinkel

CONTENTS

SERIES EDITOR'S PREFACE   v

PREFACE   xi

CHAPTER 1
MATHEMATICAL MODELS OF DETERMINISTIC DISCRETE AND CONTINUOUS DYNAMICAL SYSTEMS   1

1. Simplest typical models of discrete dynamical systems   2
2. Chaotic motions of deterministic dynamical systems   19
3. Simplest mathematical models of continuous dynamical systems   25
4. Discrete version of continuous dynamical systems   31
5. Models for locally active continuous media   37

CHAPTER 2
ORDER AND CHAOS AS TWO GENERAL BASIC TRENDS IN THE EVOLUTION OF DYNAMICAL SYSTEMS   40

1. Order and chaos: stability and instability   41
2. Time synchronization phenomena   49
3. Spatial and temporal order and chaos   52

CHAPTER 3
STOCHASTICITY TRANSFORMERS, AMPLIFIERS AND GENERATORS   57

1. Stochasticity transformers   57
2. Stochasticity amplifiers   62
3. Stochasticity generators   70
4. Is the stochasticity of stochastic and chaotic motions of deterministic dynamical systems real ?   78

CHAPTER 4
BRIEF SURVEY OF STUDIES RELATED TO THE APPEARANCE OF THE PROBLEM OF CHAOTIC AND STOCHASTIC MOTIONS AND TO TURBULENCE THEORY   81

CHAPTER 5
LOCAL PHASE PORTRAITS OF THE SIMPLEST STEADY-STATE MOTIONS AND THEIR BIFURCATIONS   97
1. Equilibrium states   97
2. Bifurcations of equilibrium states   103
3. Phase portraits in the vicinity of periodic motions   115
4. Bifurcations of periodic motions   116
5. Toroidal integral manifolds   127

CHAPTER 6
STOCHASTIC AND CHAOTIC ATTRACTORS   132
1. Auxiliary mappings and sequences of point mappings   134
2. Transition from the "negative" to the "positive" and investigation of typical appearance of chaos   145
3. Conditions for the appearance of chaotic and stochastic attractors   173

CHAPTER 7
BIFURCATIONS AND ROUTES TO CHAOS AND STOCHASTICITY   177
1. General description of the tree of possible bifurcations   179
2. Series of bifurcations   183
3. Bifurcations and the stochastic attractor in a Lorenz system   201
4. Bifurcations and the phase portrait of parametrically excited oscillator or rotator   215
5. On the appearance of chaos and stochasticity in dissipative dynamical systems   231

CHAPTER 8
QUANTITATIVE CHARACTERISTICS OF STOCHASTIC AND CHAOTIC MOTIONS. SOME UNIVERSAL PROPERTIES IN ORDER-CHAOS AND INVERSE TRANSITIONS — 241

1. Statistical characteristics — 241
2. Lyapunov exponents. Dimension and entropy of a stochastic attractor — 252
3. Synchronization threshold as a quantitative characteristic of chaotic motions — 264
4. Certain universal laws in order-chaos transitions, and analogy with phase transitions — 266

CHAPTER 9
EXAMPLES OF MECHANICAL, PHYSICAL, CHEMICAL, AND BIOLOGICAL SYSTEMS WITH CHAOTIC AND STOCHASTIC MOTIONS — 293

1. Non-linear impact negative-friction oscillator and other systems with discontinuous characteristics — 293
2. Tunel-diode generators — 295
3. Non-linear oscillators with periodic external force — 298
4. Lorenz equations and other systems of order three — 322
5. Action of a harmonic external force on periodic and chaotic oscillation generators — 351
6. Interaction of oscillation generators of various kinds — 368
7. Certain discrete models of turbulence — 373
8. Examples of models for chemical kinetics — 383
9. Systems with delay and other continuous systems — 401
10. Stochasticity in quantum systems — 430

BIBLIOGRAPHY — 443

INDEX — 499

Now we wonder if structure was originally made
Or whether the turbulence did come first.
Could it be there's a scale of cascade
And another for the nasty, tricky burst ?
[468]

PREFACE

The discovery of irregular oscillations in deterministic dynamical systems of various nature, which arise in physics, mechanics, technology, chemistry, biology or economics, has become one of the greatest sensations in science in recent years. It has initiated an enormous amount of theoretical and experimental studies. Quite many surveys and collections of papers appeared (e.g. [120, 122, 141, 143, 219, 257, 293, 313, 329, 339, 341, 413, 433, 484, 499, 525, 598, 665]), as well as several monographs [23, 157, 237, 336, 483, 655, 694, 645, 696] and popular science works [121, 168, 328, 626]. The corresponding chapters were included in several textbooks [100, 314, 317]. It is characteristic that references concerning the theory of oscillations and waves can be divided into general and specialized. The former include [15, 92, 96, 100, 111, 132, 243-245, 252, 269, 307, 314, 346, 352, 359, 360], and [157, 237] and are mostly devoted to Hamiltonian systems; to the more specialized, ones belong those mostly considering systems of concrete physical nature [89, 90, 102, 103, 107, 116, 172, 174, 193, 194, 216, 217, 251, 318, 337, 340, 345, 347-350, 358]. The book which is now offered to the reader is, in the authors' opinion, of general character, and devoted to stochastic and chaotic oscillations in dissipative systems. It is an attempt at summarizing and expressing their understanding of the problem.

The establishment and recognition of the fact that in comparatively simple systems without visible random sources complicated unpredictable motions are possible, subsequently called stochastic and chaotic, is an indisputable success first of all of the theory of oscillations. It was a break-through in the understanding of the temporal and spatial evolution of dynamical systems, as was the discovery of regular self-excited oscillations. It is therefore natural to discuss stochastic and chaotic motions of deterministic dynamical systems from the standpoint of oscillation theory, using its general ideas and methods of investigation. The most efficient method for investigating and chaotic motions of dynamical systems is that of point mappings [13, 15, 88, 100,

124, 259, 261, 267, 269, 287, 311, 312], which are widely used in the book. Since the motions in question are random, statistical methods are also important, and are treated. As a rule, chaotic and stochastic motions of concrete systems cannot be investigated without involving modern computer science. Algorithmic methods and software play a part which cannot be regarded as least important (e.g., see [283]); however, we do not discuss them in this book.

Of special interest in the study of stochastic and chaotic oscillations is in combining the efforts of researchers in various branches of science: mathematicians, mechanics specialists, physicists, chemists, or biologists.

Chapts. 1-7 were written by Yu. I. Neimark; Sect. 1, Chapt. 8, by P. S. Landa and Yu. I. Neimark; Sects. 2-4, Chapt. 8, and Chapt. 9, except Sect. 10, by P. S. Landa; and Sect. 10, Chapt. 9, on stochastic motions of quantum systems, by B. A. Grishanin.

CHAPTER 1

MATHEMATICAL MODELS OF DETERMINISTIC DISCRETE AND CONTINUOUS DYNAMICAL SYSTEMS

This chapter is introductory in nature, and describes the simplest mathematical models of deterministic dynamical systems as differential equations in phase space, thus specifying its variation with time. In the course of treating motions of dynamical systems and oscillatory phenomena phasewise, chaotic motions of deterministic dynamical systems crop up for the first time.

Mathematical models play a twofold role in oscillation theory: in idealizing real dynamical systems and in mathematically describing various oscillatory phenomena such as harmonic, divergent, damped, self-excited, stochastic, chaotic, and forced oscillations with flat and dropping characteristic, resonance, parametric excitation, a number of wave phenomena, traveling, standing or shock waves, many types of wave interaction, etc.

The world of dynamical models has become really large, and now embraces, besides astronomy, mechanical engineering and physics, also less traditional areas of application, such as, e.g., chemistry [113, 152, 199, 309, 356], biology [112, 199, 319], and economics [135]. The great variety of models can neither be described nor classified reasonably well. Nevertheless, the basic simplest and most characteristic models can be distinguished from this infinite set. The complete list of even the simplest typical models turns out to be sufficiently long, and will be different if made by a different specialist. The list of typical models discussed in the sequel make up the common part of all such lists; it includes models which are necessary and familiar to everyone studying oscillations and waves.

In addition to these simplest typical models, we here give comparatively new, and, in a sense, more complicated models of locally excitable media, describing continuous dynamical systems, and to this purpose we introduce the concepts of time generator and spatial former.

1. Simplest typical models of discrete dynamical systems

Harmonic oscillator. A load of mass m, attached to a spring of rigidity k (mechanical model), and an electrical circuit of capacitance C and self-induction L (electrical model) are ideal models of a harmonic oscillator (Fig. 1.1). Energy dissipation is not taken into account in both models, because it is assumed that there is no energy loss in the spring under deformation, no resistance of the medium to load motion, no resistance in the electrical circuit wires, no charge dissipation in the capacitor, and no electromagnetic radiation. Energy is not dissipated; it is conserved at a constant value.

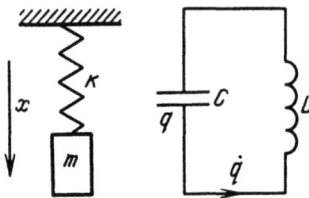

Fig. 1.1

The position x (resp. the charge q) is described in the ideal model by the second-order linear differential equation

$$\ddot{x} + \omega^2 x = 0, \qquad (1.1)$$

where the quantity ω is $\sqrt{k/m}$ (resp. $\sqrt{1/(LC)}$). If we specify the position $x(t_0) = x_0$ and velocity $\dot{x}(t_0) = x_1$ at the initial moment $t = t_0$, then we can uniquely determine the solution of (1.1)

$$x(t) = x_0 \cos \omega(t - t_0) + \frac{x_1}{\omega} \sin \omega(t - t_0), \qquad (1.2)$$

and thereby $x, \dot{x}$ at any subsequent moment $t \geq t_0$.

According to (1.2), variation of x with time for any initial conditions is represented by harmonic oscillations with the same frequency ω but with different amplitudes and phases.

Individual motions (1.2) are represented as closed phase curves (ellipses) in the

Mathematical models of deterministic discrete and continuous dynamical systems 3

phase space, i.e., the $(x, \dot{x})$-plane (the phase plane), of the harmonic oscillator (Fig. 1.2). The point $x = \dot{x} = 0$ is associated with the equilibrium position. The diagram of closed phase curves and the equilibrium point form the phase portrait.

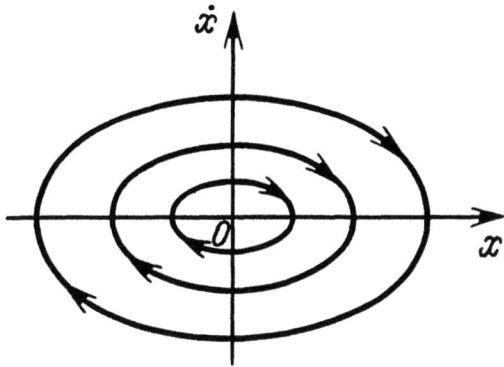

Fig. 1.2

That the phase trajectories are closed reflects the fact of energy conservation, expressed by the relation

$$\dot{x}^2 + \omega^2 x = \text{const} \tag{1.3}$$

which is derived by integrating (1.1).

We can arrive at (1.1) not only by considering the oscillation of a load attached to a spring (or the charge of a capacitor in an oscillatory circuit) but, to a certain approximation, also by considering small oscillations of a physical pendulum, building constructions, machines and mechanisms, atoms, molecules, and many other systems.

Linear oscillator. If we take energy dissipation in mechanical and electrical models into account (Fig. 1.1) as an additional term for (1.1), linear in velocity of the load (or in the current in the circuit), then we arrive at a somewhat more general mathematical model, the differential equation

$$\ddot{x} + 2\delta\dot{x} + \omega^2 x = 0 \tag{1.4}$$

called a *linear oscillator*.

Its solution can be represented as

$$x = e^{-\delta(t-t_0)} \left[x_0 \cos\Omega(t-t_0) + \frac{x_1 + \delta x_0}{\Omega} \sin\Omega(t-t_0) \right] \tag{1.5}$$

for $\omega^2 - \delta^2 = \Omega^2 > 0$, and as

$$x = \frac{\lambda_2 x_0 - x_1}{\lambda_2 - \lambda_1} e^{\lambda_1 (t-t_0)} + \frac{\lambda_1 x_0 - x_1}{\lambda_1 - \lambda_2} e^{\lambda_2 (t-t_0)} \tag{1.6}$$

for $\omega^2 - \delta^2 < 0$, where $\lambda_{1,2} = -\delta \pm \sqrt{\delta^2 - \omega^2}$ and x_0, x_1 still denote the initial values of x, $\dot{x}$ when $t = t_0$.

The term $2\delta \dot{x}$ in (1.4) takes into account energy dissipation. It follows that

$$\frac{d}{dt}\left(\frac{\dot{x}^2}{2} + \frac{\omega^2}{2} x^2\right) = -2\delta \dot{x}^2 . \tag{1.7}$$

The quantity between brackets is the total energy of the oscillator, up to a constant factor. The coefficient δ is positive; therefore, the total energy decreases with time until $\dot{x}$ vanishes. The linear oscillator then describes damped oscillations.

However, δ can also be negative. The linear oscillator then describes divergent oscillations, or non-oscillatory deviation from the equilibrium position. Fig. 1.1 represents models with $\delta < 0$, associated with so-called *negative friction* and *negative resistance*. That the parameter δ is negative, and that therefore the system builds up, is due to certain energy sources, i.e., some pumping mechanisms to compensate for falling portions of the friction or current-voltage curves; e.g., of a tunel diode, and also energy pumping mechanisms with valve-like action (pumping due to directed feedbacks making the moving oscillator exert forces amplifying the motion).

Thus, a linear oscillator describes both damping and increase of the initial perturbation, which are both of oscillatory and non-periodic nature. In the former case, the equilibrium associated with $x = \dot{x} = 0$ is stable; in the latter, unstable. The phase portraits are in Fig. 1.3 for $\omega^2 > \delta^2$.

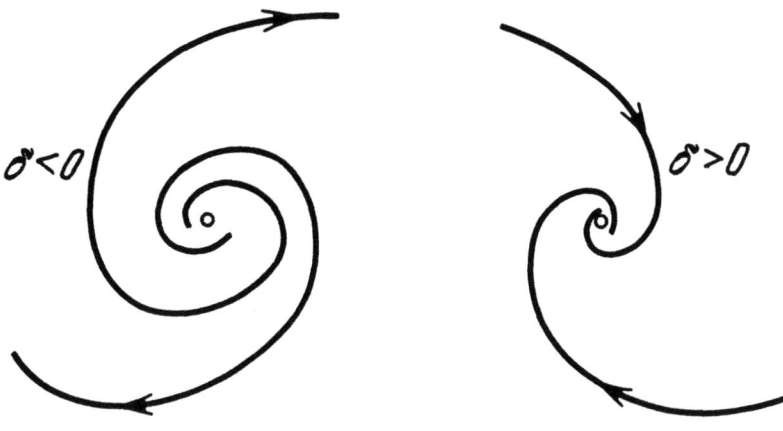

Fig. 1.3

Conservative oscillator. A linear oscillator generalizes a harmonic oscillator, which takes into account energy dissipation and pumping, as well as, in general, any process leading to exponential damping or amplification of the original perturbation. Another generalization is the conservative oscillator

$$\ddot{x} + Q(x) = 0 \qquad (1.8)$$

which takes into account the non-linear dependence elastic force of the spring on the spring's extension in the mechanical model (Fig. 1.1) and the dependence of the capacitance on the capacitor charge in an electrical model. Here, the phase trajectories are closed, since

$$\frac{\dot{x}^2}{2} + V(x) = \text{const}, \qquad (1.9)$$

where $V(x) = \int_0^x Q(\xi)\, d\xi$ is the energy integral.

For $Q'_x > 0$ and $Q(0) = 0$, the phase portrait of a conservative oscillator is of the same shape as that of a harmonic one. Phase portraits can be more complicated when using a more general function $Q(x)$; e.g., those for $Q(x) = ax + bx^3$ in Fig. 1.4. In Huygens' times, the problem of oscillation isochronism arose much interest in the independence of the oscillation period from the amplitude. Equation (1.8) (resp. (1.1)) describes non-isochronal (resp. isochronal) harmonic oscillations. Accordingly, a conservative oscillator is sometimes said to be *non-isochronal*, or *anharmonic*.

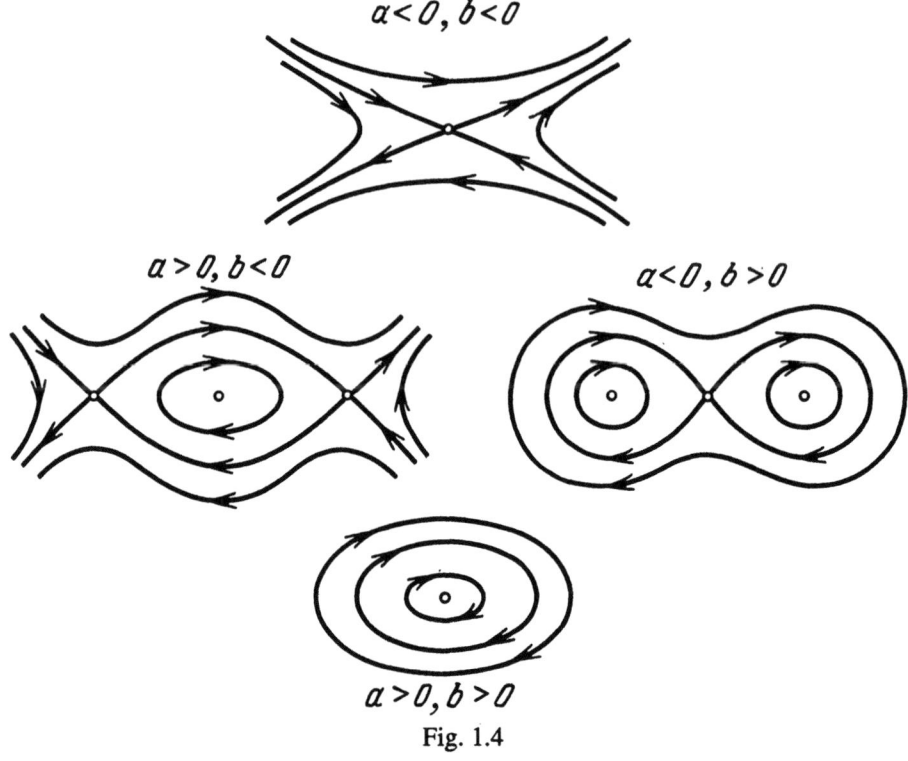

Fig. 1.4

The familiar example of (1.8) is the physical pendulum equation

$$\ddot{\varphi}+\frac{mgl}{J}\sin\varphi=0 \ . \qquad (1.10)$$

Characteristically, the function Q(φ) is periodic. It is also essential that the phase space is now a two-dimensional cylinder (Fig. 1.5), and not the (x, ẋ)-plane. The cylindricity of the phase space of the physical pendulum is not only a consequence of the periodicity of Q(φ) but also of cyclicity of the variable x (x being the angle variable φ). It may also happen that x is not the angle variable, and that the periodicity of Q(x) is related to a periodic structure of the field of forces. In Fig. 1.6a, the phase portrait of a physical pendulum is shown; in Fig. 1.6b, that of a material point moving in a periodically changing potential field of the form $Q(x) = \omega^2 \sin x$. It is clear that, physically, these are two different systems. To distinguish these, we will refer to physical pendulums as *conservative* rotors, meaning systems with cylindrical phase space, as described by an equation of form (1.8), whereas the similarly described conservative oscillator has as phase space a 2-plane. In this connection, we should stress that the mathematical model of a dynamical system is determined by its phase space and the operator of change on its phase state [269, 286]. The operator is specified for any state, i.e., any point of the phase space. Thus, it is defined on the whole phase space. In the conservative rotor case, the function Q(x) is given on the circle; in the oscillator case, on the straight line. The operator alone does not determine a dynamical system; it is also necessary to indicate its domain of definition, i.e., the state space.

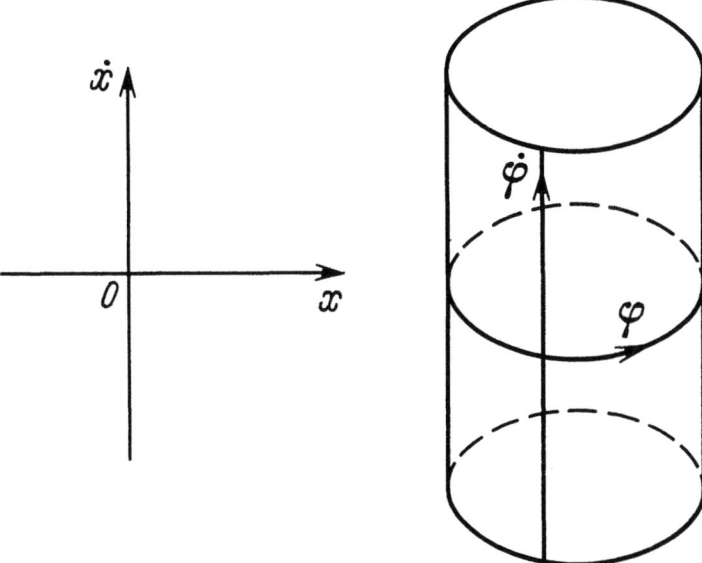

Fig. 1.5

The mathematical representation of all motions of a dynamical system is the phase portrait, not only being a geometric representation of individual motions, equilibrium states, periodic, chaotic, and stochastic motions, but also determining the "logic" of the behaviour of the dynamical system and its dependence on the parameters.

Note that, for the physical pendulum and a material point in a periodically changing field, the periodicity of Q(x) is due to absolutely different circumstances. In the former case, it is a consequence of the cyclicity of the angle variable determining the position on the circle; in the latter, of the periodicity of the field of forces.

Let us now clarify the phase portrait of the physical pendulum in Fig. 1.6a. It follows from the differential equation (1.10) that the energy integral

$$\frac{1}{2}\dot{\varphi}^2 - \omega^2 \cos\varphi = h \qquad (\omega^2 = mgl/J) \qquad (1.11)$$

is valid. The correlation between φ and $\dot{\varphi}$, determined by relation (1.11) with a certain constant h, is geometrically representable by a curve in phase $(\varphi,\dot\varphi)$-cylinder. Considering all possible h, we obtain on the cylinder the family of phase curves (Fig. 1.6a), the phase portrait of all possible motions of the physical pendulum. The phase curves degenerating into the points O and O' are associated with the lower and upper equilibrium positions of the pendulum, respectively. Those encircling O are associated with all possible motions of the periodic oscillator; those enclosing the cylinder, with all possible motions of the periodic rotator. It can be seen from the phase portrait that the different motions of the pendulum turn into each other as the energy (parameter) h varies smoothly. The minimum value $h = -\omega^2 l$ is associated with the lower equilibrium position O. As h increases, oscillations of increasing amplitude (closed phase trajectories encircling O) emerge; the value $h = \omega^2 l$ is associated with three motions: the upper equilibrium position O' and two motions S which tend to the upper equilibrium position. As h increases further, motions of the rotator in both senses emerge.

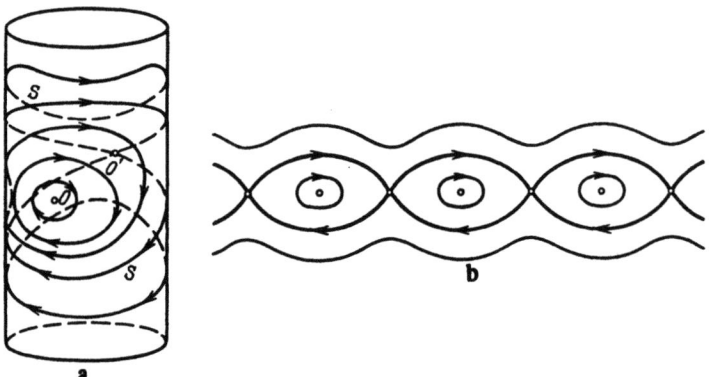

Fig. 1.6

It can immediately be seen from the phase portrait of the physical pendulum that the lower equilibrium position is stable, and the upper unstable; this follows from the pendulum being near to the lower equilibrium position for values close to $-\omega^2 l$, and far from the upper equilibrium position for those close to $\omega^2 l$.

We now consider changes due to dissipation in motion, and the phase portrait of a conservative oscillator or rotator. By this, we mean the system described by the equation

Mathematical models of deterministic discrete and continuous dynamical systems

$$\ddot{x} + 2\delta \dot{x} + Q(x) = 0 \quad . \tag{1.12}$$

Multiplying it by $\dot{x}$, we obtain

$$\frac{d}{dt}\left[\frac{1}{2}\dot{x}^2 + V(x)\right] = -2\delta\dot{x}^2 , \tag{1.13}$$

where $V(x)$ is, as before, determined by formula (1.9). Let the phase portrait of (1.12) be known for $\delta = 0$, and let it consist of the level curves of the function $W(x, \dot{x}) = \frac{1}{2}\dot{x}^2 + V(x)$. By (1.13), the phase point $(x, \dot{x})$ moves for $\delta > 0$ in such a way that it goes from level curves with a greater value to those with a lesser value of $W(x, \dot{x})$.

This means that, with dissipation, the phase portraits of conservative oscillators take the form shown in Fig. 1.7, whereas those of the physical pendulum (or conservative rotator) shown in Fig. 1.8.

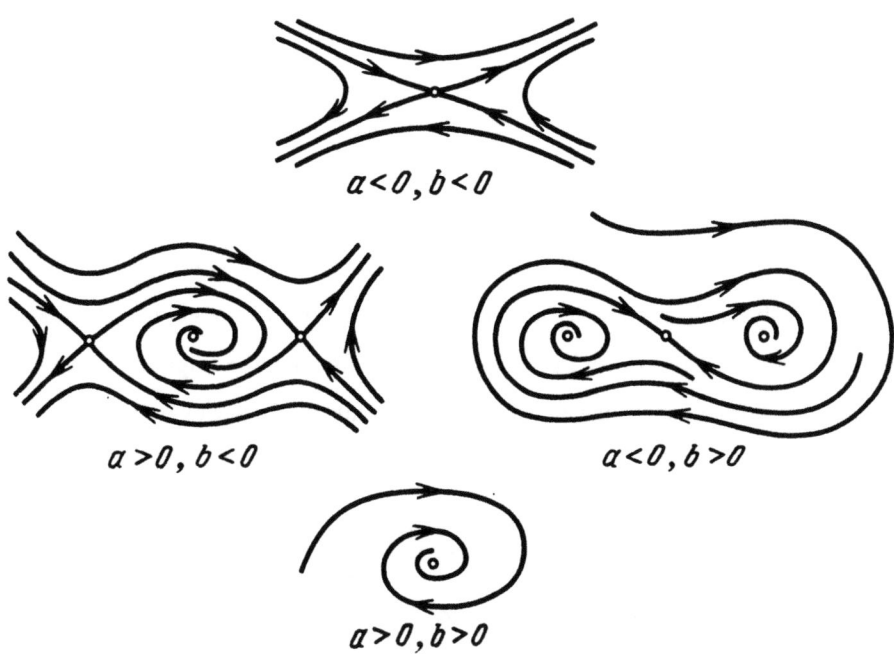

Fig. 1.7

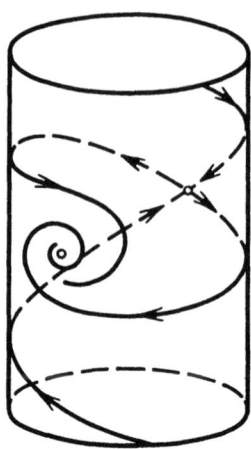

Fig. 1.8

Van der Pol oscillator. Differential equations (1.1), (1.8) describe periodic oscillations, while (1.4) describes exponentially increasing and damped oscillations. The typical model to describe the excitation processes and periodic (self-excited) oscillations is the van der Pol oscillator

$$\ddot{x} - 2\delta\dot{x}(1-\alpha x^2) + \omega^2 x = 0. \tag{1.14}$$

The differential equation is not integrable by quadratures, and has been studied by the usual numerical methods. For the most interesting case ($\delta > 0$, $\alpha > 0$), its phase portrait is represented in Fig. 1.9. It can be seen that it is a self-oscillatory system with unique globally stable periodic motion (or limit cycle). For small $\delta > 0$, the latter can be studied by the familiar van der Pol method, and we find that the amplitude of the self-excited oscillations is approximately equal to $2/\sqrt{\alpha}$, while the frequency is ω.

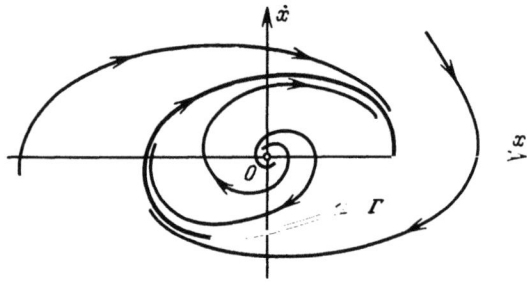

Fig. 1.9

Oscillator and rotator. A natural generalization of differential equations (1.1),

(1.4), (1.7), (1.8), (1.12) and (1.14) is

$$\ddot{x} + 2\delta(x,\dot{x})\dot{x} + Q(x) = 0. \tag{1.15}$$

When $\delta(x,\dot{x}) = 0$, this is a conservative oscillator; when $\delta(x,\dot{x}) =$ const and $Q(x) = \omega^2 x$, a linear oscillator; and, when $\delta(x,\dot{x}) = 2\delta(1-\alpha x^2)$, $Q(x) = \omega^2 x$, a van der Pol oscillator. If $\delta(x,\dot{x}) = 0$, $Q(x) = \omega^2 \sin x$ and the phase space is cylindric, this is a mathematical model of the physical pendulum, or a special case of a conservative rotator.

We call the dynamical system described by (1.15) on the phase 2-plane an *oscillator*, and the one described on the phase two-dimensional cylinder a *rotator*. Note that (1.15) is a general differential equation of second order, since any function $f(x, \dot{x})$ differentiable with respect to $\dot{x}$ can be written as

$$f(x,\dot{x}) = f(x,0) + [f(x,\dot{x}) - f(x,0)] = Q(x) + 2\delta(x,\dot{x})\dot{x}.$$

Non-autonomous oscillator and rotator. Oscillators and rotators (1.15) are autonomous dynamical systems, and do not explicitly depend on time. Non-autonomous oscillators and rotators are described by the equation

$$\ddot{x} + 2\delta(x,\dot{x},t)\dot{x} + Q(x,t) = 0; \tag{1.16}$$

particular cases are the familiar Duffing, Hill and Mathieu equations, and a linear oscillator under the action of an external harmonic force. A van der Pol oscillator with external harmonic action is of the same form.

A linear oscillator acted upon by an external harmonic force

$$\ddot{x} + 2\delta\dot{x} + \omega^2 x = F\cos\nu t, \tag{1.17}$$

and the Duffing equation

$$\ddot{x} + 2\delta\dot{x} + ax + bx^3 = F\cos\nu t \tag{1.18}$$

describe linear and non-linear resonance phenomena, respectively. Briefly, they are as follows. The general solution of (1.17) is representable as the sum of the general solution of the corresponding homogeneous equation, determined by formulas (1.5), (1.6), and a particular solution of the non-homogeneous equation (1.17), which can in complex form be written as

$$x = \frac{F}{\omega^2 - \nu^2 + 2\delta\nu i} e^{i\nu t} = A(\nu)e^{i\nu t}. \tag{1.19}$$

For $\delta > 0$, it is this particular solution that is steady-state after a sufficiently long time; it describes the steady-state forced oscillations of the linear oscillator, and their response. The modulus of the complex number $A(\nu)$ is the amplitude of forced oscillations; the argument, their phase. The resonance phenomenon lies in the infinite increase of the maximum amplitude of the forced oscillations as $\delta \to 0$ (Fig. 1.10). Meanwhile, the value of the frequency ν associated with the maximum amplitude tends to ω.

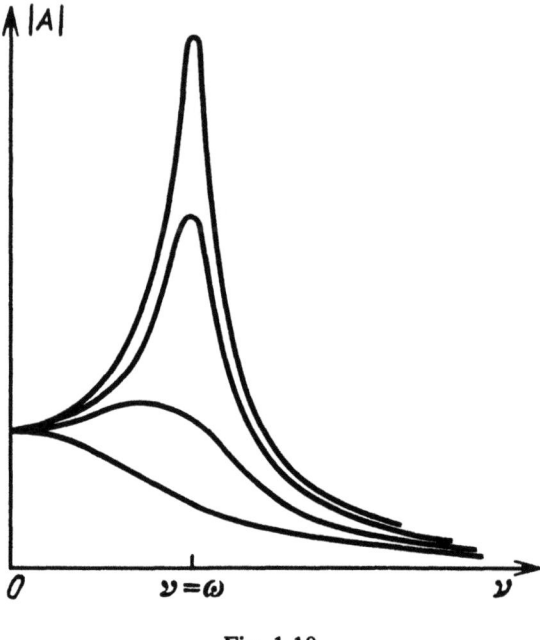

Fig. 1.10

The forced oscillations of a non-linear system described by the Duffing equation are not so simple to study. It has not been studied completely until now. Only the case of small dampings of δ and $a > 0$ can be studied without much difficulty. The response curves are of the form represented in Fig. 1.11, and differ from those for a linear oscillator (Fig. 1.10) in the slope of the "peak" and in non-uniqueness. The inclination is either to the left or right, depending on the sign of b in (1.18), which, along with the non-uniqueness, causes the familiar *hysteresis effect* of the amplitude of forced oscillations as the frequency of external force varies slowly. It consists in jumps of the amplitude, and in their occurrence for different ν as the frequency increases or decreases slowly and monotonically (Fig. 1.11).

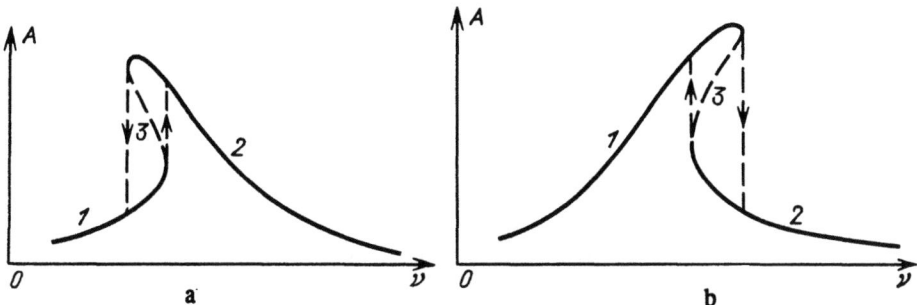

Fig. 1.11

The study of forced oscillations of the non-linear Duffing oscillator reveals another property, i.e., that resonance is possible not only at $v = \omega$ but also at $v = \omega/2$, $\omega/3$, ... ($\omega^2 = a$), the so-called *subharmonic resonances*.

Mathieu and Hill equations. They describe the phenomenon of linear parametric excitation and resonance of the harmonic oscillator. The Hill equation is

$$\ddot{x} + \left[\mu + \varphi(t)\right] x = 0, \tag{1.20}$$

where $\varphi(t)$ is an arbitrary periodic function. The Mathieu equation is associated with the particular case $\varphi(t) = v \sin t$ of (1.20). The regions of parametric excitation are unshaded in Fig. 1.12. For parameters μ and v in the unshaded regions of parametric resonance, the equilibrium state $x = \dot{x} = 0$ is unstable, a deviation from it has oscillatory nature and exponentially increasing amplitude. It is essential that, as can be seen in Fig. 1.12, such behaviour is possible with arbitrarily small amplitude v of parametric change of the oscillator

$$\ddot{x} + (\mu + v \sin t) x = 0 \tag{1.21}$$

stiffness if only frequency $\omega = \sqrt{\mu}$ is subject to the relation

$$2\omega = n \quad (n = 1, 2, 3, ...) \tag{1.22}$$

with frequency 1 of harmonic parametric action. If one of the conditions in (1.22) holds, parametric excitation and response consist just in this excitation under an arbitrarily small parametric action.

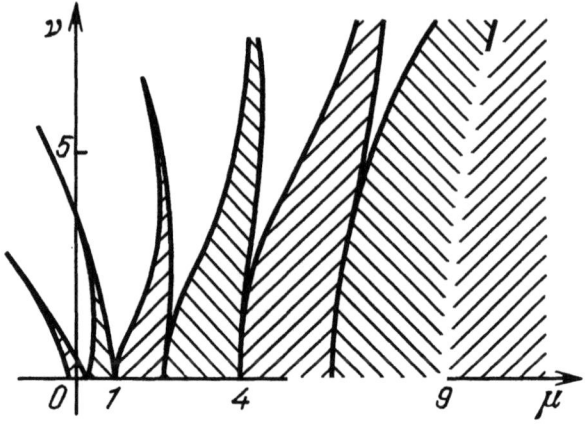

Fig. 1.12

The chaotic phenomena of non-linear response and motion for a non-linear rotator are reflected by the non-autonomous differential second-order equation

$$\ddot{x} + 2\delta \dot{x} + (\mu + \nu \sin t)\sin x = M \qquad (1.23)$$

which, in particular, describes the oscillations of an unbalanced rotor (physical pendulum) on a vibrating base with torque. Note that, also in the Duffing oscillator, chaotic forced oscillations are possible. Meanwhile, the changes in the variables x and $\dot{x}$ are chaotic and unpredictable in nature in both cases. The phase spaces of the van der Pol oscillator (1.18) and of the unbalanced rotor on a harmonically vibrating base are three-dimensional. If the phase spaces of these systems are cut by a plane, then the moving phase point leaves as trace points in the intersection with the plane. As the phase point moves periodically, there are several such intersection points, and the phase point passes through them periodically. If the motion is asymptotically periodic, there are infinitely many intersection points; however, the phase trajectory intersects the plane in several points which tend in the course of time to certain limiting positions. If the motion is chaotic, there are infinitely many such intersection points, and they fill certain sets, sometimes arranged in a very complicated way. An example is given by "clouds" of intersection points for equation (1.23); see Fig. 1.13.

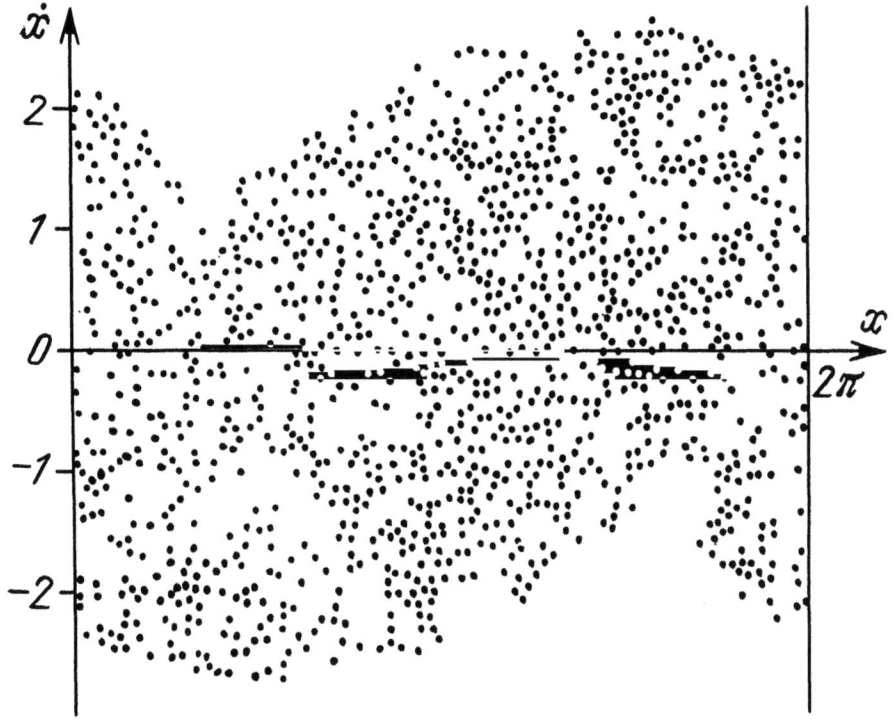

Fig. 1.13

Lorenz equations. These are [563]

$$\dot{x} = -\sigma(x-y), \quad \dot{y} = rx - y - xz, \quad \dot{z} = -bz + xy, \qquad (1.24)$$

and became widely known comparatively recently. They arise as the three-mode truncation of the 2-dimensional convection equations for parallel horizontal walls at constant but different temperatures (lower wall temperature being higher).

The phase space of such a system is three-dimensional Euclidian space. All the phase trajectories enter a bounded region, in which they can be linked together in a most intricate way. The efforts of many researchers, who made use of methods from the qualitative theory of differential equations and numerical experiments on modern computers, have shown that the complicated motions of the phase points in the Lorenz system are chaotic. One shape of phase trajectories associated with such a motion is given in Fig. 1.14. The picture was obtained on the oscilloscope display, making the projection of the phase point visible at equal time intervals.

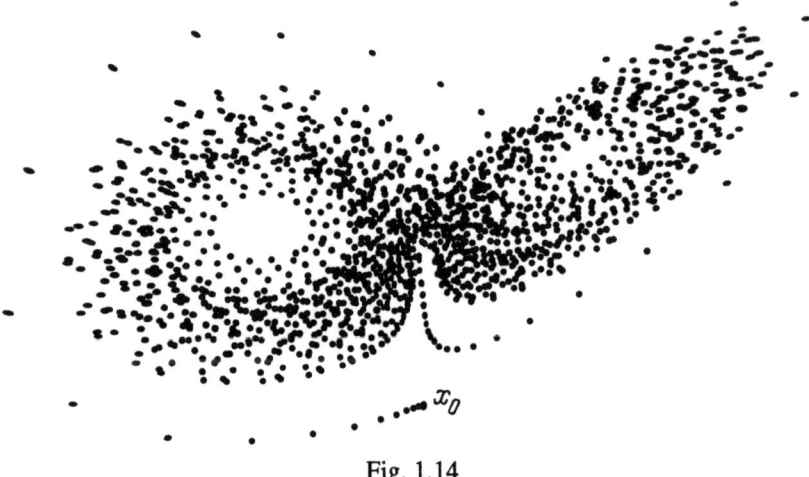

Fig. 1.14

Equations (1.24) can be reduced to the form

$$\ddot{\xi}+\varepsilon h\dot{\xi}+\xi^3+(q-1)\xi=0, \qquad \dot{q}=-\varepsilon aq+\varepsilon\beta\xi^2 \qquad (1.25)$$

$$\left(h=\frac{\sigma+1}{\sqrt{\sigma}}, \quad a=\frac{b}{\sqrt{\sigma}}, \quad \beta=\frac{2\sigma-b}{\sqrt{\sigma}}\right),$$

enabling us to treat them as a non-linear oscillator with non-linear inertial stiffness, after replacing the variables x, y, z and the time t by

$$\xi=\frac{\varepsilon x}{\sqrt{2\sigma}}, \quad q=\frac{\varepsilon^2}{\sigma}\left(\sigma z-\frac{1}{2}x^2\right), \quad \tau=\frac{\varepsilon t}{\sqrt{\sigma}},$$

where $\varepsilon=(r-1)^{-1/2}$ [11]. As $\varepsilon \to 0$, the stiffness ceases to be inertial, and the motion of any phase point satisfies the non-linear conservative oscillator equation

$$\ddot{\xi}+\xi^3+(q-1)\xi=0$$

for a certain constant value of the parameter q.

System of weakly interacting harmonic oscillators and a system with rapidly rotating phases. Until now, all models of discrete dynamical systems were at most three-dimensional. It is not always possible to confine ourselves to such simple models. Research on models of larger dimension with strong non-linearities has only started, and turns out to be sufficiently advanced for small non-linearities; certain typical models have already been formed. In this connection, we can first of all point to systems of weakly interacting oscillators and rotors, and to systems with so-called *rapidly rotating phases*.

A system of n weakly interacting harmonic oscillators can be written in the form

$$\ddot{x}_s + \omega_s^2 x_s = \varepsilon f_s(x_1, \ldots, x_n; \dot{x}_1, \ldots, \dot{x}_n) \quad (s = 1, 2, \ldots, n), \quad (1.26)$$

where ε is a small parameter. If $\varepsilon = 0$, it decomposes into non-interacting harmonic oscillators, and can be integrated easily, viz.,

$$x_s = A_s \cos(\omega_s t + \varphi_s). \quad (1.27)$$

For small ε, the solution is also representable as in (1.27); however, here, A_s and φ_s are not constant, but slowly varying functions of time. Their rates of change, i.e., $\dot{A}_s$ and $\dot{\varphi}_s$, are of order ε.

We now estimate how efficient the effect of oscillators on each other can be. Their interactions are carried out via the right-hand sides εf_s of equations (1.26). The functions f_s depend on $x_1, x_2, \ldots, x_n$ and $\dot{x}_1, \dot{x}_2, \ldots, \dot{x}_n$, while each of the variables x_s and $\dot{x}_s$ is, to a first approximation, a harmonic oscillation with frequency ω_s. Since f_s depends on $x_1, x_2, \ldots, x_n$ and $\dot{x}_1, \dot{x}_2, \ldots, \dot{x}_n$ non-linearly, all possible linear combinations of the form

$$m_1 \omega_1 + m_2 \omega_2 + \ldots + m_n \omega_n \quad (1.28)$$

with integer coefficients are involved in the spectral representation of f_s ($\omega_1, \omega_2, \ldots, \omega_n$ are the frequencies). Generally speaking, the greater the values $m_1, m_2, \ldots, m_n$, the less are the amplitudes of such harmonics; therefore, attention should be focussed on the first combination of frequencies. It is reasonable to assume that the frequencies close to $\omega_1, \omega_2, \ldots, \omega_n$ are the most important. "Close" means "different by a quantity of order ε". Since the parameter ε is arbitrarily small, strong effects for arbitrarily weak interactions can be expected only if the system has internal resonances, i.e., if relations of the form

$$\omega_s = m_1 \omega_1 + m_2 \omega_2 + \ldots + m_n \omega_n$$

with integer coefficients hold. Resonant interactions occur only if there exist integers $m_j \neq 0$ with $j \neq s$. It is only then that the jth oscillator strongly influences the sth.

Differential equations of the form

$$\dot{\varphi} = \omega(x) + \varepsilon \Phi(\varphi, x), \quad \dot{x} = \varepsilon X(\varphi, x) \quad (1.29)$$

are called *differential equations with rapidly rotating phases*. Here, φ is the vector of the angle variables $\varphi_1, \varphi_2, \ldots, \varphi_m$, x a vector of n-dimensional Euclidian space with components $x_1, x_2, \ldots, x_n$, ε a small parameter, and the functions Φ, X are periodic relative to $\varphi_1, \varphi_2, \ldots, \varphi_m$ with period 2π. The differential equations (1.26) can be reduced to the form (1.29) by putting $x_s = \rho_s \cos \varphi_s$. Meanwhile, m = n and

$$\omega(x) = (\omega_1, \omega_2, \ldots, \omega_n).$$

The system of differential equations

$$\ddot{x}_s + Q_s(x_s) = \varepsilon f_s(x_1, x_2, \ldots, x_n; \dot{x}_1, \dot{x}_2, \ldots, \dot{x}_n) \quad (1.30)$$

can be reduced to (1.29), which is more general than (1.26) and is associated to a system of n weakly interacting non-linear conservative oscillators, while

$$\omega(x) = (\omega_1(x), \omega_2(x), \ldots, \omega_n(x)).$$

Hamiltonian dynamical systems. Hamiltonian systems and those close to them are important in celestial mechanics and theoretical physics if dissipative effects are taken into account. A *Hamiltonian system* is a dynamical system whose equations of motion can by means of a unique Hamiltonian function H(q, p) be written in the form

$$\dot{q}_s = \frac{\partial H}{\partial p_s}, \quad \dot{p}_s = -\frac{\partial H}{\partial q_s} \quad (s = 1, 2, \ldots, n). \quad (1.31)$$

Hamiltonian systems have been studied in detail for a long time, both analytically and qualitatively. They include both very simple (totally integrable) and very complicated equations.

Difficulties arise even when n ≥ 2. It suffices to mention the famous restricted three body problem, the problem of the motion of a solid with a fixed point, etc.

An *integrable* Hamiltonian system with Hamiltonian function of the form

$$H(q, p) = \sum \frac{p_s^2}{2} + V(q_s) \quad (1.32)$$

is a set of non-interacting conservative oscillators or rotators, and coincides with (1.30) for $\varepsilon = 0$. Decomposition into n-dimensional integral tori with quasi-periodic or periodic coils is characteristic of the phase portrait. Equations (1.31) are then reduced to the form

$$\dot{q}_s^2 = h_s - V_s(q_s) \quad (s = 1, 2, \ldots, n).$$

Their solutions $q_s(t, h_s)$ are, under certain general assumptions, periodic functions of time with periods depending on the integration constants. The question of preserving this structure for small perturbations of the Hamiltonian function is the content of KAM theory (so called after the names of its discoverers Kolmogorov, Arnold and Moser [255]). As the perturbation increases, this structure gets destroyed, in general, and chaotic motions may appear. That such a destruction is possible has been numerically justified in the paper of M. Henon and C. Heiles [501] for a Hamiltonian function of the form

$$H = \frac{1}{2}(p_1^2 + p_2^2) + \frac{1}{2}(q_1^2 + q_2^2) + \varepsilon H_1 ,$$

where $H_1 = q_1^2 q_3 - \frac{1}{3} q_2^2$ and ε is a perturbation parameter. As ε increases, chaotic trajectories arise, forming complicated patterns in the three-dimensional subspace of fixed energy.

2. Chaotic motions of deterministic dynamical systems

Going through the list of simplest typical models of discrete deterministic dynamical systems, it is not hard to notice that two-dimensional systems do not have the chaotic motions arising when considering three-dimensional ones. This is not at all accidental, because two-dimensional smooth dynamical systems, i.e., autonomous oscillators and rotators, do not have chaotic regimes.

The unique steady-state motions in two-dimensional systems are stable equilibrium states and periodic motions. However, under periodic forced or parametric action on the oscillator or rotator, chaotic motions emerge. They also arise in changing the oscillator or rotator parameters autonomously and inertially.

Examples of chaotic motions of an oscillator under external periodic perturbations are the chaotic motions in the Duffing equation and van der Pol oscillator. An example of chaotization under periodic parametric action was given above (equation (1.23)). An example of chaotization in changing the parameter (or Lorenz equation) inertially has also been given. In Chapts. 7 and 9, these and many other examples are considered in more detail.

Dimension three is the least in which chaotization of motions in a smooth (differentiable) dynamical system is possible.

That aperiodic steady-state motions are possible in a system of third order was noted as early as 1963 by E. Lorenz's [563]. However, this became widely known only in the late seventies.

Other similar, an even earlier, examples turned out to be considerably less lucky, becoming known either much later or being kept in the dark at all. We mean the studies of the Gorky school on the theory of non-linear oscillations, conducted already in those times when computers had not yet been invented and the only practical way to discover chaotic motions was by reducing the problem to the study of the one-dimensional point mapping. We could also mention later work of the same school (i.e., theoretical studies and heuristic calculations on modern computers), enabling us to obtain an insight into the mechanism behind chaotic motions and into their general nature.

Original examples of point mappings with quite complicated (we would say *chaotic* today) iterations arose in considering concrete problems. Papers [3, 4] (1952-1957) specified complicated regimes as a consequence of applying the point mapping method to study the functioning of a two-position thermoregulator with a zone of advance. Complicated motions were also discovered in studying a magnetic interrupter model [354]. Iteration of an arbitrary piecewise linear mapping (consisting of two pieces) from a straight line onto itself was already studied in the papers [234, 235] (1959-1960). It is also necessary to draw the reader's attention to [59, 60] (1966-1967), where vibratory pile driving and unbalanced rotor motion on a harmonically vibrating base were studied on a computer.

Reading these paper now, it is necessary to recall that the terms *chaotic motion* and *strange attractor* appeared much later; such motions were then said to be *aperiodic Poisson-stable central*, or *Poisson-stable*, for short.

The terminology dates back to the book of G. Birkhoff *Dynamical Systems*, translated into Russian in 1941, and also the book by V.V. Nemytsky and V. V. Stepanov *Qualitative Theory of Differential Equations*, published in 1947. The term *Poisson-stable* itself means that the motion x(t) is bounded (as the time interval increases), and that, for any t_0, $\delta > 0$, there exist arbitrarily large moments $t_0 < t_1 < t_2 < \ldots$ for which

$$\| x(t_s) - x(t_0) \| < \delta \qquad (s = 1, 2, 3, \ldots).$$

We also refer to papers [70, 269] (1972), in which it was numerically discovered that chaotic motions are possible in the three-dimensional dissipative quasi-Hamiltonian system described by the equations

$$\dot{x} = \frac{\partial H}{\partial y} - \mu H \frac{\partial H}{\partial x} + vy \sin \omega\theta ,$$

$$\dot{y} = -\frac{\partial H}{\partial x} - \mu H \frac{\partial H}{\partial y} + \nu y \sin \omega \theta, \qquad (2.1)$$

$$\dot{\theta} = 1,$$

where $H(x, y) = (x^2 + y^2)^2 - 2a^2(x^2 - y^2)$ and μ, ν are two small parameters. These papers studied certain statistical characteristics of chaotic motions. The "cloud" of intersection points of a chaotic phase trajectory with the cutting plane $\theta = 0$ is represented in Fig. 1.15. All nearby points were attracted by the "cloud" engulfed by it. The statement was a consequence of theoretical arguments. However, exponential Lyapunov instability of motions in the "cloud" could not be proved, and numerical results regarding the aperiodic or periodic motions, but with a very large period, had to be restricted, too. Note that a similar example for a non-differentiable point mapping was constructed in 1974 [271, 279], where global contraction and local instability were established theoretically. In fact, a certain general scheme leading to such examples and manifesting their generality and frequent occurrence was indicated.

Fig. 1.15

The chaotic motions discovered in the Lorenz system are governed by this scheme, as was found later in 1977 [276, 278, 280].

The above examples of systems with chaotic motions belong to the class of non-conservative systems. Examples of chaotic motions for conservative Hamiltonian systems were known earlier; this is resistanceless coasting of a material point in a space of negative curvature [363, 485, 498], and, in particular, across the surface in Fig. 1.16, etc. We discuss this branch in Chapt.4.

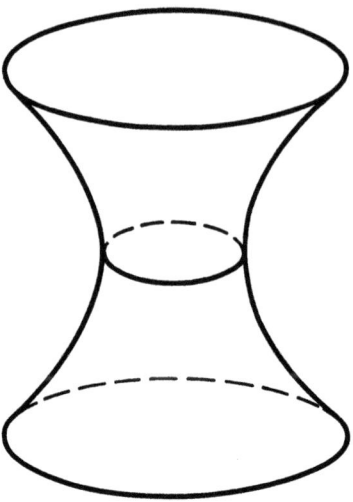

Fig. 1.16

We now return to the above-mentioned work of the representatives of the Gorky school on the theory of non-linear oscillations: A. S. Alekseyev, N. N. Leonov and Z. S. Batalova.

A. S. Alekseyev's work [4] (1955) studied the dynamics of a two-position thermoregulator with a zone of advance by the point mapping method, reducing it to the study of the point mapping from a straight line onto a straight line generated by mappings T_1, T_2, R_1, R_2, S_1 and S_2 of the half-lines $u_1 \geq 0$, $u_2 \geq 0$, $u_3 \geq 0$, and $u_4 \geq 0$ of the two-sheeted phase space (Fig. 1.17).

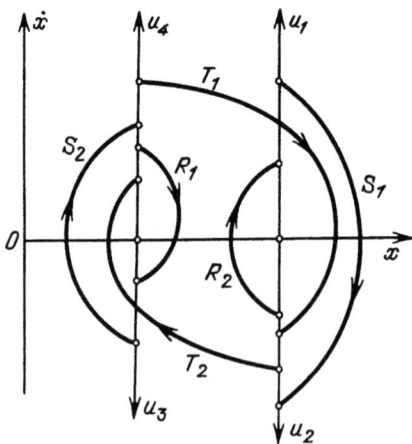

Fig. 1.17

Depending on the parameter values, the mappings may be of various form (Fig. 1.18). Despite its complexity, this mapping from a straight line onto a straight line was given a detailed study. It turned out that if the case in Fig. 1.18a is associated with a unique stable periodic motion, then, as we pass from Fig. 1.18a through Figs. 1.18b, c to Fig. 1.18d, complicated periodic motions emerge, also including asymmetric ones, despite the symmetry of the system in question; meanwhile, these motions undergo infinitely many bifurcations, so that each of the complicated steady-state regimes is associated with one of the infinitely many intervals on the parameter axis, while the complement to the intervals has the structure of the Cantor set, whose points are associated wit Poisson-stable motions (as the time interval increases). Oscillograph recordings of certain types of periodic motions are in Fig. 1.19.

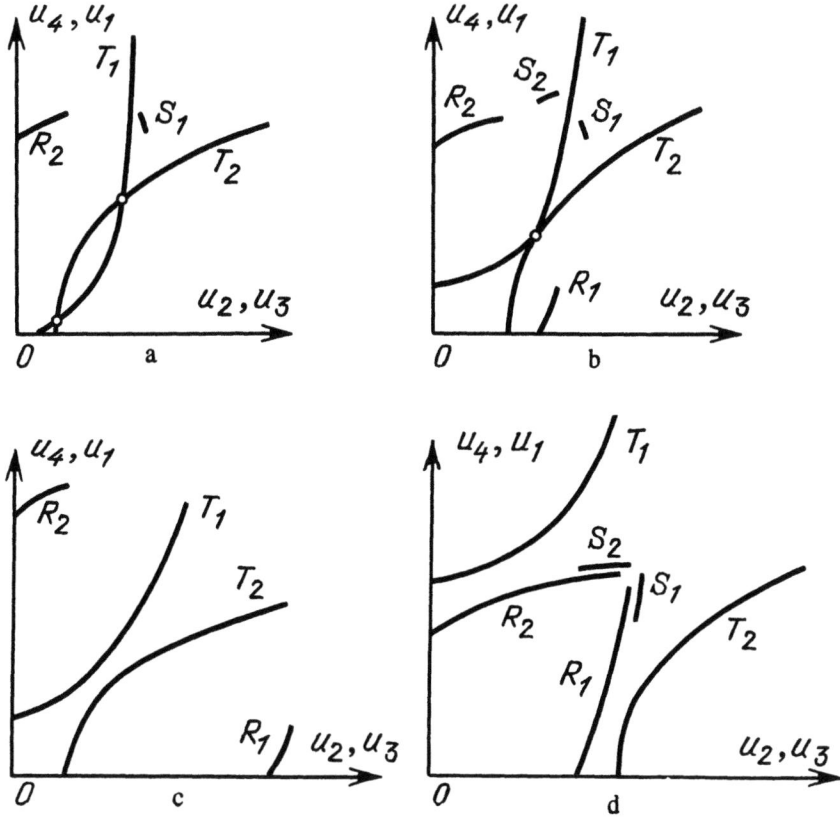

Fig. 1.18

Fig. 1.19

For a concrete automatic control system, doubling bifurcation series and other phenomena, later termed *intermittency*, were discovered in A. S. Alekseyev's paper. N. N. Leonov's papers [234, 235] contained considerations of general nature, the latter (1960) studying a non-bijective point mapping from a straight line onto a straight line, of the form

$$\bar{x} = \begin{cases} T_1 x = a + \lambda_1 x, & x \leq 0, \\ T_2 x = b + \lambda_2 x, & x > 0, \end{cases} \qquad (2.2)$$

and depending on three essential parameters λ_1, λ_2 and $\Delta = |ab^{-1}|$. A decomposition of the (λ_1, λ_2, Δ)-space was discovered, into domains associated with different types of multiple stable fixed points. The remaining part of the space is associated with transformations either leading to infinity or with Poisson-stable points.

The domains of the parameter space, associated with Poisson-stable motions, were also discovered for $|\lambda_1| > 1$, $|\lambda_2| > 1$, which, in accordance with modern terminology, is necessarily related to a stochastic attractor. One example is in Fig. 1.20. The Lorenz mapping enjoying wide popularity is its particular case $\Delta = 1$ ($a = -b$) and $\lambda_1 = \lambda_2 > 1$.

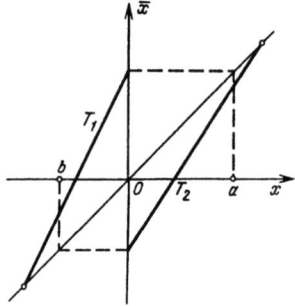

Fig. 1.20

Z. S. Batalova studied in [59, 60] (1966-1967) models for vibratory pile driving and an unbalanced rotator on a harmonically vibrating base. A computerized study was conducted by the point mapping method. That aperiodic Poisson-stable motions exist for the first system follows from the numerically determined point mapping (from a straight line onto a straight line); accordingly, there is an interval transformed into itself, by the expanding (or unstable) mapping on the segment. In the second system, the derivation was based on the aperiodic iterations, with the possible conclusion that if a fixed point exists, then its multiplicity is very large. Meanwhile, regions of Poisson-stable motion were distinguished in the parameter space, as well as stable periodic motions which get complicated by doubling their periods as the parameters vary and approach these regions.

The importance of homoclinic and heteroclinic Poincaré curves in forming aperiodic Poisson-stable motions was appraised by theoretical studies of the Gorky school on the theory of non-linear oscillations in 1964-1970 [262, 263, 266, 268, 288, 289, 370, 371, 374]; thereby, their general and essential nature in the theory of non-linear oscillations and temporal evolution of dynamical systems was clarified. We continue our discussion of these studies in Chapt.4. where we describe them after having gained certain familiarity with the basics of the question in relation to the results obtained in other branches of science. These results led (in the late 1970's and early 1980's) to the explosion of universal interest in the stochastic motions of dynamical systems.

3. Simplest mathematical models of continuous dynamical systems

We now describe mathematical models of continuous dynamical systems. Their variety is so great that we can hardly speak of any clear and convenient set of basic typical models. Still, some attracted the attention of researchers, and enabled them to make good progress in investigating wave and diffusion phenomena, laminar and turbulent hydrodynamic and convective fluid flows, etc.

Here, we briefly describe some of the models, and treat the mixed problems in mathematical physics from the viewpoint of phase: The equations are made phase state change operators. This treatment enables us to consider discrete and continuous dynamical systems from a unified standpoint, since phase state changes are described by a point of a finite-dimensional (respectively, infinite-dimensional) space.

Diffusion equation. The temporal change of the density $\rho(x, y, z, t)$ of a substance diffusing in an unbounded permeable homogeneous isotropic medium satisfies the parabolic partial differential equation

$$\frac{\partial \rho}{\partial t} = D \left(\frac{\partial^2 \rho}{\partial x^2} + \frac{\partial^2 \rho}{\partial y^2} + \frac{\partial^2 \rho}{\partial z^2} \right). \tag{3.1}$$

It enables us to find the density distribution for the given initial distribution $\rho(x, y, z, t_0)$ at any other moment $t \geq t_0$. Thus, if at the initial moment $t = t_0$, the whole mass m is concentrated at the point (x_0, y_0, z_0), then it follows from (3.1) that

$$\rho(x, y, z, t) = \frac{m}{8\pi^{3/2}(t-t_0)^{3/2} D^{3/2}} \exp\left[-\frac{(x-x_0)^2 + (y-y_0)^2 + (z-z_0)^2}{4D(t-t_0)} \right]. \tag{3.2}$$

The function $\rho(x, y, z, t)$ of these variables for fixed time t plays here the role of state, while the set of all possible non-negative functions of three variables x, y, z for the given total mass, or the given integral on the whole space, plays the part of state space.

In the case in question, $\rho(x, y, z, t) \to 0$ with time; therefore, all phase trajectories of the phase portrait of the system tend to the same point O associated with the function $\rho(x, y, z, t) \equiv 0$, which itself does not belong to the phase space.

Let the substance diffuse in a bounded region Ω, and let $\rho(x, y, z, t)$ still change according to the differential equation (3.2); however, not tending to zero with time. What is the differece from the preceding problem ? This can be seen by comparing the phase spaces. The new space is the set of all possible functions $\rho(x, y, z, t)$ which are non-negative on Ω, and with integral

$$\int_\Omega \rho(x, y, z, t) \, dx \, dy \, dz = m, \tag{3.3}$$

i.e., equal to the mass m of the diffusing substance, which remains constant; it is also required that the diffusing substance does not permeate through the boundary of Ω, or that the normal derivative of $\rho(x, y, z, t)$ on the boundary of Ω vanishes.

The latter is usually regarded as a necessary boundary condition for the diffusion equation (3.2) in order that the solution be complete and unique. If follows from the above that the boundary condition can also be treated as a consequence of the definition of phase space. A change in state, or in substance density, always satisfies the same diffusion equation, but now defined on a different phase space.

General linear partial differential equation. We now continue with considering simplest mathematical models for continuous systems, taking the partial differential equation

$$A\frac{\partial \mathbf{u}}{\partial t} + B\frac{\partial \mathbf{u}}{\partial x} + C\mathbf{u} = 0 \qquad (3.4)$$

as an example, where **u** is a vector-valued function of two variables x and t of dimension n, and A, B, C are three (n × n)-matrices.

We look for its solution in the form

$$\mathbf{u}(x, t) = \mathbf{a} e^{i(\omega t - kx)}, \qquad (3.5)$$

substitution of which in (3.4) gives

$$(i\omega A - ikB + C)\mathbf{a} e^{i(\omega t - kx)} = \mathbf{0}; \qquad (3.6)$$

therefore,

$$(i\omega A - ikB + C)\mathbf{a} = \mathbf{0}. \qquad (3.7)$$

Thus, if condition (3.7) is fulfilled, then (3.5) is the solution of equation (3.4). For (3.7) (**a** ≠ **0**) to be fulfilled, it is necessary and sufficient that

$$\text{Det}(i\omega A - ikB + C) = 0. \qquad (3.8)$$

Relation (3.8) is called the *dispersion equation*, and determines the connection between the temporal frequency ω and the wave number k of the solution (3.5), and the dependence on ω of the phase velocity of the travelling wave, which equals ω/k.

We now consider in greater detail how solutions of the form (3.7) are related to (3.8). For any real or complex k, (3.8) determines n corresponding values $\omega_1(k), \omega_2(k), \ldots, \omega_n(k)$, and each pair $(k, \omega_s(k))$ is associated with certain **a** determined by equation (3.7). Along with a solution k, ω and **a**, -k*, -ω* and **a*** are also solutions (asterisks meaning the conjugate complex number). This enables us to construct a real solution of equations (3.4) in the form of a traveling wave,

$$\mathbf{u}(x, t) = \mathbf{a} e^{i(\omega t - kx)} + \mathbf{a}^* e^{-i(\omega^* t - k^* x)}, \qquad (3.9)$$

thus specifying a temporally and spatially harmonic travelling wave if ω and k are real, and a temporally and spatially growing or evanescent wave if ω and k are complex.

Thus, expression (3.9), where ω, k and **a** are determined by equations (3.8),

(3.7), specifies a multi-parameter family of solutions of (3.4). Any solution of it can be constructed from the family by a linear superposition. This is what determines their importance and decisive role.

Superposition of waves (3.9) is not so simple as similar superposition in linear discrete systems. This is due to the relation of the temporal process with the spatial changes. Here, the key new concepts are the *envelope velocity* and *dispersion* [107, 132] describing the motion in space and the temporal changes of the wave packet, superposition of harmonic waves in a small frequency interval and corresponding wave numbers. The envelope velocity is the velocity of displacing the wave packet as a certain formation. Dispersion specifies the velocity of wave packet diffusion. With no dispersion, the wave packet keeps its shape, i.e., a travelling wave of fixed shape, or a so-called *stationary* wave. If there is dispersion, the wave packet diffuses with time. A complex ω causes exponential growth or decrease of the intensity. Thus, the envelope velocity determines the packet velocity; the dispersion, its diffusion; and the imaginary part of ω, its intensity increase or decrease. The envelope velocity is $d\omega/dk$, while the dispersion is determined by the quantity $d^2\omega/dk^2$.

Quasi-linear theory of oscillations and waves. From linear partial differential equations (3.4), we now turn to the more general non-linear equations

$$A\frac{\partial u}{\partial t} + B\frac{\partial u}{\partial x} + Cu = f\left(u, \frac{\partial u}{\partial t}, \frac{\partial u}{\partial x}\right). \quad (3.10)$$

How are the above purely linear effects altered when a non-linearity arises ? For high non-linearities, the question does not make sense, since linear effects are completely suppressed by non-linear ones; however, for small non-linearities, we can speak of a certain total action of non-linear and linear effects, and construct what is termed the *quasi-linear theory of oscillations and waves*. It is based on the deviations from the superposition principle, caused by the non-linearities. The latter are due to interaction and self-interaction of the waves, owing to the right-hand sides of equations (3.10), which involve (because of non-linearity) all possible combination frequencies, and not only the fundamental frequencies.

To obtain strong interaction for small non-linearities, as in the discrete case, it is necessary that resonance relations of the form

$$n_1\omega_1 + n_2\omega_2 + \ldots + n_m\omega_m = 0,$$

$$n_1 k_1 + n_2 k_2 + \ldots + n_m k_m = 0 \tag{3.11}$$

hold for wave numbers k, and not only for frequencies ω. The nature of resonant non-linear wave interaction is substantially different with dispersion and without it. Without dispersion, the pairs (sω, sk) (s = 2, 3, ...) satisfy the dispersion equation along with a pair (ω, k), generating (3.11), and thus pumping energy from low-order harmonics into higher. With dispersion, resonant interactions pumping energy into higher harmonics vanish, and conditions arise for the small wave ensemble (ω_1, k_1), (ω_2, k_2), ... , (ω_m, k_m) to satisfy synchronism condition (3.11), where $n_1, n_2, \ldots, n_m$ are small integers differing from zero. Such waves interact strongly, and make a certain closed group. Closedness or isolation of the ensemble is due to absence of waves with frequency ω and wave number k satisfying the synchronism conditions

$$n\omega + \bar{n}_1 \omega_1 + \ldots + \bar{n}_m \omega_m = 0,$$
$$nk + \bar{n}_1 k_1 + \ldots + \bar{n}_m k_m = 0, \tag{3.12}$$

where $n \neq 0$ and $n, \bar{n}_1, \bar{n}_2, \ldots, \bar{n}_m$ are small integers.

The simplest of such wave trains can be formed by three waves with synchronism relation of the form

$$\omega_1 + \omega_2 - \omega_3 = 0, \tag{3.13}$$
$$k_1 + k_2 - k_3 = 0,$$

Riemann wave equation. The Riemann non-linear wave equation

$$\frac{\partial u}{\partial t} + V(u) \frac{\partial u}{\partial x} = 0 \tag{3.14}$$

has been considered since Riemann's times (1860). It describes non-linear processes in a one-dimensional medium, and admits the so-called *Riemann waves*

$$u(x, t) = f(x - V(u)t) \tag{3.15}$$

as solutions. If we write a solution in the form (3.15), then we can determine u(x, t) implicitly as a solution for u. The wave shape, i.e., u as a function of the spatial coordinate x, is determined by (3.15) if time t is fixed. As time changes, the shape alters, and the variation can be tracked qualitatively, proceeding from the meaning of V(u) as the velocity with which a wave part spreads if the deviation u is fixed. The changes in the wave shape essentially depend on the correlation V(u). If V(u) = const, the shape does not change. When $\partial V / \partial u > 0$, the wave is "overturned".

Burgers and Korteweg-de Vries equations. Taking into account linear dissipation and dispersion effects leads to additional terms in the Riemann wave equation (3.15) which takes the form

$$\frac{\partial u}{\partial t} + V(u)\frac{\partial u}{\partial x} - \alpha\frac{\partial^2 u}{\partial x^2} + \beta\frac{\partial^3 u}{\partial x^3} = 0, \qquad (3.16)$$

combines the familiar Korteweg-de Vries ($\alpha = 0$) and Burgers ($\beta = 0$) equations.

When $V(u) = u$ and $\alpha = 0$, its solutions have been studied thoroughly; in particular, those of travelling-wave type. All possible shapes of travelling waves can be found as solutions of a certain ordinary differential equation. Substituting

$$u = f(\xi), \quad \xi = x - ct \qquad (3.17)$$

in (3.16) when $\alpha = 0$ and $V(u) = 0$, we arrive at the ordinary differential equation

$$\beta f''' + (f - c)f' = 0,$$

or, after integration,

$$\beta f'' + \frac{f^2}{2} - cf = h, \qquad (3.18)$$

which is associated with a non-linear conservative oscillator. Its phase portrait is represented in Fig. 1.21a, where the single wave, a soliton in the case in question, can only be associated with one phase curve γ, provided that the saddle point O is at the origin, i.e., if $h = 0$. The soliton is represented in Fig. 1.21b as a one-peak symmetric curve with exponentially decaying boundaries. The velocity with which the soliton spreads is c. It follows from the above that each value of c is associated with a soliton. The peak height is 3c, i.e., the "higher" the soliton, the greater the velocity. Other phase trajectories in Fig. 1.21a are associated with travelling waves of other shape, among which the soliton is the unique single wave with finite energy. The energy of all other waves is infinite; in particular, this testifies to the impossibility of concluding that the soliton is unstable on the basis of phase trajectories substantialy different from a phase curve γ being close to γ. The perturbation associated with this transition is not at all small, and requires infinite energy.

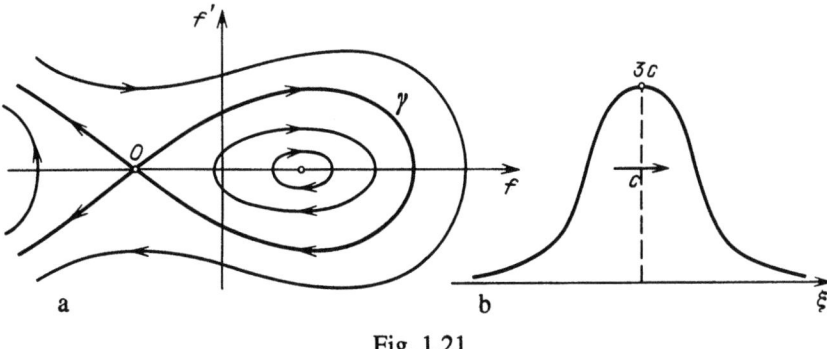

Fig. 1.21

Solitons are particular solutions of the Korteweg-de Vries equation; however, they play a very important, and, possibly, determining part, viz., any motion with finite energy decomposes after a sufficient time into a number of solitons, each of which has its own velocity and corresponding shape.

4. Discrete version of continuous dynamical systems

The above examples of mathematical models for discrete and continuous dynamical systems revealed that they are sharply different, since the description of a continuous system involves its spatial distribution, and not only its temporal dependence. This difference is especially significant in the description of discrete and continuous systems in the wave treatment of motions of continuous systems. However, Fourier standing waves can be the basis for describing such motions, next to d'Alembert travelling waves. The Fourier, Ritz, Bubnov and Galerkin methods do not stress any difference; on the contrary, they make the descriptions of motions of continuous and discrete system more alike.

We start by clarifying the possibilities and properties of a continuous dynamical system, treating the motions from the viewpoint of phase, with consideration of a concrete example of one-dimensional convective liquid flow in a vertical circular pipe heated from below (Fig. 1.22).

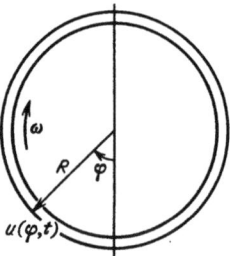

Fig. 1.22

Let $u(\varphi, t)$ be the temperature of the liquid in the pipe at the site determined by φ (where φ is the angle measured form the vertical line) at moment t, let $f(\varphi)$ be the temperature of the environment, and let ω be the liquid's velocity.

The equation for heat propagation in the pipe with one-dimensional idealization is then written in the form

$$\frac{\partial u}{\partial t} + \omega \frac{\partial u}{\partial \varphi} = -\lambda[u - f(\varphi)] + D\frac{\partial^2 u}{\partial \varphi^2}, \qquad (4.1)$$

where λ is the coefficient of heat transfer from the environment to the liquid, and D is the heat diffusion coefficient.

We inter from the law of conservation of momentum that

$$J\dot\omega = -R^2 \oint g\rho \sin\varphi \, d\varphi - h\omega, \qquad (4.2)$$

where J is the moment of inertia of the liquid in the pipe, R is the boundary radius (Fig. 1.22),

$$\rho = \rho(\varphi, t) = \rho_0[1 - \kappa u(\varphi, t)] \qquad (4.3)$$

is liquid density, and κ is its coefficient of linear expansion.

The adopted simplifications and idealization can be immediately seen from (4.1) and the relations (4.2), (4.3). In particular, hydrodynamic flow and temperature distribution are assumed to be one-dimensional, and only depending on one spatial coordinate φ. The liquid is assumed to be incompressible, so that the angular velocity ω does not depend on the angle φ. Meanwhile, its density ρ varies according to (4.3). These contradictory requirements are amenable to the familiar Boussinesq approximation, viz., heat transfer is assumed to be proportional to the difference in temperature, liquid heat capacity is constant, and (4.3) is independent of temperature.

We will look for a solution $u(\varphi, t)$ of the equations (4.1), (4.2) and (4.3) as the Fourier series

$$u(\varphi, t) = u_0(t) + \sum_{s=1}^{\infty} a_s(t) \sin s\varphi + b_s(t) \cos s\varphi . \qquad (4.4)$$

Assuming that $f(\varphi) = f(-\varphi)$, and, therefore,

$$f(\varphi) = f_1 \cos \varphi + f_2 \cos 2\varphi + \ldots ,$$

after simple calculations we arrive at the system of ordinary differential equations in $u_0(t)$, $\omega(t)$, $a_1(t)$, $b_1(t)$, ..., $a_s(t)$, $b_s(t)$, ...:

$$\begin{aligned}
\dot{u}_0 &= -\lambda u_0, \\
J\dot{\omega} &= \frac{\kappa R P}{2} a_1 - h\omega, \\
\dot{a}_s &= s\omega b_s - (\lambda + s^2 D) a_s, \\
\dot{b}_s &= -s\omega a_s - (\lambda + s^2 D) b_s + \lambda f_s .
\end{aligned} \qquad (4.5)$$

The structure of the relations among the variables u_0, ω, a_1, b_1, ... can be represented as in Fig. 1.23. Groups of interacting variables are in the rectangles, and the arrows indicate the action of some groups on others. Accordingly, we break (4.5) into several parts: First, the equation

$$\dot{u}_0 = -\lambda u_0 .$$

It follows that $u_0 \to 0$ in the course of time. Secondly, the equations

$$\begin{aligned}
J\dot{\omega} &= \frac{\kappa R P}{2} a_1 - h\omega, \\
\dot{a}_1 &= \omega b_1 - (\lambda + D) a_1, \\
\dot{b}_1 &= -\omega a_1 - (\lambda + D) b_1 + \lambda f_1 ,
\end{aligned} \qquad (4.6)$$

which can be reduced to the Lorenz equations with coefficients $b = 1$ and $r = \lambda \kappa R P f_1 / 2h(\lambda + D)^2$, $\sigma = \dfrac{h}{J(\lambda + D)}$ by the substitution

$$\tau = (\lambda + D)t, \quad x = \frac{\omega}{\lambda + D}, \quad y = \frac{\kappa R P}{2h(\lambda + D)^2} a_1$$

$$z = -\frac{\kappa R P}{2h(\lambda + D)} b_1 + \frac{\kappa R P \lambda f_1}{2h(\lambda + D)^2} .$$

The solution of the Lorenz system, and, therefore, of (4.6), can be very complicated and stochastic. Finally, the sth part is

$$\begin{aligned}
\dot{a}_s &= s\omega b_s - (\lambda + s^2 D) a_s , \\
\dot{b}_s &= -s\omega a_s - (\lambda + s^2 D) b_s + \lambda f_s .
\end{aligned} \qquad (4.7)$$

We immediately find that

$$\frac{d}{dt}(a_s^2 + b_s^2) = -(\lambda + s^2 D)(a_s^2 + b_s^2) + \lambda f_s b_s = -A_s(a_s^2 + b_s^2) + \lambda f_s b_s ;$$

therefore, the point (a_s, b_s) is after a time inside the circle with centre at the origin and of radius a little greater than $\lambda f_s / A_s$.

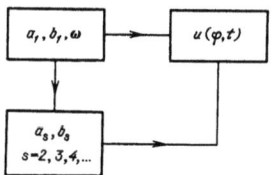

Fig. 1.23

As s increases, the radii of such circles tend to zero. The behaviour of (a_s, b_s) satisfies the equation (4.7), where ω is a certain function of time, determined by (4.6), which are equivalent to the Lorenz equations. After substituting

$$\tau = (\lambda + s^2 D)t = A_s t, \tag{4.8}$$

the equations (4.7) are reduced to the form

$$\dot{a}_s = -a_s + \frac{s\omega}{A_s} b_s, \quad \dot{b}_s = -b_s - \frac{s\omega}{A_s} a_s + \frac{\lambda f_s}{A_s} . \tag{4.9}$$

It follows (taking into account that $sA_s^{-1} \to 0$ as $s \to \infty$, and $f_s \to 0$ as $s \to \infty$) that, as s increases, the effect of ω on the variables a_s, b_s decreases, and, for very large s, a_s, b_s are close to zero.

We now turn to the representation of the distribution $u(\varphi, t)$ of the thermal field by the Fourier series (4.4). By (4.4), the thermal field is represented as infinitely many standing sine and cosine waves with amplitudes $a_s(t), b_s(t)$ ($s = 1, 2, 3, \ldots$). Recall that $u_0(t) \to 0$ with time. If solutions of (Lorenz equations) (4.6) approach an equilibrium state for $s = 1$, then all other a_s, b_s tend to constant values, and the thermal field is given a stationary distribution. If the dependence on time of the variables a_1, b_1, and ω tends to periodic, then all other variables a_s, b_s also tend to periodic changes, with the same period. By (4.4), in the course of time, this is associated with transition to a temporally periodic thermal field. Finally, if the variations of a_1, b_1, and ω are stochastic, then the temporal changes of the thermal field are of the same character. Meanwhile, the stochastic nature of the variation of the distributed thermal field is only due to stochasticity of the solutions of

Mathematical models of deterministic discrete and continuous dynamical systems 35

three differential equations, whereas the field itself is determined by solutions of an infinite system of equations, describing an infinite sequence of stochastically excited oscillators, and not only by a solution of that system of third order. This circumstance causes spatial, and not only temporal, chaotic behaviour. The slower the amplitudes decay with the variation of a_s, b_s as s increases, the more explicit the spatial chaotization is.

Thus, the system considered is an example of a continuous system whose motions are completely determined by solutions of a system of ordinary differential equations of small order. To what extent can this particular conclusion be extended to other continuous systems ? At present, a definite and comprehensive reply cannot be given easily, because the situation qualitatively (at least, within the framework of the quasi-linear theory) depends on the number of instability and freedom degrees with small damping. The problem considered has one instability degree (or one positive Lyapunov exponent). With respect to other degrees of freedom, damping increases rapidly. As is shown in the sequel, it is precisely this fact that determines the possibility of constructing a one-dimensional model as a point mapping from a straight line onto a straight line and adequately reflecting the system's temporal behaviour. However, the spatial distribution requires that we take into account considerably more wave modes. In the case in question, by a wave mode, we mean the term

$$s_s \sin s\varphi + b_s \cos s\varphi$$

involved in the one-dimensional thermal field representation by the Fourier series (4.4).

Thus, the number of degrees of freedom, necessary to describe the processes of temporal change in an acceptable manner, should be shown to differ from the number of modes required for the spatial description. For quasi-linear systems, the determining role of the number of degrees of linear instability is theoretically sound. For strongly non-linear systems, such theoretic foundation cannot be given; however, we can still believe that this number is to same degree determining for them, too. As far as the number of modes taken into account goes, it can be very large; apparently, the larger, the less the radius of spatial correlation. In describing turbulent flows, this number can reach very large values.

This is geometrically illustrated in Fig. 1.24 depicting a time generator, i.e., generally speaking, a system with a large number of degrees of freedom, excited by a time generator, and its spatio-temporal description. It is an approximate diagram; however, to make it completely precise and corresponding to the original distributed mathematical model, it suffices to take into account the reaction of the excited system on the time generator. It follows that to construct such a simplified diagram means to distinguish a time generator with a negligible action on it by the other degrees of freedom. This is only possible after transition from the description of the continuous system by partial differential equations to a discrete description, i.e., from functions u(x, t), where x are space variables and t is time, to series

$$u(x, t) = \sum_{s=1}^{\infty} a_s(t) v_s(x), \qquad (4.10)$$

and to finding systems of equations for the "amplitudes" of the modes

$$\dot{a}_s(t) = f_s(a_1, a_2, \ldots) \qquad (s = 1, 2, 3, \ldots). \qquad (4.11)$$

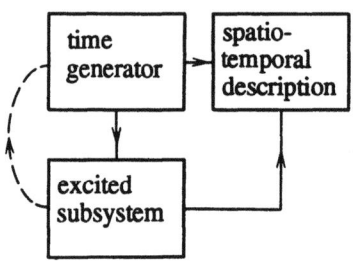

Fig. 1.24

The result, i.e., form of (4.11), substantially depends on the system of spatial modes $v_s(x)$ selected. It is difficult to give any general recipes here. However, many successful choices in various concrete problems can be indicated. One general method, though, whose realization is based on the capacity of modern computers, can still be pointed out. It consists in the following. A representative sample of solutions $u^i(x, t)$ (i = 1, 2, ... , N) of interest is found on a computer. The so-called *adapted* basis $v_1(x)$, $v_2(x), \ldots ,v_k(x)$ is constructed from the sample $u^i(x, t_j)$ (i = 1, 2, ... , N; j = 1, 2, ... , M). The adapted basis is, possibly, a small set of orthonormal functions such that all sample functions $u^i(x, t_j)$ are approximately representable by linear combinations of them. Sufficiently simple procedures for the construction of such an adapted basis are known [284], admitting practical realization for MN ≈ 10^5 and higher values even today. After $v_1(x), \ldots ,v_k(x)$ has been found, equations (4.11) are sought for in accordance with generally known schemes [170, 253].

Discrete models for continuous dynamical systems, both at the stage of equations (4.10), (4.11) and by subsequently distinguishing a time and a spatial generator, can be both partial and general. A *partial* model is one admitting the consideration of only certain of the motions of a given continuous system, or a definite class of motions, and not all. The problem of finding a class of motions admitting the construction of discrete partial models is quite complicated. In an attempt to give an answer, it is necessary to bear in mind that the requirement of stability (in some sense or other) for the distinguished class of motions with respect to all other motions close to the class is natural. This requirement can be treated as a condition for the distinguished class to form a stable integral manifold. The theory of integral manifolds exists in quite developed form; however, it does not, in fact, enable us to clarify the stability of the distinguished class of motaions of a continuous system described by partial differential equations.

The discrete model obtained via the adapted basis can be both general and partial, according to the sample used in its construction. Meanwhile, the necessity to check local stability is replaced by the requirement of closedness of the partial sample; this holds naturally.

In conclusion, we note that discrete models for continuous systems draw bridges between seemingly much different classes of dynamical systems. If the very important wave representations are dissimilar for them, then the description of continuous systems with the aid of a time and a spatial generator makes them alike, and the discrete system becomes a particular case of a continuous system reducible to one time generator. Meanwhile, the theory of discrete systems is an indispensable part of that of continuous systems, and, in a sense, its basis and core. Such an approach does not make wave representations obsolete at all, but only transforms them, by considering certain properties of a spatial generator.

5. Models for locally active continuous media

The above general argument regarding motions of a continuous system are considerably pervaded by the quasi-linear "ideology" of attaching a determining role to a certain small set of modes approximately of the same form as in a linear system. Meanwhile, totally different cases are possible. In non-linear active media, perturbations can locally arise, and, in some sense, even spontaneously. We encounter a similar situation in considering chemical reactions in liquids or on surfaces without mixing and with moderately rapid diffusion, in excitable and living environments (such as the cardiac muscle or a nutritive substrate with growing micro-organisms). Similar processes can proceed in a medium whose active and excited state is supported by permanent energy pumping from an external energy source. This localized activity can lead to the global regularization of motions of the continuous medium, or to certain ordered motions, but can also generate chaotic and disordered motions.

Apparently, the simplest mathematical model for a continuous system of the form in question first appeared in [204, 455], viz.,

$$\frac{\partial u}{\partial t} = f(u) + D\frac{\partial^2 u}{\partial x^2}. \tag{5.1}$$

More general models described by the system of equations

$$\frac{\partial u}{\partial t} = f(u, v) + D\Delta u, \quad \frac{\partial v}{\partial t} = g(u, v) \tag{5.2}$$

appeared subsequently [1, 319, 354, 356].

For $D = 0$, equations (5.1), (5.2) degenerate into the systems

$$\frac{\partial u}{\partial t} = f(u), \tag{5.3}$$

$$\frac{\partial u}{\partial t} = f(u, v), \quad \frac{\partial v}{\partial t} = g(u, v), \tag{5.4}$$

respectively, which makes natural the relations between solutions of the partial differential equations (5.1), (5.2) and the ordinary differential equations (5.3), (5.4), respectively. In particular, this is revealed by the instability of equilibria and the association of the self-oscillatory nature of systems (5.3), (5.4) with the local excitability of media described by (5.1), (5.2). The study of models of the form (5.2) is quite complicated, and far from complete. However, what we already know makes it possible to draw our attention to the important role of localized waves, of expanding rotating spirals, and, possibly, of certain others, next to that of travelling waves.

Distributed models with small-dimensional time generator and local excitability can be regarded as representatives of the "pure" cases. Very complicated distributed models in hydrodynamics are "mixed" cases.

The model for hydrodynamics, described by the Navier-Stokes equation

$$\frac{\partial v}{\partial t} + (v\nabla)v = -\frac{1}{\rho}\nabla p + v\Delta v, \quad \text{div } v = 0, \tag{5.5}$$

depending on the Reynolds number, can be both of the type associated with a small-dimensional time generator and of locally-excitable medium type. The former is characteristic of the so-called laminar flows and the initial phases of violation of their stability; the latter, of well-developed turbulence.

CHAPTER 2

ORDER AND CHAOS AS TWO GENERAL BASIC TRENDS IN THE EVOLUTION OF DYNAMICAL SYSTEMS

In the previous chapter, we have become acquainted with the simplest typical models for deterministic dynamical systems and motions described by them, viz., equilibrium states, self-excited and forced oscillations, various types of wave motions, diffusion processes and chaotic motions. All this unusual variety of motions can be separated into two basic types: order and chaos, or regularity and irregularity.

In the case of discrete systems, the division of motions into regular and chaotic only characterizes their temporal behaviour. In the continuous system case, next to temporal, we can speak of spatial order and chaos, as well as of the periodicity and the aperiodicity, the regularity and the irregularity of a spatial and not only temporal structure.

The temporal behaviour of a dynamical system has long been in the focus of attention, and the ideas of temporal order and chaos acquired a precise mathematical formalization in synchronization and stochasticity phenomena; the differences between spatial order and chaos have not, in fact, been analyzed before.

Both temporal and spatial aspects of the differences between the regular and chaotic motions are discussed below.

An adequate mathematical representation of temporal order and chaos is given by attractors, i.e., by stable equilibrium states, stable periodic motions or self-excited oscillations, and, finally, strange attractors. Saddle equilibrium states, saddle periodic motions and more complicated saddle invariant sets turned out to be an adequate mathematical representation of spatial order and chaos in the dual representation of a continuous dynamical system.

Order and chaos as two general basic trends in the evolution of dynamical systems   41

1. Order and chaos: stability and instability

As was mentioned above, the present chapter deals with the general properties of motions of discrete and continuous dynamical system, which can be classified into order and chaos. Order in temporal change is the balance of interaction, leading to stable equilibrium, synchronism of motions of individual parts of the system, entailing a periodic motion of the system as a whole. Chaos in temporal change is the absence of regularity, irregularity, unpredictability, and randomness. Spatial effects of order are spatial regularity and consistency. Spatial chaos is the absence of spatial regularity and discord.

The changes in the surrounding world can be naturally separated into these two classes. Alternation of day and night forms a regular time series with periodic structure to a first approximation; however, under a more attentive look, it is quasi-periodic (biperiodic), and reflects the Earth's rotation about its axis and the Sun. Waves spreading out from a spot where a stone was thrown into a smooth sheet of water present an example of spatio-temporal regularity. Sea-waves in the strong wind or ripple can serve as an example of chaos; but, sometimes, sea-waves form regular sequence one after the other. Day and night, summer and winter alternate periodically, whereas the weather changes in quite an intricate way, hard to predict, though some regular components are there, because +30°C cannot be registered in winter (just as -30°C in summer).

All these examples are common and customary. They are consistent with the intuitive ideas of order and chaos. But, what kind of unpredictability and randomness can appear when considering quite deterministic behaviour of solutions of differential equations, whose uniqueness has been proved ?

An argument in support of chaos on the basis of Cauchy's theorem on the uniqueness of solutions of differential equations has very recently been regarded as irreproachable; it would seem that any thought of possible random motions of deterministic dynamical systems should be banished. In any event, randomness within the framework of classical ideas was thought of as something alien. Stochastic motions of deterministic systems were only regarded as the result of random external actions. Only in problems of the nature of stochasticity of motions of gas molecules and, generally, in statistical mechanics and statistical physics, a concession was made, because there randomness was obvious. After a while, this divergence became habitual, and was erroneously believed to be due to the enormous numbers of molecules.

How surprising thus was the irrefutable discovery of unpredictability, chaos and stochasticity in simple systems, too!

The events leading to the "enlightment" are described in Chapt.4 in more detail; now, we only note and clarify the decisive role of oscillation theory in the problem.

It is not true that everyone realizes the role of oscillation theory as a front-rank science. Its great success and influence, i.e., on forming the superposition principle, the spectral approach and linear theory, the discovery and study of self-excited oscillations, and, now, of stochastic oscillations, often lose their special look and immediate relation to oscillation theory, and rapidly become an asset of all. Our book is first of all devoted to the latest achievements in oscillation theory. These will change our fundamental ideas about natural science, will lead to the discovery and study of chaotic motions of deterministic autonomous dynamical systems, to the possible generation by such systems of stochastic oscillations, to the new, and broader, approach to possible motions of dynamical systems, and to two opposite tendencies in the evolution of a dynamical system, i.e., the strive for order or chaos.

Tendencies towards order or chaos are stipulated by stability and instability. Quite recently, stability was regarded as an indispensable requirement of physical realizability. It seemed that unstable equilibrium states and periodic motions could not be physically realized in long time intervals, and were only important in mathematical investigations, since they are essential in forming the boundaries of domains of attraction of stable equilibrium states and periodic motions. The role of unstable equilibrium states and periodic motions as the boundaries of domains of attraction is reflected by the phase portrait of a certain dynamical system (Fig. 2.1) which can be called a *flip-flop*. The points O_1 and O_2 are two stable equilibrium states (one of focus, and the other of node type). The point O is a saddle-type unstable equilibrium state. It involves only two phase trajectories γ_1 and γ_2, separating O_1 and O_2 and determining two nearby domains of attraction Π_1 and Π_2. Any phase point of Π_1 tends to O_1, while any point of Π_2 tends to

O_2. Thus, depending on the initial conditions, a system with this phase portrait finds itself in one of the equilibrium states (either O_1 or O_2). It can only make the transition from one of these to the other due to a certain perturbation or jerk making it leave one domain of attraction for the other. Such a system can be called a *flip-flop, binary storage*, or a *switch*.

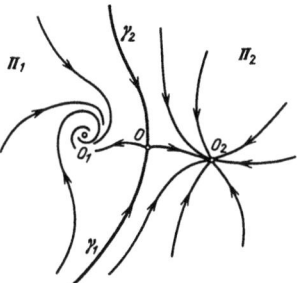

Fig. 2.1

However, the role of unstable motions is actually much greater. It turned out (unexpectedly) that unstable motions are what generate turbulences and complicated chaotic, stochastic motions of dynamical systems. The two principal kinds, or evolution tendencies, viz., synchronization (including equilibrium) and stochasticity, are nothing but a revelation of stability and instability.

The mechanism for revealing stability is habitual and clear, possibly, thanks to intuition enrooted in our minds and based on Brouwer's theorem and Banach's contractive mapping principle. Asymptotic stability always entails stable equilibria or stable periodic motions. An asymptotically stable limit motion is either a stable equilibrium state, a stable periodic motion, or a motion asymptotically approaching one of them. The mechanism for revealing instability is much more complicated and unusual. To understand it, first of all, the idea of physical realizability of motion as of the requirement of stability should be discarded, i.e., preservation of nearness of unperturbed of perturbed phase trajectories. The trajectories may not be near; moreover, they can locally diverge exponentially. Individual phase trajectories are meanwhile not realizable physically; however, they are realizable as a certain collection of motions similar to each other in some respect. It is not simple to have a clear idea of all that; therefore, the geometric representation of such phase trajectories could be possibly called a "strange attractor".

In spite of the complicated structure and unusualness of this formation, the conditions for its appearance are very simple, i.e., the combination of global contraction with local instability. A priori, it is unclear whether this combination is possible; but, if it is possible, then the existence of a strange attractor necessarily follows.

Concrete examples show that such a combination is in fact possible. Moreover, if it occurs, then the combination can be found theoretically to be typical, i.e., it is preserved for all nearby dynamical systems, and is not destroyed by small changes in the dynamical system. The answer to the question of how this all is possible is based on determining the role of so-called *homoclinic structures* which we consider in the sequel. Now, however, we confine ourselves to stating two essentially different versions of the behaviour of phase trajectories, viz.,

(1) general (global) contraction with local stability

and

(2) general (global) contraction with local instability.

The former leads to stable equilibrium states and stable periodic motions, whereas the latter to stochastic motions; the former leads to order, whereas the latter to chaos. Thus, two basic tendencies to be observed everywhere in evolution, i.e., order and chaos, are associated with two general ways of behaviour of phase trajectories: on the one hand, with general contraction and local stability, and, on the other hand, with general contraction and local instability.

By *general contraction*, we mean that a certain domain G of the phase space is sent into itself, i.e., is transformed into a domain $\overline{G} \subset G$ after a certain time.

By *local instability*, we mean divergence of phase trajectories that are initially very close, so that, being arbitrarily near to an unperturbed trajectory γ, there are perturbed trajectories which can be further than some $\varepsilon > 0$ from it after a certain time. When such a nearby phase trajectory γ has gone outside the ε-neighbourhood of γ, it can then get into the neighbourhood again; however, in general, it will necessarily leave it again, etc.

We illustrate by simple examples the behaviour of the first and second types. A phase portrait where global contraction is accompanied by local instability is represented in Fig. 2.2. Also, not only G is transformed into a region $\overline{G}$ contained in it, but any small region g is transformed into a still smaller region $\overline{g}$. Such an example is just the system with phase portrait in Fig. 2.1.

Order and chaos as two general basic trends in the evolution of dynamical systems 45

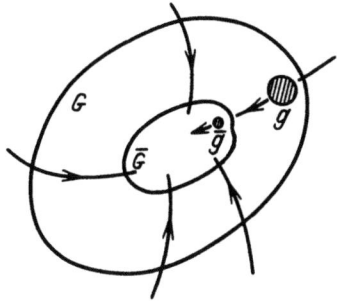

Fig. 2.2

We now turn to examples of the second type, where global contraction is companied by local instability. Take a toroidal region G, or the interior of a two-dimensional torus, elongate it, make it thinner and bend, so that it changes into the toroidal region $\bar{G}$ contained in G, as shown in Fig. 2.3 [332]. This mapping of G into $\bar{G}$ (denoted by T) can be determined analytically, for which we introduce coordinates x, y, ψ as in Fig. 2.4. Here, ψ is the angle variable $0 \le \psi < 2\pi$, and x, y satisfy the inequality $x^2 + y^2 \le 1$ in G. Not resorting to the figures, G can be defined as the set of points (x, y, ψ) for which ψ is the angle variable and $x^2 + y^2 \le 1$, i.e., as the topological product of a circumference and the unit disc. The transformation T mapping G into $\bar{G}$ can be written as

$$\bar{x} = qx + 0.5 \sin \psi, \quad \bar{y} = qy + 0.5 \cos \psi, \quad \bar{\psi} = 2\psi \,(\text{mod } 2\pi) \tag{1.1}$$

in terms of x, y, ψ, where $q < 1/2$. It also transforms the torus G into the torus $\bar{G}$ contained in G and passing through it twice (Fig. 2.3). In fact, under (1.1), any section ψ = const of G is a disc G_ψ of unit radius, and is mapped onto a domain $\bar{G}_\psi$ inside $G_{\bar\psi}$ (Fig. 2.5). As ψ varies from 0 to 2π, the section G_ψ traverses G once, and $G_{\bar\psi}$ twice. Meanwhile, $\bar{G}_\psi$ inside $G_{\bar\psi}$ "sweeps out" $\bar{G}$. $\bar{G}$ is inside G; therefore, global contraction occurs. Local instability occurs, too, since any two nearby points with coordinates ψ differing by $\Delta\psi$ are sent into two other points with coordinate difference $2\Delta\psi$, i.e., twice as much.

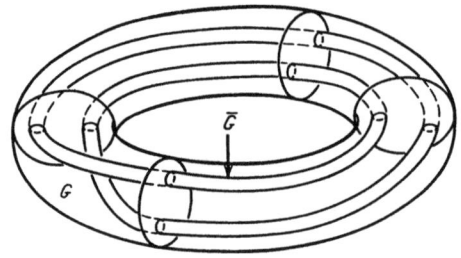

Fig. 2.3

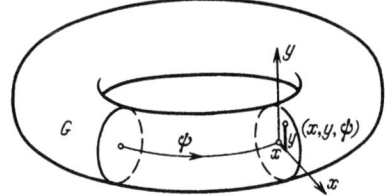

Fig. 2.4

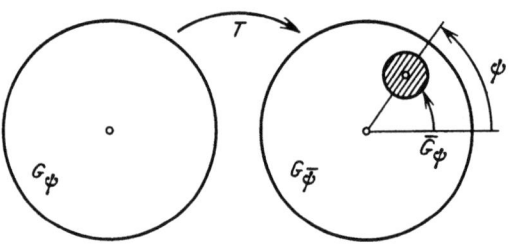

Fig. 2.5

Thus, the mapping T of G into $\bar{G}$ represented in Fig. 2.3 and determined by formulas (1.1) is globally contracting and locally unstable. Consider it in more detail. G is transformed into $\bar{G}$ which, in turn, is likewise transformed into $\bar{\bar{G}}$, etc. The toroidal region $\bar{G}$ is inside the toroidal region G. Similarly, $\bar{\bar{G}}$ is inside $\bar{G}$, etc. Meanwhile, contractions along the x- and y-coordinate occur always ($q < 1/2$), the number of turns of the image in G relative to ψ doubling with each transformation. The domains G, $\bar{G}$, $\bar{\bar{G}}$, ... approach indefinitely some limit set J invariant under T, i.e., transformed onto itself under T. For ψ, T is written as

$$\bar{\psi} = 2\psi \pmod{2\pi} \tag{1.2}$$

on this invariant set. The graph of the mapping (1.2) is in Fig. 2.6. The mapping is expanding, i.e., any small line segment of length $\Delta\psi$ is transformed into a line segment of

Order and chaos as two general basic trends in the evolution of dynamical systems 47

length $2\Delta\psi$. Consecutive transformations $\psi, \overline{\psi}, \overline{\overline{\psi}}, \ldots$ make up a periodic sequence if $\psi = 2\pi k/(2^n - 1)$, where k and n are two integers. There are countably many values of such ψ, the Lebesgue measure of this set being zero. The sequence $\psi, \overline{\psi}, \overline{\overline{\psi}}, \ldots$ with a value of ψ not belonging to the set is everywhere dense on the circle $0 \leq \psi < 2\pi$. Thus, on J, the transformation T is exponentially unstable and expanding, while the iterates of any point of J, generally speaking, are everywhere dense on J. The consecutive values of 24 transformations are represented in Fig. 2.7. Their irregular and chaotic nature can be seen explicitly. In conclusion of this example of a globally contracting and locally exponentially unstable mapping, we make a number of general remarks.

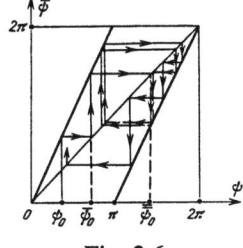

Fig. 2.6

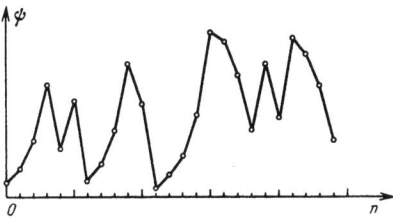

Fig. 2.7

The mapping T is one-to-one; however, the mapping of the variable ψ is not one-to-one on the invariant set J. T is three-dimensional in the above example. It can easily be generalized to the case of large dimensions; but, to construct similar examples in smaller dimensions while preserving bijectivity is not possible.

The mapping (1.1) is a simple example of a one-to-one, globally contracting and locally unstable mapping. Its iterations form, in general, a chaotic sequence. The limit set of these sequences is a certain invariant set J. Whatever the accuracy of specifying the original point x, y, ψ, its sufficiently far iterates cannot be found, since, by iteration, an unbounded and rapid exponential increase of the error occurs. Sufficiently far iterates are in this sense unpredictable. Thus, with original accuracy of order 10^{-6}, the error is, generally speaking, of order one after the 20th iteration.

This example of a three-dimensional point mapping can easily be transformed into one of a dynamical system described by differential equations but with four-dimensional phase space. For this system, the point mapping T is a Poincaré map on the 3-dimensional section plane.

Something similar to the above abstract example can also occur in concrete systems. Thus, in the phase space of the Lorenz equations (1.24) for $b = 8/3$, $\sigma = 10$, $r = 24.4$, the consecutive points where phase trajectories meet the section plane $z = r - 1$ fall into a very small neighbourhood of a curve J, remaining there and thereby generating a mapping of J onto itself. If we introduce a variable u along the curve, then this mapping is of the form shown in Fig. 2.8. It is expanding everywhere, the typical consecutive values of u being chaotic. We will speak of Lorenz systems and their strange attractors in more detail in Sect.4, Chapt.7.

A somewhat different situation also leading to chaotic motions can occur in the section plane of the system discussed in the preceding chapter.

Just as in the case of the Lorenz system, consecutive transformation points, starting with some of them, approximately to a high degree of accuracy, are all on a certain curve J on the section plane; this generates on J a mapping as in Fig. 2.8 for the Lorenz equations, and mappings as in Fig. 2.9a-d for the non-autonomous rotator (1.23) discussed in the preceding chapter. The mapping in Fig. 2.8 is very much similar to that in Fig. 2.6, also being expanding. The mappings in Figs. 2.9a-d are of another kind. They map the disc onto itself, and are not expanding everywhere: They always have parts with dilatation coefficient less than one, reaching zero at certain points. Figs. 2.9a-d are associated with the parameter values $v = 0, 0.7, 0.9, 1.32$, and 2.2. As can be seen from the diagrams (e.g., if the successive mappings for $0 \leq v \leq 2.2$ are constructed from them), both simple and multiple stable fixed points and chaotic sequences are possible. The former are related to rotor motions synchronous with the frequency of parametric effect; the latter, with chaotic motions. As v increases, these types of motions alternate (which can be seen in Figs. 2.9a-d), the synchronization regions with respect to v decrease, while the chaotic motion ranges expand. A more complete consideration of the phase portrait and its changes (bifurcations) for this dynamical systems is given in

Sect.5, Chapt.7.

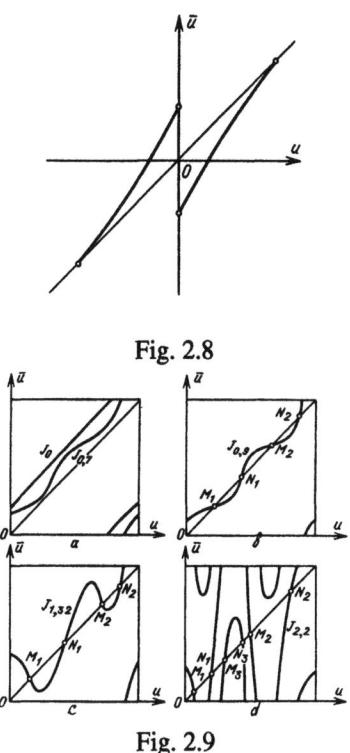

Fig. 2.8

Fig. 2.9

We now turn to more complete descriptions of synchronism and then temporal and spatial chaotization phenomena. We recall once more that synchronization understood as temporal regularity or temporal order is the effect of general contraction and local stability. Temporal chaotization is due to general contraction and local instability. Spatial regularity or chaos can be related to properties of the structure of solutions of the dual discrete representations of a continuous system's motions.

2. Time synchronization phenomena

Man got familiar with the synchronization phenomenon very long ago. All his and environmental life-cycles were synchronous with the seasons and alternations of day and night. He not only observed them, but actively strove for possibly better synchronization of his actions, as a hunter, angler, cattle-breeder, and ploughman, with alternations of seasons and day and night. The calendar and clock are consequences of this striving. Man observed synchronization during mass dances, collective songs and movements to the music or drum beat.

However, the science of synchronization originated form the observation of incomparably simpler phenomena.

Synchronization of clock pendulum oscillations. The first known description of this simple synchronization phenomenon was made by Christian Huygens, who invented a magnificent clock with a very accurate rate. In his observations, Huygens discovered an extremely striking phenomenon, and described it as follows: The pendulum was 9 inches long, with half a pound of weight attached. The mechanism set in motion by the weights encased together. The box was 4 ft long, and loaded with at least 100 lb of lead, so that the whole mechanism might be as vertical as possible. Two such timepieces hung from the same beam on two supports. Both pendula always moved in opposite directions, whereas the oscillations coincided precisely and never diverged. The ticking of both clocks was simultaneous.

If this coincidence was disturbed artificially, then it restored itself after a short time.

Synchronization of organ oscillations. Much later, in the 19th century, the synchronization phenomenon in acoustic and electro-acoustic systems was discovered by Lord Rayleigh, who found that two organ pipes placed side by side sounded in unison with small detuning. Each was a self-oscillatory system generating sound of a certain frequency ω. If these frequencies are little different, then both pipes start emitting sound of the same frequency and in unison. Sometimes, this is expressed by complete disappearance of sound. This is associated with generation by the pipes of sound of precisely the same frequencies of opposite phases (their oscillation phases being π different).

Synchronization of electric generator rotations. The synchronization phenomenon first became considerably important in applications when a number of a. c. electric generators were set in one circuit, i.e., were given the same load. As first electric power-stations appeared, disastrous accidents often occurred. As a matter of fact, the phases of the generated voltages did not coincide when working with the same load, and instead of sending a current to the common circuit together, the generators let the current pass through each other, and burned out the coils. To prevent this, they had to rotate with the same frequency (frequency coincidence should be very strictly observed) and the same (sufficiently close) phases. How can that be achieved if the generators are set in rotation by different turbines ? Even the most exact control of turbine speed cannot provide for a complete coincidence of frequencies and phases, since, for arbitrarily small difference $\Delta\omega \neq 0$ of their rotation frequencies, the phase difference $\Delta\omega T$ accumulates with time T, and, after $T = \pi/\Delta\omega$, their phases are π different.

In this seemingly hopeless situation, the synchronization phenomenon proved helpful. It turned out that, under certain conditions, for not very considerable frequency detunings, the electric generators synchronize themselves, and rotate synchronically to a very high accuracy.

Synchronization in radio reception. A not less considerable role is given the synchronization phenomenon by radioengineering and radiophysics. We can say without any exaggeration that the synchronization phenomenon for oscillations of a radio receiver by an external weak radio signal is the basis for receiving radio signals.

The basic mathematical model considered in direct relation to the problem of reception of a weak radio signal by a regenerative receiver was the van der Pol equation with E cos vt as the right-hand side. Its study showed that synchronization of it can occur for an arbitrarily small amplitude of external effect if only the frequency detuning is sufficiently small. Otherwise, the biperiodic regime emerges.

Phase synchronization system. The synchronization phenomenon is the basis for various numerous phase synchronization systems widely used in communication engineering and measurement theory.

One phase synchronization problem consists in making powerful, quite stable high-frequency oscillators. The problem can be solved by synchronizing a high-power oscillator with a low-power one of high stability in terms of frequency; e.g., a crystal-controlled generator.

Synchronization of mechanical vibrators. The synchronization phenomenon finds many applications in various vibrational technological devices. Quite efficiently it is used in the joint operation of rotors of a vibrational machine, rotating simultaneously and not coupled mechanically [89, 90].

Synchronization of planetary and satellite motion. The synchronization phenomenon is used for orienting in space artificial satellites of the Earth. Synchronization itself of heavenly body motions was first discovered for the planets and satellites of the solar system in the form of remarkable relationships between their periods of revolutions and rotations. The explanation is very complicated, and has not been completed even today. The discovered properties are simple integer relations between the frequencies of orbital and proper rotations of the planets and satellites, viz.,

$$n_1\omega_1 + n_2\omega_2 + \ldots + n_s\omega_s = 0 \qquad (2.1)$$

[79, 256], where $n_1, n_2, \ldots, n_s$ are small integers and $\omega_1, \omega_2, \ldots, \omega_s$ the frequencies of orbital or proper motions. Certainly, the relations can also be accidental. Moreover, since the frequency values are known approximately, the relations are approximate, with small deviations. However, this assumingly accidental coincidence can only be very rare, with probability estimate of order 10^{-11}. To believe that an event with this negligible probability can occur is hard. Besides, certain of these relations were given a theoretical foundation.

Synchronization of systoles. The rhythmic contractions of the heart are due to synchronization. Any disturbance in the normal synchronization of cardiac-muscle fiber contractions leads to arrhythmias, including such a terrible one as cardiac fibrillation.

The first mathematical model for cardiac contraction synchronization was suggested by van der Pol. It was influenced by the synchronization phenomena studied at that time in radio engineering, and was nothing but three coupled oscillators. Many arrhythmias could be successfully explained with its help, and even a number of new ones detected. However, to see the reason for fibrillations to appear, a more complicated model for a conducting excitable medium was required. This gave an account of the reasons and mechanism for cardiac fibrillation when, instead of the synchronous rhythmic contractions, irregular waves of contractions start running across the cardiac muscle.

3. Spatial and temporal order and chaos

In spite of the variety of concrete manifestations of time synchronization, they all are made up of consistent changes in individual subsystems of a dynamical system with external periodic effect, leading to periodic change in the state of the whole system, no matter whether the system is discrete or continuous. Spatial phenomena have been investigated considerably less, and have not been used so extensively as time synchronization. Moreover, time synchronization has been defined rigorously [89, 90], and there is no similar definition as far as spatial order is concerned; everything is confined to a comparatively modest, and only partly theoretically studied, set of concrete examples, such as Hale-Shaw or Bénard cells in convective flows of a liquid, Taylor vortices in a viscous liquid between rotating cylinders, and certain systems for which an explicit spatial structure is observed experimentally, such as stable, self-excited standing waves, Karman vortices behind a streamlined body, contractions of excitable cardiac-muscle fiber, spatio-temporal biological cell organization, etc. In the latter cases, a definite spatial structure and self-adjustment are discussed, and not only spatial order; in accordance with this treatment, *synergetics* is spoken of as a new science of self-adjustment [355, 356, 487].

Still, what is temporal and spatial order ? Of course, we have in mind a continuous system, and resort to its discrete representation of the form

$$u(x, t) = \sum a_s(t) v_s(x),$$
$$\dot{a}_s = f_s(a_1, a_2, \ldots) \quad (s = 1, 2, \ldots),$$
(3.1)

considered in Chapt.1. From the standpoint of this representation, temporal order is the periodicity of motions of its time generator or of the whole infinite system for variables $a_1, a_2, \ldots$. And what is spatial order ? E.g., the periodicity of $u(x, t)$ with respect to x and any t can be meant, or somewhat more general regularity with respect to x. However, these properties of $u(x, t)$ are related to the functions $v_1, v_2, \ldots$ and $a_1, a_2, \ldots$ in a sufficiently complicated way. We therefore turn from a discrete representation of the motions of a continuous system as in (3.1) to a representation of the form

$$u(x, t) = \sum a_s(t) v_s(x),$$
$$\frac{dv_s}{dx} = g_s(v_1, v_2, v_3, \ldots) \quad (s = 1, 2, \ldots),$$
(3.2)

which is called *dual*. Besides (3.2), the functions $v_1, v_2, v_3, \ldots$ must also satisfy certain boundary conditions on the boundaries of the spatial domain, which in the one-dimensional case can be written in the form

$$K_1(v_s)|_{x=l_1} = 0, \quad K_2(v_s)|_{x=l_2} = 0 \quad (s = 1, 2, \ldots). \tag{3.3}$$

Note that, in (3.2), v_s are vector-valued functions of a certain dimension m, so that the total number of the conditions (3.3) is m, where p are determined by the first and q by the second condition ($p + q = m$).

For the sake of simplicity, we confine ourselves to finitely many (e.g., N) functions $v_1, v_2, \ldots, v_N$, and accordingly consider the truncated system (3.2) with only the corresponding part of (3.3) for $s = 1, 2, \ldots, N$.

First, we consider all possible solutions of (3.2) without boundary conditions for all x on the infinite interval $-\infty < x < +\infty$, and construct its "phase portrait". We construct two surfaces K_1 and K_2, of codimensions qN and pN, respectively, associated with (3.3). The problem of searching for a solution of (3.2), satisfying (3.3), can then be formulated as that for a segment of the phase trajectory of (3.2), whose endpoints associated with $x = l_1$ and $x = l_2$ are on K_1, K_2, respectively.

The problem of appearance of spatial order is the most interesting in the case where the distributed problem is homogeneous with respect to x, which is associated with the variable x not being explicitly involved in (3.2). If the variable is treated as time, then the system is autonomous. For a solution of a boundary-value problem, the length of the interval between $x = l_1$ and $x = l_2$ is important, and not the numbers l_1, l_2 themselves; more precisely, it is not important that the boundary conditions are specified for $x = l_1$ and $x = l_2$. Denote this length by $l = l_2 - l_1$, and consider solutions existing as $l \to \infty$ and continuously changing as l varies. Such solutions arise if there are a saddle equilibrium state O in the phase space and invariant manifolds S^+, S^- with non-empty intersections with K_1, K_2, respectively. This situation is made clear in Fig. 2.10. The dimensions of K_1 and K_2 should be consistent [334] with those of S^+, S^-; otherwise, a small change in the position of K_1, K_2, S^+, or S^- can make the intersection of K_1 with S^+ (or K_2 with S^-) empty. As $l \to \infty$, in the limit, the required segment L of the phase trajectory is made up of two segments, γ^+ and γ^-, of the phase trajectories lying on S^+ and S^-. γ^+ joins the point where K_1 meets S^+ and O; γ^- joins the point O and the point where S^- intersects K_2. For large l, L approaches γ^+ and γ^- consecutively (Fig. 2.11a). All the above can be repeated also when a saddle periodic motion Γ (Fig. 2.11b) or another more general invariant saddle set replaces the saddle equilibrium.

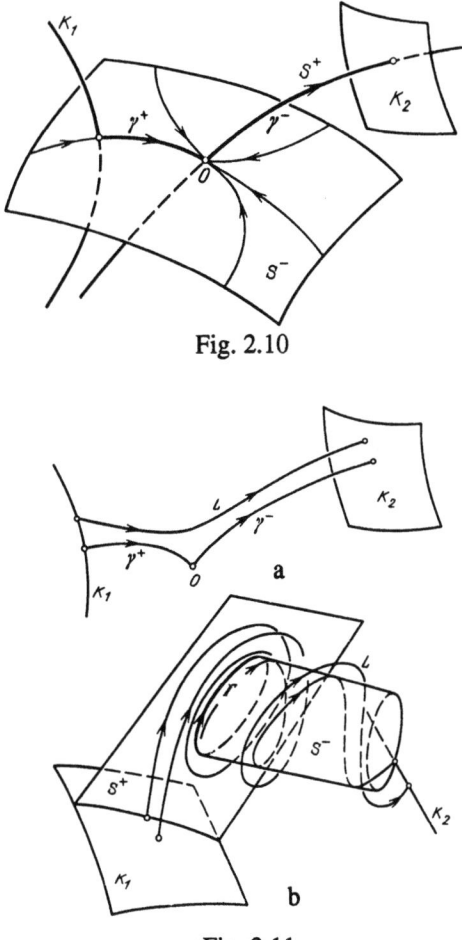

Fig. 2.10

Fig. 2.11

In the above situation, the appearance of spatial (equilibrium, periodic or chaotic) structures is explained naturally in relation to the saddle equilibrium O, a saddle periodic motion Γ or a saddle chaotic motion J. Appearance of a periodic or a chaotic spatial structure is similar to the appearance of periodic or stochastic self-excited oscillations, the difference being that where stability was essential before, the solution should now be of saddle-point type and consistent with the boundary conditions. The consistency lies in the fact that the sum of the dimensions of K_1 and S^+, and K_2 and S^-, respectively, should be not less than the dimension mN of the space. We also note that, in order that equilibrium, periodic or chaotic spatial structures should appear in the above situations, it is also necessary that the spatial extent l be sufficiently large.

The above also enables us to obtain a similar and sufficiently natural treatment of temporal and spatial order phenomena, as well as of tendencies to time and spatial chaos, which are opposite to the former. Apparently, in addition to these two types of evolution of dynamical systems, i.e., synchronization and stochasticity, another type of evolution, self-adjustment, can, and should, be regarded. Self-adjustment phenomena should hardly be reduced to temporal and spatial order. On the contrary, it is apparent that the new sense given by self-adjustment should be stressed, i.e., the possibility of processes leading to the appearance of structures which are stable with respect to more or less considerable and diverse changes in environment and, finally, which are capable of increase and expansion. To reduce self-adjustment to some form of temporal and spatial order should not possibly be carried out also because of their qualitatively different basis: Synchronization is the manifestation of stability in interacting subsystems, while self-adjustment is that of control and organizing feedback functions, which can expediently be regarded from an information point of view.

In conclusion, it should be stressed that the separation of motions of deterministic dynamical systems into regular and chaotic is schematic: intermediate cases are possible. Moreover, any chaotic motion is to some degree endowed with certain regularity, temporal properties and spatial structure. The problem of studying chaotic motions is primarily the discovery and description of their temporal and spatial properties of deterministic nature in some of their parts, and of random nature in others. Chaotic motions have both regular and stochastic features; e.g., the flow behind a streamlined body that moves sufficiently fast is, as a whole, turbulent and chaotic. However, the separation of Karman vortices from the streamlined body is clearly a regular and periodic process. Meanwhile, the flow velocities are random, with more or less periodic components.

CHAPTER 3

STOCHASTICITY TRANSFORMERS, AMPLIFIERS AND GENERATORS

Above, all possible motions of a deterministic dynamical system were separated into regular and irregular, i.e., associated with order and chaos. Meanwhile, irregular chaotic motions were represented as the joint effect of regular and random components, or as the perturbation of some regular component by a certain random component. Here, we discuss the nature of the random components of irregular chaotic motions.

1. Stochasticity transformers

Let a linear oscillator be subject to a random action $\xi(t)$, so that change in its state x, $\dot{x}$ satisfies the equation

$$\ddot{x} + 2\delta\dot{x} + \omega_0^2 x = \xi(t). \qquad (1.1)$$

Solving (1.1) with arbitrary initial conditions $x = x_0$, $\dot{x} = \dot{x}_0$ for $t = t_0$, we obtain

$$x(t) = e^{-\delta(t-t_0)} \left\{ x_0 \cos\Omega(t-t_0) + \frac{\dot{x}_0 + \delta x_0}{\Omega} \sin\Omega(t-t_0) \right\} +$$

$$+ \frac{1}{\Omega} \int_{t_0}^{t} e^{-\delta(t-\tau)} \xi(\tau) \sin\Omega(t-\tau)\, d\tau, \qquad (1.2)$$

where $\Omega^2 = \omega_0^2 - \delta^2 > 0$. Let $\delta > 0$. Then it follows from (1.2) that

$$x(t) = \frac{1}{\Omega} \int_{-\infty}^{+\infty} e^{-\delta(t-\tau)} \xi(\tau) \sin\Omega(t-\tau)\, d\tau \qquad (1.3)$$

as $t_0 \to -\infty$. According to (1.3), the random action $\xi(t)$ is also associated with the random change $x(t)$. However, if $\xi(t)$ is a stationary random process with zero expectation and given autocorrelation function $K_{\xi\xi}(\tau)$, then $x(t)$ is also a stationary random process with zero expectation, and with autocorrelation function

$$K_{xx}(\tau) = \frac{1}{\Omega^2} \int_0^\infty \int_0^\infty K_{\xi\xi}(\tau - v + \mu) e^{-\delta(v+\mu)} \sin \Omega v \sin \Omega \mu \, dv \, d\mu. \quad (1.4)$$

The relationship between the spectral density $S_{\xi\xi}(\omega)$ of the random action $\xi(t)$ and the spectral density $S_{xx}(\omega)$, associated with (1.4), is known to be of the form

$$S_{xx}(\omega) = \frac{S_{\xi\xi}(\omega)}{(\omega^2 - \omega_0^2)^2 + 4\delta^2 \omega^2}. \quad (1.5)$$

Thus, a linear oscillator with an external random action $\xi(t)$ can be regarded as a transformer of $\xi(t)$ into a new random process $x(t)$. Formula (1.3) determines the realizations of $x(t)$ in terms of that of $\xi(t)$, while (1.4), (1.5) establish the relationships between the autocorrelation functions and spectral densities in the case where $\xi(t)$ is a stationary random process, enabling us to find the variance of the random variable $x(t)$, viz.,

$$Dx = K_{xx}(0) = \frac{1}{\Omega^2} \int_0^\infty \int_0^\infty K_{\xi\xi}(\mu - v) e^{-\delta(v+\mu)} \sin \Omega v \sin \Omega \mu \, dv \, d\mu =$$

$$= \frac{1}{\pi} \int_0^\infty \frac{S_{\xi\xi}(\omega) \, d\omega}{(\omega^2 - \omega_0^2)^2 + 4\delta^2 \omega^2}. \quad (1.6)$$

The vanishing of $D\xi$ entails that of Dx, as expected. Recall that it follows from $D\xi = 0$ that $K_{\xi\xi}(\tau) = 0$, i.e., if the action $\xi(t)$ is not random, then the thereby generated change in the coordinate $x(t)$ of the linear oscillator is not random. If the random action $\xi(t)$ is replaced by the sum of a certain given action $\xi_0(t)$ and the same random action $\xi(t)$, then, if the random component of $\xi(t)$ vanishes, the output quantity $x(t)$ does not; however, it loses its randomness, and becomes deterministic; i.e., as expected, the randomness of $x(t)$ is only due to that of $\xi(t)$.

We now make another remark regarding the nature of the transformation of the random component of the input action $\xi(t)$ to the output quantity $x(t)$. Due to (1.6), a transformation of the spectrum of $\xi(t)$ into that of the output $x(t)$ occurs in accordance with the resonant properties of the linear oscillator, or its response curves represented in Fig. 1.10. Meanwhile, the random harmonic components far from the resonance are suppressed; on the contrary, the nearby ones increase.

Stochasticity transformers, amplifiers and generators

Now, let the random perturbation $\xi(t)$ affect the van der Pol oscillator, and not a linear one, so that

$$\ddot{x} - 2\delta\dot{x}(1-x^2) + \omega^2 x = \xi(t) \ . \tag{1.7}$$

We will consider $\xi(t)$ as the input action as before, whereas $x(t)$ denotes the generated output. In contrast to the above case where $\xi(t)$ vanishes, the output $x(t)$ does not; on the contrary, it is different from zero ($\delta > 0$), and in the course of time tends to a certain periodic process $x^*(t)$. As before, we see that randomness in the output disappears if the action is not random; however, the output itself does not vanish at all.

If $\xi(t)$ and $\delta > 0$ are small, the problem posed admits a comparatively simple approximate solution. With respect to new variables A and φ related to the old phase variables x and $\dot{x}$ by

$$x = A\cos(\omega t + \varphi), \quad \dot{x} = \omega A \sin(\omega t + \varphi), \tag{1.8}$$

equation (1.7) is written in the form

$$\dot{A} = \delta A\left(1 - \frac{A^2}{4}\right) - \frac{1}{\omega}\xi \sin(\omega t+\phi), \quad \dot{\phi} = -\frac{1}{\omega A}\xi \sin(\omega t+\phi), \tag{1.9}$$

whose phase portrait for $\xi = 0$ is depicted in Fig. 3.1. All other phase points, except the unstable equilibrium point O, asymptotically approach the circle Γ consisting of equilibrium states. Small random actions ($\xi \neq 0$) lead to random walks of the phase point in the vicinity of Γ, i.e., the oscillation amplitude A is close to two, whereas the phase varies slowly, and can accumulate its changes. In the steady state, the probability density $p(A, \varphi)$ does not depend on the angle φ, and is represented by a surface of the form as in Fig. 3.2. Thus, the input random action ξ is transformed by the van der Pol oscillator into output fluctuations of the oscillation amplitude and random drift of the phase φ. To find the corresponding probability density, the familiar Fokker-Planck partial differential equation can be formed. The surface in Fig. 3.2 and the transition probability densities from one state A, φ into another $\overline{A}$, $\overline{\varphi}$ in time τ [216, 310, 320, 342] can be found by means of this equation, as can the steady-state probability distribution itself, i.e., the equation of the surface. This probability density $p(A, \varphi; \overline{A}, \overline{\varphi}; \tau)$ has the steady-state probability density $p(\overline{A})$ as its limit when $\tau \to \infty$.

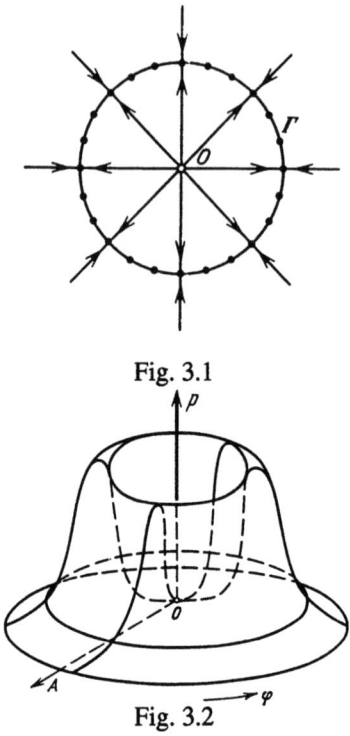

Fig. 3.1

Fig. 3.2

The above examples illustrate the presence of "stochasticity transformers" in the dynamical systems of a linear and a van der Pol oscillator. Meanwhile, external noise is transformed into random changes of x, ẋ or A, φ. Randomness in dynamical systems has only been considered in this respect until recently. Randomness was acquired by a dynamical system from the outside in the form of certain random action. It is this action that caused the randomness of change in the state of the system and in any other quantities determined, which enabled us to represent a dynamical system as a certain transformer of random inputs into random outputs. This standpoint is especially natural in considering complicated dynamical systems in the form of less complicated interacting subsystems. Each of the subsystems, distinguished physically or ideally, experiences some random actions and acts on other subsystems with its output actions. It was earlier regarded as something indisputable that randomness of the output actions was only due to that of the input actions, and that each subsystem was a certain stochasticity transformer, or a transformer of randomness into some other, and that, as input noise vanished, e.g., as its variance tended to zero, the randomness at the output and its variance also vanished.

Stochasticity transformers, amplifiers and generators

However, it was found that, sometimes, as the input variance tended to zero, the output variance remained finite. The account and study of this feature is the principal goal and contents of the present monograph. It turned out that, along with stochasticity transformers, there are dynamical systems that act as stochasticity generators. It also turned out that, according to the input-output coupling, dynamical systems could be divided into stochasticity transformers and generators. The latter are different from the former in the possibility of output stochasticity with finite variance for arbitrarily small input variance. This rough division could be expediently specified in more detail by introducing a certain intermediate case, called a *stochasticity amplifier*. The difference between a stochasticity transformer and a stochasticity amplifier is not so explicit as that between a stochasticity transformer and a stochasticity generator. Thus, a stochasticity amplifier with moderate amplification factor is a stochasticity generator, whereas a stochasticity amplifier with infinite or very large amplification factor is already a syochastic amplifier somewhat resembling a stochasticity generator.

Thus, in accordance with the above, we will distinguish between stochasticity transformers, stochasticity amplifiers, and stochasticity generators, judging by the nature of relation between input stochasticity and output stochasticity [100, 272, 278, 280, 281]. We have already discussed the stochasticity transformer. Summarizing, we note that it transforms a random action at its input into some other random action, so that the probabilistic description of the randomness of the output, at least after a certain time, is determined by that of the input; if the input is not random, then the output is also not random. It is this, and only this, case that has been studied until recently [104, 193, 194, 216, 299, 310, 320, 342]. It was studied intensively in considering random oscillations of mechanical and radio-engineering systems. The above examples are all of this sort. The main problem in the studies was the search for characteristics and the description of randomness at the output for given characteristics and stochastic descriptions of the input. This could be solved simply in some cases. Thus, within the framework of the theory of correlation of stationary random processes, the spectral output function $S_{xx}(\omega)$ of a linear dynamical system is related to the input spectral function $S_{\xi\xi}(\omega)$ by

$$S_{xx}(\omega) = |K(i\omega)|^2 S_{\xi\xi}(\omega), \qquad (1.10)$$

where $K(i, \omega)$ is the gain factor for the linear system. In some cases, these problems are

2. Stochasticity amplifiers

In principle, a stochasticity amplifier is not different from a stochasticity transformer. The difference is in the amplification factor. If it is vary large, then output stochasticity cannot be controlled, since it requires unattainable accuracy in the random input actions. The simplest example is a threshold element whose output η is related to an input ξ by a correlation of the form

$$\eta = \begin{cases} 1 \text{ for } \xi > 0, \\ 0 \text{ for } \xi \leq 0. \end{cases} \qquad (2.1)$$

This threshold transformer changes a random process $\xi(t)$ of arbitrarily small magnitude with $M\xi = 0$ and $D\xi \ll 1$ into a random sequence of 1 and 0. Another example is given by a conservative oscillator with as phase portrait a family of nested closed curves (Fig. 3.3a). For arbitrarily small random actions, the phase point starts a random walk from one closed phase curve to another, and its total displacement in the course of time can be arbitrary. The amplification of a small random effect to a finite, and even large, displacement occurs on account of the "perturbation build-up" effect. A similar stochasticity amplifier is a system whose phase portrait is a two-dimensional torus with periodic or quasi-periodic coil (Fig. 3.3b). A different example is given by a dynamical system with phase portrait as in Fig. 3.4. Without perturbations, depending on the initial conditions, the phase point either moves along the phase curves γ_1, α, or along γ_1, β, or along γ_2, α, or, finally, along γ_2, β. As an arbitrarily small perturbation appears, the situation changes sharply; from time to time, the phase point can make random transitions in the vicinity of the points O_1, O_2, continuing to move either along γ_1 or γ_2 and either along the phase curve α or β, respectively. Which of the motions it selects depends on arbitrarily small random actions, and is, therefore, random.

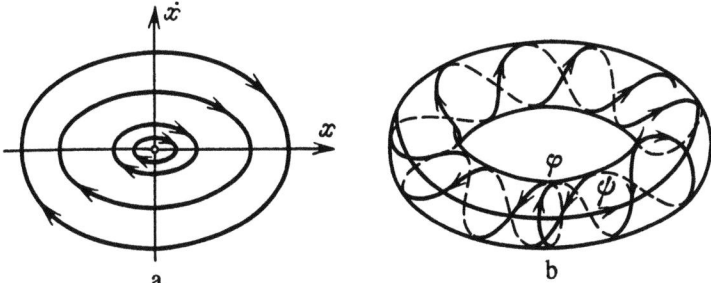

Fig. 3.3

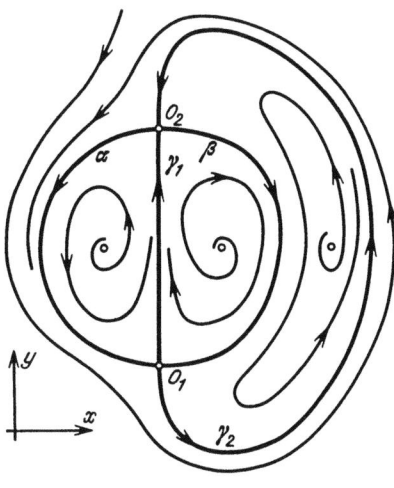

Fig. 3.4

The above examples are associated with special and limiting cases. The appearance of a very small insensitivity zone in a threshold element radically changes the above, since the value of ξ should be sufficiently large and exceed the threshold. Systems with phase portraits such as a centre and periodic or quasi-periodic coils of a two-dimensional torus are generally destroyed under arbitrarily small non-conservative perturbations, and are *systèmes grossiers* in the sense of Andronov-Pontryagin (are structurally unstable). Consequently, the above examples may seem to be far-fetched, and associated only with certain ideal models and structurally unstable systems. The same could also be said about the example having phase portrait as in Fig. 3.4, since, under a slight perturbation of the dynamical system, the separatrices α, β, γ_1, and γ_2 do not lead from one saddle equilibrium into another, and the phase portrait can take, e.g., the form as in Fig. 3.5. For similar jumps, the random actions must be not very small.

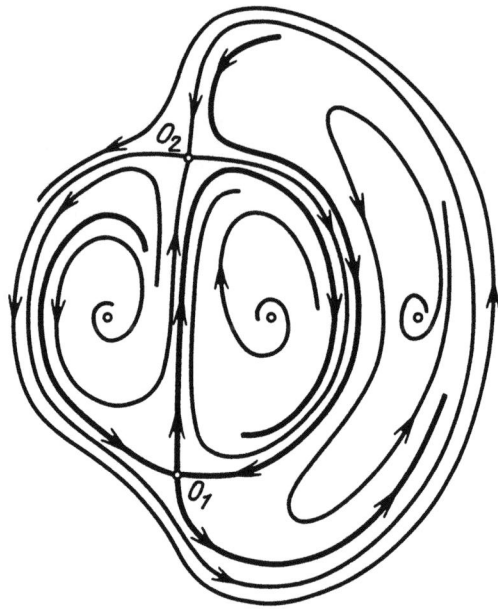

Fig. 3.5

Although, in the above examples, the slightest perturbation excludes the existence of a stochasticity amplifier with indefinitely large amplification factor, as is shown in the sequel, in somewhat generalized form, such mechanisms can generate chaotic motions of the dynamical system. Meanwhile, small perturbations only change the capacity of the system as regards stochasticity increase, but do not eliminate it at all.

We start with a very simple example. Take the dynamical system determined by the following point mapping T of the interval [0, 1] into itself (Fig. 3.6):

$$\bar{x} = Tx = \begin{cases} ax & \text{for} \quad 0 \leq x \leq \frac{1-\varepsilon}{2}, \\ \frac{1-\varepsilon}{2} a & \text{for} \quad \frac{1-\varepsilon}{2} \leq x \leq \frac{1+\varepsilon}{2}, \\ -a(x-1) & \text{for} \quad 1 \geq x \geq \frac{1+\varepsilon}{2}. \end{cases} \quad (2.2)$$

In (2.2), a and ε are two parameters (generally speaking, ε is a small positive number and $1 \leq a \leq 2$). Any point x_0 in [0, 1] falls, after successive application of T, into a subinterval containing ε, and then transforms cyclically with a period of N transformations. The number N depends both on a and ε in a complicated way; however, generally speaking, as ε decreases, N increases indefinitely. For large N, the points $x_1, x_2, \ldots x_N$ of this periodic sequence are closely packed on [0, 1], and alternate (see Fig. 3.7). Near each x_i, a neighbourhood δ_i transformed into the subinterval containing ϵ can be constructed. This neighbourhood is of width ϵa^{-N+s} near x_s. $\delta_1, \delta_2, \ldots, \delta_N$ do not overlap, but are close to each other, and random in the general case. The mean distance between them is of order N^{-1}; therefore, for large N, even very small random actions violate the consecutive transitions from δ_i into δ_j, δ_j into δ_k, etc. These violations occur as soon as the random action exceeds ϵa^{-N}, and, as the random action increases, the chaotization of transition increases. It is essential to stress that the noise amplitude threshold is very small; e.g., for $\epsilon = 10^{-2}$, $a = 2$ and $N = 30$, it is of order 10^{-11}.

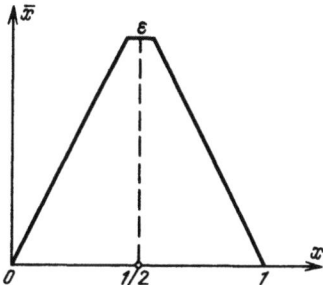

Fig. 3.6

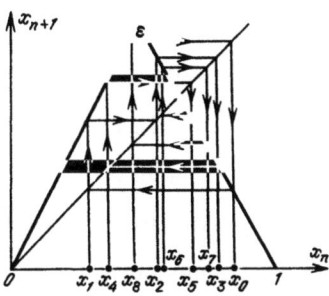

Fig. 3.7

We encounter a similar situation in the case of the familiar and well-studied point mapping

$$\bar{x} = ax(1-x) \quad (0 < a \leq 4) \tag{2.3}$$

whose graph is given in Fig. 3.8. The role of the neighbourhood containing ϵ is played by a small neighbourhood of the point associated with the parabola vertex. The number N can be as large as we please. Moreover, infinitely many values of the parameter a are known to exist, and N increases indefinitely as they are approached. These increases occur by doubling N indefinitely. In Fig. 3.9, the bifurcation diagram of (2.3) is given, and two limit values for these infinite series are indicated.

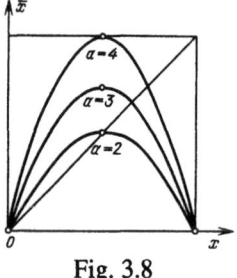

Fig. 3.8

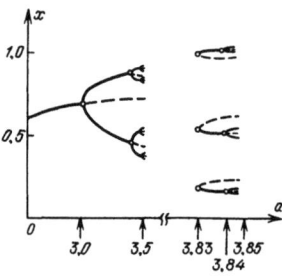

Fig. 3.9

The above examples already show that the mechanism for chaotization of the motions of a dynamical system is sufficiently general, provided the system is in a stable periodic motion with a very small domain of attraction. The size of this domain at the "narrowest" places is in the example of order εa^{-N}, and it is this smallness and the considerable changes in motion after leaving the neighbourhood that causes chaotization of motion. Meanwhile, despite the random jumps, the consecutive images lie in small neighbourhoods of points $x_1, x_2, x_3, \ldots, x_N$ making up a stable N-cycle of fixed points; these jumps only violate the cyclic order of displacement across the neighbourhoods, and cause slight blurring in each. Even with twice greater accuracy, the necessary accuracy to provide for cyclicity for large N considerably exceeds a computer's capacity, which leads to chaotic, and not strictly periodic, sequences at the output. Thus, chaotization of motions can arise due to extreme narrowness of the domain of attraction of a stable periodic motion, when jumps from one domain of attraction into another occur duly. Meanwhile, the transitions should not completely take the phase point out of the whole domain of attraction of this stable motion. In other words, on the one hand, since the domain of attraction is narrow, the phase point leaves part one of it, and gets into another under small random actions; on the other hand, it must always stay in this domain in order not to fall into another part of the phase space and to find itself in the domain of attraction of another steady-state motion. It is clear that the concepts of "domain of attraction" and "leaving" require some sharpening, consisting in the necessity to distinguish between global and local domains of attraction. The global domain of attraction $\Pi(\Gamma)$ of an asymptotically stable periodic motion Γ, regarded as a phase curve, is the set of phase points whose phase curves have Γ as their limit set if the time interval increases indefinitely. In other words, for any point $M \in \Pi(\Gamma)$, the phase trajectory $x(M,t)$ emanating from M asymptotically approaches Γ, so that

$$\lim_{t \to \infty} \rho(\Gamma, x(M, t)) = 0. \tag{2.4}$$

The domain $\Pi(\Gamma)$ can be obtained from a sufficiently small neighbourhood of Γ by time reversal (as $t \to -\infty$). By a sufficiently small neighbourhood we mean, e.g., an ε-neighbourhood of Γ so small that, for each of its points M, there is a point M_0 on Γ such that for certain K and the same κ, for all $t > 0$,

$$\rho(x(M, t), x(M_0, t)) < K e^{-\kappa t} \rho(M, M_0) \tag{2.5}$$

(Fig. 3.10).

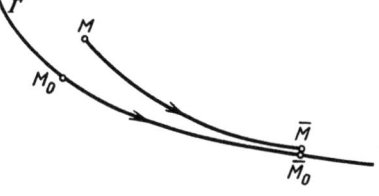

Fig. 3.10

The point M_0 on Γ is unique for M. Denote this one-to-one mapping by F, so that $M_0 = F(M)$. It is obvious that $F(M_0) = M_0$. If the point M is sent to $\overline{M}$ in time τ, and M_0 to $\overline{M}_0$, then it is obvious that $F(\overline{M}) = \overline{M}_0$. For points M sufficiently close to Γ, the function $\Gamma(M)$ is continuous. As M moves father from Γ, discontinuity is possible. If M_p is a point of discontinuity, then also all points of the phase trajectory emanating from M_p are associated with discontinuity points, i.e., the set of discontinuity points of the function

F consists of phase trajectories. At M_p, F is not defined, whereas the value of K increases indefinitely as M approaches M_p. Generally speaking, the point $M_0 = F(M)$ changes jumplike in passing through a point of discontinuity, e.g., from M_0^1 to M_0^2. This means that if the point $x(M, t)$ is close to the point $x(M_0^1, t)$ before the discontinuity, then, after the discontinuity point, $x(M, t)$ is close to $x(M_0^2, t)$. Thus, an arbitrarily small displacement of M in the course of time turns into a finite displacement. It is in this that the indefinite increase in small perturbations consists if only they send M across a point of discontinuity of F. The above is illustrated in Fig. 3.11, where M' and M'' are two nearby points "separated" by a discontinuity point M_p. Despite the nearness of M' and M'', the phase trajectories emanating from them diverge strongly. The function F sends M' and M'' to two different points M_1 and M_2 on the phase curve Γ.

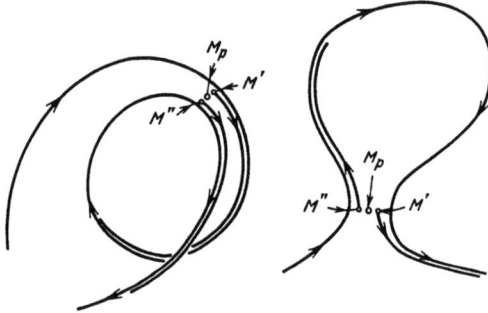

Fig. 3.11

Thus, in order that the above-mentioned mechanism of stochasticity amplification arise, on the one hand, it is necessary that almost all points of a certain not very small neighbourhood of an asymptotically stable periodic motion Γ should belong to global domain of attraction $\Pi(\Gamma)$ of Γ, and, on the other hand, that the discontinuity points M_p of $F(M)$ should be contained in a very small neighbourhood of Γ, associated with a periodic motion. The latter requirement always holds if Γ is of large length, and belongs to a certain bounded domain. Note that it is the length of a closed phase curve which is important, and not the period during which the phase point traverses it. Certainly, the period is also large; however, this alone does not provide for large length of the phase curve, because Γ may pass near saddle equilibrium states. If this is excluded, then both requirements are equivalent.

Similar mechanisms for stochasticity amplification are possible in the general case of several asymptotically stable periodic motions $\Gamma_1, \Gamma_2, \Gamma_3, \ldots$. Here, amplification arises not only because of "jumps" along Γ_s, but also due to the transitions from some periodic motions to others (Fig. 3.12).

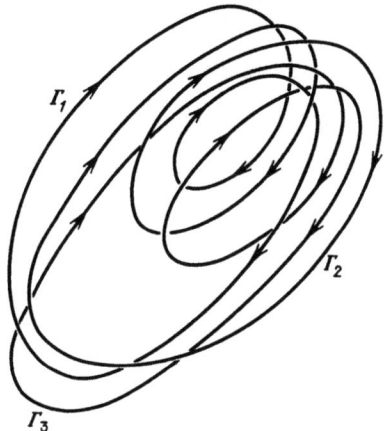

Fig. 3.12

An estimation of the form

$$\rho < c \left(\frac{V}{L}\right)^{1/(n-1)}, \qquad (2.6)$$

where V is the total volume of a domain containing the curves $\Gamma_1, \Gamma_2, \ldots$, L their total length, c a certain constant, and n the dimension of the phase space, can be proved for the minimum distance ρ of discontinuity points M_p from $\Gamma_1, \Gamma_2, \ldots$. It can be seen that, as L increases indefinitely, the quantity ρ decreases and tends to zero. The same estimate also holds for each part of $\Gamma_1, \Gamma_2, \ldots$ with total length L.

Thus, when regarding a dynamical system as a stochasticity transformer, we may distinguish the case where this transformation occurs with infinitely large amplification factor.

Such stochasticity transformers are called *stochasticity amplifiers*. Possessing an infinitely large amplification factor, stochasticity amplifiers have a certain threshold, which need not be defined in the "ideal" cases.

3. Stochasticity generators

Qualitatively, a stochasticity generator is different both from a stochasticity transformer and a stochasticity amplifier. It is similar to the ideal amplifier in that the randomness of its output does not disappear if the random input component decreases and disappears, and is different in that the stochastic characteristics of its output negligibly little depend on sufficiently small random inputs, and are completely determined by the dynamical properties of the system. In the limit, as the random actions approach zero, randomness of the output does not disappear, and, irrespective of the stochastic characteristics of the input, tends to the same stochastic output. The statistical description of the latter can always be found from the description of the dynamical system.

We begin with a very simple example. Let the state of the system vary according to

$$x_n = 2x_{n-1} - [2x_{n-1}], \qquad (3.1)$$

where $[z]$ is the entier of z. The graph of this point mapping is the same as in Fig. 2.7. There are no stable fixed points. The mapping is unstable everywhere, and makes any interval twice as long. The whole interval $[0, 1]$ is mapped by (3.1) to $[0, 1]$, covered twice. At $x = 1/2$ the point mapping (3.1) is discontinuous. Let the initial value x_0 be random, and have an arbitrary smooth probability density $p_0(\xi)$. By (3.1), all subsequent values $x_1, x_2, x_3, \ldots$ are also random, and their probability densities $p_1(\xi), p_2(\xi), \ldots$ are related by

$$p_n(\xi) = \frac{1}{2}\left[p_{n-1}\left(\frac{\xi}{2}\right) + p_{n-1}\left(\frac{\xi+1}{2}\right)\right] \qquad (3.2)$$

[100].

A remarkable property of transformation (3.2) is that

$$\lim_{n \to \infty} p_n(\xi) = 1 \qquad (3.3)$$

irrespective of the initial probability density $p_0(\xi)$, i.e., the limiting probability density is uniform. The proof is not complicated. Indeed, differentiating (3.2) with respect to ξ, we find

$$p'_n(\xi) = \frac{1}{4}\left[p'_{n-1}\left(\frac{\xi}{2}\right) + p'_{n-1}\left(\frac{\xi+1}{2}\right)\right];$$

hence,

$$\max_{\xi} |p'_n(\xi)| \le \frac{1}{2} \max_{\xi} |p'_{n-1}(\xi)|,$$

and $p'_n(\xi) \to 0$ as $n \to \infty$, which means that $p_n(\xi)$ approaches a constant, equal to one due to the normalizations. Thus, the probability density of the state established in the system in question does not depend on the original probability density, and is determined by the properties of the system only. Randomness can be added to the state by a random external action. If the external action then disappears, a uniform distribution emerges. However, if the random actions continue, but are quite small, a close-to-uniform distribution emerges. That the established distribution is uniform is a property of the simple example considered; but the fact that there is a limiting probability distribution irrespective of the original distribution is common to both.

As another example, we consider a linear impact oscillator with negative friction, described by the differential equation

$$\ddot{x} - 2\delta\dot{x} + \omega_0^2 x = -\sum p\delta(t-t_s), \qquad (3.4)$$

[280], where t_s, $s = 1, 2, 3, \ldots$, are consecutive moments at which $x = 0$ and $\dot{x} \ge a > 0$. The parameters ω_0, δ, p, a, and $\omega^2 = \omega_0^2 - \delta^2$ are positive. This is an autonomous system of order two with two-dimensional phase space, or phase plane cut along the ray $x = 0$, $\dot{x} \ge a$, with identified boundaries, so that $x = 0$, $\dot{x}$ is identified with the point $x = 0$, $\dot{x} - p$ (Fig. 3.13). It can be seen from the phase portrait in the diagram and from equation (3.4) that the consecutive values $\dot{x}_s$ and $\dot{x}_{s+1}$ of the velocity immediately after an impact are related by

$$\dot{x}_{s+1} = \begin{cases} q\dot{x}_s & \text{for } q\dot{x}_s < a, \\ q\dot{x}_s - p & \text{for } q\dot{x}_s \ge a, \end{cases} \qquad (3.5)$$

where $q = \exp 2\pi\delta\omega^{-1}$. Relation (3.5) can be regarded as a point mapping sending the point $\dot{x}_s$ to $\dot{x}_{s+1}$. The possible substantially different forms of its graph are depicted in Fig. 3.14. In Fig. 3.14a, unrestricted oscillator build-up occurs. The amplitude of the oscillator velocity $\dot{x}$ turns out to be greater than $p(q-1)^{-1}$ from a certain moment onwards, and then monotonically and indefinitely increases. This is valid when the point M of the graph in Fig. 3.14a lies above the point N, i.e., for

$$p|q-1|^{-1} < a. \qquad (3.6)$$

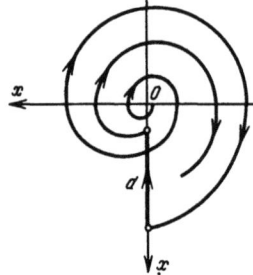

Fig. 3.13

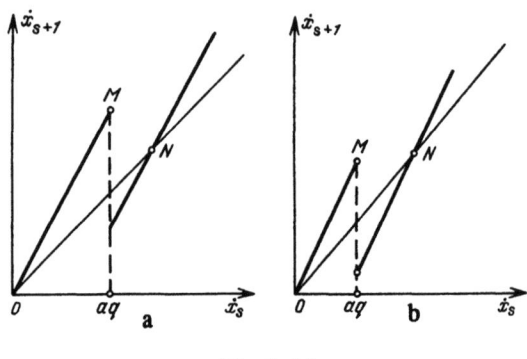

Fig. 3.14

If the reverse inequality holds (Fig. 3.14b), two totally different types of behaviour are possible, depending on the initial conditions: if $\dot{x}_0 > |pq-1|^1$, divergent oscillations; if $\dot{x}_0 < |pq-1|^1$, restricted chaotic oscillations. In the latter case, the consecutive images of the point $\dot{x}_0$ (i.e., $\dot{x}_1, \dot{x}_2, \ldots$) are everywhere dense in the interval J with ends at a-q and a. The mapping (3.5) is expanding on J, so that there can be no stable fixed points (of any multiplicity). The expansion and restricted general motion make this case similar to the preceding example: Any smooth initial probability distribution of the initial velocity $\dot{x}_0$ tends in the course of time to a certain limiting distribution, now non-uniform. Therefore, the above linear unstable oscillator with impacts damping its oscillations is a stochastic oscillator. The oscillogram of variation of its coordinate x is given in Fig. 3.15.

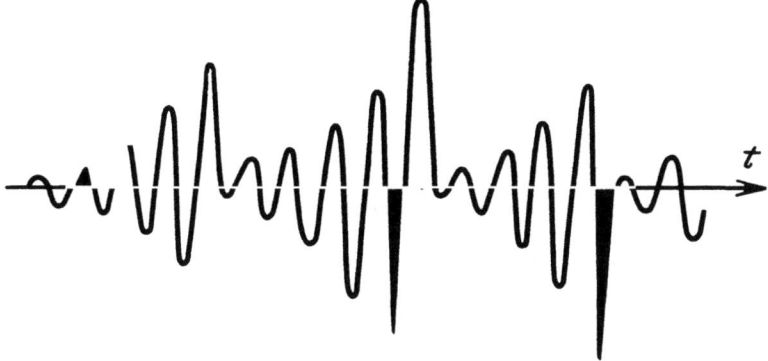

Fig. 3.15

We have already mentioned above that dimension three is the least in which a dynamical system can possess chaotic motions. The damping impact oscillator is however two-dimensional, and, nevertheless, capable of chaotic oscillations. There is no contradiction, since the statement that three is that least dimension concerns smooth dynamical systems, and motions of impact oscillators may have phase point jumps. Still, we note that the jumps can be eliminated if we cut the $(x, \dot{x})$-plane along the ray $x = 0$, $\dot{x} \geq a$, and glue the left boundary to the right along the ray $x = 0$, $\dot{x} \geq a - p$. The phase point moves smoothly across the cut phase plane with the cut boundaries identified (glued together); however, not being two-dimensional, the phase space is not a plane any more, which is sufficient for chaotic motions to arise.

We now draw our attention to the transition from chaotic motions to those extending at infinity. If the point M of the graph in Fig. 3.14 lies above the point N, then only motions with unrestricted build-up are possible. If the relation is reverse, both motions with unrestricted build-up and restricted chaotic oscillations occur. However, if M is just a little higher than N, then restricted chaotic oscillations are possible for a very long time, making the transition into divergent oscillations at some unpredictable moment.

Consider the two-dimensional mapping

$$\begin{cases} \bar{x} = 2x - 1 \\ \bar{y} = \frac{1}{3}y + \frac{2}{3} \end{cases} \text{ for } x \geq 0, \quad \begin{cases} \bar{x} = 2x + 1 \\ \bar{y} = \frac{1}{3}y - \frac{2}{3} \end{cases} \text{ for } x < 0. \qquad (3.7)$$

The points $O_1(1, 1)$ and $O_2(-1, -1)$ are its fixed saddle points. The square $G(|x| \leq 1, |y| \leq 1)$ is sent into itself by the mapping (3.7), i.e., into the two regions G_1, G_2 equal to the images of $0 \leq x \leq 1$, $|y| \leq 1$ and $-1 \leq x < 0$, $|y| \leq 1$, respectively, which make up G (Fig. 3.16). Each of G_1, G_2 is in G. Thus, for (3.7), a global contraction occurs, G is transformed into itself, and local instability holds, for (3.7) is (twice) expanding with respect to x. In G, the transformation (3.7) (we call it T) can have no stable fixed points, since any two arbitrarily close points different in the coordinate x diverge at a finite distance under iteration by T, so that, in G, the transformation is chaotic. To grasp this fact better, consider the successive images of G. Under the mapping applied once, G is transformed into G_1, G_2 (Fig. 3.16a). Then each of G_1, G_2 is sent into two domains, G_{11}, G_{12} and G_{21}, G_{22}, respectively (Fig. 3.16b). Meanwhile, the inclusions

$$G_{11} \subset G_1, \ G_{12} \subset G_2, \ G_{21} \subset G_1, \ G_{22} \subset G_2$$

hold. Similarly, each G_{ij} (Fig. 3.16c), i, j = 1, 2, is mapped into two regions G_{ij1}, G_{ij2} with

$$G_{ij1} \subset G_{j1} \ , \ G_{ij2} \subset G_{j2} .$$

By iteration, we arrive at all possible regions

$$G_{i_1 i_2 \ldots i_n} \quad (i_1, i_n = 1, 2),$$

with

$$G_{i_1 i_2 \ldots i_n} \subset G_{i_2 \ldots i_n} \subset \ldots \subset G_{i_{n-1} i_n} \subset G_{i_n} . \tag{3.8}$$

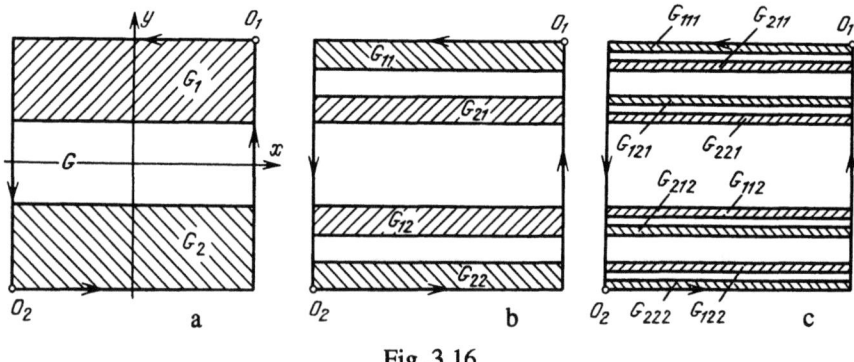

Fig. 3.16

Each rectangular region $G_{i_1 i_2 \ldots i_n}$ is of width 2 with respect to x, and of width 3^{-n} with respect to y. It follows that these regions contract to a certain set J as $n \to \infty$, a continuum of straight line segments placed along the x-axis. The limit set J is sent onto itself by T, so that x undergoes the transformation

$$\bar{x} = \begin{cases} 2x-1 & \text{for } x>0, \\ 2x+1 & \text{for } x<0, \end{cases}$$

which can also be written as the mapping (3.1), replacing x by x-1. Note also that a mapping very similar to (3.7) is generated by the phase trajectories of the Lorenz equations (1.24), Chapt.1, on the section 2-plane z = r-1.

The above examples can be extended; the last chapter is devoted to them completely. Now, we will try to answer the problem of the reason for the stochastic behaviour of the above systems, and give general conditions for their appearance. It has already been noted that stochasticity generation is impossible in the stability case, since only stable equilibrium states and stable periodic motions can be steady-state motions. In these cases, even if the initial condition is random, randomness disappears with time, so that the probability density as $t \to \infty$ vanishes everywhere outside an equilibrium state or a closed curve associated with a periodic motion. If the motion is periodic, some trace of randomness in the initial condition is still there, as the random phase of periodic oscillations. Meanwhile, the probabilistic description of phase depends on the initial condition's probability distribution, and is determined by it. Thus, to generate stochasticity, we need local instability, global boundedness of the motion, and a certain global contraction or, at least, no expansion.

H. Poincaré himself pointed to instability as the reason for unpredictability [608]. Stronger arguments can be found in N. S. Krylov [207] and M. Born [410], and lead to the unpredictability of the events arising because of unstable motions, since, whatever accurate the initial conditions are specified (accuracy being practically restricted), the slightest error leads to quite a palpable difference after a sufficiently long time. Applied to long-term weather forecast, the effect of unusually sensitive dependence on small perturbations was noted by E. Lorenz, and called the *butterfly effect* in connection with the events in a story by R. Bradbury. Instability imparts indeterminacy to future evolution of the dynamical system, making the future unpredictable to a certain extent. Restrictions on indeterminacy were given by A. N. Kolmogorov, and R. Adler and A. Konheim (metric and topological entropy). Both entropies somewhat differently characterize the quantitative measure of the divergence rate of phase trajectories and the degree of indeterminacy arising. Thus, the unpredictability and stochasticity of motions of a dynamical system are generated by their instability under global contraction. Still, how does this happen in reality, and, in particular, how can local instability go along with global contraction ?

We will try to answer the question by analyzing the phase portrait. Let γ be a phase trajectory, and x_0, x_t two of its points. Assume that a small spherical neighbourhood δ_0 of x_0 is transformed into a neighbourhood δ_t of x_t after time $t > 0$. In the general case, as long as δ_t is small, the variation of δ_t is determined by the linearized equations with respect to deviations ξ from γ, as

$$\dot{\xi} = \frac{\partial X}{\partial x}\bigg|_\gamma \xi \tag{3.9}$$

when the original differential equations are

$$\dot{x} = X(x). \tag{3.10}$$

In some directions, contractions occur; in others, extensions. Consequently, the original spherical neighbourhood δ_0 is transformed for large t into a neighbourhood δ_t of ellipsoid form, extended along some axes and contracted along others (Fig. 3.17). There are as many ellipsoid axes with longwise extension as instability degrees. The variation of the volume of the ellipsoid with time t is determined by div $X(x)$. For div $X(x) > 0$, the volume increases; if the inequality is reverse, the volume decreases.

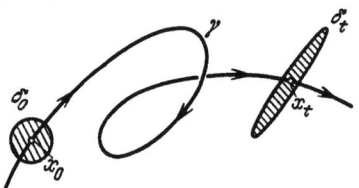

Fig. 3.17

Each ball in δ_t is in the course of time, in turn, transformed into a prolate ellipsoid, etc.; therefore, the originally spherical neighbourhood δ_0 still more and more extends in the instability direction, and contracts in the stability direction. However, the expanded neighbourhood δ_t should remain in a bounded domain due to global contraction, and, therefore, must bend and fold somehow. For three-dimensional smooth dynamical systems described by differential equations, the only mechanism for this extension, bending and folding is a homoclinic structure, where a sequence of transformations of δ_t and its parts in neighbourhoods of saddle periodic motions and equilibrium states are carried out. The metamorphoses occurring in the originally spherical domain near a saddle equilibrium and a saddle periodic motion are in Fig. 3.18. Sucking the neighbourhood into this "transformer" occurs along the surface S^+; throwing out, along S^-. After the sucking-in and throwing-out are done, the whole process can repeat as soon as part of the neighbourhood discarded falls into the range of entrainment of some surface S^+, etc. It is this infinite sequence of transformations that can occur in a homoclinic structure.

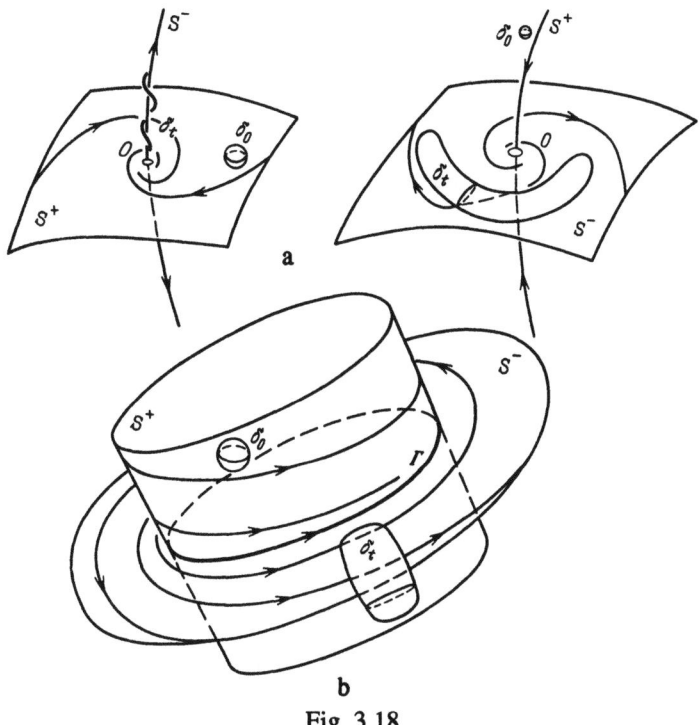

Fig. 3.18

The transformation (3.7) considered above is the simplest example of this infinite sequence of transformations in neighbourhoods of fixed saddle points O_1, O_2. A further description of homoclinic structures and their role in forming chaotic motions is in Chapts. 4, 6; Chapts. 7, 9 list various examples.

For dynamical systems of dimension higher then three, the mapping (1.1) of Chapt.2 yields another example of extending indefinitely and covering. The characteristic property is covering without folds, requiring dimension at least three for the point mapping; for the corresponding dynamical system described by differential equations, the dimension should be four at least. This fact itself is already an argument for possible substantial difference of four-dimensional systems from three-dimensional, and not only three-dimensional from two-dimensional. How real this difference is cannot be determined at present easily, because four-dimensional systems are very little studied. We also note here that, in four-dimensional systems, bifurcations are possible; they are absent in three-dimensional ones, as can be seen in the sequel. It is quite feasible that the difference between four-dimensional and three-dimensional systems is most considerable.

It can be seen from the above that complicated transformations in the phase space's stochasticity generator and related chaotization of motion do not have any external random actions as the cause for their appearance. Everything occurs in accordance with the deterministic equations of motion for the dynamical system, and is generated by the latter itself. This is where the difference lies between a stochasticity generator and a stochasticity amplifier, whose stochasticity is generated by, and substantially depends on, small random, possibly, unaccountable, perturbations. Meanwhile, a stochasticity generator has no randomness either if we do not assume that there are even negligibly small random perturbations. The statistical characteristics of a stochasticity generator are independent of these uncontrolled random perturbations; however, the latter , though being possibly arbitrarily small, are necessary for randomness. It is hard to say whether this treatment is erroneous in reality; it is an inevitable consequence of the present ideas.

4. Is the stochastic and chaotic motions of deterministic dynamical systems real ?

In conclusion, it is necessary to touch upon the very complicated question of whether the stochasticity of deterministic dynamical systems is "real", to be studied by probability theory. The formulation should be sharpened from the start. In fact, it is hardly possible to deny that the stochasticity of the surrounding world is generated by deterministic dynamical systems in the fluctuating quantum microcosmos. The question is to what extent and for what reasons the stochasticity of macroscopic deterministic systems, in classical physics, is associated with canonical axiomatics of probability theory. This form of question is alien to probability theory, which postulates randomness and probability, and does not inquire into their origin (possibly, because the empirical basis of probability theory left no hope for a solution). Probably, the situation is somewhat different now. Moreover, the development of the theories of non-linear oscillations and dynamical systems leads just to the question of the origin of randomness and probability, opening new ways and approaches to the solution of arising problems.

The random sequences generated by digital computers are commonly said to be *quasi-random*. The same name is often given the stochastic motions of dynamical systems. We can agree to the former, since after repetition a computer produces a sequence which reflects certain properties of a random sequence, but is not random to the full. It is not possible to justify as simply the quasi-randomness of stochastic motions of dynamical systems. Let us sharpen the conditions for kinds and realization of stochastic motions. If the conditions are thought to be the same as in the computer case, then the motions should be quasi-random; however, they are not. The finiteness of computer word length enables us to reproduce the initial conditions exactly, while small interference cannot affect the computational result. Both the former and the latter do not hold for a continuous dynamical system, since the initial conditions cannot be reconstructed, and even very small interference can affect the motion. These new circumstances can be somewhat reflected by a mathematical model of the form

$$\varepsilon \dot{\mathbf{x}} = f(\mathbf{x}, \mathbf{y}, \xi), \quad \dot{\mathbf{y}} = g(\mathbf{x}, \mathbf{y}), \qquad (4.1)$$

where **x**, **y** are phase variables of the dynamical system on two different scales, which we call *macro-* and *microdescriptions*, ξ (in the cases of interest) is a very small random variable reflecting the probabilistic nature of motions in the microcosmos, and, generally speaking, ε is a very small parameter, so that **x** varies rapidly while **y** varies slowly. There may bet more than two such different levels of variation rate.

Since **y** varies slowly, it can be taken constant in short time intervals. We assume that the lengths of these constancy intervals are sufficient for the variation of **x** to be steady-state in time. There arises a probability density for **x**, depending on **y**. If the dynamical system describing the variation of **x** for constant **y** is a stochasticity transformer, then we can neglect the action ξ, due to its smallness. Conversely, if it is a stochasticity amplifier or generator, then neglecting ξ is inadmissible. Note that, in the latter case, where the variation of **x** is described by a stochastic generator, the steady-state probability density does not depend on the statistical characteristics of the random variable ξ, and is completely determined by the dynamical system

$$\varepsilon \dot{\mathbf{x}} = f(\mathbf{x}, \mathbf{y}, 0) \qquad (4.2)$$

for constant **y**.

If the time interval is of order ε, then the variation of **x** is of deterministic nature; however, on the scale of duration of change in **y**, the change in **x** is random. We can assume, e.g., that it is already random for time intervals of order $\sqrt{\varepsilon}$. If so, then, in the second equation of (4.1), **x** can be regarded as a random variable with very short correlation time (of white-noise type); consequently, **y** is a Markov process, while the components of **y** are non-Markov processes. This occurs if the system with respect to **y** is a stochasticity transformer. It can also be a stochasticity generator, and its stochasticity is then of another sort, characterized by the fact that the probability density of state transition is the delta-function. We stress that the density is the delta-function during substantial changes in the components of **y** for small effect **x**.

Thus, what is a possible answer to the question in the Sections's headline ? Stochastic and chaotic motions of deterministic dynamical systems are just actual randomness of the environment. We have only begun to study the randomness in this new aspect, and it is to this new approach that we owe the "discovery" of stochastic and chaotic motions of deterministic dynamical systems.

CHAPTER 4

BRIEF SURVEY OF STUDIES RELATED TO THE APPEARANCE OF THE PROBLEM OF CHAOTIC AND STOCHASTIC MOTIONS AND TO TURBULENCE THEORY

Until quite recently, all phenomena in the world were separated into deterministic and random. The theory of differential equations, mathematical physics, classical mechanics, and electro-dynamics studied the former; probability theory, the theory of random processes, statistical physics, and quantum mechanics, the latter. This sharp borderline seemed unquestionable, and raised no doubts, though certain discrepancies did arise: The probabilistic nature of classical statistical physics laws was not justified accurately, and was accounted for by the very large number of particles and degrees of freedom.

It turned out, however, that even the motion of one particle within the framework of Newton's laws of motion could be unpredictable and random. Moreover, it was found that many, even the simplest, deterministic autonomous dynamical systems could have stochastic motions. Today's scientific community, shaken by the avalanche of new facts and developments, has assimilated the discoveries as something extraordinary, and affecting our profoundly intuitive ideas.

Especially burning was the question of finding mechanisms for stochasticity to appear in the theory of turbulence in liquids, gases and plasma. Possibly, it is because of this fact that mechanics and physics specialists occupied with the theory of waves and turbulence confronted this new understanding of the mechanisms with great partiality and ardour. First, somewhat unexpected, familiarity with the new outlook occurred at the schools on oscillations and waves in 1972, 1973 and 1975 [291, 292], seminars on phase synchronization in 1973 [273], the seminar-conference on dynamical systems and control processes in 1974 [271], the school of specialists in mechanics in 1975, and schools on non-linear hydrodynamic stability theory in 1976 and 1978. It all began with the lectures [154, 270, 366] at the first school on oscillations and waves in March 1972. Subsequent events were recorded in numerous publications. If it was not observed earlier that motions of an autonomous dynamical system could be random, then now random motions are discovered everywhere.

Similar "familiarity" and subsequent reaction occurred in a number of other branches, too, but somewhat later. We can point out the seminars in 1976-1977 on turbulence and the Navies-Stokes equation, and the seminar in 1973 on point mappings and applications in Toulouse. If the impetus was mostly given by lectures on the basis of studies of the Gorky school on non-linear oscillation theory (already discussed in Chapt.1) and the Siberian group of physicists, whose representatives were introduced to each other for the first time at the schools on oscillations and waves (1972-1973), then the pivot of the seminars in 1976-1977 on turbulence and the Navier-Stokes equation was Lorenz' 1963 paper [563] on the non-periodic nature of motions of a 3-mode model of convective turbulence and the paper of D. Ruelle and F. Takens [627], containing new hypotheses on the nature of turbulence. The works were reprinted, and it is on them that many subsequent papers were focussed.

Until a certain time in the past, problems on the structure of a phase portrait, its possible bifurcations, symbolic dynamics, ergodic theory, and the chaotization and stochastization of motions of deterministic dynamical systems had been studied only by a small circle of mathematicians and few specialists in oscillation theory.

The above and other schools, congresses, conferences, symposia, partial publication of the lectures and reports, and a number of surveys brought the earlier unknown papers to light, while the new ideas caused an explosion of interest in chaotic motions of dynamical systems, resulting in a multitude of scientific papers (see the Bibliography).

This is the outside of intrinsically much more complicated events during this comparatively short period of time, which was preceded by little known and scarcely noticeable investigations for many years. We will discuss the internal motivation and

Problem of chaotic and stochastic motions

results in the sequel, briefly list their importance, and contrast the prior approach to determinism and randomness with the new one. The principal goal of our treatment is to show the internal coherence of the problem of randomness in deterministic dynamical systems to the development of oscillation theory, topological dynamics and ergodic theory, and other branches of modern mathematics, mechanics and physics.

It is natural that the discussion of the above problems in the present monograph is not detailed or complete. It is necessarily brief and schematic.

The earlier approach to the nature of randomness was substantially based on the following ideas:

1. Probability theory admits an axiomatic structure when the existence of probability is postulated, but the question of its origin is not posed. Note that geometry, serving as a model for this axiomatization, did focus on its foundations, as developed by N. I. Lobachevsky, F. Gauss, B. Rieman, D. Hilbert, F. Klein.

2. Motions of a deterministic dynamical system cannot be random, since, by Cauchy's theorem, they are uniquely determined by the initial conditions.

3. Randomness of the environment, and, for specialists in mechanics and physics, of objects in statistical physics, was usually accounted for by the incomprehensibly large number of particles and degrees of freedom, incomplete data and, finally, complicated and intricate motions.

4. Oscillation theory, the qualitative theory of differential equations and dynamical systems succeeded in fully investigating two-dimensional systems only, while stochastic self-oscillations are only possible in systems of dimension not less than three. Poincaré's perturbation method [243, 244] and the asymptotic methods of N. M. Krylov and N. N. Bogolyubov [92], which could be applied to systems of any dimension, did not make it possible to discover stochastic motions if the generating system did not have them, due to the non-exponential infinitesimality of domains of existence of stochastic motions with respect to a small parameter.

We see that the earlier general ideas about motions of deterministic dynamical systems left no room for chaotic and stochastic motions.

But, possibly, this is not the only reason. Examples of conservative dynamical systems with quite complicated behaviour of phase trajectories (which we would call today *chaotic* and *stochastic* without the slightest hesitation) were known rather long ago as individual instances of non-conservative systems reducible to point mappings with chaotic behaviour of iterations. Moreover, G. Birkhoff [88] suggested a general classification of motions of dynamical systems, taking the above complicated motions into account. The scheme was given in A. A. Andronov's book *Mathematical Problems in Self-Excited Oscillation Theory* of 1933 [12]; in particular, it stressed that the totality of all motions could make up a complicated system. Reading really prophetic lines in the monograph and examining G. Birkhoff's classification, it is hard to understand what actually hampered the making of another, apparently quite small, step towards the recognition of chaotic and stochastic motions. Certainly, they were some reasons for that. Chaotic motions were regarded as something exceptional and ephemeral, not realizable physically: If they are possible, then in a different, conservative, world of stellar and atomic dynamics, and not in structurally stable dissipative systems.

Random motions are also objects belonging to another world separated by a high "barrier" from deterministic dynamical systems. Seemingly hard to refute or ignore, the theorem on the unique solution of the Cauchy problem, stating that the solution of a system of differential equations is uniquely determined by the initial conditions, is taken as the barrier. "Therefore," reasoned the adherents of tradition, "no randomness can be spoken of". Judging formally, this simple argument not only applies to stochastic motions of simple dynamical systems with few degrees of freedom, but also testifies against the whole of classical statistical mechanics and physics. However, the argument is not discarded in statistical mechanics, which resorts to a trick, viz., reference to the very large number of particles. This leaves the feeling of dissatisfaction, somehow enabling one to agree to the contradiction.

The new outlook on the nature of randomness is based on the following achievements of modern mathematics.

1. *The development of ergodic theory and the "bridging of gaps" between the deterministic and stochastic description of dynamical systems.* Ergodic theory dates back to the ergodic hypothesis suggested by L. Boltzman himself, according to which time averaging in statistical mechanics can be replaced by ensemble averaging [56]. Further attempts to justify the hypothesis led to creating complicated and ramified ergodic theory,

whose main stages of development are related to the names of G. Birkhoff, J. von Neumann, E. Hopf, M. Morse, A. Ya. Khinchine, A. N. Kolmogorov, V. I. Arnold, J. Moser, Ya. G. Sinai, etc. It is within the framework of these ideas that the concepts of *well-stirredness*, KAM, (Kolmogorov-Arnold-Moser) theory of existence of integral tori in Hamiltonian dynamical systems, the *entropy* of a dynamical system and symbolic description of its motions, and *topological Markov chain* arose, which have paved the way to statistically describing deterministic dynamical systems.

Ergodic theory naturally led to numerical studies of concrete Hamiltonian dynamical systems of small dimension and to the discovery of complicated chaotic motions in them. We should first of all point out to the papers of M. Henon and C. Heiles [501], G. M. Izrailev, G. M. Zaslavsky, B. V. Chirikov [154-158, 160-164, 366, 418], G. Walker and J. Ford [672], J. Ford and G. Lanford [456], etc.

A decisive role in the transition in statistical mechanics from systems with number of particles of order 10^{23} to those with few degrees of freedom, besides the paper of E. Fermi, J. Pasta and S. Ulam [450], belongs to papers on the motion of a free material point on a Riemanian manifold with negative curvature (i.e., the geodesic problem, originating in J. Hadamard's [485] of 1889), and also so-called *billiard problems* [88, 205, 326].

A billiard problem is concerned with the motion of a material point in the domain bounded by a certain curve, with elastic rebounds from the boundary as a perfectly elastic wall (see the convex billiard in Fig. 4.1a, and the concave one in Fig. 4.1b). A *convex billiard* was studied by G. Birkhoff on the basis of the familiar Poincaré's geometric theorem. A *concave billiard* possesses totally different properties, all its motions being unstable and divergent, since the originally small difference between particle motions increases after each rebound (Fig. 4.1b).

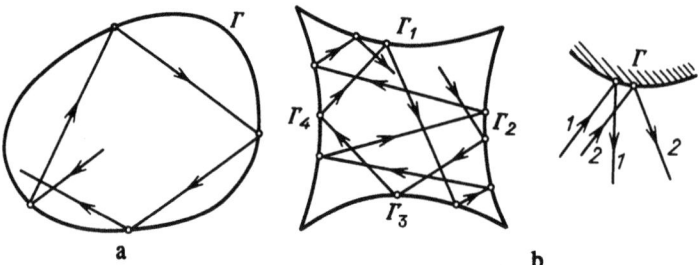

Fig. 4.1

The familiar Sinai system of N perfectly elastic balls inside a perfectly reflecting sphere and a *stadium billiard* can be regarded as further generalizations of the above billiard [315]. In the former case, for N = 1, we have a convex billiard. However, already for N = 2 or larger, motions of balls are similar to those in a concave billiard, due to mutual collisions when each ball is repelled by the "concave boundary" of another.

2. The development of the general topological theory of smooth dynamical systems. The basic results of principal importance were obtained by S. Smale and his disciples [6-9, 39-40, 175, 295, 330-332, 653], who distinguished the class of dynamical systems of Morse-Smale type, a direct multidimensional analog to the *systèmes grossiers* in the sense of Andronov-Pontryagin [20]. Their unique steady-state motions can only be equilibrium states and periodic motions. Recall that the dynamical system

$$\dot{x} = X(x, y), \quad \dot{y} = Y(x, y) \tag{1}$$

is said to be *structurally stable* (*grossier* by Andronov-Pontryagin) if its phase portrait is topologically equivalent to any sufficiently close perturbed system

$$\dot{x} = X(x, y) + p(x, y), \quad \dot{y} = Y(x, y) + q(x, y). \tag{2}$$

Meanwhile, smallness of the perturbations p(x, y), q(x, y) means that of their partial derivatives with respect to x, y, and not only smallness of themselves. The requirement that the derivatives should be small cannot be dropped, since without it structurally stable systems do not exist at all. We should also probably include the requirement for the phase space to be compact. The phase plane does not satisfy it, and, in the two-dimensional case, we should also assume additionally that only a bounded part is considered with all phase curves crossing the boundary, and not the whole phase plane.

Topological equivalence means the existence of a one-to-one and bi-continuous

mapping
$$\bar{x} = \xi(x, y), \quad \bar{y} = \eta(x, y) \tag{3}$$
carrying the phase trajectories of the unperturbed system (1) to those of the perturbed (2), or, for short, sending the unperturbed system's phase portrait to the perturbed one's. We should also add the requirement that the homeomorphism (3) should be continuously transformed into the identity transformation as the perturbing contributions p(x, y), q(x, y) decrease indefinitely (along with the derivatives).

The definition of a structurally stable system can be carried over to a multidimensional system without any substantial corrections. For this purpose, we somewhat geometrize the definition. Let G be the space of dynamical systems
$$\dot{x} = X(x), \tag{4}$$
where x is a multidimensional vector, and X(x) a smooth function of x. By *nearness* in G, we mean nearness of the right-hand sides of X(x) and their partial derivatives, i.e., the matrices dX/dx. The dynamical system associated with a point X(x) in G is structurally stable if its phase portrait is topologically equivalent to that of all points in a sufficiently small neighbourhood. Necessary and sufficient conditions for two-dimensional systems (1) to be structurally stable were formulated by A. A. Andronov and L. S. Pontryagin. Similar, but only sufficient, conditions for multidimensional systems were given by S. Smale, and dynamical systems satisfying them said to be *Morse-Smale systems*. It turned out that so-called *hyperbolic* multidimensional systems, substantially more complicated than Morse-Smale systems, could be structurally stable [175, 332]. The hyperbolicity property means the instability of motions, and is expressed by exponential extension of the mobile phase volume element in some directions, while by contraction in other directions. As a simple example of a hyperbolic dynamical system, we give that determined by the point mapping
$$\bar{x} = 2x + y \pmod{1}, \quad \bar{y} = x + y \pmod{1} \tag{5}$$
on a two-dimensional torus (Fig. 4.2). It is linear, and has characteristic equation
$$\begin{vmatrix} 2-\lambda & 1 \\ 1 & 1-\lambda \end{vmatrix} = \lambda^2 - 3\lambda - 1 = 0,$$
with one root less than one (contraction occurring in the corresponding direction). In the other direction associated with the second root greater than one, extension occurs.

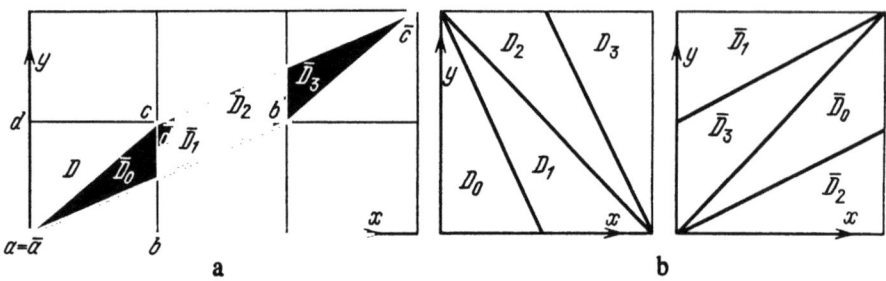

Fig. 4.2

A hyperbolic point map, subsequently called the *Smale horseshoe*, i.e., structurally stable, and, at the same time, with infinitely many different saddle (unstable) fixed points, became widely known.

Dynamical systems with hyperbolic structures are similar to systems considered earlier in symbolic dynamics [88, 588], and, first of all, to those describing coasting of a material point in a Riemann manifold of negative curvature [363]. However, the mobile point's phase volume is not necessarily preserved, possibly decreasing, and the system can be dissipative.

Still more unexpected than the existence of totally new kinds of structurally stable systems in the multi-dimensional case was the discovery of typical structurally unstable systems [331].

It seemed earlier that structurally unstable systems in the parameter space of dynamical systems were only limiting cases between structurally stable systems, i.e., surfaces of codimension one or higher. It became known after Smale's work that structurally unstable systems could fill whole regions in this parameter space, or that, for multidimensional systems, the concept of structural stability in the sense of Andronov-Pontryagin has a limited area of application.

3. *The study of homoclinic structures and the clarification of their role in forming complicated chaotic and stochastic motions of deterministic dynamical systems.* Curves called by H. Poincaré *homoclinic* and *heteroclinic* [312] were discovered in the restricted three-body problem, of motion of three material points attracted to each other by Newton's law, assuming that the motion is planar, one of the masses is vanishingly

small, and does not affect the motion of the two others. (The problem repeatedly drew the attention of researchers also in the post-Poincaré period.)

Homoclinic and heteroclinic curves attracted Poincaré's attention because the phase trajectories nearby, reflecting motions of the system, have a very complicated behaviour.

Seeing this, he wrote:

"*Que l'on cherche à se représenter la figure formée par ces deux courbes et leur intersections en nombres infinis dont chacune correspond à une solution doublement asymptotique, ces intersections forment une sorte de treillis, de tissu, de réseau à mailles infiniment serrées; chacune des deux courbes ne doit jamais se recouper elle-même, mais elle doit se replier sur elle-même d'une manière très complexe pour venir recouper une infinité de fois toutes les mailles do réseau .*

"On sera frappè de la complexité de cette figure, que je ne cherche même pas à tracer. Rien n'est plus propre à nous donner une idée de la complication du problème des trois corps ..." [1]

This complicated picture was described in 1967 [263, 266]. It was noted that certain combinations of homoclinic and heteroclinic curves could form structures generating quite complicated motions of stochastic nature, of interest if we want to describe the turbulence of fluid flows. The complicated formations themselves were later called *homoclinic structures* [268].

A homoclinic γ and a heteroclinic γ' phase trajectory are represented in Figs. 4.3, 4.4, where Γ, Γ' are saddle periodic motions. As t changes from $-\infty$ to $+\infty$, γ

---

[1] If one seeks to represent the figure formed by these two curves and their infinitely many intersections, each of which corresponds to a doubly asymptotic solution, then these intersections will be seen to form a kind of lattice, of web, of net with infinitely contracted looping; each of the two curves must never intersect itself again, but it must fold on itself in a very complicated manner to intersect infinitely many times all the net loops.

One will be startled by the complexity of this figure, which I do not even attempt to trace. Nothing is more proper to give us an idea of complications of the three-body problem ...

winds off, and then again on, Γ. Meanwhile, γ' winds off Γ, and then on Γ'. Intersect Γ with a plane Σ as in Fig. 4.5. Let T be the Poincaré point map sending the previous intersection point of Σ with the phase curve to the subsequent intersection point. The intersection points with Γ, Γ' are fixed under T. The homoclinic and heteroclinic phase trajectories meet Σ at certain infinite sequences of points ... , M_{-1}, M_0, M_1, ... Since γ tends to the closed curve Γ both as t → - ∞ and t → + ∞, the phase curve lies on both integral manifolds S^+, S^- of Γ. Similarly, since γ' tends asymptotically to Γ as t → - ∞, and to Γ' as t → + ∞, γ' lies on S^- of Γ and on S^+ of Γ'. The traces of the intersection of S^+, S^- with Σ are in Fig. 4.6 (only small parts are represented). It is rather complicated to extend them, the difficulty being mentioned by H. Poincaré himself. Instead, we confine ourselves to a schematic representation (Figs. 4.7, 4.8). By means of these quite simplified schemes only demonstrating how the phase trajectories enter and leave neighbourhoods of saddle periodic motions, an idea of a homoclinic structure and the generated complicated motions can be given.

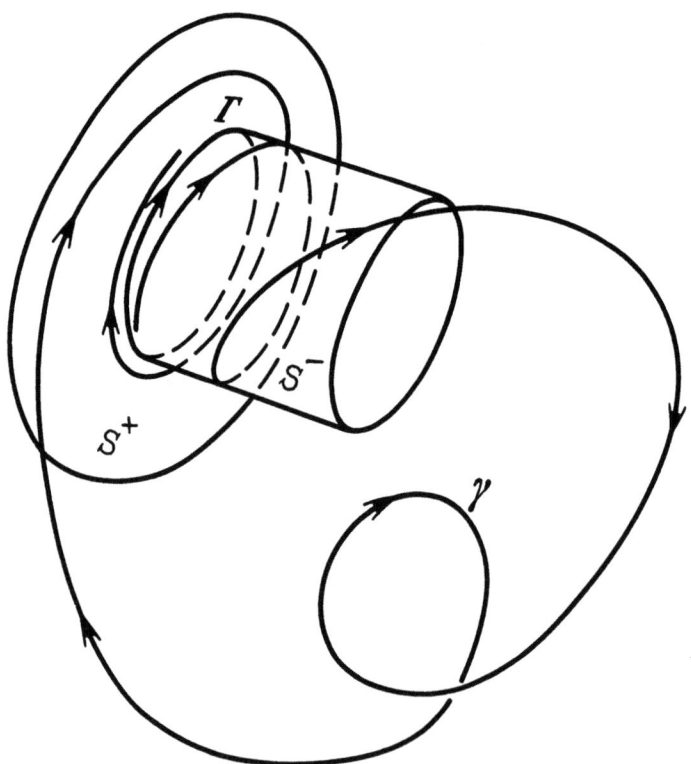

Fig. 4.3

Problem of chaotic and stochastic motions

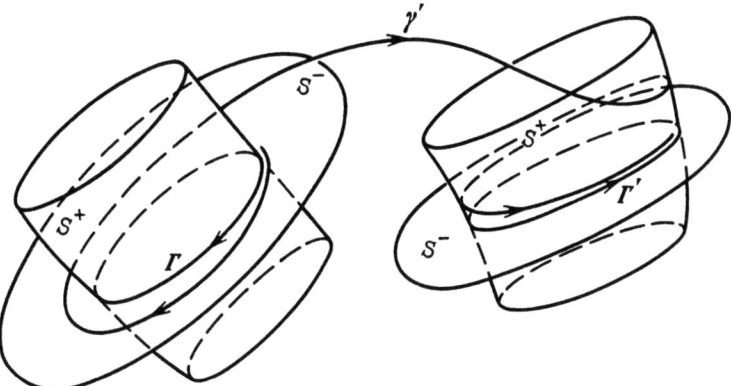

Fig. 4.4

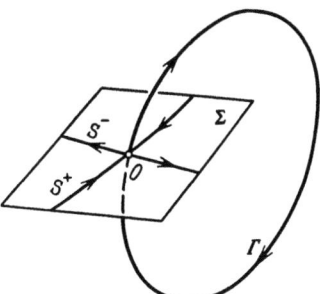

Fig. 4.5

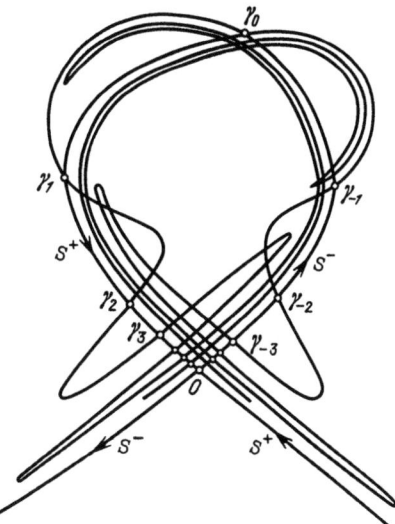

Fig. 4.6

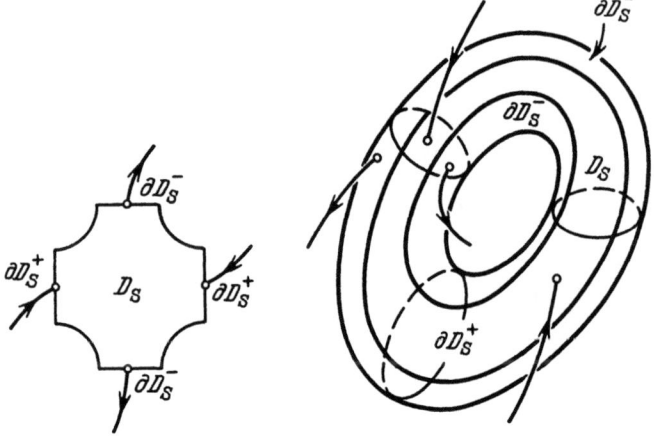

Fig. 4.7

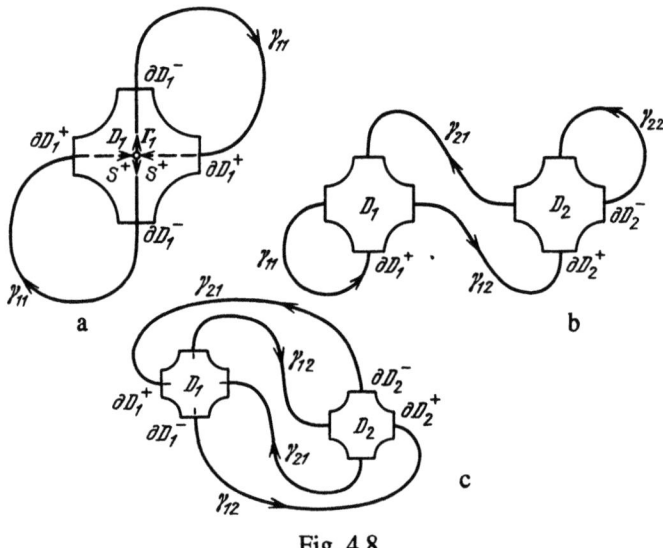

Fig. 4.8

A homoclinic structure can be schematically represented as in Fig. 4.8. Inside the regions D_s, saddle periodic motions Γ_s are located. In their vicinity, phase space elements are extended and contracted. Extension occurs along the surface S^-; contraction, along S^+.

The phase trajectories in some part ∂D_s^+ of the boundary (s, j = 1, 2) enter D_s, and leave it in some other part ∂D_s^-. The arrow at γ_{sj} means that the set of phase trajectories leaves D_s, crosses its boundary part ∂D_s^-, and then enters D_j, crossing ∂D_j^+. It may happen that there are no phase trajectories which, leaving any region D_s, do not find themselves in any region with in-going γ_{sj}, in which case each of the phase trajectories gets in some D_s across ∂D_s^+, and then either gets stuck in some D_j or always leaves some $D_1, D_2, \ldots$ for others. The motion of the phase point can be described by the sequence of subscripts of D_j which the point passes through. It stays for some time in each D_j, shifting in the vicinity of the saddle periodic motion Γ_j. This sojourn can be specified by the number of rotations n_j which the point makes in the toroidal neighbourhood. Thus, the motion of the phase point can be described by the sequence of pairs

$$\ldots, j_0, n_{j_0}, j_1, n_{j_1}, j_2, n_{j_2}, \ldots . \qquad (6)$$

It turns out that the sequence of subscripts j_s and of rotation numbers n_{j_s} uniquely, and, therefore, completely, determines the phase trajectory. Furthermore, we can indicate possible and impossible sequences, and thereby supply a full description of the phase trajectories always staying in D_j either as $t \to +\infty$ or $t \to -\infty$. It remains to indicate what the randomness of the motion of the phase point consists in. It is in unpredictability, or that and randomness of (6), from a certain moment onwards, and the randomness of transitions from one D_j to another as well as of sojourn (rotation numbers) in each. We also stress that the phase point moves in all D_j near a saddle periodic motion; therefore, motion is unstable, originally close points diverge, and move perfectly differently after some time.

Recall also that a simple example of this motion has already been considered. It is a homoclinic structure that generates stochasticity of motion (Chapt.1) in the Lorenz system (1.23).

4. The impasses in completing the foundations of statistical mechanics, and in the development of turbulence theory. Statistical mechanics and physics are based on a number of general, totally unjustified, non-substantiated but well checked, efficient postulates. By non-substantiation, we mean that the postulates should follow from the universal laws of mechanics and physics; meanwhile, no one could as yet derive them from the laws. This situation has persisted for a ling time; it is habitual and quite acceptable from the standpoint of the physicist who is unwilling to rummage in the foundations. In some sense, this situation does not worry the theorist either, for the simple reason that classical mechanics and physics are not more than the limit approximation of quantum theory, which postulates statistical properties, though in no way related to the problem of stochastic motion of deterministic dynamical systems. Accordingly, the discovery of stochasticity, albeit possibly promising some advancement in the problem of substantiating classical statistical mechanics and physics, caused no noticeable enthusiasm.

The matter is totally different for problems in hydrodynamic and plasma turbulence. First, turbulence theory should be fully based on classical macroscopic equations such as the Navier-Stokes, gas-dynamic, magnetohydrodynamics, and plasma equations; however, to derive the basic characteristics of turbulent motion from the macroscopic equations is not feasible at present, and we have to resort to other arguments. Turbulence theory has developed extraordinarily, but its path is thorny and difficult. The theory is forced to resort to a semiempirical, rather doubtful, approach, and cannot make out today even the simplest types of flows. It satisfies itself with quite scarce theoretical results concerning stability loss, and involves numerical computations not soundly supported. This unsatisfactory situation occurred not only because fluid mechanics and its equations turned out to be very complicated, while the number of degrees of freedom is depressingly large, but also because it was perfectly unclear in which direction to move, or, at least in principle, how such a theory could be constructed.

The new views on the stochastic motions of a dynamical system (ideas in non-linear oscillation, dynamical system and bifurcation theories, new for fluid mechanics) raised hopes for creating the long-awaited theory, a way out of the impasse in the solution of a number of long-standing problems. So bright prospects could not help in inducing enthusiasm which, as can be foreseen, dies down after realizing how colossal the difficulties are. However, the attack has begun, and will sooner or later lead to certain results.

Until recently, the Landau hypothesis of 1944 [230, 231] reigned supreme in the theory of how turbulence appears. In 1948, similar ideas were advanced by E. Hopf [511]. Essentially, Landau's theory was the first attempt since its discovery by O. Reynolds in 1883 to account for turbulence as a parameter, subsequently called the *Reynolds number*, increases [613]. Landau's theory relates turbulence to instability, which is undoubtedly true; still, how this is done requires much sharpening. From the new standpoint, Landau's theory is incomplete, but valid. It only points out to one possible version of how turbulence appears, and obviously not at all the most important. Modern bifurcation theory suggests many other ways. Some are more or less corroborated by experiments. These new routes to chaotization and stochastization of fluid motion are in contrast with Landau's theory.

According to Landau, turbulence appears as the result of the equilibrium state's, a periodic, doubly periodic motion, etc., gradual stability loss. (The motion becomes multiperiodic of the form

$$v(t) = f(\omega_1 t, \omega_2 t, \ldots, \omega_m t), \tag{7}$$

where $\omega_1, \omega_2, \ldots, \omega_m$ are frequencies and the function $f(\varphi_1, \varphi_2, \ldots, \varphi_m)$ is periodic with respect to each variable with period 2π.). D. Ruelle and F. Takens [627] paid attention to the fact that the way indicated by Landau is not universal, and that the general situation is the formation of a strange attractor. However, they did not study the latter's appearance. These ideas have their own prehistory. G. S. Gorelik spoke of the self-oscillatory nature of turbulence in the early fifties [217]. The *Fifth Mathematical Summer School* in 1967 discussed the possible relation of turbulence to homoclinic structures and the generated complicated motions [266].

Getting familiar now with the earlier propositions, one cannot help noticing that they underwent considerable changes according to the general ideas and development of the theories of non-linear oscillations and dynamical systems. Landau's theory is a direct consequences of the contemporary ideas of how multiperiodic oscillations arise. G. S. Gorelik's ideas follow from A. A. Andronov's theory of self-excited oscillations, developed at the time. Yu. I. Niemark's statements on turbulence result from his study of the homoclinic structures discovered by H. Poincaré himself, while the new term *strange attractor* and the hypothesis of D. Ruelle and F. Takens concerning the nature of turbulence are directly influenced by S. Smale.

We should possibly stress another, now forgotten, direction in the attempts to account for turbulence, stemming from the absence of theorems on the existence of solutions to the Navier-Stokes equations, where turbulence is treated as a consequence of temporal inextendability and destruction of their solutions. The reasoning dates back to J. Leray, and was developed later [633].

Now, what can be added ? First of all, that all statements above only proceed from the temporal treatment of motions of a dynamical system and its phase portrait, whereas the fluid is a spatially distributed medium, and its motion description involves a spatial component as well as the temporal. Turbulence is also spatial chaos, and not only temporal. Certainly, time and spatial chaos are interrelated, but irreducible to each other; generally speaking, time chaos and spatial order or temporal order and spatial chaos can co-exist. Generally speaking, turbulence is both time and spatial chaos. The double nature of chaos in turbulence was taken up by A. S. Monin in his survey [257].

Meanwhile, it is apparent that the cases of only time, or only spatial, chaos can be referred to turbulence.

Time chaos has a strange attractor as its adequate geometric representation; spatial chaos, a consistent saddle (hyperbolic) invariant set. Apparently, further sharpening is required.

Finally, it is also necessary to note that the only geometric representation of time chaos until very recently was a strange attractor. It now becomes clear that there are more representations. The geometric representation of temporal turbulence can also be one or several stable periodic motions with transitions from one to another or to itself on account of small fluctuations, i.e., turbulence can be a consequence both of stochasticity generation and amplification.

CHAPTER 5

LOCAL PHASE PORTRAITS OF THE SIMPLEST STEADY-STATE MOTIONS AND THEIR BIFURCATIONS

The present chapter deals with the simplest steady-state motions, equilibrium states and periodic motions. The classification of equilibrium states and periodic motions is discussed, the principal types of bifurcation are given and investigated. Unstable saddle equilibria and periodic motions are considered along with those are stable and periodic. If the latter are actually the main simplest steady-state motions, then the former are basic in determining the boundaries of their domains of attraction, chaotic or stochastic motions, and the whole phase portrait.

1. Equilibrium states

The equilibrium states satisfy the differential equation

$$\dot{x} = X(x) \tag{1.1}$$

of the form $x = x^*$, where x^* is a constant vector to be found from

$$X(x^*) = 0. \tag{1.2}$$

In the general case, the equilibrium states form a set of isolated points. In the singular cases, manifolds of equilibrium states, of some dimension, can arise.

Introducing the new variable $\xi = x - x^*$, we write (1.1) in the vicinity of $x = x^*$ as

$$\dot{\xi} = X(x^* + \xi) = \frac{\partial X}{\partial x}\bigg|_{x=x^*} \xi + R(\xi), \tag{1.3}$$

where $R(\xi)$ is an infinitesimal of order at least two with respect to ξ. Omitting $R(\xi)$, we arrive at the equations of motion,

$$\dot{\xi} = \frac{\partial X}{\partial x}\bigg|_{x=x^*} \xi = A\xi, \qquad (1.4)$$

linearized in a vicinity of the equilibrium state, where A is a certain constant square matrix. Generally speaking, when all eigenvalues of A are different, there exists a non-singular, linear and, generally complex, change of variables

$$\eta = L\xi \qquad (1.5)$$

after which the differential equation (1.4) assumes the form

$$\dot{\eta} = \Lambda\eta,$$

where
$$\Lambda = LAL^{-1} \qquad (1.6)$$

is a diagonal matrix whose diagonal entries are roots of the characteristic equation

$$\chi(\lambda) = \mathrm{Det}\,(A - \lambda E) = 0, \qquad (1.7)$$

Det denotes the determinant, and E the identity matrix. In scalar form, the differential equations (1.6) are written as

$$\dot{\eta}_s = \lambda_s \eta_s \quad (s = 1, 2, \ldots, n). \qquad (1.8)$$

The roots can be either real or complex. In the latter case, along with each λ_s, there is also the complex conjugate root $\lambda_{s+1} = \lambda_s^*$. Putting $\lambda_s = \sigma_s + i\omega_s$, $\lambda_{s+1} = \sigma_s - i\omega_s$ and introducing the new real variables

$$u_s = \frac{1}{2}(\eta_s + \eta_{s+1}), \quad v_s = \frac{1}{2i}(\eta_s - \eta_{s+1}),$$

we rewrite (1.8) in the form

$$\dot{\eta}_s = \lambda_s \eta_s \qquad (1.9)$$

if λ_s is real, and in the form

$$\dot{u}_s = \sigma_s u_s - \omega_s v_s, \quad \dot{v}_s = \omega_s u_s + \sigma_s v_s \qquad (1.10)$$

if $\lambda_s = \sigma_s + i\omega_s$ is complex.

The differential equations (1.9), (1.10) can be solved easily, so that

$$\eta_s = \eta_s^0 e^{\lambda_s t} \qquad (1.11)$$

and
$$\begin{aligned} u_s &= e^{\sigma_s t}(u_s^0 \cos \omega_s t - v_s^0 \sin \omega_s t) \\ v_s &= e^{\sigma_s t}(u_s^0 \sin \omega_s t + v_s^0 \sin \omega_s t) \end{aligned} \qquad (1.12)$$

where η_s^0, u_s^0, v_s^0 are the initial values of the variables for $t = 0$.

We exclude the cases $\lambda_s = 0$ and $\sigma_s = 0$. Then, according to (1.11), (1.12), any of η_s, u_s, v_s either increases indefinitely with time or tends to zero. Denote the vector of variables from the former group by **v**, and from the latter by **u**. They can be determined by the equations

Local phase portraits of the simplest steady-state motions and their bifurcations

$$\dot{\mathbf{u}} = A\mathbf{u}, \quad \dot{\mathbf{v}} = B\mathbf{v}, \tag{1.13}$$

where A, B are matrices. The diagonal of A contains either negative numbers λ_s or matrices of order two,

$$\begin{pmatrix} \sigma_s & -\omega_s \\ \omega_s & \sigma_s \end{pmatrix}, \tag{1.14}$$

with negative σ_s; similarly, the diagonal of B either contains positive λ_s or matrices (1.14) with positive σ_s.

Let p, q be the dimensions of **u**, **v**, respectively. Then A is a $p \times p$ matrix, while B is a $q \times q$ matrix. The sum p+q coincides with the total dimension n of the vector **x**. We denote by $O^{p,q}$ the equilibrium states with the numbers p, q of the roots $\lambda_1, \lambda_2, \ldots, \lambda_n$ to the left and right of the imaginary axis.

The linearized equations (1.13) are remarkable for the fact that **u**, **v** can be separated, and behave differently, with all components of **u** tending to zero as $t \to +\infty$ and all those of **v** increasing indefinitely. Thus, in a neighbourhood of $O^{p,q}$, the equations (1.1) can, under the above assumptions, be written as

$$\begin{pmatrix} \dot{\mathbf{u}} \\ \dot{\mathbf{v}} \end{pmatrix} = \begin{pmatrix} A & 0 \\ 0 & B \end{pmatrix} \begin{pmatrix} \mathbf{u} \\ \mathbf{v} \end{pmatrix} + F(\mathbf{u}, \mathbf{v}), \tag{1.15}$$

where the function $F(\mathbf{u}, \mathbf{v})$ is an infinitesimal of order not les than two with respect to the variables **u**, **v**.

The latter can be separated in (1.15) in the linear terms; in the higher-order ones, generally speaking, they cannot. The question arises: How successfully can the variables be separated in the terms of higher order, too ? As it turns out, the answer depends on the eigenvalues $\lambda_1, \lambda_2, \ldots, \lambda_n$. The following remarkable theorem, dating back to H. Poincaré, holds. We give it as stated by A. D. Bryuno [95, 96].

The system of differential equations

$$\dot{x}_s = \lambda_s x_s + f_s(\mathbf{x}) \quad (s = 1, 2, \ldots, n), \tag{1.16}$$

where $f_s(\mathbf{x})$ are analytic functions of **x** of order not less than two with respect to **x**, can be reduced to the so-called *normal form*

$$\dot{y}_s = \lambda_s y_s + y_s \sum_{Q \in m_s} g_{sQ} \mathbf{y}^Q \tag{1.18}$$

by a change of variables

$$x_s = y_s + g_s(y) \quad (s = 1, 2, \ldots, n), \tag{1.17}$$

where $g_s(y)$ is of order not les than two with respect to y, Q the vector with components $q_1, q_2, \ldots, q_n$, $y^Q = y_1^{q_1} y_2^{q_2} \ldots y_n^{q_n}$, the set m_s determined as that of all possible Q for which

$$q_1, q_2, \ldots, q_{s-1}, q_{s+1}, \ldots, q_n \geq 0, \quad q_s \geq -1,$$
$$\|Q\| = q_1 + q_2 + \ldots + q_n \geq 1, \tag{1.19}$$
$$(\Lambda, Q) = \lambda_1 q_1 + \lambda_2 q_2 + \ldots + \lambda_n q_n = 0,$$

and g_{sQ} are, generally speaking, non-zero coefficients. The theorem expresses the formal possibility to arrive at (1.18). Meanwhile, the series determining the transformation, and the normal form itself, can diverge. The normalization can be preserved for the terms of order less than N. Instead of (1.18), we then obtain

$$\dot{y}_s = \lambda_s y_s + y_s \sum_{Q \in m_s} g_{sQ} y^Q + Y_{sN}, \tag{1.20}$$

where Y_{sN} are terms of order not less than N with respect to y. The questions of convergence then fall off, and not more than N+1 times differentiability of the right-hand sides is required. The condition

$$\|Q\| < N - 1 \tag{1.21}$$

determines an at most finite set of vectors

$$Q^1, Q^2, Q^3, \ldots, Q^L,$$

together with inequalities (1.19). m_s contains only those Q^i for which the last condition in (1.19)v holds. Therefore, m_s is empty if

$$(\Lambda, Q^1)(\Lambda, Q^2) \ldots (\Lambda, Q^L) \neq 0. \tag{1.22}$$

Therefore, under general assumptions regarding the roots $\lambda_1, \lambda_2, \ldots, \lambda_n$, m_s are empty, and (1.20) can be written as

$$\dot{u} = Au + f(u, v), \quad \dot{v} = Bv + g(u, v) \tag{1.23}$$

with respect to the real variables u, v, where the functions f, g are infinitesimals of order not lower than N.

We now assume that there are also zero and purely imaginary roots. Then, with the other roots in general position, we distinguish the variables w associated with the zero and purely imaginary roots, and write the differential equations (1.20) in the form

Local phase portraits of the simplest steady-state motions and their bifurcations 101

$$\dot{u} = Au + uP(w) + f(u, v, w),$$
$$\dot{v} = Bv + vQ(w) + g(u, v, w), \qquad (1.24)$$
$$\dot{w} = Cw + wR(w) + h(u, v, w),$$

where the functions f, g, h are still of order not less than N with respect to u, v, w, and P(w), Q(w), R(w) polynomials in w, determined by the fact that (1.19), (1.21) hold for the zero and purely imaginary roots. Note that distinguishing only the linear terms also reduces the differential equations (1.1) to the form (1.24); meanwhile, N = 2.

Other useful arguments which can be used to give (1.24) a convenient form in a neighbourhood of an equilibrium state are the theorems on the existence, smoothness and smooth dependence on parameters of the integral manifolds of the differential equations [264, 265, 269, 285], originating from Hadamard's and Perron's work. Accordingly, (1.24) admit two integral manifolds S^+ ($v = f^*(w, u)$) and S^- ($u = g^*(w, v)$) in a neighbourhood of the equilibrium, whose intersection yields the integral manifold

$$J(u = \varphi^*(w), v = \psi^*(w)). \qquad (1.25)$$

To represent S^+, S^- and J visually, we first consider them in the case where there are neither purely imaginary, zero roots, nor non-linear terms. Then the surfaces S^+, S^-, J have the equations

$$v = 0, \quad u = 0 \quad \text{and} \quad u = v = 0,$$

respectively. Therefore, J is an equilibrium point, S^+ consists of all phase trajectories tending to J as $t \to +\infty$, and S^- of all phase trajectories tending to J as $t \to -\infty$ (Fig. 5. 1). When the non-linear terms f, g appear in equations (1.24), $S^+(v = 0)$, $S^-(u = 0)$ deform slightly, remaining tangent to the manifolds determined by $v = 0$, $u = 0$ at the point J.

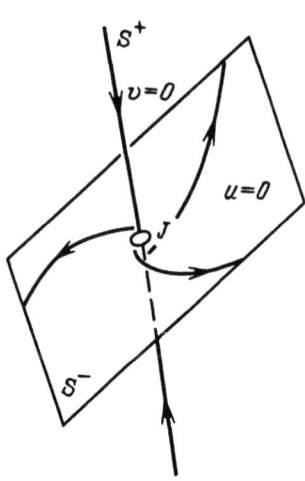

Fig. 5.1

Let there be no non-linear terms. Suppose, however, that there are roots on the imaginary axis. Then S^+, S^-, J have the equations $\mathbf{v} = 0$, $\mathbf{u} = 0$ and $\mathbf{u} = \mathbf{v} = 0$, respectively. But now the manifold J is not an equilibrium point but an integral manifold of dimension equal to the number of roots of the characteristic equation on the imaginary axis.

As non-linear terms appear, S^+, S^-, J deform, remaining tangent to the surfaces $\mathbf{v} = 0$, $\mathbf{u} = 0$ and $\mathbf{u} = \mathbf{v} = 0$, respectively. The phase trajectories on the manifold satisfy the differential equation

$$\dot{\mathbf{w}} = C\mathbf{w} + \mathbf{w}R(\mathbf{w}) + h(\varphi^*(\mathbf{w}), \psi^*(\mathbf{w}), \mathbf{w}) \qquad (1.26)$$

according to the last equation in (1.24), and (1.25) for J. The behaviour of the phase trajectories is thereby described fully on S^+, S^- and J. The phase points approach J along S^+, and then move farther from J along S^- when $p \neq 0$, $q \neq 0$. If $p = 0$ or $q = 0$, all phase trajectories move farther from J, or, conversely, approach it. The case $q = 0$ is of special interest: S^- does not exist, S^+ coincides with the whole phase space (a certain neighbourhood of the equilibrium point $O^{p,0}$), and the phase trajectories exponentially tend to J. If they do not leave this small neighbourhood in the course of time, then each tends exponentially to a phase trajectory on J. Their asymptotic behaviour near $O^{p,0}$ is therefore determined by that of J only; it is in this sense that the phase portrait of a neighbourhood of $O^{p,0}$ is determined by that on J. If $p+q = n$ and $p \neq 0$, $q \neq 0$, then the equilibrium state is of saddle type. When $q = 0$, it is stable; when $p = 0$, unstable. In all cases, the behaviour

Local phase portraits of the simplest steady-state motions and their bifurcations 103

of phase trajectories was described above. We regard the corresponding phase portraits for equal p, q as identical. It can be established that such phase portraits are topologically isomorphic, i.e., can be transformed onto each other by a one-to-one and bicontinuous mapping.

With respect to the new variables

$$\bar{u} = u - g^*(w, v), \quad \bar{v} = v - f^*(w, u), \quad \bar{w} = w, \tag{1.27}$$

the differential equations (1.24) are written as

$$\begin{aligned}
\dot{\bar{u}} &= A\bar{u} + \bar{u}P(\bar{w}) + \bar{u}\bar{f}(\bar{u}, \bar{v}, \bar{w}), \\
\dot{\bar{v}} &= B\bar{v} + \bar{v}Q(\bar{w}) + \bar{v}\bar{g}(\bar{u}, \bar{v}, \bar{w}), \\
\dot{\bar{w}} &= C\bar{w} + \bar{w}R(\bar{w}) + \bar{w}\bar{h}(\bar{u}, \bar{v}, \bar{w}),
\end{aligned} \tag{1.28}$$

while the integral manifolds S^+, S^-, J are determined by $\bar{v} = 0$, $\bar{u} = 0$, and $\bar{u} = \bar{v} = 0$, respectively. The non-linear functions $\bar{u}\bar{f}$, $\bar{v}\bar{g}$, and $\bar{w}\bar{h}$ are still infinitesimals of orders not less than N - 1 with respect to $\bar{u}, \bar{v}, \bar{w}$; the form of the function R is unaltered.

The question remains about the role of the terms in the normal form, for which condition (1.19) holds. No complete answer is known. It can only be noted that, as is shown in Sect.2, these, so-called *resonant terms,* are sometimes quite important.

2. Bifurcations of equilibrium states

Below, we investigate the dependence of the phase portrait in a neighbourhood of an equilibrium state on the parameters μ on the right-hand sides of the describing differential equations

$$\dot{x} = X(x, \mu). \tag{2.1}$$

The investigation mostly consists in considering the phase portrait and its variations for a continuous change of μ along a curve γ in the parameter space. It turns out that, in passing through certain points on γ, the phase portrait gets restructured qualitatively. These were called its *bifurcation points*, while the associated parameter values μ *bifurcation values*. Many different curves γ can pass through one point of the parameter space, and it is never known a priori whether a change in the phase portrait does not depend on the particular curve of parameter variation. Therefore, the concept of bifurcation also depends on the parameter space, i.e., new bifurcations can be discovered by extending the space, while lost by restricting. The clarification of each particular case requires sharpening when studying some or other dynamical system.

Bifurcations can be considered in the space, or some class, of all dynamical systems; e.g., of order two, with polynomial right-hand sides, or described by Hamiltonian equations, etc.

A point μ^* of the space of dynamical systems or in its parameter space is called a *bifurcation point* if, in an arbitrarily small neighbourhood there are points associated with dynamical systems with qualitatively different phase portraits. This definition is sufficiently comprehensive; however, for the present, it remains unclear what *qualitatively different* means. Usually, topologically non-equivalent phase portraits are meant. In certain cases, this definition can, however, turn out to be excessively detailed; e.g., in the case where bifurcation values prove to be everywhere dense. Qualitative difference may then tentatively be determined otherwise.

In studying bifurcations, it is desirable that the parameter space should be of smallest dimension possible, while the bifurcation points form easily assimilated geometric images. How can it be done ? Let M be the space of parameters or dynamical systems of interest, and μ^* a bifurcation point. The simplest case is of points associated with only two qualitatively different phase portraits in a small neighbourhood of μ^*, so that we can choose one parameter μ in this neighbourhood, the surface of separation being $\mu = 0$ (i.e., $\mu > 0$ for one type of phase portraits, while $\mu < 0$ for the other) (Fig. 5.2). We will then say that the bifurcation at μ^* is *1-parametric*, while the bifurcation surface $\mu = 0$ is associated with systems of structural instability of degree one in the sense of Andronov-Leontovich [18, 19][2].

---

[2]Another treatment of the problems considered, in which the concepts of structural instability and structural instability degrees are not involved, can be found in [43].

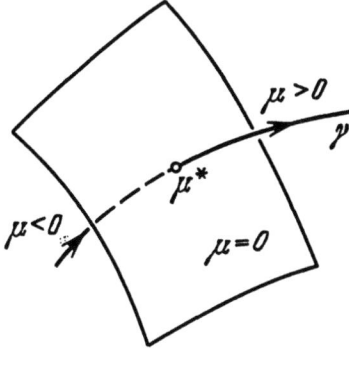

Fig. 5.2

A more complicated case is possible where there are four types of phase portraits in a neighbourhood of μ^*, separated by two surfaces. Selecting parameters μ_1, μ_2 so that the surfaces may be associated with $\mu_1 = 0$, $\mu_2 = 0$, and $\mu_1 > 0$ and $\mu_2 > 0$, $\mu_1 < 0$ and $\mu_2 > 0$, $\mu_1 > 0$ and $\mu_2 < 0$, $\mu_1 < 0$ and $\mu_2 < 0$ in the four regions of phase portraits of different types, we arrive at the 2-parameter representation of bifurcations in a neighbourhood of μ^* (Fig. 5.3a). The latter may also be of more complicated form; e.g., as in Fig. 5.3b, in which case there are only three regions associated with different types of phase portraits, and not four. The regions correspond to the three combinations of parameter signs $\mu_1 > 0$ and $\mu_2 < 0$, $\mu_1 > 0$ and $\mu_2 > 0$, and $\mu_1 < 0$ and $\mu_2 < 0$. Thus, a 2-parameter representation is also possible here. The intersection "curve" of the bifurcation surfaces $\mu_1 = 0$ and $\mu_2 = 0$ is in all cases associated with structurally unstable systems of second degree of structural instability.

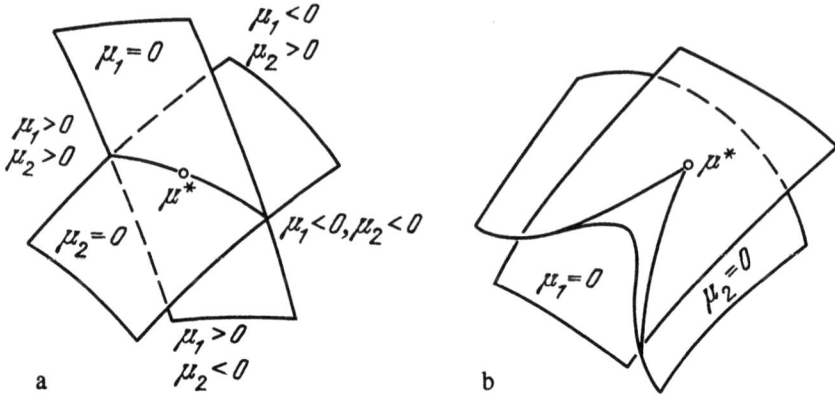

Fig. 5.3

These simple examples enable us to give a general definition of representability of bifurcations by r parameters $\mu_1, \mu_2, \ldots, \mu_r$; viz., in a neighbourhood of the bifurcation point μ^* in the space M of parameters or dynamical systems, representability by means of r parameters occurs if a choice of r variables is possible so that each type of phase portrait is associated with one or several combinations of signs.

We can now determine representability of bifurcations also in a region $\tilde{M}$ of M, and not only in the vicinity of a bifurcation point. Bifurcations in $\tilde{M}$ are representable by $\mu_1, \mu_2, \ldots, \mu_r$ if there exists a finite of infinite set of piecewise smooth functions $f_1(\mu_1, \mu_2, \ldots, \mu_r), f_2(\mu_1, \mu_2, \ldots, \mu_r), \ldots$ such that, at any two points of M associated with different types of dynamical systems (or their phase portraits), at least two among $f_1, f_2, \ldots$ are of opposite signs. We stress that the phase portrait of a dynamical system, and its different types, can be regarded globally, in the whole phase space, or locally, in some part only; e.g., in a small neighbourhood of an equilibrium state.

We now have all the information necessary to give the definition of an *isolated bifurcation point* μ^*. A point μ^* is associated with an isolated bifurcation if, in a sufficiently small neighbourhood, at most finitely many $f_1, f_2, \ldots$ vanish, and there are at most finitely many connected components of the space of dynamical systems of different types. Otherwise, the point is *non-isolated*, and an accumulation point of infinitely many regions associated with phase portraits of different types. The accumulation around a bifurcation point μ* may be unilateral or multilateral.

That there are accumulation points was first discovered explicitly by considering structurally stable maps of a circle into itself [242]. Other mechanisms for generating non-isolated bifurcation points were discovered in connection with studying homoclinic structures [262, 268]. In later, American, works, this phenomenon was called *omega-explosion* [295]. It was already stressed above that the opportunity for an everywhere dense covering of non-isolated bifurcation points was established by S. Smale [331]. Infinite series of period doubling bifurcations recently became generally known [444-448]. We consider them in Sect.2, Chapt.7.

We now retrace our steps to the differential equation (2.1) and its equilibrium states. Let $O(x = x^*)$ be one of them, $\lambda_1, \lambda_2, \ldots, \lambda_n$ the roots of its characteristic equation, M the space of all possible smooth functions $X(x)$, and μ* the point in M, associated with the system (2.1) in question.

THEOREM 5.1. *The point μ* is a bifurcation point of the equilibrium state O if there are no roots $\lambda_1, \lambda_2, \ldots, \lambda_n$ on the imaginary axis.*

In fact, an equilibrium state O exists not only for $\mu = \mu^*$, but for all μ sufficiently near μ*, since the Jacobian of (2.1),

$$\text{Det} \left.\frac{\partial X}{\partial x}\right|_{x=x^*}, \qquad (2.2)$$

is other than zero. That it vanishes would mean, according to (1.7), that the root of the characteristic equation is zero.

For all μ sufficiently near to μ*, the equilibrium state does not alter its type (i.e., dimensions of its integral manifolds S^+, S^-, and the numbers p, q); it therefore remains of type $O^{p,q}$.

It follows from Theorem 5.1 that the basic simplest bifurcation points of an equilibrium state are μ* for which the characteristic equation has root zero or two purely imaginary conjugate roots. Each case in M is associated with surfaces N_0, N_ω of codimension one. Their equations can be written in the form

$$\text{Det} \left.\frac{\partial X}{\partial x}\right|_{x=x^*} = 0 \qquad (2.3)$$

and

$$\text{Det} \left\{ \left.\frac{\partial X}{\partial x}\right|_{x=x^*} - i\omega E \right\} = 0 \quad (0 < \omega < +\infty), \qquad (2.4)$$

respectively.

Expanding the determinant, we write characteristic equation (1.7) in the form

$$\chi(\lambda) = \lambda^n + a_1 \lambda^{n-1} + \ldots + a_{n-1}\lambda + a_n = 0. \quad (2.5)$$

The equations for N_0, N_∞ can be written as

$$a_n = 0, \quad \Delta_{n-1} = \begin{vmatrix} a_1 & 1 & 0 & 0 & 0 & \ldots \\ a_3 & a_2 & a_1 & 1 & 0 & \ldots \\ a_5 & a_4 & a_3 & a_2 & a_1 & \ldots \\ \cdot & \cdot & \cdot & \cdot & \cdot & \end{vmatrix} = 0, \quad (2.6)$$

respectively, in terms of the coefficients a_1, a_2, ..., a_n, which are functions of the parameter μ, where Δ_{n-1} is the penultimate Routh-Hurwitz determinant of (2.5). Recall that the last is $\Delta_n = a_n \Delta_{n-1}$, so that its vanishing determines both bifurcation boundaries N_0, N_∞.

Let $\mu^* \in N_0$. The equations of motion for $\mu = \mu^*$ can be written as (1.28). We focus our attention on

$$\dot{w} = c_2^* w^2 + c_3^* w^3 + \ldots + h(\mathbf{u}, \mathbf{v}, w)$$

with $\mu = \mu^*$, written as

$$\dot{w} = c_0 + c_1 w + c_2 w^2 + c_3 w^3 + \ldots + h(\mathbf{u}, \mathbf{v}, w) \quad (2.7)$$

for μ close to μ^*, where the coefficients c_0, c_1, c_2, c_3, ... and the function h depend smoothly on μ, and $c_0 = c_1 = 0$ for $\mu = \mu^*$. In the simplest case, $c_2 = c_2^* \neq 0$ for $\mu = \mu^*$; and $c_2^* = 0$, $c_3^* \neq 0$ follows. We consider just these two cases. That the term $h(\mathbf{u}, \mathbf{v}, w)$ is small, together with the non-singularity of the matrices A, B, enables us to confine ourselves to the equation

$$c_0 + c_1 w + c_2 w^2 + c_3 w^3 + \ldots = 0 \quad (2.8)$$

in seeking the equilibrium state for μ close to μ^*. Solutions vanishing for $\mu = \mu^*$ are of special interest. Generally speaking, the parameter μ is of arbitrary dimension. To simplify the argument, we therefore assume that an arbitrary straight line passes through $\mu^* \in M$, and that ν varies along the line, vanishing at μ^* (Fig. 5.4). Consider the basic simplest cases

(a) $c_0^* = c_1^* = 0$, $c_0' = \dfrac{\partial c_0}{\partial v}\bigg|_{v=0} \neq 0$, $c_2^* \neq 0$,

(b) $c_0^* = c_1^* = c_2^* = 0$, $c_0' \neq 0$, $c_1' \neq 0$, $c_3^* \neq 0$. (2.9)

In the former, according to (2.8),

$$w = \sqrt{-\dfrac{c_0'}{c_2^*}} v^{1/2} + \ldots ; \qquad (2.10)$$

in the latter,

$$w_1 = \sqrt[3]{-\dfrac{c_0'}{c_3^*}} v^{1/3} + \ldots ; \qquad w_{2,3} = w_1 + \sqrt{-\dfrac{c_1'}{c_3^*}} v^{1/2}, \qquad (2.11)$$

which is associated with the graphs of the dependence on v of the coordinate w of the equilibrium state, represented in Fig. 5.5a. To plot the bifurcation diagrams, the corresponding equilibrium states should also be pointed out on the branches of the curves. By substituting the found values of w in (2.7), we find all possible substantially different diagrams; they are represented in Fig. 5.5b.

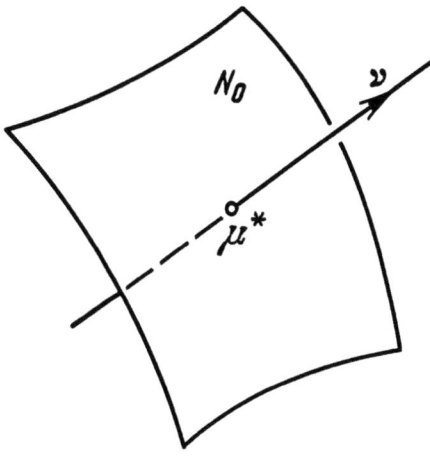

Fig. 5.4

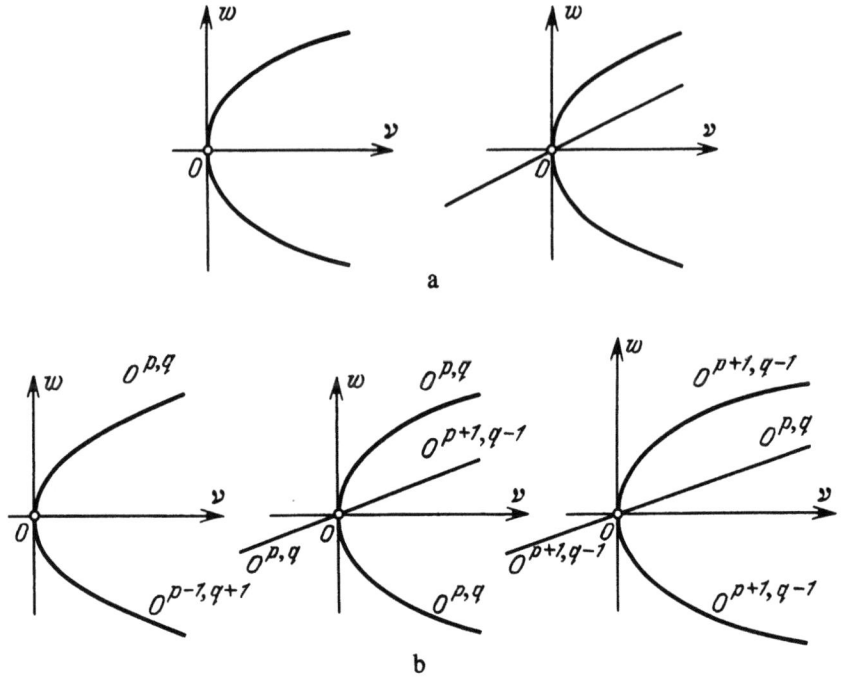

Fig. 5.5

Accordingly, the following holds.

THEOREM 5.2. *In cases (2.9), up to the direction of the w- and v-axes, the bifurcation diagrams in Fig. 5.5 hold, where the values of w as functions of v are determined by formulas (2.10), (2.11), respectively.*

Changes in the three-dimensional phase portrait for the first and second of the above bifurcations are represented in Fig. 5.6, with the assumption that the original equilibrium state for $v < 0$ is stable, or of type $O^{3,0}$.

Local phase portraits of the simplest steady-state motions and their bifurcations

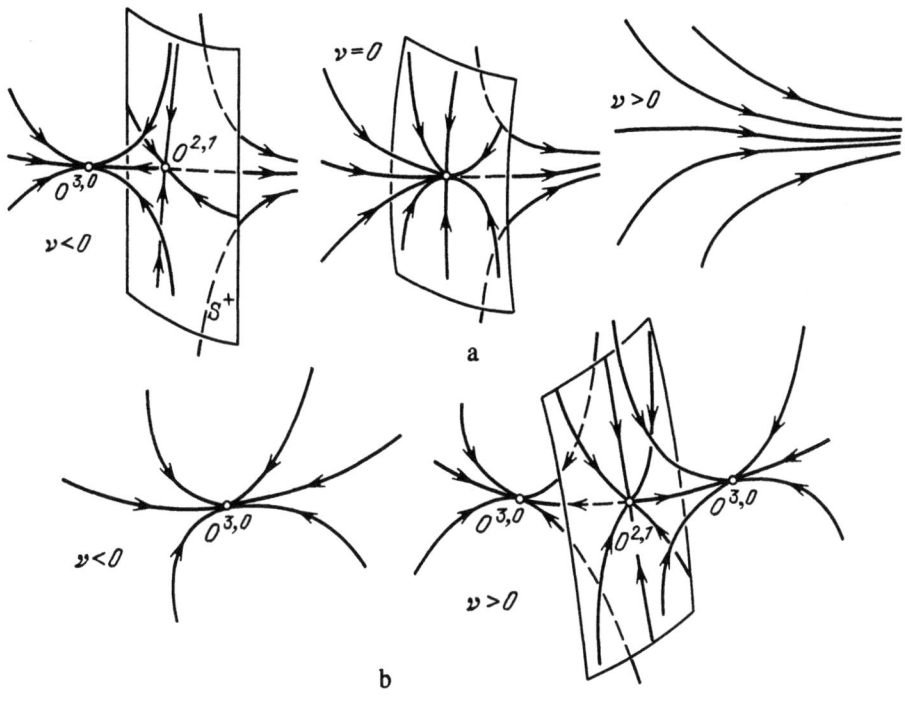

Fig. 5.6

Let $\mu^* \in N_\omega$. In contrast to the above, as μ varies, the equilibrium state is preserved and is unique in its small neighbourhood for all μ close to μ^* (no zero root and non-zero Jacobian). This enables us to consider the last of equations (1.28) on the integral manifold $\mathbf{u} = \mathbf{v} = 0$, which becomes

$$\dot{\mathbf{w}} = C\mathbf{w} + \mathbf{w}R(\mathbf{w}) + \mathbf{w}h(0, 0, \mathbf{w}) \qquad (2.12)$$

on the 2-manifold J ($\mathbf{u} = \mathbf{v} = 0$). For $\mu = \mu^*$, the matrix C has two purely imaginary roots $\pm i\omega$ ($\omega \neq 0$), while the lowest resonant term is such that equation (2.12) can be written as

$$\dot{w}_1 = \sigma w_1 - \omega w_2 + g(w_1^2 + w_2^2)w_1 + \ldots$$
$$\dot{w}_2 = \omega w_1 + \sigma w_2 + g(w_1^2 + w_2^2)w_2 + \ldots$$

componentwise, where $\sigma = 0$ for $\mu = \mu^*$ and the dots denote infinitesimals of order higher than any term written out. Multiplying the first equation throughout by w_1, the second by w_2, and adding together, we have

$$\frac{1}{2}\frac{d}{dt}(w_1^2 + w_2^2) = \sigma(w_1^2 + w_2^2) + g(w_1^2 + w_2^2)^2 + \ldots,$$

or, introducing the polar coordinate r and denoting r^2 by ρ, we can rewrite the equation obtained in the form

$$\frac{1}{2}\dot{\rho} = \sigma(v)\rho + g(v)\rho^2 + \ldots . \qquad (2.13)$$

Taking into account that $\sigma(0) = 0$ and putting $\sigma = \sigma'(0)v + \ldots$, where $\sigma'(0) \neq 0$, depending on the signs of $\sigma'v \neq 0$ and $g(0) \neq 0$, we obtain one of the phase portraits (Fig. 5.7). Meanwhile, r equals $\sqrt{\rho}$ (oscillation amplitude) for a periodic motion Γ that is stable relative to w_1, w_2 for $\sigma'v < 0$, $g > 0$, and unstable for $\sigma'v > 0$, $g < 0$, and, by (2.13), is thus of order $\sqrt{v}$; more precisely,

$$r = \sqrt{-\frac{\sigma'(0)v}{g(0)}} + \ldots , \qquad (2.14)$$

where the dots denote infinitesimals in v of order higher than $v^{1/2}$.

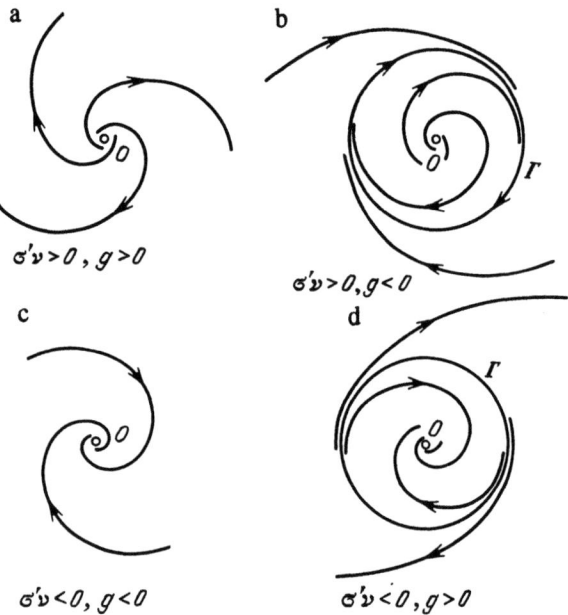

Fig. 5.7

Let $\sigma'v \neq 0$, and a direction of change of v be chosen so that $\sigma'v > 0$. Then only the following two cases are possible for $g(0) \neq 0$.

THEOREM 5.3. *For $g(0) > 0$, as v increases, the equilibrium state $O^{p,q}$ changes into $O^{p-2, q+2}$, and the periodic motion $\Gamma^{p-1, q+2}$ merges with it. For $g(0) < 0$, $O^{p,q}$ changes into $O^{p-2, q+2}$; however, the periodic motion $\Gamma^{p, q+1}$ gets separated.*

The meaning of p, q is similar to the notation of equilibrium states, and is sharpened in Sect. 3.

By Theorem 5.3, the stable periodic motion $\Gamma^{n,1}$ is generated by stability loss of the equilibrium state $O^{n,0}$ for $g(0) < 0$; for $g(0) > 0$, stability is lost when the state merges with the unstable periodic motion $\Gamma^{n-1,2}$.

The phase portrait if the stable periodic motion $\Gamma^{3,1}$ is generated is in Fig. 5.8.

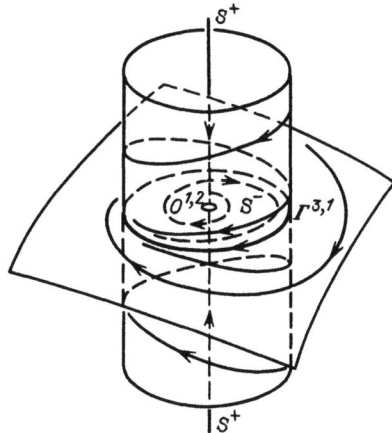

Fig. 5.8

Theorem 5.3 was first established in the two-dimensional case by A. A. Andronov and E. A. Leontovich (1939) [16, 17], and generalized to arbitrary dimensions in the papers [259, 260]. E. Hopf considered the multidimensional case in 1942, but the result is less complete, since the sign of the Lyapunov quantity g is not taken into account. For two-dimensional and three-dimensional systems, g was eventually calculated in [72, 73] on the basis of [259, 260]; for multidimensional ones in [351].

Table 1.

No	Bifurcation formula	Codimension	Boundary crossed	Stability loss	Phase portrait	Theorem
1	$O^{n,0} + O^{n-1,1} \to \emptyset$	1	N_0	hard	5.6, a	5.2
2	$O^{n,0} \to O^{n-1,1} + 2O^{n,0}$	2	N_0	soft	5.6, b	5.2
3	$O^{n,0} + 2O^{n-1,1} \to O^{n-1,1}$	2	N_0	hard		5.2
4	$O^{n,0} \to O^{n-2,2} + \Gamma^{n,1}$	1	N_ω	soft	5.7, c b	5.3
5	$O^{n,0} + \Gamma^{n-1,2} \to O^{n-2,2}$	1	N_ω	hard	5.7, d a	5.3
6	$O^{p,q} + O^{p-1,q+1} \to \emptyset$	1	N_0			5.2
7	$O^{p,q} \to O^{p+1,q-1} + 2O^{p,q}$	2	N_0			5.2
8	$O^{p+1,q-1} \to 2O^{p+1,q-1} + O^{p,q}$	2	N_0			5.2
9	$O^{p,q} \to O^{p-2,q+2} + \Gamma^{p,q+1}$	1	N_ω			5.3
10	$O^{p,q} + \Gamma^{p,q+2} \to O^{p-2,q+2}$	1	N_ω			5.3

Table 1 gives all the results about bifurcations of equilibrium states.

Numbered 1-5 are bifurcations of stable equilibria of type $O^{n,0}$; e.g., the first formula means that the saddle equilibrium $O^{n-1,1}$ merges with $O^{n,0}$, and then both vanish. In the next column, the bifurcation codimension is given (i.e., generally speaking, the number of conditions in order that bifurcation should occur). Which boundary of bifurcation is crossed is indicated next, followed by the column which specifies the nature of stability loss, and indicates what occurs with phase points earlier close to $O^{n,0}$ under stability loss when the parameter very slowly passes through zero. *Soft* transition means that the phase points are very slightly displaced, since a stable equilibrium or periodic motion is separated from an equilibrium which became unstable. In contrast, *hard* transition means that the points considerably shift to some other unknown steady-state motion. The last two columns give the representations of these bifurcations in the two-dimensional or three-dimensional case, and the numbers of related theorems.

Local phase portraits of the simplest steady-state motions and their bifurcations 115

3. Phase portraits in the vicinity of periodic motions

The study of the phase portrait associated with the differential equations (2.1) is in a neighbourhood of a closed phase curve reduced to the investigation of the point mapping

$$\bar{x} = F(x, \mu) \qquad (3.1)$$

via the Poincaré map in a neighbourhood of the corresponding fixed point O (Fig. 5.9). As for an equilibrium state, by linearization, theorems on reducing a point mapping to normal form, and theorems on invariant manifolds, this equation can be reduced to

$$\begin{aligned}\bar{u} &= Au + uP(w) + uf(u, v, w), \\ \bar{v} &= Bv + vQ(w) + vg(u, v, w), \\ \bar{w} &= Cw + wR(w) + wh(u, v, w).\end{aligned} \qquad (3.2)$$

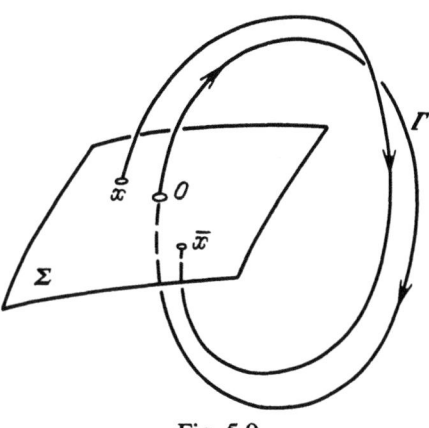

Fig. 5.9

In this notation, the invariant manifolds S^+, S^-, and J have the equations $v = 0$, $u = 0$, and $u = v = 0$, respectively.

As before, the resonant terms associated with the roots of the characteristic equation on the unit circle are collected in the polynomial R(w). The resonance conditions similar to (1.19) are of the form

$$\lambda_1^{q_1} \lambda_2^{q_2} \ldots \lambda_n^{q_n} = 1, \quad q_k \geq 0 \ (k \neq s), \quad q_s \geq -1, \quad \sum q_i \geq 1, \qquad (3.3)$$

where $\lambda_1, \lambda_2, \ldots, \lambda_n$ are the roots of the characteristic equation for a fixed point $O(x = x^*)$:

$$\chi(\lambda) = \operatorname{Det}\left\{\frac{\partial F}{\partial x}\bigg|_{x=x^*} - \lambda E\right\} = 0. \qquad (3.4)$$

The eigenvalues of the matrix A are in modulus less than one; those of B, greater; those of C lie on the unit circle.

We also indicate the general form of the blocks of A, B and C in (3.2), assuming that the matrices are reducible to block-diagonal form.

A real root λ_s on the diagonal of A or B is associated the element λ_s. If $\lambda_s = 0$, a zero is placed at the corresponding place on the diagonal of C. Two complex conjugate roots $\rho e^{\pm i\phi}$ ($\rho \neq 1$) are associated the second-order matrix

$$\begin{bmatrix} \rho \cos \phi & -\rho \sin \phi \\ \rho \sin \phi & \rho \cos \phi \end{bmatrix} \quad (3.5)$$

on the diagonal of A ($\rho < 1$) or B ($\rho > 1$).

If $\rho = 1$, (3.5) is placed on the diagonal of C.

The type of a simple fixed point is determined by the numbers p, q of roots of (3.4) outside and inside the unit circle, respectively; meanwhile, we assume that there are no roots on this circle. For a fixed point $O^{p,q}$ of type p, q, S^+, S^- are of dimensions p, q, respectively, and consist of points tending to $O^{p,q}$ if the mapping (3.2) and, respectively, the inverse one are repeated indefinitely. $O^{p,q}$ is where the manifolds S^+ and S^- meet, and coincides with the invariant manifold J. If there are roots of the characteristic equation on the unit circle, then the dimension of J equals the number of roots on it.

The qualitative behaviour of (3.1) in the vicinity of the fixed point O determines the behaviour of phase trajectories near the closed curve Γ associated with the periodic motion. Meanwhile, the closed phase trajectory Γ is associated to O; and the integral manifolds S^+, S^-, of dimensions p+1, q+1, of the periodic motion Γ are associated to the invariant manifolds S^+, S^- of dimensions p, q of the fixed point. Accordingly, p+1, q+1 determine the type of Γ; this is denoted by $\Gamma^{p+1,q+1}$. Along the integral surfaces S^+, S^-, the phase trajectories tend asymptotically to Γ as $t \to +\infty$ and $t \to -\infty$, respectively. The remaining phase trajectories in the vicinity of Γ first approach Γ along S^+, and then diverge from it along S^-.

4. Bifurcations of periodic motions

By definition, the type of a fixed point can be altered only when one of the roots

$\lambda_1, \lambda_2, \ldots, \lambda_n$ crosses the unit circle; therefore, only those parameter values $\mu = \mu^*$ are bifurcation values for which there is a root on the unit circle. Accordingly, the principal simplest bifurcation surfaces are N_{+1}, N_{-1} and N_φ, associated with one root, $+1$ or -1, respectively, and two complex conjugate roots $e^{\pm i\varphi}$.

THEOREM 5.4. *A point $\mu^* \in M$ is a bifurcation point if and only if (3.4) has a root on the unit circle, i.e., if μ^* is on one of N_{+1}, N_{-1} or N_φ.*

If we write the characteristic equation in the form

$$\chi(\lambda) = \lambda^n + a_1 \lambda^{n-1} + \ldots + a_{n-1}\lambda + a_n = 0, \tag{4.1}$$

then N_{+1}, N_{-1} and N_φ are determined by the equations

$$1 + a_1 + a_2 + \ldots + a_n = 0, \tag{4.2}$$

$$1 - a_1 + a_2 - \ldots + (-1)^n a_n = 0, \tag{4.3}$$

$$e^{in\varphi} + a_1 e^{i(n-1)\varphi} + \ldots + a_n = 0 \quad (0 < \varphi < \pi), \tag{4.4}$$

respectively. The equation of the latter surface N_φ is parametric. All N_{+1}, N_{-1}, N_φ are of codimension one.

Up to now, this is similar to the case of two equilibrium states: N_{+1} and N_φ of a fixed point are similar to the bifurcation surfaces N_0 and N_ω of an equilibrium state; N_{-1} is new. However, not all possible bifurcations of a periodic motion are exhausted by this. Also, a bifurcation of a periodic motion Γ is possible on account of the vanishing of the motion, in the following three scenarios: Γ loses closedness at infinity, an equilibrium state appears on Γ, or Γ is contracted to a point. There are no other possibilities; more precisely, there are no chances to stop the periodic motion Γ that are not related with moving across N_{+1}, N_{-1} or N_φ. Consider each case.

Bifurcation with Γ at infinity can be treated as drift of a point of Γ at infinity as the parameter μ varies.

Appearance of an equilibrium state O on Γ means that O acquired a phase trajectory γ tending to this state both as $t \to +\infty$ and $t \to -\infty$. Therefore, the curve γ is the intersection of the integral surfaces S^+ and S^- of O. Thus, a bifurcation of Γ consists in transforming the latter into a doubly asymptotic curve of an equilibrium state O where S^+ meets S^-.

μ changes continuously, and the closed curve Γ contracts to a point, which should be a bifurcation equilibrium state. Those states were considered, including equilibrium states' bifurcations, for which a closed phase curve merges with the state.

In the next chapter, we will continue our consideration of possible ways to make a closed phase trajectory Γ vanish; now, we return to the bifurcations N_{+1}, N_{-1}, and N_φ. The study of bifurcations caused by traversing a certain curve across N_{+1}, N_{-1}, and N_φ is only on J reduced to the study of the point mapping (3.2) in the form

$$\overline{w} = Cw + wR(w) + wh(0, 0, w). \tag{4.5}$$

If there is only one root on the unit circle, either $+1$, -1, or only the two complex conjugate roots $e^{i\varphi}, e^{-i\varphi}$, (4.5) becomes

$$\overline{w} = w + a_2 w^2 + a_3 w^3 + \ldots, \tag{4.6}$$

$$\overline{w} = -w + a_3 w^3 + a_5 w^5 + \ldots, \tag{4.7}$$

$$\overline{w} = e^{i\varphi} w + w \sum_{q_1, q_2} a_{q_1 q_2} w^{q_1} w^{*q_2} + \ldots, \tag{4.8}$$

respectively, where the dots denote infinitesimals of higher order, and all possible combinations of the integers q_1 and q_2 are, according to (3.3), determined by the equation

$$e^{iq_1\varphi - iq_2\varphi} = 1, \tag{4.9}$$

where $q_1 \geq -1$, $q_2 > 0$, $q_1 + q_2 \geq 1$.

For φ incommensurate with π, (4.9) has no solutions different from $q_1 = q_2 = q \geq 1$; therefore, (4.8) can be written as

$$\overline{w} = e^{i\varphi} w + a_3 w^2 w^* + a_5 w^3 w^{*2} + \ldots. \tag{4.10}$$

For $\varphi = 2\pi k/m$, where k and m are coprimes, other resonant terms appear. They are determined by the equation

$$\frac{2\pi k}{m}(q_1 - q_2) = 0 \pmod{2\pi}, \tag{4.11}$$

or, taking into account that $\varphi \neq 0$, $\varphi \neq \pi$,

$$q_1 - q_2 = m, \quad m = \pm 3, \pm 4, \ldots. \tag{4.12}$$

For $|m| = 3$, $|m| = 4$, the resonant terms to be added are essential; they need not be taken into account for $|m| \geq 5$ in considering the principal case $a_3 \neq 0$. The first two equalities are associated with $\varphi = 2\pi/3$ and $\varphi = \pi/2$; therefore, there are singular bifurcation curves associated with $\varphi = 2\pi/3$ and $\varphi = \pi/2$ on N_φ. Their codimension is two. In Fig. 5.10, N_{+1}, N_{-1}, N_φ, $N_{2\pi/3}$ and $N_{\pi/2}$ are given. Note that only the surfaces N_0

and N_ω are in a similar diagram for an equilibrium state (Fig. 5.11).

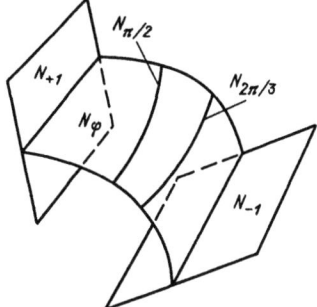

Fig. 5.10

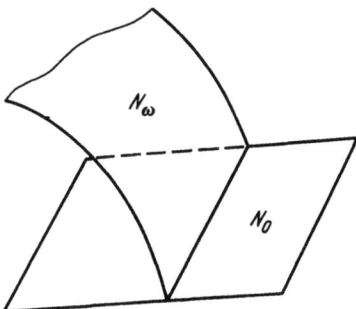

Fig. 5.11

We now continue our consideration of the principal case N_φ. Formula (4.10) yields the form of transformation of the manifold J for $\mu = \mu^*$. Similarly to the case of bifurcations of an equilibrium state, we introduce the parameter v, and write the perturbed mapping (4.10) in the form

$$\overline{w} = \lambda(v)w + a(v)|w|^2 w + \ldots = \lambda(v)[1 + a(v)|w|^2]w + \ldots, \qquad (4.13)$$

or, using polar coordinates ρ, ϑ,

$$\overline{\rho}^2 = |\lambda(v)|^2 [1 + 2\operatorname{Re} a(v)\rho^2]\rho^2 + \ldots$$
$$\overline{\vartheta} = \vartheta + \arg \lambda(v) + \operatorname{Im} a(v)\rho^2 + \ldots \qquad (4.14)$$

where the variable ρ has been separated up to inessential terms. Therefore, the first equation of (4.14) can be regarded as a one-dimensional mapping with fixed points

$$\rho_1^2 = 0, \quad \rho_2^2 = -[\operatorname{Re} a(v)]^{-1} \frac{d|\lambda(v)|}{dv}\bigg|_{v=0} v + \ldots . \qquad (4.15)$$

The possible diagrams of point mappings are given in Fig. 5.12.

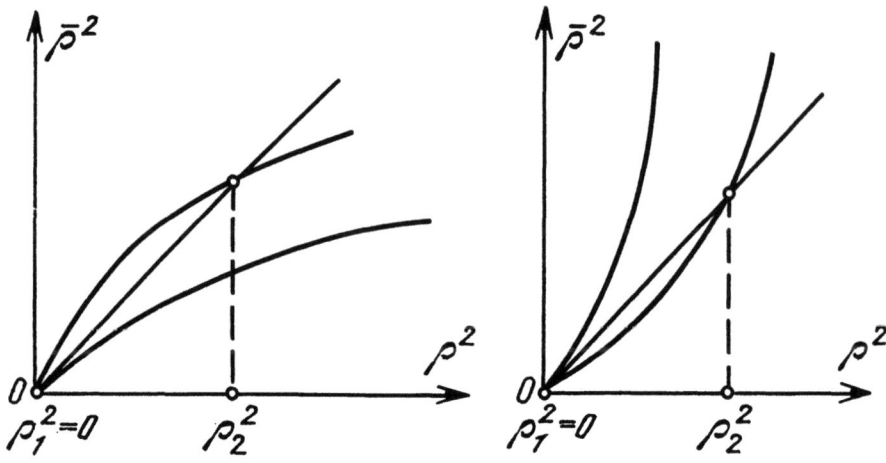

Fig. 5.12

The types of ρ_1 and ρ_2 (meaning stability and instability) are determined by the derivatives

$$\left.\frac{d\bar{\rho}^2}{d\rho^2}\right|_{\rho=\rho_1} = |\lambda(v)|^2, \quad \left.\frac{d\bar{\rho}^2}{d\rho^2}\right|_{\rho=\rho_2} = 1 - 4\frac{d|\lambda(v)|}{dv}v + \ldots . \qquad (4.16)$$

Putting $|d\lambda(v)/dv|_{v=0} > 0$ for definiteness, we arrive at the bifurcation diagrams in Fig. 5.13a, where the small circles denote stability, and crosses instability.

Local phase portraits of the simplest steady-state motions and their bifurcations 121

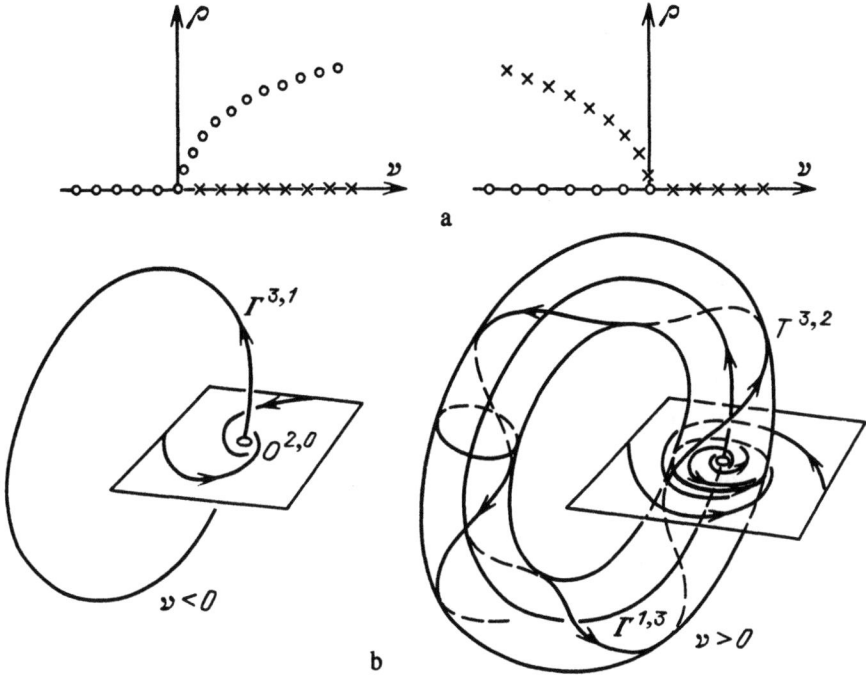

Fig. 5.13

In traversing across N_φ in the direction of forming two new roots of (3.4) exceeding one in modulus, the following thus holds [58, 260, 265, 629].

THEOREM 5.5. *The fixed point $O^{p,q}$ is sent to $O^{p-2,q+2}$. Simultaneously, either the invariant closed curve $\Gamma^{p-1,q+2}$ merges with $O^{p-2,q+2}$, or the invariant curve $\Gamma^{p,q+1}$ separates. The former occurs for* Re $a(0) > 0$; *the latter, for* Re $a(0) < 0$.

The following is accordingly true for the corresponding periodic motion.

THEOREM 5.5'. *The periodic motion $\Gamma^{p+1,q+1}$ is transformed into $\Gamma^{p-1,q+3}$. Simultaneously, depending on the sign of* Re $a(0) \neq 0$, *either the toroidal invariant manifold $T^{p,q+3}$ merges with $\Gamma^{p-1,q+3}$ or the invariant toroidal manifold $T^{p+1,q+2}$ separates* (Fig. 5.13b).

The particular case where $p+1 = n$, $q = 0$ is of special interest; the periodic motion $\Gamma^{n,1}$ loses stability, and a stable integral 2-manifold $T^{n,2}$ is generated. This occurs when Re $a(0) < 0$. If the inequality is reversed, $\Gamma^{n,1}$ loses stability, merging with the unstable two-dimensional integral torus $T^{n-2,4}$. Theorem 5.5' does not say anything about the phase trajectories on two-dimensional tori. We only note that, as $v \to 0$, the Poincaré rotation number of these phase trajectories tends to $\varphi/2\pi$.

When crossing the boundary of N_{+1}, the point mapping is of the form (4.6) for bifurcation at $\mu = \mu^*$. Studying the mapping, we obtain results completely analogous to bifurcation of equilibrium states when the boundary of N_0 is crossed. The following is valid [259].

THEOREM 5.6. *In crossing the boundary of N_{+1}, either the fixed points $O^{p,q}$, $O^{p-1,q+1}$ merge under (4.6) for the principal case $a_2 \neq 0$, and vanish, or such a pair is generated*

Accordingly, the following holds for a periodic motion.

THEOREM 5.6'. *In crossing the boundary of N_{+1} in the principal case $a_2 \neq 0$, either the periodic motions $\Gamma^{p+1,q+1}$, $\Gamma^{p,q+2}$ merge and then vanish, or are generated simultaneously.*

The case $a_2 = 0$, $a_3 \neq 0$ and $\mu = \mu^*$ is similar to the above for equilibrium states. Note that the particular case $a_2 = 0$, $a_3^* \neq 0$ may become principal if the system in question is symmetric. Just because of this fact, we make an exception for this case, and consider it along with the general one.

As bifurcations of fixed points in crossing the boundaries of N_{+1} and N_φ are similar to bifurcations of equilibrium states in crossing the boundaries of N_0 and N_ω, crossing the boundary of N_{-1} is accompanied by a new type of change, without any similarity to an equilibrium state [259, 260]. Consider the type in greater detail. For $\mu = \mu^*$, the mapping of the 1-manifold J is of form (4.7). When μ is close to μ^*, the point $O^{p,q}$ in question is preserved (the Jacobian on which this fixed point depends vanishes only on N_{+1}); therefore, for small v, (4.7) is

$$\overline{w} = -a_1(v)w + a_2(v)w^2 + a_3(v)w^3 + \dots , \qquad (4.17)$$

where $a_1(0) = -1$, $a_2(0) = 0$ and $a_3(0) \neq 0$.

We iterate this mapping as follows:

$$\begin{aligned}
\overline{\overline{w}} &= -a_1(v)[-a_1(v)w + a_2(v)w^2 + \dots] \\
&\quad + a_2(v)[-a_1(v)w + a_2(v)w^2 + \dots]^2 \\
&\quad + a_3(v)[-a_1(v)w + a_2(v)w^2 + \dots]^3 + \dots \\
&= a_1^2(v)w + [-a_1(v)a_2(v) + a_2(v)a_1^2(v)]w^2 \\
&\quad + [-a_1(v)a_3(v) - 2a_2^2(v)a_1(v)]w^3 + \dots \\
&= [1 + 2a_1'(0)v]w - a_3(0)w^3 + \dots . \qquad (4.18)
\end{aligned}$$

Iterating twice, we find the three fixed points

$$w_1 = 0, \quad w_{2,3} = \sqrt{\frac{2a_1'(0)}{a_3(0)}} v^{1/2} + \ldots, \tag{4.19}$$

w_1 also being fixed for the original mapping (4.17), while w_2 and w_3 associated with a cycle of fixed points of multiplicity two. We assume that $a_1'(0) \neq 0$ and that an increase in v is associated with the root crossing through -1 as a_1 decreases, so that $a_1'(0) < 0$. Then, depending on the sign of $a_3(0)$, the cases below are possible.

THEOREM 5.7. *In crossing the boundary of bifurcation of* N_{-1}, *when the fixed point* $O^{p,q}$ *loses one root of the characteristic equation, less than one in modulus, and is sent to the fixed point* $O^{p-1,q+2}$, *the cycle of double fixed points of type* $O^{p,q}$ *or* $O^{p-1,q+2}$, *respectively, separated from* $O^{p,q}$ *for* $a_1'(0)a_3(0) > 0$, *while they coincide for* $a_1'(0)a_3(0) < 0$.

THEOREM 5.7'. *For* $a_1'(0) < 0$ *and* $a_3(0) < 0$, *the periodic motion* $\Gamma^{p+1,q+1}$ *is transformed into* $\Gamma^{p,q+2}$, *with* $\Gamma^{p+1,q+1}$ *separating simultaneously, its period being twice as large. For* $a_1'(0) < 0$ *and* $a_3(0) > 0$, *where* $\Gamma^{p+1,q+1}$ *is replaced by* $\Gamma^{p,q+2}$ *again, the former merges with the latter, of period twice as large.*

The corresponding bifurcation diagrams for the point mapping (4.17) are in Fig. 5.14. The consecutive replacement of phase portraits in the three-dimensional case for $p = 2$, $q = 0$ and $a_1'(0) < 0$, $a_3(0) < 0$ is represented in Figs. 5.15 and 5.16.

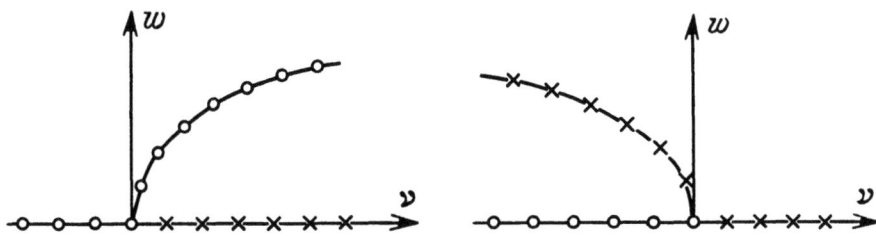

Fig. 5.14

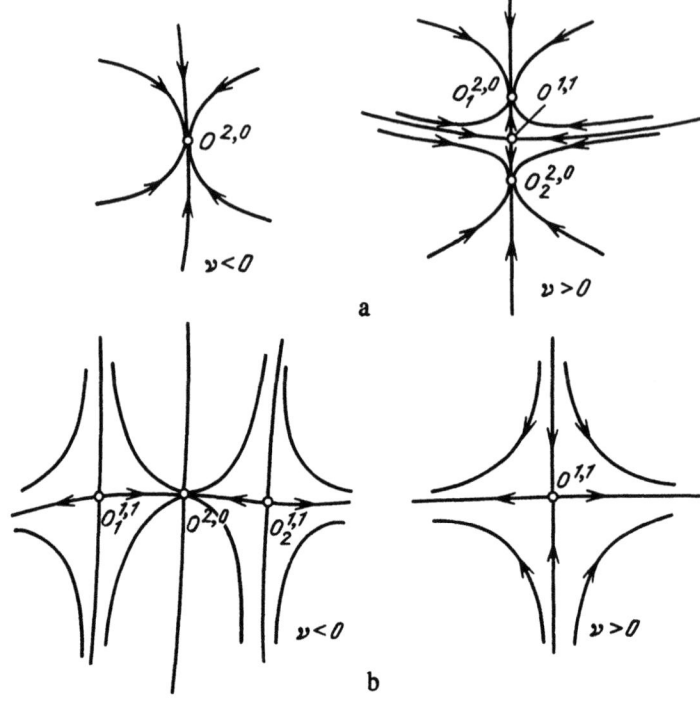

Fig. 5.15

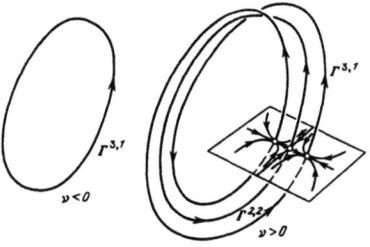

Fig. 5.16

We now clarify why there is no analog of the bifurcation N_{-1} for a fixed point in an equilibrium state. Let $x = x^*$ be an equilibrium state of the system of differential equations

$$\dot{x} = X(x), \qquad (4.20)$$

and $x = f(t, x_0)$ its solution satisfying the initial condition $x(0) = x_0$. Fix $t = \tau$, and consider the point mapping

$$\bar{x} = f(\tau, x). \qquad (4.21)$$

The point x^* is fixed. (4.21) is particular, since its Jacobian is non-negative; therefore, the characteristic equation of the fixed point cannot have an odd number of negative roots. Moreover, for $\tau = 0$, (4.21) is the identity, and all roots of its characteristic equation are +1. Note that the roots $\lambda_1, \lambda_2, \ldots$ of the characteristic equation of the equilibrium state and the roots $z_1, z_2, \ldots, z_n$ of the characteristic equation of the corresponding fixed point for the point mapping satisfy the equality $z_s = e^{\lambda_s \tau}$; therefore, the simple root -1 cannot emerge as τ increases. Then, what does the root -1 of the characteristic equation of the Poincaré map mean, with fixed point O related to a periodic motion Γ? To give the answer, we clarify how the invariant manifolds of a fixed point of a point mapping are associated with the same manifold J for Γ. We confine ourselves to the visual three-dimensional case. As can be seen in Fig. 5.17a, for $z = +1$, the integral manifold J is a two-sided surface; for $z = -1$ (Fig. 5.17b), a one-sided Möbius strip.

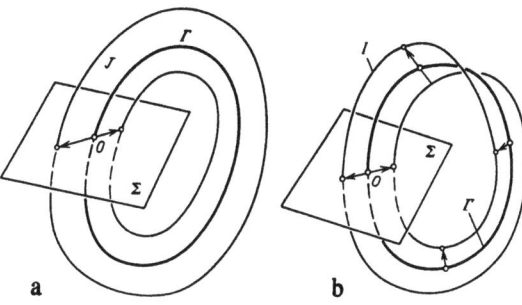

Fig. 5.17

In considering bifurcations of fixed points and periodic motions, the singular cases of N_φ-type with $\varphi = 2\pi/3$, $\varphi = \pi/2$ of codimension two passed unnoticed. Generation of a two-dimensional torus for the general bifurcation of N_φ-type, and the bifurcation of doubling of N_{-1}-type ($\varphi = \pi$) were discovered in 1959 [259, 260]. The consideration of singular bifurcations for $\varphi = 2\pi/3$, $\varphi = \pi/2$ still continues, and cannot be assumed complete, in spite of the considerable number of papers by different researchers [42, 58, 195-197, 280, 290]. A purely analytic study is impossible; there is a great variety and complexity of phase portraits which can arise in crossing the boundary of N_φ with $\varphi = 2\pi/3$, $\varphi = \pi/2$. Depending on the parameter values, both two-dimensional tori and periodic motions of saddle type with intersecting invariant manifolds S^+, S^-, and stable periodic motions of period three times and four times as large, respectively, can be generated. The singular bifurcations can lead to chaotic motions. Note that, for the case $\varphi = \pi$, the role of the bifurcation N_{-1} for chaotic motions has been clarified considerably.

In conclusion, we summarize the theorems proved for bifurcations of fixed points and periodic motions (Table 2). The meaning of the bifurcation formulas and related columns is similar to that in the table of bifurcations of equilibrium states, given above.

Table 2

Bifurcation formula	Codimension	Boundary crossed	Stability loss	Phase portrait	Theorem
Fixed point bifurcations					
$O^{n,0} + O^{n-1,1} \to \varnothing$	1	N_{+1}	hard		5.6
$O^{n,0} \to O^{n-1,1} + 2O^{n,0}$	2	N_{+1}	soft		5.6
$O^{n,0} + 2O^{n-1,1} \to O^{n-1,1}$	2	N_{+1}	hard		5.6
$O^{n,0} \to O^{n-1,1} + (O_1^{n,0}, O_2^{n,0})$	1	N_{-1}	soft	5.15, a	5.7
$O^{n,0} + (O_1^{n-1,1}, O_2^{n-1,1}) \to O^{n-1,1}$	1	N_{-1}	hard	5.15, b	5.7
$O^{n,0} \to O^{n-2,2} + \Gamma^{n,1}$	1	N_φ	soft		5.5
$O^{n,0} + \Gamma^{n-1,2} \to O^{n-2,2}$	1	N_φ	hard		5.5
$O^{p,q} + O^{p-1,q+1} \to \varnothing$	1	N_0			5.6
$O^{p,q} + 2O^{p-1,q+1} \to O^{p-1,q+1}$	2	N_0			5.6
$O^{p,q} \to 2O^{p,q} + O^{p-1,q+1}$	2	N_0			5.6
$O^{p,q} \to O^{p-1,q+1} + (O_1^{p,q}, O_2^{p,q})$	1	N_{-1}			5.7
$O^{p,q} + (O_1^{p-1,q+1}, O_2^{p-1,q+1}) \to O^{p-1,q+1}$	1	N_{-1}			5.7
$O^{p,q} \to O^{p-2,q+2} + \Gamma^{p,q+1}$	1	N_φ			5.5
$O^{p,q} + \Gamma^{p-1,q+2} \to O^{p-2,q+2}$	1	N_φ			5.5

Local phase portraits of the simplest steady-state motions and their bifurcations 127

Periodic motion bifurcations

$\Gamma^{n,1} + \Gamma^{n-1,2} \to \varnothing$	1	N_0	hard		5.6'
$\Gamma^{n,1} \to \Gamma^{n-1,2} + 2\Gamma^{n,1}$	2	N_0	soft	5.16	5.6'
$\Gamma^{n,1} + 2\Gamma^{n-1,2} \to \Gamma^{n-1,2}$	2	N_0	hard		5.6'
$\Gamma^{n,1} \to \Gamma^{n-2,3} + T^{n,2}$	1	N_φ	soft	5.13	5.5'
$\Gamma^{n,1} + T^{n-2,4} \to \Gamma^{n-2,3}$	1	N_φ	hard		5.5'
$\Gamma^{n,1} \to \Gamma^{n-1,2} + \Gamma^{n,1}$	1	N_{-1}	soft		5.7'
$\Gamma^{n,1} + \Gamma^{n-1,2} \to \Gamma^{n-1,2}$	1	N_{-1}	hard		5.7'
$\Gamma^{p,q} + \Gamma^{p-1,q+1} \to \varnothing$	1	N_0			5.6'
$\Gamma^{p,q} \to \Gamma^{p-1,q+1} + 2\Gamma^{p,q}$	2	N_0			5.6'
$\Gamma^{p,q} + 2\Gamma^{p-1,q+1} \to \Gamma^{p-1,q+1}$	2	N_0			5.6'
$\Gamma^{p,q} \to \Gamma^{p-2,q+2} + T^{p,q+1}$	1	N_φ			5.5'
$\Gamma^{p,q} + T^{p-1,q+2} \to \Gamma^{p-2,q+2}$	1	N_φ			5.5'
$\Gamma^{p,q} \to \Gamma^{p-1,q+1} + \Gamma^{p,q}$	1	N_{-1}			5.7'
$\Gamma^{p,q} + \Gamma^{p-1,q+1} \to \Gamma^{p-1,q+1}$	1	N_{-1}			5.7'

5. Toroidal integral manifolds

Toroidal integral 2-manifolds naturally arise in bifurcation of a periodic motion as the surface N_φ is traversed. It follows from the preceding section (Theorem 5.5) that, under certain conditions, traversal of N_φ is accompanied by the separation of a toroidal 2-manifold from the periodic motion. A toroidal integral 2-manifold may carry a great variety of phase portraits on its surface, possible bifurcations not being accompanied by torus destruction. Besides, there are bifurcations possible under which the torus vanishes as a smooth integral surface. The torus can be destroyed in a variety of ways. Cases where the torus gets destroyed as a whole are especially interesting. Bifurcations of a torus as a whole are similar to those of a periodic motion of types N_{+1}, N_{-1} and N_φ. However, studying the bifurcations similarly to periodic motions meets with a new difficulty, since the reduction of equations in a neighbourhood of the torus to normal form assumes the reducibility of linearized equations in the vicinity of the torus to linear differential equations with constant coefficients. Another approach to considering bifurcations of a torus as a whole is also possible, using reduction of the two-dimensional torus bifurcation problem to the problem of bifurcations of an invariant closed curve of the point mapping; for that, we have to cut the torus by a plane to obtain closed curve Γ. The phase trajectories on the torus and in its vicinity generate on Γ and nearby a point mapping. We assume that the latter is representable in a neighbourhood of Γ by differential equations, i.e., can be regarded as a shift operator of the equations. This arbitrarily exact representation has been established for close-to-identity mappings [269]. A similar situation holds in many other cases. This is how the matter stands for a point mapping T without intersecting invariant manifolds of fixed saddle points, and with invariant phase curves asymptotically tending to stable and unstable fixed points, respectively. For a two-dimensional mapping, the arguments justifying the statement are formulated in [100], and can easily be extended to the multi-dimensional case.

We now retrace our steps to the two-dimensional integral torus T cut by a plane Σ. Let it have a quasi-periodic coil. Then a degree of the mapping of T to Σ on a closed curve Γ is close to the identity transformation, and representable by differential equations on Σ. For the latter, Γ is associated with a periodic motion, possibly affected by bifurcations of types N_{+1}, N_{-1}, and N_φ for various parameters. The question of bifurcations of the torus as a whole is thereby reduced to the above ones of periodic motions. Bifurcations of types N_{+1}, N_{-1}, and N_φ of the periodic motion Γ are associated with those of the two-dimensional torus as a whole. Meanwhile, the stable two-dimensional torus accordingly merges with the unstable one or they both vanish (for N_{+1}); the two-dimensional torus loses stability, and simultaneously, either a new stable "double" torus is generated or it merges with the existing unstable double saddle torus (for N_{-1}), the two-dimensional torus loses stability; either a three-dimensional stable torus is generated from it or it merges with the three-dimensional unstable torus. We assumed above that, as parameters vary, the motions remain quasi-periodic on the torus. Satisfying of this requirement gets more complicated by the association of the general coil with one or several stable periodic motions separated by unstable ones. Which case (quasi-periodic or periodic) holds depends on the Poincaré rotation number. If it is rational, then periodic motions and those asymptotic to the latter occur, separated by unstable periodic ones; if

Local phase portraits of the simplest steady-state motions and their bifurcations 129

irrational, then quasi-periodic ones. For each rational value, the number continuously depends on the parameters as a piecewise constant function. The torus coil associated with a Poincaré rotation number of the form p / q, where p and q are two coprimes and the sum |p|+|q| is sufficiently large, is quite close to the quasi-periodic one; in particular, in the sense that, in directions not tangential to the torus, the characteristic Lyapunov exponents of motions on it are close. If the coil is quasi-periodic, then all these exponents are equal. The same holds for the roots of the characteristic equation of the fixed points of the mapping generated by phase trajectories on the cutting surface Σ, as in Fig. 5.18, where Σ is three-dimensional. The numbers of stable and unstable fixed points on the invariant curve Γ are rq, where r is a certain integer (usually, r = 1; sometimes, r = 2).

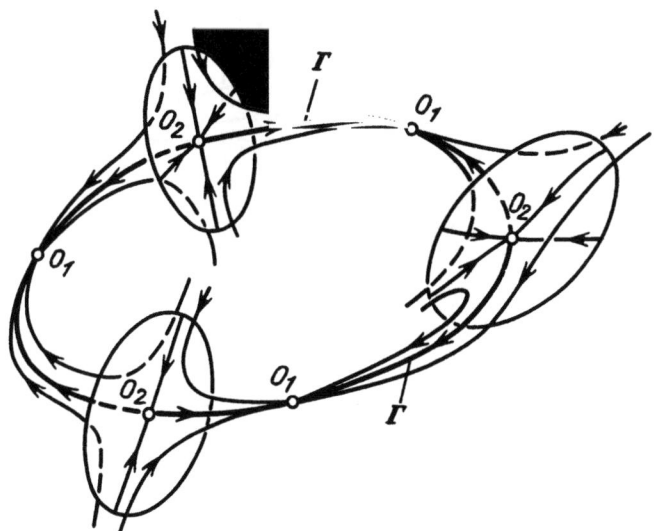

Fig. 5.18

We now consider how the phase portraits of a point mapping vary in the vicinity of a closed curve Γ for bifurcations of types N_{+1}, N_{-1}, and N_φ. We first neglect the small difference in the roots of the characteristic equation for fixed points in different cycles, then take it into account, and estimate the changes introduced. For bifurcations of type N_{+1}, the fixed points on and outside Γ merge and vanish; this is associated with the merging of the stable torus with the unstable one or with their vanishing. For a bifurcation of type N_{-1}, by Theorem 5.7, either new fixed points of multiplicity twice as large emerge from each fixed point or merge with it. One of the cases is in Fig. 5.19. But, it is necessary to bear in mind that they are only possible if the dimension of the original phase space is at least four, and that of Σ at least three, respectively. The restriction is due to a necessarily even number of negative real roots, since the Jacobian of the point mapping should be positive.

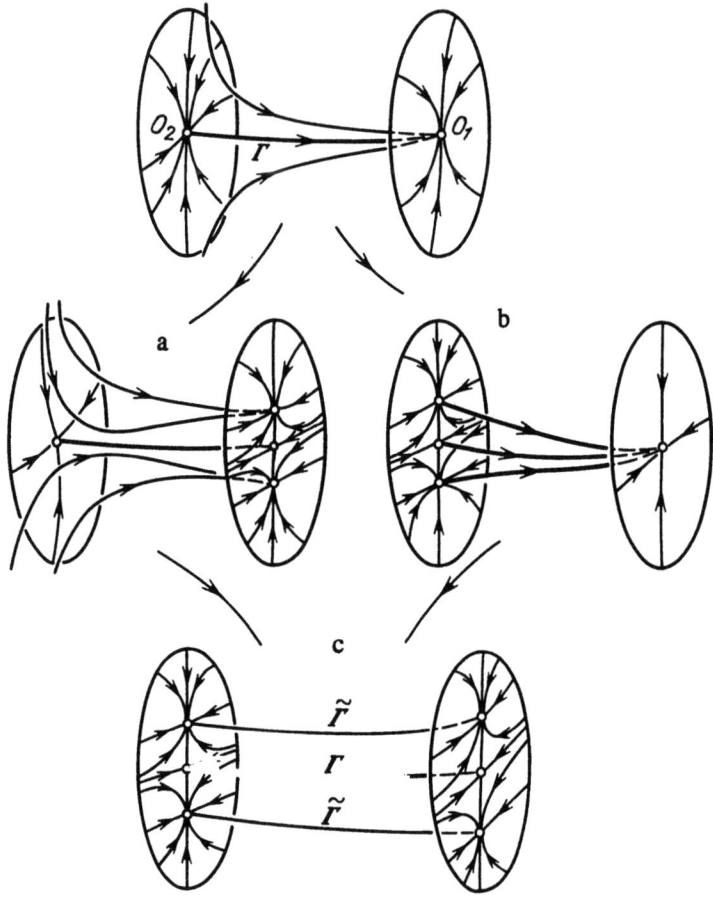

Fig. 5.19

For a bifurcation of type N_φ and also requiring the dimension of the original phase space to be at least four, by Theorem 5.6, two different cases are possible. In one, a three-dimensional stable torus is generated (which is associated with generation of a two-dimensional invariant stable torus from a closed curve Γ on Σ).

The above is obvious if motions on the two-dimensional integral torus are quasi-periodic. Little changes for large p, q and exactly coinciding bifurcation moments of individual periodic motions on the torus. If the moments are little different, then changes

Local phase portraits of the simplest steady-state motions and their bifurcations 131

occur, illustrated for a bifurcation of type N_{-1} in Fig. 5.19. If the difference is substantial, then changes not leading to the "double" torus can occur.

A similar argument is applicable to a three-dimensional integral torus, and leads to bifurcations of it as a whole, of types N_{+1}, N_{-1}. However, now with increasing dimension, changes on the torus itself can become still more important. They can cause chaotization and stochastization of motions if the torus is preserved as a stable manifold. In the two-dimensional case, the change cannot chaotize the motions on the torus, but can lead to its destruction. The merging of a stable periodic motion on the torus (N_{+1}) with an unstable one and then their vanishing should be regarded as such a bifurcation. One example will be considered in Chapt.6. It should be borne in mind that this bifurcation does not always lead to torus destruction, and everything can end in a change of Poincaré rotation numbers for phase trajectories on the torus. Torus destruction can be a consequence of bifurcations of individual periodic motions on it, of types N_{-1}, and N_φ. First of all, this can refer to cases where the periodic motion under bifurcation does not cover the torus in a sufficiently dense way. A bifurcation of type N_φ can lead to subsequent formation of a homoclinic structure via the contact of integral manifolds S^+ and S^- of saddle motions, previously lying on the torus.

CHAPTER 6

STOCHASTIC AND CHAOTIC ATTRACTORS

Among the basic important properties of a dynamical system are its steady-state motions, to which after a certain time any other of its motions leads. A steady-state motion is limiting, occurring asymptotically in the system isolated from all uncontrollable or random actions, and can be regarded as ideal deterministic.

Any motion and any phase trajectory, if bounded[3], tend to a certain limiting motion or motions. x_ω is an accumulation point for a phase trajectory $x(t)$ if an indefinitely increasing sequence of time intervals $t_1, t_2, t_3, \ldots$ can be indicated so that

$$\lim_{s \to \infty} x(t_s) = x_\omega.$$

It is not difficult to see that the set X_ω of accumulation points of a bounded phase trajectory $x(t)$ is non-empty, and consists of phase trajectories. However, physically, i.e., taking into account unavoidably small perturbations, the phase trajectory can tend to a limit set X_ω if X_ω is limiting both for $x(t)$ and all others close to X_ω. If X_ω possesses this asymptotic stability property, then it is an attractor. The simplest attractors are asymptotically stable equilibrium states and periodic motions.

The Banach contraction mapping principle enables us to regard contractibility of the phase space as a reason for existence of a stable equilibrium state, whereas contractibility on the cutting surface is a reason for existence of a periodic motion. If a phase region D contracts in the course of time into a region $\tilde{D}$ inside it, and subsequently $\tilde{D}$ into a still smaller region $\bar{\tilde{D}} \subset \tilde{D}$, etc., then it is intuitively clear that D contracts to a point as $t \to \infty$, i.e., a stable equilibrium state. Contractibility can be understood not only as justification for an equilibrium, but also as its natural reason. Similarly, if D on the cutting

---

[3]It is assumed that a bounded set is compact.

plane Σ is mapped into $\bar{D} \subset D$, $\bar{D}$ inside itself, etc. (Fig. 6.1), after intersection with phase trajectories again, and if, as before, contraction occurs, i.e., any two points are closer to each other after transformation, then there is a unique fixed point (sent to itself) in D, and a closed phase trajectory passes through it. Contraction is again understood as the natural reason for stable periodic motion.

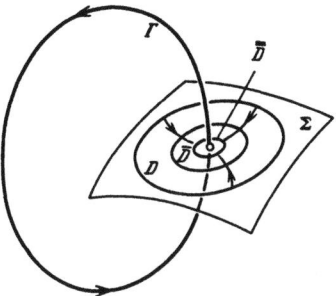

Fig. 6.1

These visual and sound arguments have a solid theoretical foundation of the familiar, widely used Brouwer and Banach theorems, and their subsequent generalizations [206, 239].

Properties of a dynamical system, e.g., energy dissipation, can also be connected with the contractibility property. Relations of this kind are sufficiently strong. A totally different situation arises as soon as expansion accompanies contraction. It is just to these not only contracting, but also expanding, mappings that stochasticity in dynamical systems is intimately related. We have already said before that stochasticity is a consequence of global contraction with local instability.

The subsequent discussion of the auxiliary mapping method enables us to formulate a new approach as a visual basis for intuitive ideas and theoretical study of non-contractive mappings. This approach makes it possible to study consistently and visually all familiar simplest typical situations which lead to chaotizing and stochastizing of motions of deterministic dynamical systems. In the sequel, a simple homoclinic Poincaré loop and a loop with contact, the Smale horseshoe, the Lorenz attractor, a mapping of an annulus into itself, a loop of a saddle-focus equilibrium state and of a saddle-node, a loop of a fixed singular saddle node, a homoclinic figure-eight, and stochastic synchronism are investigated just in this way. The machinery of auxiliary mappings and sequences of point mappings was studied comparatively recently especially for homoclinic structures [262, 269, 282]. We now turn to its treatment. From necessity, it is to be brief. Additional information can be found in [100, 269].

1. Auxiliary mappings and sequences of point mappings

We begin with a very simple example. The point mapping

$$\bar{x} = \frac{1}{2}x, \quad \bar{y} = 3y \tag{1.1}$$

transforms the square $G(|x| \leq 1, |y| \leq 1)$ into the rectangle $\bar{G}(|x| \leq 1/2, |y| \leq 3)$, contracting G along the x-axis and extending it along the y-axis (Fig. 6.2a). There is a unique fixed point, where the square intersects the rectangle. To make this obvious, we write (1.1) in the form

$$\bar{x} = \frac{1}{2}x, \quad y = \frac{1}{3}\bar{y}, \tag{1.2}$$

and regard it as a mapping of the point $x, \bar{y}$ to the point $\bar{x}, y$. Meanwhile, the rectangle $\tilde{G}(|x| \leq 1, |\bar{y}| \leq 3)$ is transformed into the rectangle $\bar{\tilde{G}}(|x| \leq 1/2, |y| \leq 1)$ (Fig. 6.2b). The latter lies in the former, the mapping is contractive; hence, there is a unique fixed point x^*, $\bar{y}^*$ sent to itself. But then it is also fixed for the original transformation (1.1).

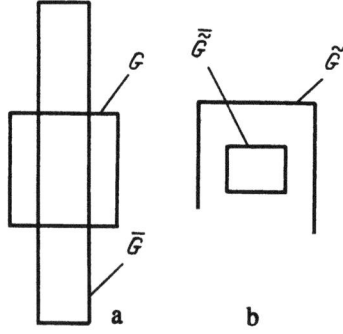

Fig. 6.2

It may seem that the fixed point is a consequence of G intersecting with $\bar{G}$. However, it can easily be seen that the intersection is not enough, while the inclusion of $\bar{\tilde{G}}$ in $\tilde{G}$ already suffices.

This simple argument can be generalized to any mapping of the form

$$\bar{u} = f(u, v), \quad \bar{v} = g(u, v) \tag{1.3}$$

Stochastic and chaotic attractors 135

that is contractive with respect to the variables **u** and expanding with respect to **v**. Solve the second equation with respect to **v**, i.e.,

$$\mathbf{v} = \tilde{g}(\mathbf{u}, \bar{\mathbf{v}}), \tag{1.4}$$

and substitute **v** in (1.3). Then

$$\bar{\mathbf{u}} = f(\mathbf{u}, \tilde{g}(\mathbf{u}, \bar{\mathbf{v}})) = \tilde{f}(\mathbf{u}, \bar{\mathbf{v}}). \tag{1.5}$$

We regard (1.4), (1.5) as a point mapping carrying a point $(\mathbf{u}, \bar{\mathbf{v}})$ into the point $(\bar{\mathbf{u}}, \mathbf{v})$. The original mapping is contractive with respect to **u** (i.e., in the transition from **u** to $\bar{\mathbf{u}}$, contraction occurs), while expanding with respect to **v** (i.e., in the transition from **v** to $\bar{\mathbf{v}}$, expansion occurs). If follows that the point mapping determined by formulas (1.4) and (1.5) is contractive with respect to both variables **u** and $\bar{\mathbf{v}}$, since then **u** is sent to $\bar{\mathbf{u}}$, and $\bar{\mathbf{v}}$ to **v**. Therefore, if the mapping (1.4), (1.5) transforms a certain region $\tilde{G}$ into a region $\overline{\tilde{G}} \subseteq \tilde{G}$, then it has a unique fixed point $(\mathbf{u}^*, \bar{\mathbf{v}}^*)$ such that

$$\mathbf{u}^* = \tilde{f}(\mathbf{u}^*, \bar{\mathbf{v}}^*), \quad \bar{\mathbf{v}}^* = \tilde{g}(\mathbf{u}^*, \bar{\mathbf{v}}^*). \tag{1.6}$$

It follows from the way the functions $\tilde{f}$ and $\tilde{g}$ were obtained that

$$\mathbf{u}^* = f(\mathbf{u}^*, \bar{\mathbf{v}}^*), \quad \bar{\mathbf{v}}^* = g(\mathbf{u}^*, \bar{\mathbf{v}}^*) \tag{1.7}$$

due to relations (1.6), i.e., the point $(\mathbf{u}^*, \bar{\mathbf{v}}^*)$ is also fixed under the original non-contractive mapping (1.3).

The goal of the sequel is to develop this simple argument, and apply it to the questions of appearance of chaotic and stochastic motions as well as of attractors as geometric representations of motions in the phase space.

1. *Definition of an auxiliary mapping and conditions for its being contractive.* Let the point mapping

$$(\bar{\mathbf{u}}, \bar{\mathbf{v}}) = T(\mathbf{u}, \mathbf{v}) \tag{1.8}$$

be defined in a region $G(\mathbf{u}, \mathbf{v})$, where **u** and **v** are r- and s-dimensional vectors, respectively. The mapping can also be written in the form (1.3). Suppose that it maps $G(\mathbf{u}, \mathbf{v})$ to a region $\overline{G}(\bar{\mathbf{u}}, \bar{\mathbf{v}})$.

Fix **u** and $\bar{\mathbf{v}}$, and find all pairs $\bar{\mathbf{u}}$, **v** for which (1.8), or, in another form, (1.3), holds. Possible values of **v** can then be found on the basis of

$$\bar{\mathbf{v}} = g(\mathbf{u}, \mathbf{v}), \tag{1.9}$$

which is solvable for **v**. For each $\bar{\mathbf{v}}$ found, the corresponding value of $\bar{\mathbf{u}}$ is determined by

$$\bar{u} = f(u, v). \tag{1.10}$$

The (possibly, multivalued) mapping determining all possible pairs $\bar{u}$, v for u and $\bar{v}$ is said to be *auxiliary* for a mapping T. We will denote it by $\tilde{T}$, and write it either as

$$(\bar{u}, v) = \tilde{T}(u, \bar{v}) \tag{1.11}$$

or as

$$\bar{u} = \tilde{f}(u, \bar{v}), \quad v = \tilde{g}(u, \bar{v}). \tag{1.12}$$

By the definition of an auxiliary mapping, if it associates a point $(u, \bar{v})$ with a point $(\bar{u}, v)$, then $(u, v) \in G$, while $(\bar{u}, \bar{v}) \in \bar{G}$.

If the auxiliary mapping $\tilde{T}$ is finitely or countably valued, then its regular branches are denoted by $\tilde{T}_j$, $j = 1, 2, 3, \ldots$.

We illustrate this by a two-valued auxiliary mapping. The mapping T

$$\bar{u} = \frac{1}{2}u, \quad \bar{v} = -4 + v^2, \tag{1.13}$$

defined in the region $G(0 \le u \le 1, |v| \le 2)$, has the two-valued auxiliary mapping

$$\bar{u} = \frac{1}{2}u, \quad v = \pm\sqrt{\bar{v} + 4}. \tag{1.14}$$

Each regular branch $\tilde{T}_1$, $\tilde{T}_2$ is defined in the region $\tilde{G}(0 \le u \le 1, -4 \le \bar{v} \le 0)$, and maps it into the regions $\bar{G}_1(0 \le \bar{u} \le 1/2, 0 \le v \le 2)$ and $\bar{G}_2(0 \le u \le 1/2, -2 \le v \le 0)$, respectively. $\tilde{T}_1$ and $\tilde{T}_2$ coincide on the curve $v = -4$, and differ at all the remaining points of $\tilde{G}$.

One can think that the two-valuedness of $\tilde{T}$ is related to that of the inverse mapping T^{-1}. However, this is not so. The mapping

$$\bar{u} = u_0 + a(v - v_0), \quad \bar{v} = bu + c(v - v_0)^2 \tag{1.15}$$

is one-to-one, while its auxiliary mapping

$$\bar{u} = u_0 \pm a\sqrt{\frac{\bar{v} - bu}{c}}, \quad \bar{v}_0 = v_0 \pm \sqrt{\frac{\bar{v} - bu}{c}} \tag{1.16}$$

is two-valued.

We call an auxiliary mapping *contractive* if, for certain $0 \le q < 1$,

$$\|\delta\bar{u}\| + \|\delta v\| < q(\|\delta u\| + \|\delta\bar{v}\|), \tag{1.17}$$

where δu and $\delta\bar{v}$ are small changes in the variables u and $\bar{v}$, and $\delta\bar{u}$, δv are the corresponding changes in the variables $\bar{u}$, v, determined from u, $\bar{v}$ according to (1.4) or

(1.5), so that
$$\delta\bar{u} = \vec{f}'_u \delta u + \vec{f}'_v \delta v, \quad \delta\bar{v} = \vec{g}'_u \delta u + \vec{g}'_v \delta v \qquad (1.18)$$
in a linear approximation. As usual, the sign $\|\cdot\|$ denotes the norm, which can be Euclidian or maximum-modulus for one component.

That $\tilde{T}$ is contractive means that the original mapping T is contractive with respect to **u**, and expanding with respect to **v**. In fact, for $\delta\bar{v} = 0$, it follows from (1.17) that $\|\delta\bar{u}\| < q\|\delta u\|$, while, for $\delta u = 0$, that $\|\delta\bar{v}\| > \frac{1}{q}\|\delta v\|$. An auxiliary mapping is contractive in the sense of (1.17) if
$$\max\{\|\vec{f}'_u\|, \|\vec{g}'_v\|\} + \max\{\|\vec{f}'_v\|, \|\vec{g}'_u\|\} \leq q < 1. \qquad (1.19)$$

Let a mapping T be of the form
$$\bar{u} = Au + f(u, v), \quad \bar{v} = Bv + g(u, v), \qquad (1.20)$$
where A, B are two linear operators such that
$$\|A\| \leq p_1, \quad \|B^{-1}\| \leq p_2, \qquad (1.21)$$
the functions f, g in the region $G(\|u\| \leq a, \|v\| \leq a)$ satisfying the conditions
$$\max\{\|f\|, \|g\|\} \leq K_0, \quad \max\{\|f'_u\|, \|f'_v\|, \|g'_u\|, \|g'_v\|\} \leq K_1, \qquad (1.22)$$
where the constants K_0, K_1, p_1, p_2, a satisfy the inequalities
$$K_0 < (1 - p_1)a, \quad p_2 K_1 < 1. \qquad (1.23)$$

The mapping T then has a one-valued auxiliary mapping $\tilde{T}$ in the region $\tilde{G}(\|u\| < a, \|\bar{v}\| \leq a)$, which is contractive in the sense of (1.17) if
$$\max\left\{\frac{p_1 + (1 + p_2)K_1}{1 - (1 + p_2)K_1}, \frac{p_2}{1 - (1 + p_2)K_1}\right\} < q < 1. \qquad (1.24)$$

A proof can be found in [269]. We will repeatedly use the theorem on the existence of $\tilde{T}$ for a mapping as in (1.20). It is important that such $\tilde{T}$ exists and is contractive in a sufficiently small neighbourhood of the point $u = 0$, $v = 0$ if only T is saddle-like, which is reflected by the conditions (1.21). That all other conditions of (1.22) and (1.23) hold is guaranteed by taking a sufficiently small neighbourhood of the point.

2. *Properties of auxiliary mappings.* Some simple lemmas on auxiliary mappings [269]:

(1) The mappings T, $\tilde{T}$ have (simple) fixed points in common.

(2) The fixed points of a mapping TR are in one-to-one correspondence with

those of the mapping $\tilde{T}\tilde{R}$, and coincide with those of the mapping TR.

(3) If $\tilde{T}$ is contractive, then its fixed point is saddle for T.

(4) If auxiliary $\tilde{T}$ and $\tilde{R}$ are defined and contractive in a region $\tilde{G}(\|u\| \leq a, \|v\| \leq b)$, and send $\tilde{G}$ into the region $\tilde{G}(\|\bar{u}\| \leq a, \|v\| \leq b)$, then the mapping $\tilde{R}T$ is defined, contractive and single-valued in $\tilde{G}$, transforming it into a region $\overline{\tilde{G}} \subseteq \tilde{G}$.

(5) Let $\tilde{T}, \tilde{R}$ be two mappings satisfying the conditions of (4), and $\tilde{S}$ defined in $\tilde{G}$ and satisfy the Lipschitz condition with a Lipschitz constant K for which

$$\frac{2q^2K}{1-2\alpha K} + \alpha < 1, \quad 2\alpha K < 1. \tag{1.25}$$

Then the mapping RST has the auxiliary mapping $\widetilde{RST}$, which is contractive in $\tilde{G}$, and takes $\tilde{G}$ into $\overline{\tilde{G}} \subseteq \tilde{G}$.

3. *Auxiliary mappings and fixed points.* Let a mapping T have the auxiliary mapping $\tilde{T}$, and one regular branch of $\tilde{T}$ reduce a certain region into itself. Then $\tilde{T}$ and, therefore, T have a fixed point.

The fixed points of T, associated with those of the regular branches of the auxiliary mapping, are necessarily different if the images of the domains of existence do not meet. On the contrary, if $\tilde{T}$ have no fixed points, e.g., the domains of existence of individual regular branches and their images are disjoint, then T does not have fixed points either.

If the auxiliary mapping is contractive, then the fixed-point statement admits sharpening, i.e., states the uniqueness in case of existence, and not only existence or non-existence. In the sequel, the above statements on auxiliary mappings and the relation of fixed points of auxiliary mappings to those of the original mappings are applied in various concrete situations. Below, we formulate theorems for two, in a certain sense, limiting, cases.

THEOREM 6.1. *Let mappings* T_i ($i = 1, 2, \ldots, n$) *have single-valued auxiliary mappings* $\tilde{T}_i$ *in a region* $\tilde{G}(\|u\| \leq a, \|v\| \leq b)$, *each transforming* $\tilde{G}$ *into* $\overline{\tilde{G}}(\|\bar{u}\| \leq a, \|v\| \leq b)$. *Then each mapping*

$$T_{i_m} T_{i_{m-1}} \ldots T_{i_2} T_{i_1}, \tag{1.26}$$

where $i_1, i_2, \ldots, i_m$ *are arbitrary numbers from 1 to n, has a fixed point* $x^*_{i_1 i_2 \ldots i_m}$ *in the region* $G(\|u\| \leq a, \|v\| \leq b)$. *If the auxiliary mappings are contractive, then the* *fixed point is unique.*

THEOREM 6.2. *Let $\tilde{T}_1, \tilde{T}_2, \ldots, \tilde{T}_n$ be the regular branches of the mapping $\tilde{T}$ which is auxiliary for T, defined in the region $\tilde{G}(\|\tilde{u}\| \leq a, \|\tilde{v}\| \leq b)$, and let any $\tilde{T}_s$ transform $\tilde{G}$ into a region $\overline{\tilde{G}}_s$ lying inside $\tilde{G}(\|\tilde{u}\| \leq a, \|v\| \leq b)$. The regions $\overline{\tilde{G}}_1, \overline{\tilde{G}}_2, \ldots, \overline{\tilde{G}}_n$ are assumed pairwise disjoint. Then each sequence of numbers $i_1, i_2, \ldots, i_m$ $(1 \leq i_s \leq n)$ is associated with the fixed point $x^*_{i_1 i_2 \ldots i_m}$ of the mapping T^m, the fixed points $x_{i_1 i_2 \ldots i_m}$ associated with different sequences $i_1, i_2, \ldots, i_m$ being different.*

The point $x^*_{i_1 i_2 \ldots i_m}$ is fixed under the mapping $T_{i_m} T_{i_{m-1}} \ldots T_{i_1}$, where $T_{i_1}, T_{i_2}, \ldots, T_{i_m}$ are the restrictions of T, associated with the regular branches of $\tilde{T}_{i_1}, \tilde{T}_{i_2}, \ldots, \tilde{T}_{i_m}$ for $\tilde{T}$. If $\tilde{T}_1, \tilde{T}_2, \ldots, \tilde{T}_n$ are contractive, then T^m has no fixed points different from $x^*_{i_1 i_2 \ldots i_m}$; for each $T_{i_m} \ldots T_{i_1}$, this is a unique fixed point.

4. *Sequences of point mappings.* Here, the above argument for fixed points and auxiliary mappings is generalized to sequences of point mappings, infinite in both directions.

Let
$$\ldots, T_{-2}, T_{-1}, T_0, T_1, T_2, \ldots \tag{1.27}$$

be a sequence of point mappings, infinite in both directions, each of which is defined in a region $G_s(\|u_s\| \leq a_s, \|v_s\| \leq b_s)$, and let

$$\ldots, (u_{-1}, v_{-1}), (u_0, v_0), (u_1, v_1), \ldots \tag{1.28}$$

be a sequence of points, infinite in both directions, such that each subsequent point, (u_s, v_s), is obtained from the one immediately preceding it, (u_{s-1}, v_{s-1}), by the transformation

$$T_s(u_{s-1}, v_{s-1}) = (u_s, v_s), \tag{1.29}$$

(Fig. 6.3).

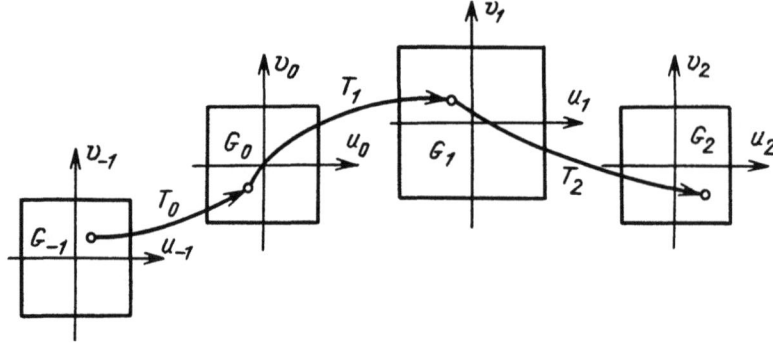

Fig. 6.3

(1.27) can be related to the unique mapping T of the direct product of the regions ... , G_{-1}, G_0, G_1, ... , sending the point (**u**, **v**), where $\mathbf{u} = (..., u_{-1}, u_0, u_1, ...)$, $\mathbf{v} = (..., v_{-1}, v_0, v_1, ...)$, to $(\bar{\mathbf{u}}, \bar{\mathbf{v}}) = (..., \bar{u}_{-1}, \bar{u}_0, \bar{u}_1, ... ; ..., \bar{v}_{-1}, \bar{v}_0, \bar{v}_1, ...)$, where

$$(\bar{u}_s, \bar{v}_s) = T_s(u_{s-1}, v_{s-1}). \tag{1.30}$$

If (1.29) holds, then

$$\bar{u}_s = u_s, \quad \bar{v}_s = v_s \tag{1.31}$$

for all s, while the point (**u***, **v***) associated with (1.28) is fixed under T.

Thus, the point sequence (1.28) related by (1.29) is associated with a fixed point y of T, and the question of existence or non-existence of this sequence is reduced to that of y. We consider the statement in somewhat greater detail, illustrating it by the phase trajectories of a dynamical system.

The question arises whether there exists a phase trajectory continually making the transition from a region A into a region B, from B into a region C, and then returning to A; the process should repeat itself. Suppose this behaviour occurs also in negative time, and not only positive, i.e., if time is negative, the phase point from A passes into C, from C into B, from B into A, etc. Let T_{AB}, T_{BC} and T_{CA} be mappings sending the phase point from A into B, from B into C, and C into A, respectively (Fig. 6.4).

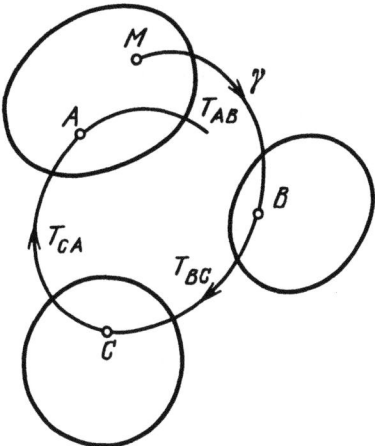

Fig. 6.4

The simplest case is a periodic motion traversing consecutively A, B and C. If T_{AB}, T_{BC} and T_{CA} are contractive or if their product is, then only this case is possible. However, if they are not, then more complicated cases are also possible. The above makes their study possible at least when the auxiliary mappings $\tilde{T}_{AB}$, $\tilde{T}_{BC}$ and $\tilde{T}_{CA}$ (or the auxiliary mapping $T_{CA}T_{BC}T_{AB}$) satisfy one of the above theorems.

Meanwhile, if $\tilde{T}_{AB}$, $\tilde{T}_{BC}$, $\tilde{T}_{CA}$ (at least two of them) are multivalued, then, along with different saddle periodic motions, many aperiodic ones also possess the required behaviour (there being a continuum of these).

A similar question may be asked for all displacements of the phase point across any sequence of regions, and not only for a periodically repeating displacement.

Numerous concrete examples using this argument are given in the subsequent section; now, we complete the consideration of (1.27) and the mapping T defined on sequences (1.28).

T is defined in the infinite-dimensional normed space of sequences (1.28) with the norm

$$\|(\mathbf{u}, \mathbf{v})\| = \sup_{s} \{\|(\mathbf{u}_s, \mathbf{v}_s)\|\}. \tag{1.32}$$

We now make the transition from T to the possibly contractive auxiliary mapping $\tilde{T}$, so that

$$\|\delta\overline{\mathbf{u}}\| + \|\delta\overline{\mathbf{v}}\| < q(\|\delta\mathbf{u}\| + \|\delta\mathbf{v}\|) \tag{1.33}$$

for certain $0 \leq q < 1$. This means in the case of (1.32) that each auxiliary mapping $\tilde{T}_s$ is contractive. In the general case, if $\tilde{T}$ transforms a certain convex region G into itself, the $\tilde{T}$ has a fixed point in G. In addition, if $\tilde{T}$ is contractive, then the fixed point is unique, and of saddle (i.e., hyperbolic) type. G is determined by the inequalities

$$\|\mathbf{u}_{s-1}\| \leq a_{s-1}, \quad \|\mathbf{v}_s\| \leq b_s \quad (s = \ldots, -1, 0, 1, \ldots), \tag{1.34}$$

is convex, and sent into itself by $\tilde{T}$ if each $\tilde{T}_s$ transforms the region $\tilde{G}_s(\|\mathbf{u}_{s-1}\| \leq a_{s-1}$, $\|\mathbf{v}_s\| \leq b_s)$ into $\overline{\tilde{G}}_s(\|\mathbf{u}_s\| \leq a_s, \|\mathbf{v}_{s-1}\| \leq b_{s-1})$ (Fig. 6.5). $\tilde{T}$ can be constructed from $\tilde{T}_s$, since specifying the relations

$$(\mathbf{u}_s, \mathbf{v}_{s-1}) = \tilde{T}_s(\mathbf{u}_{s-1}, \mathbf{v}_s) \tag{1.35}$$

for all s is just what defines the auxiliary mapping.

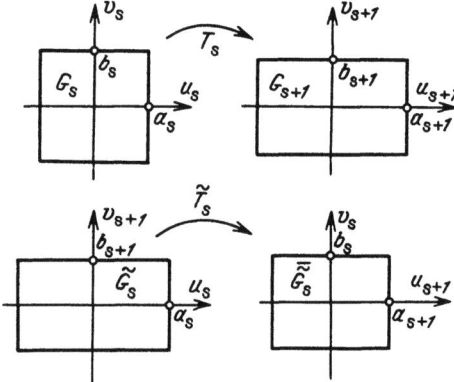

Fig. 6.5

Summarizing, we have the following.

THEOREM 6.3. *Fixed points of the mapping T defined in a region G are associated with sequences (1.28) satisfying relations (1.29).*

The mapping $\tilde{T}$ exists and is defined in the region $\tilde{G}$, transforming it into itself, if the same holds for all $\tilde{T}_s$; $\tilde{T}$ is contractive if all $\tilde{T}_s$ are.

This statement assumes implicitly that all $\tilde{T}_s$ are single-valued. However, this assumption is not at all necessary, and can be neglected if we only require that all regular branches of each multivalued $\tilde{T}_s$ should satisfy the same restrictions placed on single-valued $\tilde{T}_s$. The multivaluedness of $\tilde{T}_s$ also makes $\tilde{T}$ multivalued. Multivalued $\tilde{T}$ admit more than one fixed point of T even if the regular branches of $\tilde{T}$ are contractive.

The appearance of more than one fixed point of T means the existence of more than one sequence (1.28) satisfying conditions (1.29). The above can be formulated as follows.

THEOREM 6.4. *Let each mapping* T_s *(s = ... , -1, 0, 1, ...) have auxiliary mapping* $\tilde{T}_s$, $\tilde{T}_{sj_s}$ *(j_s = 1, 2, ... , m_s) being its regular branches,* T_{sj_s} *the associated restriction of* T_s, *and let each auxiliary mapping* $\tilde{T}_{sj_s}$ *be defined in the region* $\tilde{G}_s(\|u_{s-1}\| \leq a_{s-1}, \|v_s\| \leq b_s)$, *mapping it into* $\overline{\tilde{G}}_s(\|u_s\| \leq a_s, \|v_{s-1}\| \leq b_{s-1})$. *Then any sequence of mappings*

$$\ldots, T_{-1j_{-1}}, T_{0j_0}, T_{1j_1}, \ldots \quad (1.36)$$

is associated with at least one sequence of points

$$\ldots, (u_{-1j_{-1}}, v_{-1j_{-1}}), (u_{0j_0}, v_{0j_0}), (u_{1j_1}, v_{1j_1}), \ldots, \quad (1.37)$$

so that the equalities

$$(u_{sj_s}, v_{sj_s}) = T_{sj_s}(u_{s-1j_{s-1}}, v_{s-1j_{s-1}}) \quad (1.38)$$

hold for all s. If all auxiliary mappings in (1.36) are contractive, then a sequence (1.37) satisfying (1.38) is unique.

This theorem, with its general statement on the existence of the required fixed point or sequence, covers all the above. However, it does not contain any assertion enabling us to establish the absence of some or other fixed points or sequences. Meanwhile, it has already been noted that there is another possibility; viz., if at least one auxiliary mapping $\tilde{T}_{sj_s}$ of the mapping T_{sj_s} involved in (1.36) sends $\tilde{G}_s$ into a region $\overline{K}_{sj_s}$ not intersecting $\tilde{G}_s$, and not into $\tilde{G}_s$, then the required sequence of points (1.37) satisfying relations (1.38) does not exist.

In the two-dimensional and three-dimensional cases, the transition from original to auxiliary mappings can be given a visual geometric form. To stress the simplification and ease in the achieved understanding and investigation of the point mapping, the transition is said to be from the *negative to the positive*: The negative is the original point mapping; the positive, the corresponding auxiliary one [282]. Transitions from the positive to the negative can be conveniently used to analyze situations of different kinds, arising in the phase portraits of dynamical systems. In the subsequent section, the situations are indexed 1-10; now, we demonstrate the transition from the negative to the positive by the following hyperbolic (saddle) mapping T of the two-dimensional torus onto itself:

$$\bar{x} = 2x + y \pmod 1, \quad \bar{y} = x + y \pmod 1; \quad (1.39)$$

it has already been used in Chapt.4. Its negative is represented in Fig. 4.2, according to which this linear one-to-one mapping sends the square $D(0 \leq x < 1, 0 \leq y < 1)$ with corresponding side identifications into itself, so that each of its regions D_s (s = 0, 1, 2, 3) is mapped into the region $\overline{D}_s$. Having written this mapping in the form

$$\overline{x} = 2x + y - i, \quad \overline{y} = x + y - j, \qquad (1.40)$$

we see that the integers i, j depend on the membership of the point (x, y) in D_s; viz., for $(x, y) \in D_0$, $i = j = 0$; for $(x, y) \in D_1$, $i = 1, j = 0$; for $(x, y) \in D_2$, $i = j = 1$; for $(x, y) \in D_3$, $i = 2, j = 1$.

Regular branches of the auxiliary mapping can be written as mappings $\tilde{T}_1$:

$$x = \frac{\overline{x} - y}{2} + s, \quad \overline{y} = \frac{\overline{x} + y}{2}, \qquad (1.41)$$

where s = 0 for $\overline{x} - y \geq 0$, s = 1 for $\overline{x} - y < 0$; and $\tilde{T}_2$:

$$x = \frac{\overline{x} - y}{2} + \frac{1}{2}, \quad \overline{y} = \frac{\overline{x} + y}{2} + \frac{1}{2} - s, \qquad (1.42)$$

where s = 0 for $\overline{x} - y \geq 0$, s = 1 for $\overline{x} - y < 0$. The auxiliary mappings $\tilde{T}_1$, $\tilde{T}_2$ are contractive (with characteristic roots $1/2 \P i/2$ equal to $1/\sqrt{2}$ in modulus), send each the square $\tilde{D}(0 \leq \overline{x} < 1, 0 \leq y < 1)$ into itself, and their images are disjoint. By Theorem 6.4, any sequence of mappings

$$\ldots, \tilde{T}_{i_{-1}}, \tilde{T}_{i_0}, \tilde{T}_{i_1}, \tilde{T}_{i_2}, \ldots \qquad (1.43)$$

is associated with the unique sequence of points

$$\ldots ; x_{-1}, y_{-1} ; x_0, y_0 ; x_1, y_1 ; x_2, y_2 ; \ldots \qquad (1.44)$$

related by

$$(x_s, y_s) = T(x_{s-1}, y_{s-1}). \qquad (1.45)$$

The converse is also true, i.e., any sequence (1.44) related by (1.45) is associated with a unique sequence of auxiliary mappings (1.43); this yields the full description of all phase trajectories, both periodic and aperiodic, of the point mapping T. The description is in the spirit of so-called *symbolic dynamics*.

We will dwell on it briefly. We represent individual fragments of (1.43) by the letters $\alpha, \beta, \gamma, \ldots$. If (1.43) is periodic, i.e., of the form $\ldots \alpha \alpha \alpha \ldots$, then it is associated with a periodic phase trajectory (1.44). Sequences of the form

$$\beta \alpha \alpha \alpha \ldots \text{ and } \ldots \alpha \alpha \alpha \gamma$$

are related to phase trajectories tending to (1.44) as $s \to +\infty$ and $s \to -\infty$, respectively.

Stochastic and chaotic attractors 145

Their sets form the manifolds S^+ and S^- of the periodic motion ... ααα ... .

The phase trajectory associated with a sequence of the form ... αα ... αγα ... αα ... is doubly asymptotic to the periodic trajectory ... ααα ... . The phase trajectory ... αα ... αγββ ... β ... asymptotically tends to the periodic trajectory ... ααα ... and ... βββ ... as $s \to -\infty$ and $s \to +\infty$, respectively.

Let $\alpha_1, \alpha_2, ...$ be any, possibly countable, set of periods. It is then not difficult to indicate sequences associated with phase trajectories which arbitrary closely approach any periodic trajectory of the set. It can be shown that almost all phase trajectories are everywhere dense.

One may think that this nice set of motions is very fragile, and only associated with linear mapping (1.39). However, this is not so. Actually, we face a very solid formation, which cannot be destroyed even by a very, and not so small, perturbation of (1.39). This immediately follows from the non-destruction of the contractability of $\tilde{T}_1, \tilde{T}_2$ under the perturbations, with the consequence that the whole phase portrait structure is complicated.

2. Transition from the "negative" to the "positive" and investigation of typical appearance of chaos.

SITUATION 1. *Simple homoclinic Poincaré loop.* Consider a homoclinic Poincaré loop (Fig. 6.6), where O is a fixed point, S^+, S^- are the in- and out-going invariant manifolds, and M is the point at which they meet (one of the intersections of the curves S^+ and S^-). In this situation, the mapping T, after any sufficiently many transformations m, sends a thin region A located along the line S^+ to a thin region $\overline{A}^m$ along the curve S^-. Furthermore, since S^- meets S^+ at M, after a certain number m* of transformations T, a region B is mapped to a thin region $\overline{B}$ intersecting A.

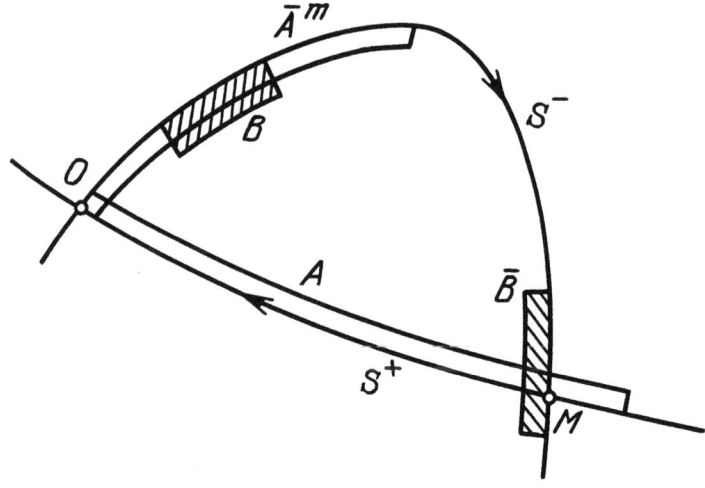

Fig. 6.6

This is how the "negative" is described. But what can be deduced from it? There are no regions mapped into themselves, and no contractibility. It is not simple to orient ourselves in this situation.

To make the transition from the "negative" to the "positive", for better representation, we illustrate the transitions from A to $\overline{A}^m$, and from B to $\overline{B}$, separately (Fig. 6.7). The mapping sending A to $\overline{A}^m$ is T^m; the one mapping B to $\overline{B}$ is T^{m^*}, and will be denoted by L (m* being fixed). The points a, b, O, c of A in Fig. 6.7 are sent to $\overline{a}$, $\overline{b}$, O, $\overline{c}$ of $\overline{A}^m$, while the points d, e, f, g of B, to $\overline{d}$, $\overline{e}$, $\overline{f}$, $\overline{g}$ of $\overline{B}$. We select coordinates u, v as in Fig. 6.7 (along the rectifiable curves S^+, S^- in Fig. 6.6).

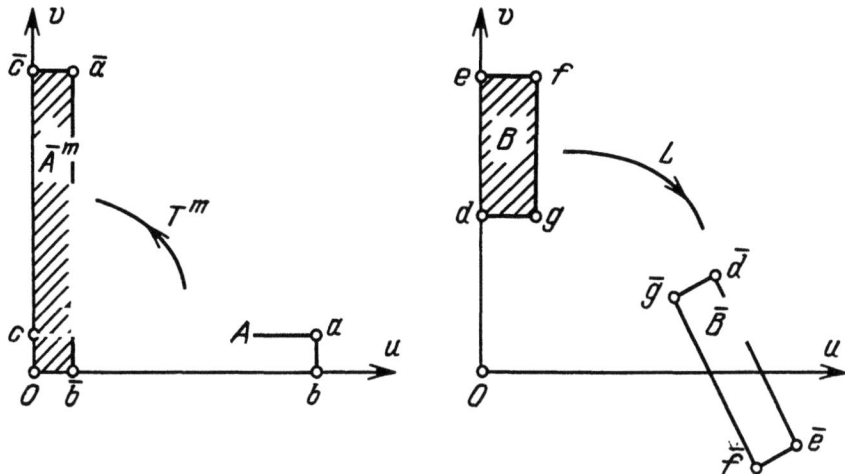

Fig. 6.7

The auxiliary mapping $\tilde{T}^m$ transforms the region A_1 into the region $\overline{A}_1^m$ (Fig. 6.8), which follows, by the contractibility of $\tilde{T}^m$, the choice of A, m, the smallness of |v| for A, and that of |u| for $\overline{A}^m$. At the same time, the variables u, v are within finite limits for A, $\overline{A}^m$, respectively. For sufficiently large m, $\tilde{T}^m$ is contractive. It is single-valued, and its coefficient of contraction tends to zero as m → ∞. Thus, $\tilde{T}^m$ maps A to itself, and is contractive with contraction coefficient tending to zero as m → ∞. The point u = 0, v = 0 is fixed. This is the "positive" for the mapping T^m.

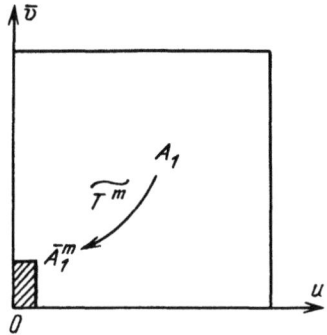

Fig. 6.8

We now turn to the "positive" for the mapping $L = T^{m^*}$; the "negative" is in Fig. 6.7. The auxiliary mapping $\tilde{L}$ transforms a certain region B_1 to a region $\overline{B}_1$ in A_1 if only A is sufficiently extended with respect to u and m sufficiently large. $\tilde{L}$ is single-valued, because the curve S⁻ meets S⁺, and has a bounded Lipschitz constant, which immediately follows from the fact that the image of a u-curve in a region B intersects a $\bar{v}$-curve in the region $\bar{B}$ at one point, without tangency. This is the "positive" for L (Fig. 6.9).

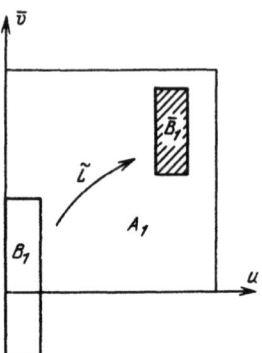

Fig. 6.9

We now make the "positives" for the mappings T^m and L coincident, and construct the auxiliary mapping $\check{L}\check{T}^m$. However, it is considerably simpler to construct the product of $\tilde{T}^m$ and $\tilde{L}$, and not $\check{L}\check{T}^m$ (which is also possible): $\tilde{T}^m$ transforming A_1 into the region $\overline{A}_1^m$, and strongly contracting the latter, $\tilde{L}$ transforming $\overline{A}_1^m$ as part of B_1 into a region $\overline{B}_1$, and, therefore, in A_1. Hence, for each $m \geq \overline{m}$, the mapping $\tilde{L}\tilde{T}^m$ has a fixed point, and, according to the preceding section, the mapping $LT^m = T^{m^*+m}$ also has a fixed point for any $m \geq \overline{m}$.

This is the conclusion from the contraction mapping principle for the "positive": In a neighbourhood of a point M, there are infinitely many fixed points of the point mapping T, while, starting with certain multiplicity $m \geq m^*$ (where $m^*+\overline{m}$ is replaced by m^*), they are saddle, and, for each m, the saddle point of some multiplicity is unique.

Making use of the transition to infinite sequences of mappings, the above statement can be formulated as the existence of points (u_s, v_s) ($s = 0, \pm 1, \pm 2, \ldots$) such that

$$(u_s, v_s) = LT^{m_s}(u_{s-1}, v_{s-1}) \qquad (2.1)$$

in the neighbourhood of M for any sequence of numbers

$$\ldots, m_{-1}, m_0, m_1, m_2, \ldots \quad (m_s \geq m^*). \qquad (2.2)$$

The mapping T in question can be interpreted as the point mapping on the

Stochastic and chaotic attractors 149

cutting plane, generated by the phase trajectories of a three-dimensional dynamical system with phase portrait as in Fig. 6.10, where Γ is a saddle periodic motion associated with a fixed point O on the cutting plane, and γ a doubly asymptotic phase curve tending to Γ as $t \to +\infty, t \to -\infty$. The point M is a point of intersection of γ and the cutting plane. Recall that γ is a homoclinic curve in the sense of H. Poincaré. Choose a small neighbourhood of Γ, γ. According to the above, all possible phase curves belong to it, rotating about Γ several times and simultaneously passing along γ. In particular, there is a closed phase trajectory of saddle type, traversing Γ $m \geq m^*$ times, and than encircling γ. There is a closed curve of saddle type, traversing along Γ $m_1 \geq m^*$ times, passing along γ, again passing along Γ $m_2 \geq m^*$ times, and, finally, encircling γ, etc. Moreover, for any infinite sequence of numbers (2.2), there exist a unique phase trajectory passing along Γ just the required number of times, and simultaneously passing along γ. All the phase trajectories together make up an invariant saddle set J of continuum power. J completely belongs to the neighbourhood of Γ, γ. As m^* increases, the phase curves are located in a still smaller neighbourhood δ of the phase curves Γ, γ. Two invariant manifolds S^+, S^- pass through each of the phase curves $\gamma_{...m_{-1}m_0m_1...}$ from J. The collection of all possible surfaces S^+ forms the bundle of phase trajectories entering δ, and remaining there as $t \to +\infty$. All possible surfaces S^- are made up of bundles of phase trajectories entering δ and remaining there as $t \to -\infty$. The sets of all possible manifolds S^+ and S^- of individual phase trajectories from J are called the *manifolds* S^+ and S^- of J. We cannot help noticing a simple analogy between J and its manifolds S^+, S^-, and the saddle motion Γ and its invariant manifolds S^+, S^-. In particular, in this more general case, too, all the remaining phase trajectories enter δ along S^+, and emanate along S^-. Meanwhile, as they enter δ, the sojourn in δ and motion of the phase point there can be quite different, strongly varying even if the initial position of the phase point changes very little.

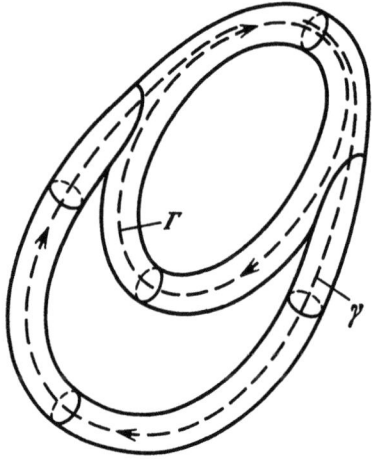

Fig. 6.10

All the above occurs in the neighbourhood δ of the saddle periodic motion Γ and the phase trajectory γ doubly asymptotic to it. In particular, among all possible phase trajectories in this small neighbourhood of Γ and γ, there are all possible doubly asymptotic ones for sequences of numbers

$$\infty, m_1, m_2, \ldots, m_k, \infty, \quad (2.3)$$

and also all doubly asymptotic phase trajectories γ_ω for any saddle periodic motion Γ_ω associated with a periodic sequence of m_s ($s = \ldots, -1, 0, 1, \ldots$) of the form

$$m_1, m_2, \ldots, m_k \quad (2.4)$$

with any period ω. In a neighbourhood of any phase trajectories Γ_ω, γ_ω, in turn, there are infinitely many ones similar to those in the vicinity of Γ, γ. A similar pattern occurs for any saddle periodic motion Γ and all possible doubly asymptotic phase trajectories associated with all possible sequences of the form (2.3). One cannot help seeing a hierarchy of nested structures, where the same structure of general form is a subset of the preceding, and the situation repeats infinitely many times.

SITUATION 2. *Poincaré loop with contact*. This is different from the above in that the curves S^+, S^- do not intersect, but touch each other at a certain point M (Fig. 6.11a). We select a region A again. The image $\overline{A}^m$ of A is the same. As before, we select a region B; however, now, its image $\overline{B}$ touches S^+ at M after several iterations of T (Fig. 6.11), which makes the "positive" for the mapping L different. The auxiliary mapping $\tilde{L}$ is two-valued, and defined in the region B_1, transforming it into two regions

$\bar{B}_{11}$, $\bar{B}_{12}$. Such is the "positive" for L. In fact, we have already met it in considering an example of auxiliary mappings.

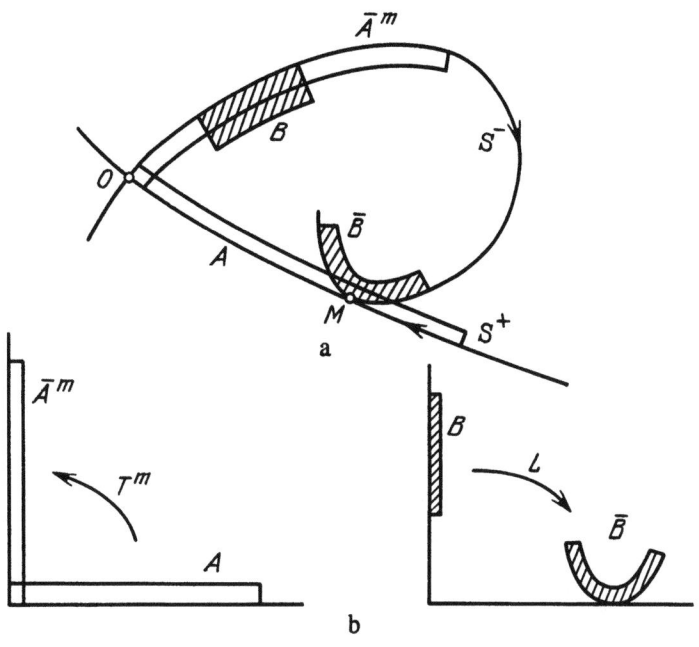

Fig. 6.11

Thus, $\tilde{L}$ transforms B_1 into $\bar{B}_1 = B_{11} \cup B_{12}$, and the mapping $\tilde{T}^m$ sends the region $\bar{B}_1$ to the region $\bar{\bar{B}}_1$. For increasing m, these regions are represented in Fig. 6.12. Judging by the way the region $\bar{\bar{B}}_1^m$ approaches the point O as $m \to +\infty$, we can distinguish between two cases of approaching O along the axis of ordinates or abscissas. In the former, from a certain m onwards, $\bar{\bar{B}}_1^m$ turns out to be inside B_1; in the latter, necessarily outside, which means that all possible sequences of mappings

$$\ldots, T^{m_s}L_{j_s}, T^{m_s+1}L_{j_{s+1}}, \ldots \qquad (2.5)$$

are realized, where $m_s \geq m^*$, $j_s = 1, 2$, and L_1 denotes the restriction of L, associated with one regular branch of $\tilde{L}$, while L_2 with the other; there are no such sequences for $m_s \geq m^*$. If λ, ν are roots of the characteristic equation of the fixed point O, then, as can easily be seen, the former case holds for $\lambda\nu < 1$; the latter, for $\lambda\nu > 1$. All possible motions associated with (2.5) for $\lambda\nu < 1$ also form a certain set J. The Lipschitz constant of the mapping $\bar{L}$ tends to ∞ on the boundary separating the regions B_{11}, B_{12}; therefore, the statement of Item 5 on the properties of auxiliary mappings, used to prove that the mappings (2.5) are of saddle type, cannot be applied immediately. Nevertheless, considering the "positive" in Fig. 6.12, one can believe that they are of saddle type. A more complete study in support of these intuitive ideas can be found in [119, 137], which makes Situation 2 similar to Situation 1 above, since, in both cases, there are saddle sets of continuum power, and associated manifolds S^+, S^-. However, the situations are substantially different in the global behaviour of phase trajectories; we shall deal with this fact in the sequel.

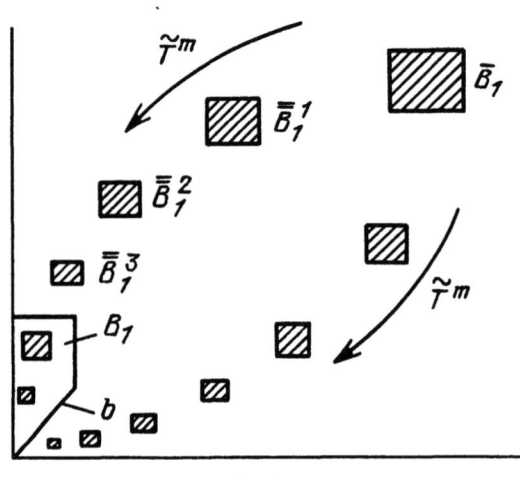

Fig. 6.12

The situation in question is marginal for two close intersections of the curves S^+, S^-, and for no intersection. The "positives" associated with all three cases are illustrated in Fig. 6.13.

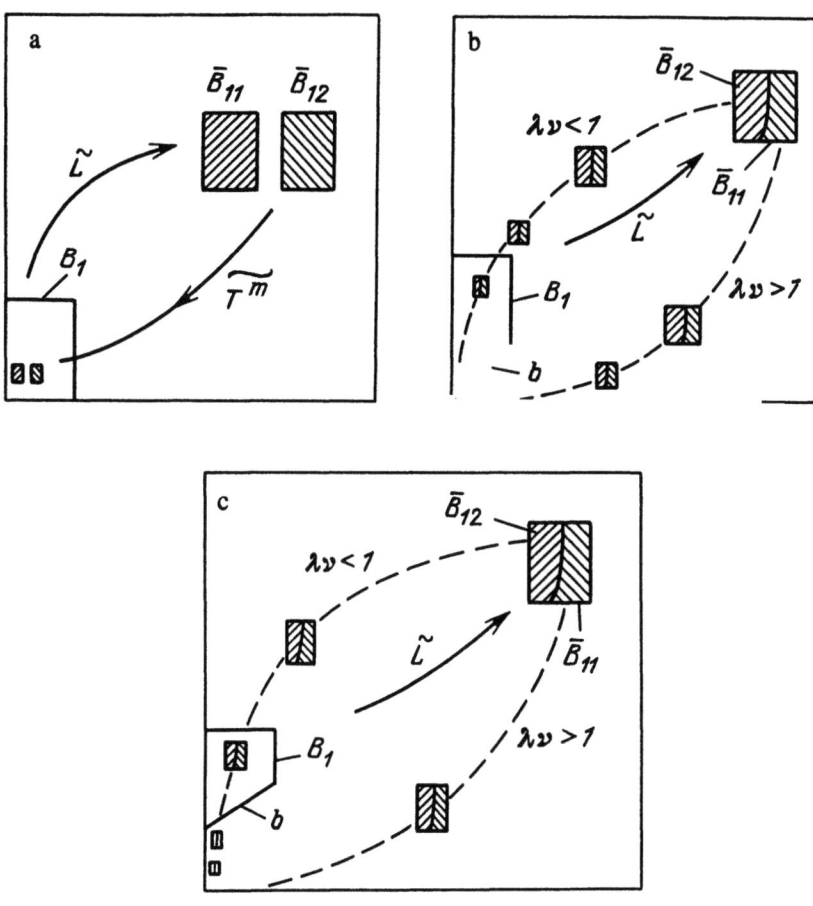

Fig. 6.13

We can gather from them how the set J and the corresponding admissible sequences of transformations vary, and which part is associated with saddle motions. The visual ideas are developed in Chapt.7.

SITUATION 3. *Smale horseshoe*. A Smale horseshoe is a certain mapping T (and its image) transforming a bar abcd into a bar $\overline{abcd}$, bending the former somewhere in the middle. The mapping from abcd to the horseshoe $\overline{abcd}$ is obtained by contracting the rectangle along the sides ad, bc, extending it along ab, cd, and bending to get a horseshoe shape (Fig. 6.14). One may notice that we are, in fact, familiar with the mapping. If we consider only one of the horseshoe intersections with the bar, then we are in Situation 1; if we lift the horseshoe, so that the bending is in the bar, then we are in Situation 2. This immediately enables us to draw its "positive" for the horseshoe "negative", representing the former in Fig. 6.15. The auxiliary mapping $\tilde{T}$ transforms the region $\tilde{G}$ into two small regions $\overline{\tilde{G}}_1$, $\overline{\tilde{G}}_2$ inside, because $\tilde{T}$ is two-valued and contractive. It can directly be seen from the "positive" that the mapping T has two different fixed saddle points; T^2, four; T^3, eight, etc. Situation 3 is governed by Theorem 6.2. Let T_1, T_2 be two restrictions of T, associated with regular branches of the auxiliary mapping $\tilde{T}$. In the case in question, T_1, T_2 represent the mapping T in the parts G_1, G_2 of G (or bar abcd). T transforms G_1, G_2 into regions $\overline{G}_1$, $\overline{G}_2$. By Theorem 6.4, any sequence ..., i_{-1}, i_0, i_1, i_2, ... ($i_s = 1, 2$) is associated with a unique sequence of points transformed by T into each other, so that they are in the region

$$\ldots, G_{i_{-1}}, G_{i_0}, G_{i_1}, G_{i_2}, \ldots \tag{2.6}$$

one after another. All the fixed points associated with periodic sequences (2.6) are saddle, as are all the phase trajectories corresponding to sequences (2.6) taken arbitrarily.

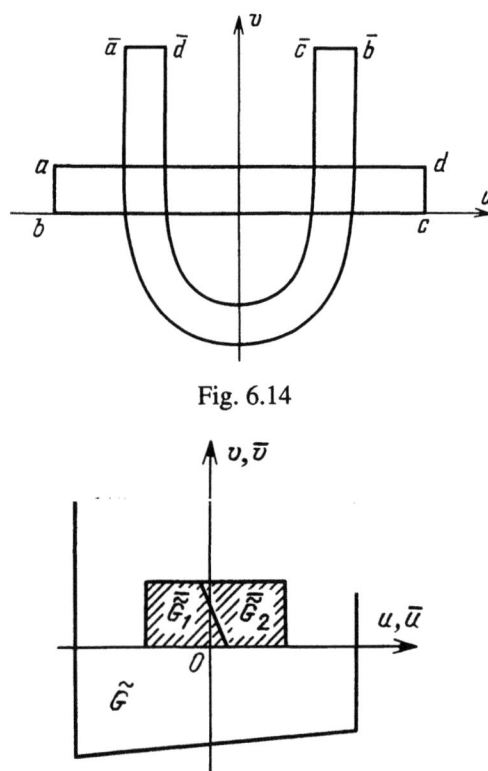

Fig. 6.14

Fig. 6.15

SITUATION 4. *Lorenz attractor*. This is the Poincaré map T on the cutting surface $z = r-1$ of the Lorenz equation as the strange attractor appears and after it appeared. The corresponding "negatives" are in Fig. 6.16, where O_1, O_2 are two fixed saddle points, S_1^-, S_1^+, S_2^-, S_2^+ their invariant curves, and R is the line of discontinuity of the mapping, T being smooth and continuously extendable to R on opposite sides of it. The "negative" is symmetric with respect to its centre. Extended on opposite sides, T sends R to two points M_1, M_2, respectively. In accordance with the above, G_1, G_2 in Fig. 6.16 are mapped to two wedges $\overline{G}_1$, $\overline{G}_2$, and the "negative" is thus described.

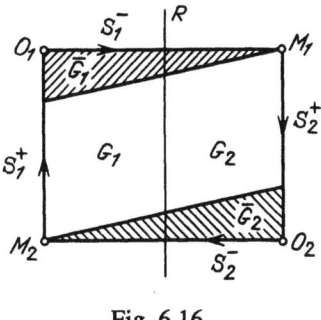

Fig. 6.16

The "positive" is described by the region $\tilde{G}$ transformed into two regions $\tilde{\overline{G}}_1$, $\tilde{\overline{G}}_2$ (Fig. 6.17) almost as in the Smale horseshoe, with substantial difference that $\tilde{\overline{G}}_1$, $\tilde{\overline{G}}_2$ are adjacent to the boundaries of $\tilde{G}$, while O_1, O_2 are fixed stable points of the auxiliary mapping $\tilde{T}$. The structure of the invariant set J follows immediately. It remains to indicate that J is an invariant saddle set, and a section of the strange attractor of the Lorenz equations. A property of J is that its invariant set S^- is contained in J. S^+ attracts "near" phase points along its invariant set S^-.

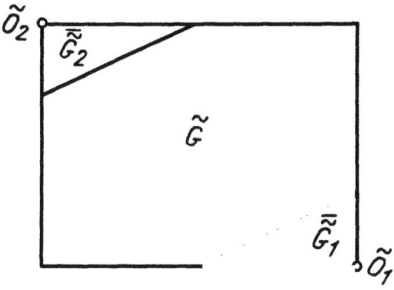

Fig. 6.17

SITUATION 5. *Mapping of an annulus into itself.* If an annulus is transformed into itself, and the mapping is contractive along the radii, then the existence of a limiting closed curve to which the whole annulus is contracted seems natural. However, this is not always so: The limiting set can also be very complicated, of the type in Situations 1–4. To see this, we make the transition from the "negative" to the "positive" as before.

Let a mapping T of the annulus $G(r_0 \leq r \leq R_0, 0 \leq \varphi < 2\pi)$ into itself be given by

$$\overline{\varphi} = f(\varphi, r), \quad \overline{r} = g(\varphi, r), \tag{2.7}$$

and let it be contractive along radii, so that

$$|g'_r(\varphi, r)| < q < 1. \tag{2.8}$$

Denote by $\tilde{T}_k$ the mappings which are auxiliary for the restrictions of T defined in the regions G_k ($\varphi_{1k} \leq \varphi \leq \varphi_{2k}$) (k = 1, 2, ... , m). Assume that we can indicate regions G_k and $\tilde{G}_s$ ($\tilde{\varphi}_{1s} \leq \varphi \leq \tilde{\varphi}_{2s}$) (s = 1, 2, ... , n), and regular branches $\tilde{R}_i$ (i = 1, 2, ... , n) of the auxiliary mappings $\tilde{T}_k$ such that each of the distinguished single-valued auxiliary mappings $\tilde{R}_i$ is defined in a certain region $\tilde{G}_s$, transforming it into a region $\tilde{G}_j$ (Figs. 6.18 and 6.19). Meanwhile, the subscript i determines the possible subscripts s, and s all possible subscripts j, so that

$$s = \alpha(i), j = \beta(s), \tag{2.9}$$

which determines some possible pairs of subscripts i, j, where i, j = 1, 2, ... , n. Construct the graph with vertices $\Gamma_1, \Gamma_2, \Gamma_3, \ldots, \Gamma_n$ and edges γ_{ij} leading from a vertex Γ_i to Γ_j, where i, j are all possible subscript pairs (Fig. 6.20). The path from vertex to vertex along γ_{ij} is called a *path in the graph*. Each path in the graph is associated with a certain sequence of its vertices Γ_i. Each vertex can be associated with a point mapping $\tilde{R}_i$, and then each path in the graph is associated with a sequence of mappings $\tilde{R}_i$, said to be *admissible*. By Theorem 6.4, each admissible sequence of mappings

$$\ldots, \tilde{R}_{i_{-1}}, \tilde{R}_{i_0}, \tilde{R}_{i_1} \tag{2.10}$$

is associated with a realizable sequence of mappings

$$\ldots, R_{i_{-1}}, R_{i_0}, R_{i_1}, \ldots \tag{2.11}$$

of the annulus into itself, where each R_{i_s} is the restriction of T, associated with the single-valued auxiliary mapping $\tilde{R}_{i_s}$. If the set of all possible paths in the graphs is not reduced to finitely many closed paths or their periodic repetion, then all possible sequences (2.11) are associated with a complicated invariant set J. It is saddle if the auxiliary mappings $\tilde{R}_i$ ($i = 1, 2, \ldots, n$) are contractive.

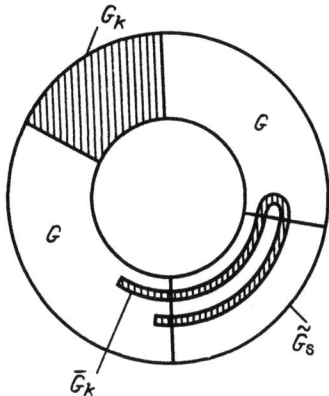

Fig. 6.18

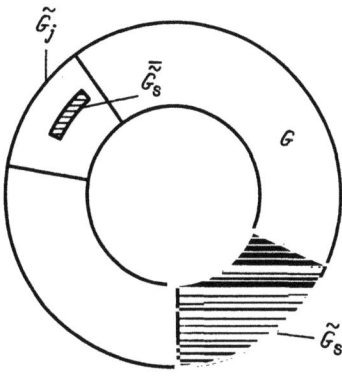

Fig. 6.19

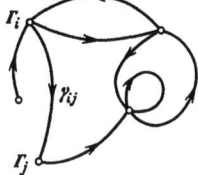

Fig. 6.20

In conclusion, we give an example of a point mapping of an annulus into an annulus, represented in Fig. 6.21, with a non-trivial invariant set J. The graph of admissible sequences of mappings is represented in the same diagram. The mapping T sends the region G_1 to the region $\bar{G}_1$. The auxiliary mapping $\tilde{T}_1$ is three-valued. Each of its regular branches is defined in $\tilde{G}_1$, and sends $\tilde{G}_1$ inside G_1, which is contained in $\tilde{G}_1$. The existence of a complicated invariant set J follows.

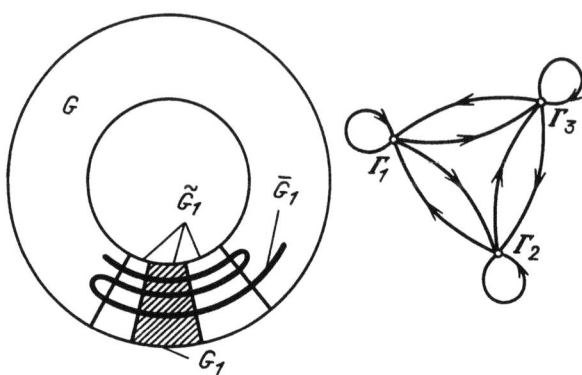

Fig. 6.21

Note that the above consideration of the two-dimensional mapping of an annulus into itself can be immediately generalized to a mapping of an n-dimensional toroidal region G ($\|r\| \leq r_0$, $0 \leq \varphi < 2\pi$) which is the topological product of the (n-1)-ball $\|r\| \leq r_0$ and the one-dimensional circle. An example of such a mapping for n = 2 was considered in Sect.1, Chapt.2.

SITUATION 6. *Saddle focus equilibrium state loop.* Since non-linear terms not essential in the sequel are neglected, in a neighbourhood of the saddle focus $O^{2,1}$, the equations of motion of a phase point u, v, z can be written as

$$\begin{aligned} \dot{u} &= -\sigma u - \omega v, \\ \dot{v} &= \omega u - \sigma v, \\ \dot{z} &= \lambda z, \end{aligned} \quad (2.12)$$

where $\sigma > 0$, $\lambda > 0$. Integrating over the time interval τ gives

$$\begin{aligned} \bar{u} &= e^{-\sigma\tau}(u \cos \omega\tau - v \sin \omega\tau), \\ \bar{v} &= e^{-\sigma\tau}(u \sin \omega\tau + v \cos \omega\tau), \\ \bar{z} &= e^{\lambda x} z. \end{aligned} \quad (2.13)$$

Cut the bundle of phase trajectories approaching the point O by the plane Ω^+

$$u^2 + v^2 = \delta^2 \quad (0 \leq z \leq \delta_1), \quad (2.14)$$

and the bundle of divergent phase trajectories by the plane Ω^-

$$z = \varepsilon (|u| + |v| \leq \varepsilon_1) \quad (2.15)$$

(Fig. 6.22). To find the relation between the coordinates of the intersection points (u, v, z) and ($\bar{u}$, $\bar{v}$, $\bar{z}$) of the phase trajectory with Ω^+, Ω^- (Fig. 6.23), we take into account (2.14), and put

$$u = \delta \cos \varphi, \quad v = \delta \sin \varphi. \tag{2.16}$$

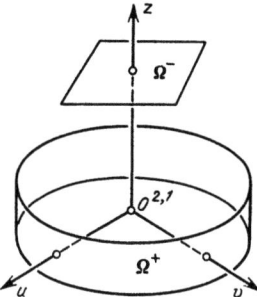

Fig. 6.22

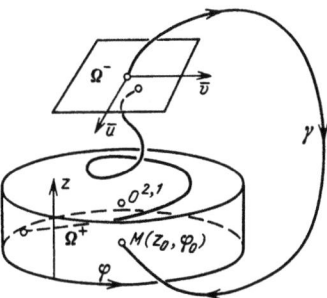

Fig. 6.23

On the basis of (2.15), we derive from the last equation in (2.13) that

$$\tau = \frac{1}{\lambda} \ln \frac{\varepsilon}{z}.$$

It then follows from the other two equations that

$$\bar{u} = \delta \left(\frac{z}{\varepsilon}\right)^{\sigma/\lambda} \cos\left(-\frac{\omega}{\lambda} \ln \frac{z}{\varepsilon} + \varphi\right),$$

$$\bar{v} = \delta \left(\frac{z}{\varepsilon}\right)^{\sigma/\lambda} \sin\left(-\frac{\omega}{\lambda} \ln \frac{z}{\varepsilon} + \varphi\right). \tag{2.17}$$

For $\sigma > \lambda$ and small z, the mapping is contractive; for $\sigma < \lambda$, it is not. Solving equation (2.17) for z and $\bar{v}$, we rewrite it in the form

$$z = f_k(\bar{u}, \varphi), \quad \bar{v} = g_k(\bar{u}, \varphi), \quad (\bar{u}, \varphi) \in \tilde{D}_k, \tag{2.18}$$

where $\tilde{D}_k$ are annular regions of the form as in Fig. 6.24, and f_k, g_k single-valued functions to be discussed below. As k increases, the regions are of still smaller and smaller height. From $\bar{u}$ and φ as determined by the first equation, we can find z by intersecting the curve

$$x = \delta\left(\frac{z}{\varepsilon}\right)^{\sigma/\lambda} \cos\left(-\frac{\omega}{\lambda} \ln \frac{z}{\varepsilon} + \varphi\right) \tag{2.19}$$

with the straight line $x = \bar{u}$ (Fig. 6.25).

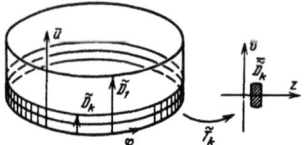

Fig. 6.24

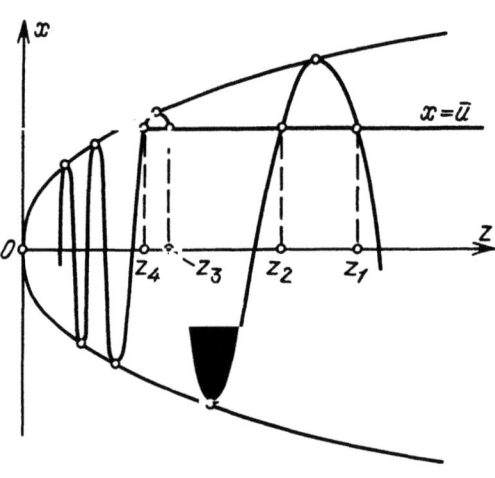

Fig. 6.25

The values $z = z_k$ determine the function $f_k(\bar{u}, \varphi)$. The function $g_k(\bar{u}, \varphi)$ is obtained via substitution $z = z_k$ in the second equation of (2.17).

The mapping (2.18) is contractive if

$$\max\left\{\left|\frac{\partial z}{\partial \bar{u}}\right| + \left|\frac{\partial \bar{v}}{\partial \bar{u}}\right|, \left|\frac{\partial z}{\partial \varphi}\right| + \left|\frac{\partial \bar{v}}{\partial \varphi}\right|\right\} < q < 1, \tag{2.20}$$

which occurs for small z outside a small neighbourhood of the upper boundary of $\tilde{D}_k$. The mapping (2.18) is auxiliary for (2.17). Denote the regular kth branch of the auxiliary mapping by $\tilde{T}_k$. If a point $(\bar{u}, \varphi) \in \tilde{D}_s = \tilde{D}_{s+1}$ does not belong to $\tilde{D}_{s+2}$ any more, then the auxiliary mapping at this point has $s+1$ images in the $(z, \bar{v})$-plane, as shown in Fig. 6.26;

this all holds for σ < λ. For σ > λ, the mapping is contractive, and determines the "positive" itself.

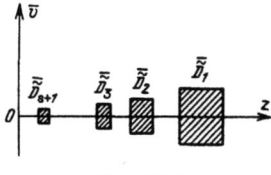

Fig. 6.26

Assume that the phase trajectory γ emanating from the point $\bar{u} = \bar{v} = 0$ of Ω^- arrives at a certain point M of Ω^+ (Fig. 6.23). The phase trajectories close to γ generate a point mapping L of Ω^- into Ω^+, of the form

$$z = h(\bar{u}, \bar{v}), \quad \varphi = k(\bar{u}, \bar{v}), \tag{2.21}$$

where $h(0, 0) = z_0$, $k(0, 0) = \varphi_0$ and z_0, φ_0 are the coordinates of M on Ω^+.

To construct the "positives" of the mapping T of Ω^+ into Ω^-, and of L of Ω^- into Ω^+, L should be written in the form

$$\bar{u} = \tilde{h}(z, \bar{v}), \quad \varphi = k(z, \bar{v}). \tag{2.22}$$

Assume that the mapping is single-valued and $h'_{\bar{u}}(0, 0) \neq 0$. We are now ready to construct the "positive" represented in Fig. 6.27. In its upper part, the "positive" of the mapping from Ω^+ into Ω^-, generated by the phase trajectories passing near the saddle focus, is represented. The mapping is multivalued, and even infinite-valued on the axis $\bar{u} = 0$; as we move farther from $\bar{u} = 0$, the number of images decreases.

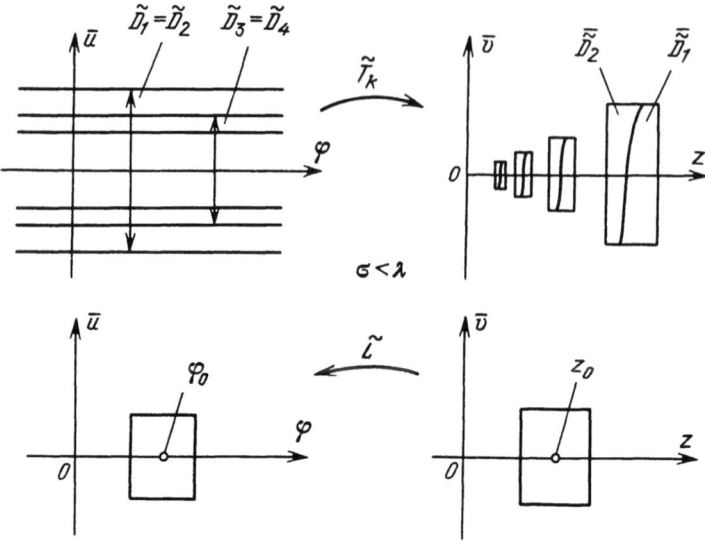

Fig. 6.27

In the lower part, a mapping $\tilde{L}$ is represented, transforming a neighbourhood of the point $(z_0, 0)$ into that of $(\varphi_0, 0)$. It can be seen from the "positive" that if $\sigma < \lambda$, then, in a neighbourhood of the saddle focus loop γ, there is a set of all possible trajectories forming an invariant set J, of continuum power. As z_0 moves farther from the v-axis, J gets thinner and vanishes, which occurs for $\sigma < \lambda$. For $\sigma > \lambda$, everything is considerably simpler: The "positive" coincides with the "negative", and represented in Fig. 6.28.

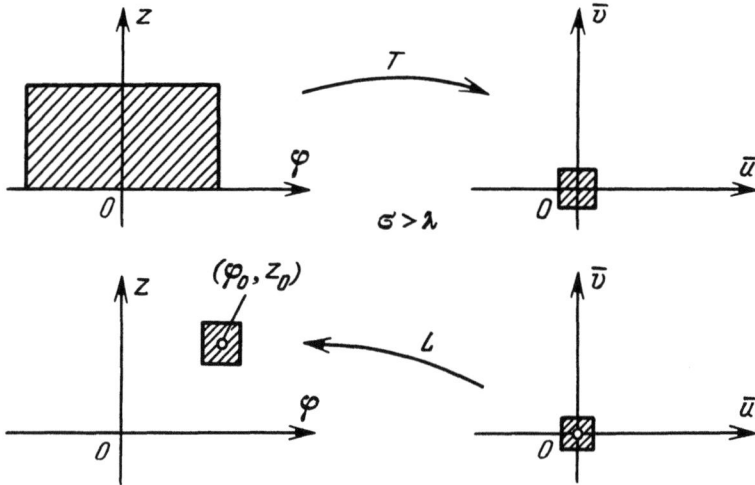

Fig. 6.28

Stochastic and chaotic attractors 163

It can be seen that if z_0 is sufficiently near to the z-axis, then there is a steady-state periodic motion coinciding with γ as z_0 vanishes, and vanishing for $z_0 < 0$.

SITUATION 7. *Saddle node loop.* Consider a situation similar to the above (Fig. 6.29) with the difference that the equilibrium state is not a saddle focus, but a saddle node, and described by the equations

$$\dot{x} = -\lambda x, \quad \dot{y} = -\nu y, \quad \dot{z} = \mu z \tag{2.23}$$

in a linear approximation, where the coefficients λ, ν, μ are positive, and $\nu > \lambda$ for the sake of being specific. The cutting surfaces Ω^+ and Ω^- have the equations $x = \delta$ and $z = \varepsilon$. The point mapping from Ω^+ into Ω^- can be written in the form

$$\bar{x} = \delta\left(\frac{z}{\varepsilon}\right)^{\lambda/\nu}, \quad \bar{y} = \left(\frac{z}{\varepsilon}\right)^{\mu/\nu} y, \tag{2.24}$$

is contractive for $\mu > \nu$, and coincides with the auxiliary one. For $\mu < \nu$, the mapping

$$\bar{y} = \left(\frac{\bar{x}}{y}\right)^{\nu/\mu} y, \quad x = \varepsilon\left(\frac{\bar{x}}{\delta}\right)^{\lambda/\mu} \tag{2.25}$$

is taken as auxiliary. The latter is contractive for small x, since $\nu > \mu$ and $\lambda > \mu$.

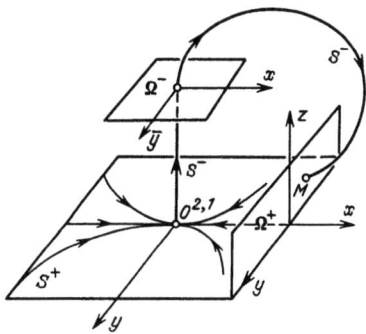

Fig. 6.29

As before, we denote the mapping from Ω^+ into Ω^- by T, and that from Ω^- into Ω^+ by L.

The "positives" for T, L are represented in Fig. 6.30. The coordinates of the point M at which the curve S^- meets Ω^+ are denoted by y_0, z_0. It can be seen that, as z_0 increases, starting with $z_0 = 0$, a periodic motion arises near the original loop, being steady-state for $\mu > \nu$ and saddle for $\mu < \nu$. Comparing the situations with the loop for the saddle focus and saddle node, we discover that, for $\sigma > \lambda$ and $\mu > \nu$, steady-state periodic motions arise in both cases; for $\sigma < \lambda$ and $\mu < \nu$, saddle motions. However, for the saddle node, this is a periodic motion; for the saddle focus, a complicated invariant set.

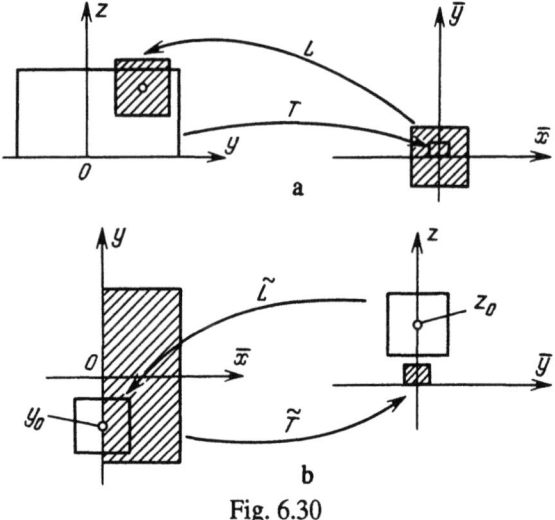

Fig. 6.30

SITUATION 8. *Loop of a singular fixed saddle node point*. The situation in question is connected with a point mapping which can arise on a cutting surface. Meanwhile, the stable and saddle periodic motions merge in the phase space of the system intersected. As the parameters responsible for intersection of the bifurcating surface N_{+1} vary further, the complicated saddle-node periodic motion vanishes; it arose due to the merger. There is a complicated fixed saddle node point O on the cutting surface, arising because the saddle and fixed nodes merge. Besides, it is assumed that, emanating from the point O, the invariant curve S^- again returns to it, forming a loop. This effect can occur differently, by entering O smoothly (Fig. 6.31a), or bending (Fig. 6.31b), or intersecting S^+ and bending (Fig. 6.31c). In the latter case, it is not quite safe to say that S^- returns to O; rather, it intersects S^+, and forms a Poincaré loop with the property that O is now a singular saddle node which can vanish or bifurcate again into a saddle point and a fixed node under slight changes in the parameters. All three cases lead to the study of the point mapping of an annulus into itself. How the latter is selected is seen in Fig. 6.32. It is assumed meanwhile that the annulus can be chosen to be sufficiently thin so that a certain degree of the mapping T in question may transform it into itself. A mapping of the annulus into itself was considered in Situation 5. The argument can also be applied to the second ant third cases, leading to a complicated invariant set. The first leads to an invariant, smooth and closed curve generated in the annulus from a closed loop S^- as the singular saddle node vanishes. The "negatives" associated with the second and third cases immediately after the singular fixed saddle node point has vanished are represented in Fig. 6.33. In the "negative", the region G is transformed after sufficiently many iterations into the region $\overline{G}$, contracted and extended in the directions of S^+, S^- respectively. G is than transformed into $\overline{G}$, their intersection being in Fig. 6.33. By analogy with the above situations, the intersections lead to a complicated invariant saddle set.

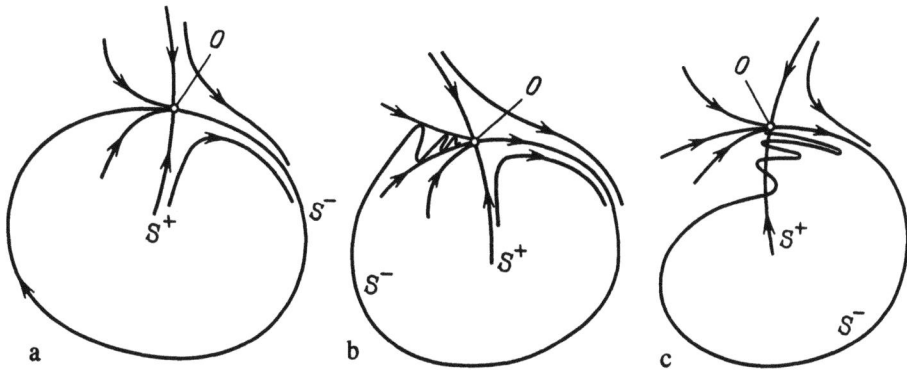

Fig. 6.31

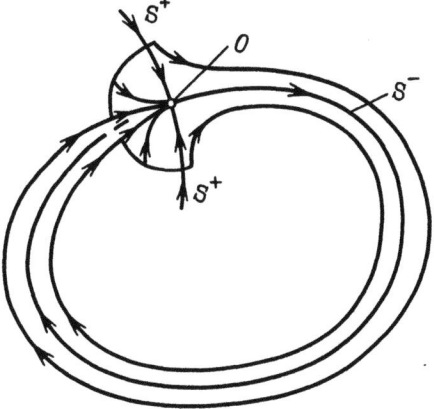

Fig. 6.32

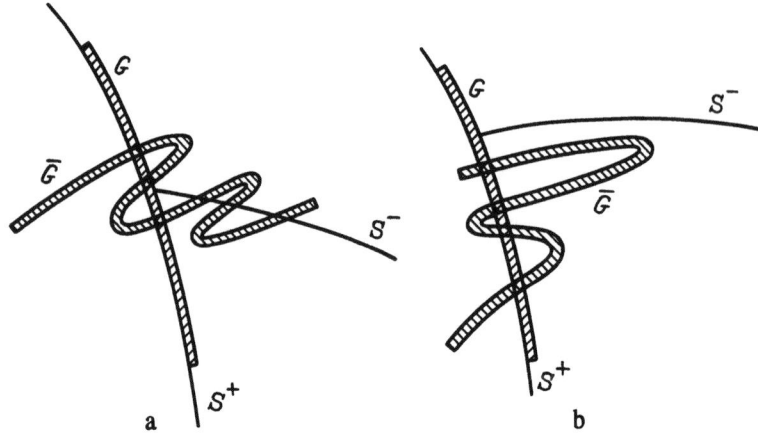

Fig. 6.33

SITUATION 9. *Figure eight.* A figure eight is the phase portrait of a point mapping from a plane into a plane with fixed saddle O and two separatrix loops $S^- \equiv S^+$ (Fig. 6.34). Even for a slight perturbation of the point mapping, the figure eight collapses, and contacts or intersections of the invariant curves S^-, S^+ can occur (Fig. 3.35). Individual loops were already considered in Situations 1, 2; therefore, it is known which particular motion occurs nearby. In particular, under contact, the saddle quantity $\lambda\nu$ turns out to be very important. A complicated invariant set inside the loops exists only for $\lambda\nu < 1$. However, in addition to motions near each loop, those nearby and with transition from one to another are possible in the figure eight. The goal is now to consider all motions, including the one near each loop as a special case.

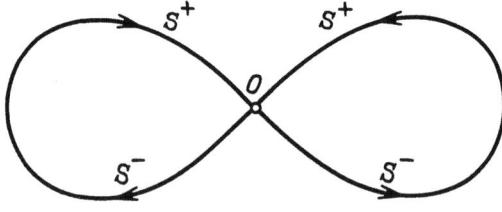

Fig. 6.34

Stochastic and chaotic attractors 167

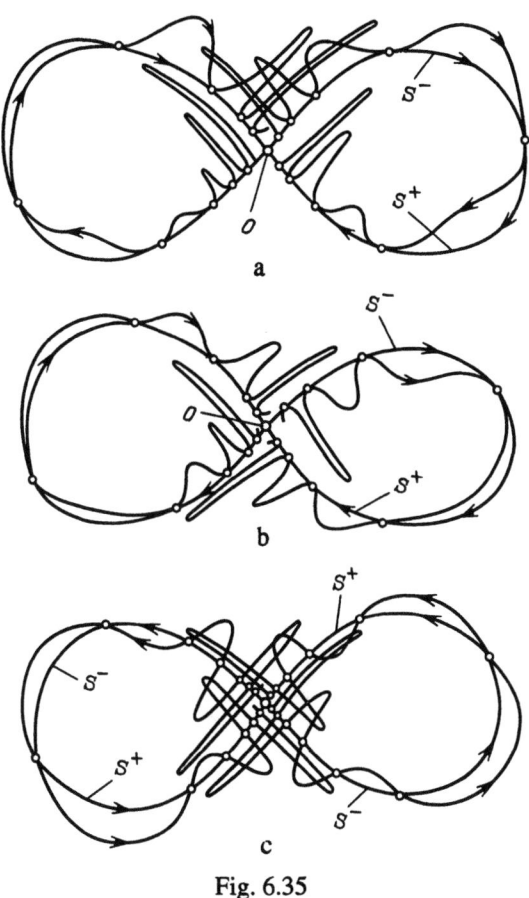

Fig. 6.35

It is natural that this consideration should not be so detailed as in Situations 1, 2. In particular, we do not represent the "positives" and "negatives" of individual transformations, but give their general scheme instead. Introduce the regions A_1, B_1, C_1, D_1 and A_2, B_2, C_2, D_2 as in Fig. 6.36. The mappings transforming A_1 into B_1, A_2 into D_2, C_1 into D_1, and C_2 into B_2 are certain powers of the mapping T; e.g., T^k, T^l, T^m, and T^n. The numbers k, l, m, n are arbitrary, but greater than a certain value N (in order that all the regions could be regarded as thin, and contraction under the auxiliary mappings $\tilde{T}^k$, $\tilde{T}^l$, $\tilde{T}^m$, and $\tilde{T}^n$ as sufficiently strong). For certain powers of the transformation T, the parts of the regions $B_1 \cup B_2$, $D_1 \cup D_2$, distinguished by shading in Fig. 6.37, are transformed into the "bent" regions $\overline{B}_1 \cup \overline{B}_2$, $\overline{D}_1 \cup \overline{D}_2$. Denote by L, M the transformations performing the above. We can now represent the general scheme of the transformations in Fig. 6.38. The

168   Chapter 6

transition from the transformation scheme associated with the "negative" to the one associated with the "positive" consists in only replacing mappings by auxiliary ones, and reindexing the regions accordingly, with those new regions now transforming into each other and with mappings that are eventually contractive. Thus, the scheme is that of admissible sequences of transformations or of different traverses along each loop and of the transitions from traversal of one loop to that of another. All possible sequences making these transitions possible are associated with all possible paths in the graph in Fig. 6.38.

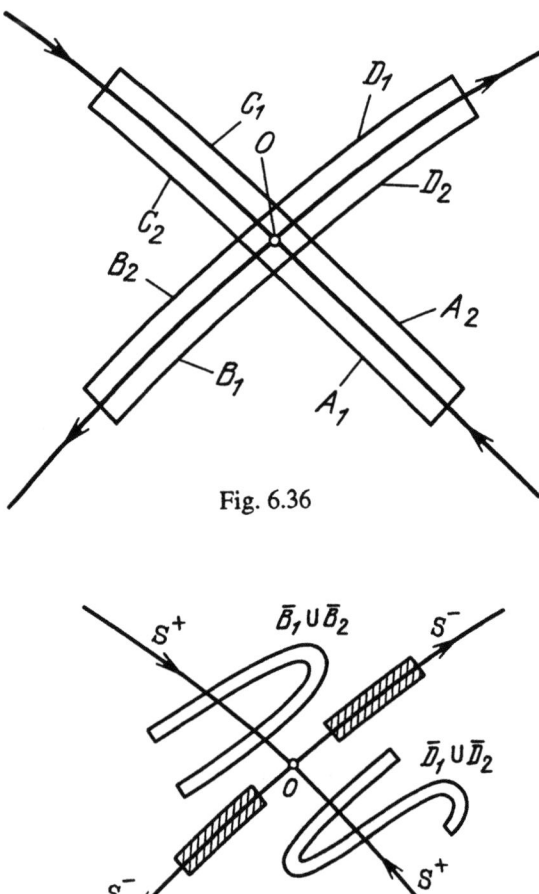

Fig. 6.36

Fig. 6.37

Stochastic and chaotic attractors

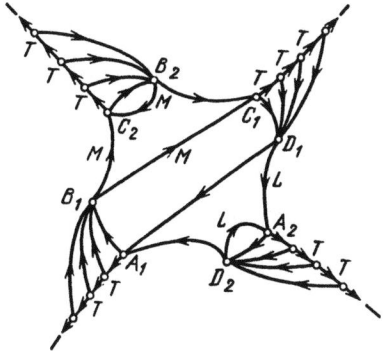

Fig. 6.38

SITUATION 10. *Stochastic synchronism*. A *synchronism* is the phase portrait of a point mapping as in Fig. 6.39. Under the mapping T, the points $O_1, O_2, \ldots, O_p$ and $O'_1, O'_2, \ldots, O'_p$ are sent into each other cyclically, so that the point indexed by i is carried to the one indexed by j, an accordance with

$$j = i + q \pmod{p}, \qquad (2.26)$$

where p, q are relatively prime. It follows that all $O_1, O_2, \ldots, O_p$ and $O'_1, O'_2, \ldots, O'_p$ are fixed points of multiplicity p; $O_1, O_2, \ldots, O_p$ are fixed saddles of multiplicity p; their invariant manifolds may, or may not, intersect, being tangent in the limiting case. For the sake of simplicity, the manifolds are represented in Fig. 6.39 as being coincident. Slight changes can lead to any of the above possibilities.

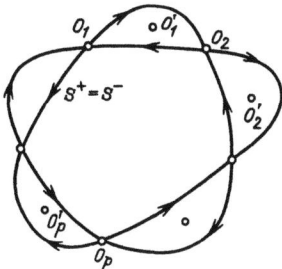

Fig. 6.39

In intersection, contact, or in cases close to tangency but without intersection, a complicated invariant saddle set can arise. Its appearance may occur in the case of intersection in general, and in the other two cases only if the saddle quantity $\lambda\nu$ is less than one. The general role of the latter is studied in the sequel.

Let the phase portrait of a synchronism be related to differential equations, and not a point mapping. It is then easy to complete the portrait to two basic simple versions as in Fig. 6.40. Meanwhile, $O_1, O_2, \ldots, O_p$ are equilibrium states of saddle type; therefore, to make them analogous to those of a point mapping, we should consider T^p instead of T. However, there are differences, too. Thus, for the differential equations, S^+ cannot intersect S^-, which on the contrary, is possible for a point mapping. In the phase portrait case in Fig. 6.40a, a stable equilibrium is inside each loop; in Fig. 6.40b, an unstable one. It is intersecting to note that, in the former case, the saddle quantity is greater than one; in the latter case, less. To see that this is actually true is easy from the meaning of the saddle quantity as the value of the Jacobian of T near fixed saddles.

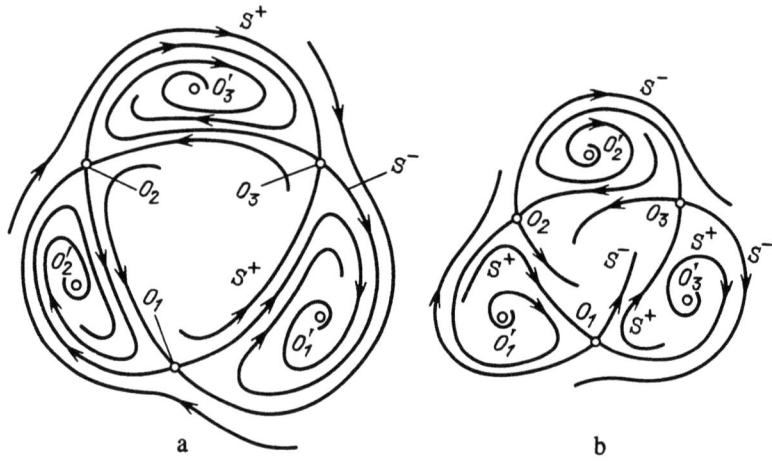

Fig. 6.40

It is clear from the above that, in a neighbourhood of the fixed points $O_1, O_2, \ldots, O_p$ and of their invariant curves in the point mapping case, complicated invariant saddle sets can exist. In the case of differential equations, only the coinciding curves S^+, S^- can be an analogue of such a set. If the merger collapses, then stable periodic motions can occur either inside or outside the loops. The same phase portrait for a point mapping on a cutting surface would be associated with the appearance in the original system of toroidal integral manifolds. Does this possibility of emergence of a complicated invariant saddle set contribute anything new to this picture ? We see that it does. To make the difference concrete, we consider variables on the cutting plane as the difference of phases with certain external periodic force, and as oscillation amplitude resultant due to the external force. Meanwhile, the transition to a differential equation can be treated, e.g., as an application of the averaging method. If we speak of the phase portrait of a differential equation, then general cases are possible, i.e., either phase synchronism, or amplitude constancy (stable equilibrium states), or periodic change in the phase difference and amplitude.

With this complicated invariant set, if the phase point could remain in a small neighbourhood of it, or, still better, tend to the set asymptotically, a new possibility would arise: chaotic variation of the phase difference, and, accordingly, of the oscillation amplitude. It is natural to call this *stochastic synchronism*. Its stochasticity manifests itself on the phase portrait of the mapping T^p in slow chaotic wandering of the phase point in a neighbourhood of $O_1, O_2, \ldots, O_p$ and S^-, S^+ (Fig. 6.41).

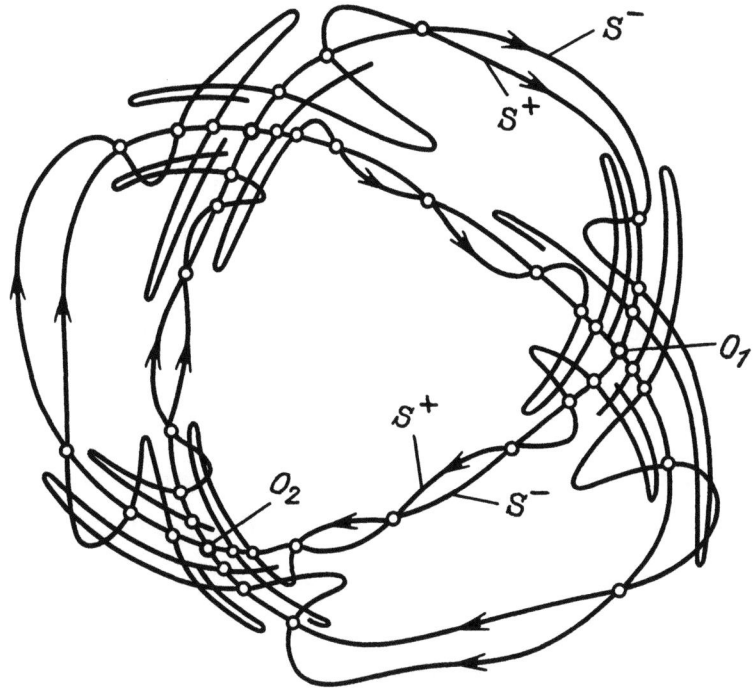

Fig. 6.41

All the above situations admit straightforward multidimensional generalizations. Their consideration is little different from the above, but certainly less visually representable. Note that the simplified notation of certain mappings does not restrict generality, being based on properties of the auxiliary mapping not destroyed by negligible non-linear terms. Meanwhile, it should be remembered that there are situations leading to complicated invariant saddle sets which can only be realized when the dimension of the point mapping exceeds two. One of these situations was described in Chapt.2. We note that, without the requirement that the transformation be one-to-one, this can be realized even for dimension one. In investigating this situation, transition from the "negative" to the "positive" can also be used.

We now retrace our steps to the mapping (1.1) of Chapt.2. Its contractive auxiliary mapping is two-valued, while regular branches are given by

$$\tilde{T}_1: \bar{x} = qx + 0.5 \sin \frac{\overline{\psi}}{2}, \quad \bar{y} = qy + 0.5 \cos \frac{\overline{\psi}}{2}, \quad \psi = \frac{\overline{\psi}}{2},$$
$$\tilde{T}_2: \bar{x} = qx - 0.5 \sin \frac{\overline{\psi}}{2}, \quad \bar{y} = qx - 0.5 \cos \frac{\overline{\psi}}{2}, \quad \psi = \frac{\overline{\psi}}{2} + \pi.$$
(2.27)

Each of the mappings $\tilde{T}_1, \tilde{T}_2$ transforms the region $\tilde{G}$ ($x^2 + y^2 \leq 1$, $0 \leq \psi < 2\pi$) into itself, where the images $\tilde{T}_1\tilde{G}, \tilde{T}_2\tilde{G}$ are disjoint. Hence, each sequence

$$\ldots, \tilde{T}_{i_{-1}}, \tilde{T}_{i_0}, \tilde{T}_{i_1}, \tilde{T}_{i_2}, \ldots \quad (i_s = 1, 2)$$
(2.28)

is associated with a unique sequence of points

$$\ldots, (x_{-1}, y_{-1}, \psi_{-1}), (x_0, y_0, \psi_0), (x_1, y_1, \psi_1), \ldots$$
(2.29)

such that

$$(x_s, y_s, \psi_s) = T(x_{s-1}, y_{s-1}, \psi_{s-1}).$$
(2.30)

All possible sequences (2.29), or phase trajectories of the mapping T, make up a complicated invariant set J. It is saddle, for each (2.29) is saddle (because $\tilde{T}_1, \tilde{T}_2$ are contractive).

Different situations leading to a non-trivial saddle set were considered above in a unified approach and in transition from the "negative" to the "positive". They include both those discovered comparatively long ago by H. Poincaré himself, and those discovered quite recently, a little more than ten years ago. Many were rediscovered repeatedly; e.g., the situation with the complicated saddle node loop was subsequently interpreted both as intermittence and as appearance of chaotic behaviour if the two-dimensional integral torus collapsed. The first three situations should be regarded as due to H. Poincaré. G. Birkhoff studied a simple loop, and found that there were infinitely many periodic motions in its small neighbourhood. The result was generalized to multidimensional systems in [372]. Further results were obtained by the method of auxiliary mappings in [263, 266, 275]. Survey [262] announced the study of loops with contact for the first time; it was completed in [100, 117, 119, 137]. The Smale horseshoe was investigated by S. Smale.

The saddle focus equilibrium loop was considered in [262, 288, 370]. The fixed singular saddle node loop has a long history dating back to A. A. Andronov and A. A. Vitt [14], and A. A. Andronov and E. A. Leontovich [16], where its analogue for differential equations was given and considered for the first time. This case was then investigated in [49, 50, 236, 271]. The figure eight was already mentioned in Sect.2, Chapt.1. Stochastic synchronism as a kind of analogue of usual synchronism but with randomly wandering phase point was first noted in [273].

Finally, we note that the situations considered are based on the lectures [282].

3. Conditions for the appearance of chaotic and stochastic attractors

When can a complicated saddle set J, known to arise in many different situations, be attracting, and turn out to be a strange attractor ? First of all, it is clearly necessary that the characteristic exponents of motions of J should provide for phase volume contraction. That this is a necessary condition is beyond doubt; however, we cannot be satisfied. In a neighbourhood of an individual motion, phase volume can contract, even very strongly. Still, an individual saddle motion cannot be an attractor, since phase points first approach the motion, and then move farther. (The possibility to approach across the surface S^+ is not to be taken into account, since the measure of S^+ is zero.) Note that a sufficient test for phase volume contraction for the dynamical system described by the differential equation

$$\dot{x} = X(x) \tag{3.1}$$

is

$$\text{div } X(x) < 0. \tag{3.2}$$

Meanwhile, if

$$\text{div } X(x) \leq \delta < 0,$$

the phase volume V at time τ is not greater than $Ve^{\delta\tau}$. Thus, $\delta = -\sigma - b - 1 < 0$ for the Lorenz equations (1.24), Chapt.1, and phase volume contracts very rapidly and strongly, becoming about 10^3-10^5 times as small per unit time. Despite phase volume contraction, an isolated saddle periodic motion alone cannot be an attractor. However, certainly, the motion can accelerate contraction via other stable and little-contracting periodic motions. In other words, a peculiar "mutual assistance" is possible; viz., saddle motions first decrease phase volume, which is then absorbed by "very feeble" stable motions with overall contraction only in their small neighbourhood. This "assistance" can, and does, occur; however, saddle motions themselves can form an attractor; it is only necessary that there should be many of them. The Lorenz attractor shows that the latter is possible. There are other examples, too. It is desirable that the general scheme covering them should be clarified; e.g., it is possible that each saddle motion of a certain set attracts or repels phase volume elements, decreases their volume, passes them over to another saddle motion, which, in turn, contracts the phase volume of the phase particle received, and sends it to another saddle motion, etc. For this scheme to be realized, the invariant surfaces S^- of all saddle motions should also be saddle motions. Recall that we encountered just this situation in the Lorenz attractor. Does this requirement, i.e., $S^- \subseteq J$, suffice ? Apparently, the answer is "yes" if we take into account that, after each traversal near a saddle motion, the distance from the phase point to the surface S^- decreases, the points approach the set S^- still nearer and nearer, and, since $S^- \subseteq J$, also nearer J. Thus, a strange attractor is an invariant saddle set J such that the manifold S^- belongs to J, and the saddle quantity σ is less that one for the motions of J. We call this strange attractor *stochastic*. Meanwhile, "strange" attractors are possible, containing both saddle and stable motions, which, during this peculiar mutual assistance, contract phase volume and "annihilate" the latter on "weak" stable motions, thus playing the role of phase volume "drains". Where do all these weak stable motions come from ? One of the mechanisms for their appearance can be indicated, viz., tangency or very close location of the invariant curves S^+ and S^-. If contact occurs, then weak stable motions are inevitable. The reason can be understood in considering Situation 2 in the preceding section, and is caused by the fact that the derivative of the auxiliary mapping L tends to infinity on a certain line. We will consider this question in greater detail in Chapt.7, and now confine ourselves to stressing the possible existence of these symbiotic strange attractors, with extraordinary smallness of local neighbourhoods of attraction of stable periodic motions entering these

Stochastic and chaotic attractors 175

neighbourhoods. We call such attractors *chaotic*. As far as attractors are concerned for which either a decrease or an increase in phase volume is possible and natural, the above should be substantially sharpened in the Hamiltonian system case. The phase volume in Hamiltonian systems is preserved, since

$$\text{div } X(x) \equiv 0 . \tag{3.3}$$

If the condition $S^- \subseteq J$ held for a Hamiltonian system, the saddle quantity σ being one, then we could not prove that the phase points approach S^-, but nothing would prevent the invariant set J from being of measure other than zero; this was impossible before, since the measure of J should have been variable. Thus, the structure of an invariant saddle set for a Hamiltonian system is substantially different. An attractor of a Hamiltonian system is an invariant saddle set with nonzero Lebesgue measure. Meanwhile, as in the case of dissipative systems, the condition $S^- \subseteq J$ should hold. Replacing t by -t does not make the dynamical system non-Hamiltonian; however, interchange of the invariant set S^+, S^- cannot help making us think that $S^+ \subseteq J$ for Hamiltonian systems; otherwise, J could suck in outer phase volume, which is impossible without throwing phase volume out.

CHAPTER 7

BIFURCATIONS AND ROUTES TO CHAOS AND STOCHASTICITY

In the previous chapter, we have considered the simplest typical situations leading to complicated non-trivial invariant saddle sets J. If, in addition, these sets are attracting, then they are also strange attractors with the local instability property, but are stable as a whole. It is of interest together with the "static" study of J to clarify how such sets arise in dynamics with varying parameters. The "static" and "dynamical" approaches make it easy to discover stochastic and chaotic motions, and not only to study them more thoroughly.

Usually, a dynamical system depends on varying parameters; a previously stable equilibrium state gets unstable, while the motions become chaotic. How this can happen is clarified in the sequel. Both a physical treatment and a more direct approach based on geometrization by a phase portrait are possible. Each method has a right to exist. They do not exclude, but complete each other.

The physics viewpoint proceeds from analyzing local instability, the reasons leading to divergent oscillations, those for bifurcation of the process, and finally leading to global contraction. The specific conditions for chaotic and stochastic oscillations, in contrast with those for periodic ones, lay in the different mechanisms of contraction. For periodic self-excited oscillations, the oscillations should be smoothly bounded; for chaotic oscillations, there are relatively abrupt collapses or transitions to other motion regimes. The reasons for instability can be the same in both cases of periodic and stochastic oscillations.

A geometric viewpoint based on considering the phase portrait and its changes focuses the attention on bifurcations, or, in modern terms, *catastrophes*. The approach is founded on the idea that qualitative changes in the phase portrait occur for specific parameters, bifurcation values, and are not smooth. Ideas of possible bifurcations have recently changed strongly. It was assumed previously that bifurcations were isolated and localized on the phase portrait. In reality, everything turned out to be considerably more complicated. Bifurcation points can condense and form complicated sets, while the associated phase portrait may change in as much as whole regions of the phase space. Bifurcations are not independent, so that we can speak of bifurcation sequences or sets as of a certain new and complicated bifurcation. It turned out to be expedient to study the dependence of a phase portrait on whole parameter groups, and not only on one parameter. The above is not new, and this approach for obtaining more complete possible bifurcation diagrams was also used earlier in solving different concrete problems. However, the ideas of the smallest parameter group related to the bifurcation in question were sharpened lately; we have discussed the matter in Chapt.5.

Meanwhile, in spite of all the complexity, bifurcations of equilibrium states, periodic motions and their integral manifolds S^+ and S^- are still basic. In addition to the precise laws of bifurcations of equilibrium states and periodic motions, new laws for bifurcation sequences were discovered, as well as their relations to so-called *nested structures*, tangency of the invariant manifolds S^+, S^-, and specific dependence of the Poincaré rotation number on parameters.

R. Feynman's lectures [353] contain a very vivid description of turbulence as the Reynolds number increases. As compared to earlier descriptions, the picture and its increased complexity cannot be better associated with the process of increasing complexity and independence of ideas in bifurcation theory. The following aims at clarifying all possible metamorphoses of the phase portrait, which could be linked with the transition of a laminar flow into turbulent, and, in general, of a stable equilibrium state into chaos. The discussion is not final, and only gives an outline. After the tree of possible bifurcations has been described, bifurcation sequences are considered in more detail. Bifurcations in two concrete and sufficiently well studied dynamical systems are described: the Lorenz system and the parametrically excited nonlinear oscillator or rotator. The examples enable us to track the routes to order and chaos in a sufficiently detailed manner.

Bifurcations and routes to chaos and stochasticity 179

1. General description of the tree of possible bifurcations

Suppose there is initially a stable equilibrium state $O^{n,0}$.

It can lose stability or vanish under continuous change of parameters in only three ways, quite definite and described in Chapt.5:

(1) It can become unstable, turning into an equilibrium state of $O^{n-2,2}$ -type, accompanied by simultaneous separation of a stable periodic motion $\Gamma^{n,1}$ or by merging with an unstable periodic motion $\Gamma^{n-1,2}$.

(2) It can lose stability, turning into an equilibrium state of $O^{n-1,1}$ -type, accompanied by simultaneous separation of two stable equilibrium states of $O^{n,0}$ -type or by merging with two unstable equilibrium states of $O^{n-1,1}$ -type.

(3) It can vanish, merging with a saddle equilibrium state of $O^{n-1,1}$ -type.

As a stable periodic motion or stable equilibrium states separate from an equilibrium state $O^{n,0}$, a soft transition occurs from the prior steady-state periodic motion (equilibrium state) to new ones (steady-state periodic motion or one of stable equilibrium states). On the contrary, as an unstable periodic motion, an unstable equilibrium or equilibria merge with an equilibrium state $O^{n,0}$, the transition to a new steady-state motion is abrupt. Local bifurcation theory does not indicate to which new steady-state motion the abrupt transition leads. This can be an equilibrium, a periodic, or chaotic, or stochastic self-oscillation. It can also tend to infinity. Note that only the bifurcations (1), (3) are general, while (2) is general only in the case of symmetry in the dynamical system (as it often happens). We stress that all of them were already considered in Chapt.5. They are now all collected together, and represented in the tree of possible bifurcations (Fig. 7.1), associated with transitions across the boundaries of bifurcation, N_0 and N_ω.

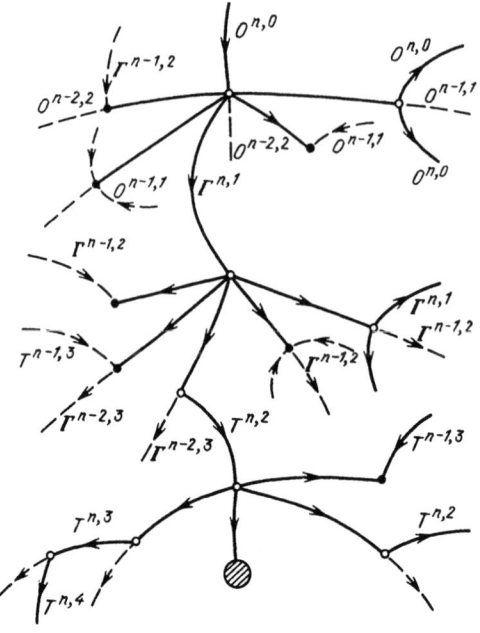

Fig. 7.1

Furthermore, if a stable equilibrium state arises again because of the bifurcations, then only the same bifurcations can naturally occur. Therefore, we only consider bifurcations of a stable periodic motion $\Gamma^{n,1}$.

We now describe the basic bifurcations of $\Gamma^{n,1}$ that are possible under a continuous change of parameters.

1. $\Gamma^{n,1}$ vanishes, merging with the saddle periodic motion $\Gamma^{n-1,2}$.

2. $\Gamma^{n,1}$ loses stability, separating a stable periodic motion with period twice as large or merging with an unstable periodic motion with the same period. In both cases, the original motion $\Gamma^{n,1}$ turns into a periodic one of $\Gamma^{n-1,2}$-type.

3. $\Gamma^{n,1}$ loses stability, separating a stable two-dimensional toroidal manifold $T^{n,2}$ or merging with an unstable two-dimensional toroidal manifold $T^{n-1,3}$.

These bifurcations are associated with transitions in the parameter space across the bifurcating surfaces N_{+1}, N_{-1}, and N_ϕ with ϕ other than 0, $\pi/2$, π, or $2\pi/3$. The values $\phi=0$, $\phi=\pi$ are associated with N_{+1}, N_{-1}, while $\phi=\pi/2$, $\phi=2\pi/3$ with special resonant cases of stability loss by $\Gamma^{n,1}$, studied in a great number of works [42, 43, 58, 195–197, 274, 280, 290]. We only note that, in the resonant cases, stable period-tripling, quadrupling and chaotic motions can emerge from the periodic motion $\Gamma^{n,1}$ losing stability.

Bifurcations for which the closed phase curve associated with $\Gamma^{n,1}$ contracts to an equilibrium state, or if such a state appears on the curve and the curve turns into a loop leading from this state to the same again, should be added to the above. These are caused by transitions across the bifurcating surfaces N_{+1}, N_{-1}, and N_ϕ. We already spoke of the former, i.e., bifurcation (1) of an equilibrium state, and considered the second in Chapt.4 as Situation 7 ("saddle node loop"). We now also indicate that the phase trajectory associated with a periodic motion can tend to infinity.

Above, we have described the basic bifurcations associated with codimension one boundaries of bifurcation. If the dynamical system is symmetric, then the bifurcation for which a stable periodic motion $\Gamma^{n,1}$ turns into a saddle one, $\Gamma^{n-1,2}$, can be basic, separating simultaneously two new stable periodic motions of $\Gamma^{n,1}$-type. This bifurcation occurs in crossing the boundary N_{+1}.

The loss of stability by (or the vanishing of) a periodic motion leads to a replacement of the steady-state motion, which can be soft or hard. If the change is soft, and the parameters vary smoothly, then the replacement occurs smoothly and gradually. If the change is hard, then, in spite of the smooth change in parameters, the replacement occurs in a jump.

The parts of the boundaries of bifurcation associated with a soft change in the steady-state motion can be said to be *safe*; those with hard change, *dangerous*. The concepts of *soft* and *hard* occurrence of self-excited oscillations were introduced by A. A. Andronov; of dangerous and safe parts of boundaries of regions of stability of equilibrium states, by N. N. Bautin [72, 73]. The notions are further generalized to periodic motions in [259], on whose basis a general definition, and a general picture of soft and hard change of steady-state motions can be given.

Let J be a steady-state motion, i.e., an equilibrium state, a periodic or a stochastic motion; more exactly, the geometric representation of the motion in the phase space. J has region of attraction $\Pi(J)$ with J inside. If the parameter μ of the dynamical system varies continuously, then both the steady-state motion and $\Pi(J)$ vary. Two different cases are possible: First, for a bifurcation value $\mu=\mu^*$, there exist arbitrarily small ε-neighbourhoods of J, transforming in the course of time strictly inside themselves, and, second, there are no such neighbourhoods, and any sufficiently small ε-neighbourhood of J is transformed into a region with points outside the neighbourhood. The limiting case is discarded if there are arbitrarily small neighbourhoods of J that are transformed into themselves. The former case is associated with the soft regime of replacement of a steady-state motion; the latter, with the hard one. In the former case, the steady-state motion occurring again is separated from the prior motion prior J, with the region of attraction generated continuously from that of J. On the contrary, in the latter case, J loses stability or vanishes, and the phase points from its neighbourhood in the course of time pass to a new steady-state motion, as a rule existing even before the bifurcation of J in question. In the tree of possible bifurcations (Fig. 7.1), those of steady-state motions, accompanied by a hard replacement, are marked by small black circles. If the replacement is soft, then local bifurcation theory indicates the new type of steady-state motion; if hard, then the new steady-state motion is not indicated. But there are some exceptions when the non-local consideration and clarification of this new regime are theoretically possible. This is bifurcation of a stable equilibrium state through a complicated saddle node if a loop is there, and the vanishing of both a stable motion and a saddle one by merging on the stable two-dimensional torus. Both bifurcations were, in fact, considered in the preceding chapter in Situations 6, 7 and 8.

We now describe bifurcations of the stable two-dimensional torus $T^{n,2}$. They can be separated into those of the torus properly as an invariant stable manifold, those of the phase portrait on the torus, and, finally, those of individual periodic motions on the torus. If the phase space is of dimension greater than three, then torus bifurcations are possible (as well as of a periodic motion), of N_{+1}-, N_{-1}- and N_ϕ-types. In the three-dimensional phase space, only bifurcations of N_{+1} are possible. The stable torus $T^{n,2}$ then merges with the saddle torus $T^{n-1,3}$, and vanishes. In the bifurcation of N_{-1}, the stable torus "doubles", and a saddle torus, $T^{n-1,3}$, separates. If N_ϕ bifurcates, then $T^{n,2}$ gets unstable simultaneously, a three-dimensional stable torus separates, or $T^{n,2}$, becomes unstable, and merges with the three-dimensional saddle torus.

Bifurcations on a two-dimensional torus can be caused by variation in the

Bifurcations and routes to chaos and stochasticity 183

Poincaré rotation number of its coil. If the number is rational, then the torus coil is periodic; more precisely, there are stable periodic motions on the torus, and the other phase trajectories tend to them, except for as many unstable periodic motions, which play the part of separating boundaries of local attraction regions of stable periodic motions. If the number is irrational, then the coil is quasi-periodic. The Poincaré rotation number as a parameter function is generally step-like; as the function varies, the torus coil, or phase portrait on the torus, bifurcates. Bifurcations of individual periodic motions on the torus do not differ from the above bifurcations of periodic motions.

The bifurcations tree (Fig. 7.1) only describes bifurcations of equilibrium states, periodic motions, and two-dimensional tori. In addition to these relatively well known ones, variations with respect to the location of the integral manifolds S^+, S^- of saddle equilibria and of periodic motions, not immediately connected with bifurcations of stable equilibrium states or periodic motions, can lead to chaotic or stochastic motions; e.g., the strange attractor in the Lorenz system (1.16), Chapt.1, arises just in this way as the parameter r increases.

2. Series of bifurcations

We already noted above that the idea of a set of bifurcations points as a certain surface in the parameter space was not justified fully. The next natural step in making this primitive picture more complicated is the idea of bifurcation series as infinite sets of surfaces accumulated to form a certain limiting surface or surfaces, in turn, making up infinite sets. Here, we will consider some of the series.

1. *Bifurcation series connected with the Poincaré rotation number.* Apparently, the first object making the prior idea of a bifurcation set obsolete was the one-to-one point transformation of the circle into itself of the form

$$\bar{\theta} = f(\theta) \pmod{2\pi}, \tag{2.1}$$

where $f(\theta)$ is a smooth periodic function of period 2π of angle variable θ, for which $f'(\theta) > 0$.

Let $\theta_0, \theta_1, \theta_2, \ldots$ be an infinite sequence of points

$$\theta_{k+1} = f(\theta_k) \quad (k=1,2,3,\ldots) \tag{2.2}$$

sent into each other, and $\Delta\theta_k$ the length of the arc joining θ_k and θ_{k+1} together. H. Poincaré established that the limit

$$\lim_{n\to\infty} \frac{1}{2\pi n} \sum_{k=0}^{n-1} \Delta\theta_k = \omega \qquad (2.3)$$

existed independently of the initial point θ_0 [311]. It is just this limit ω that was called the *Poincaré rotation number*. It can be seen from proof that the number is a topological invariant, remaining invariable under any bicontinuous change of θ. It was found much later by A. G. Mayer in [242] that the Poincaré rotation number continuously depends on the parameter, i.e., if the function $f(\theta)$ depends on μ continuously, then the rotation number ω is a continuous function of μ. Furthermore, he clarified the question of general kinds of mapping the circle into the circle. It was discovered by H. Poincaré himself that if ω is irrational, then the mapping (2.1) has no periodic points; also, if (2.1) has periodic points, then ω is rational (if $\omega=p/q$, where p and q are relatively prime integers, then the periodic points of (2.1) are of multiplicity q). It turns out that structurally stable, and thereby generic, can only be mappings of the circle with rational ω. For $\omega=p/q$, they have generally 2r, r=1,2,..., cycles of fixed points of multiplicity q, with rq stable and rq unstable. Those unstable separate the stable. The phase portraits of (2.1) are given in Fig. 7.2 for p=3, q=1, r=1; p=3, q=2, r=1; and p=3, q=1, r=2, respectively.

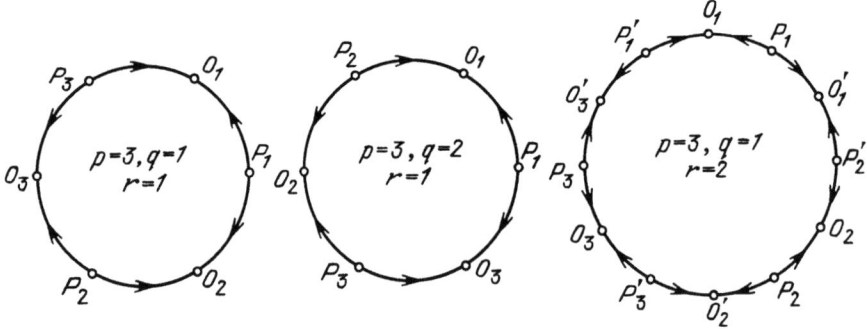

Fig. 7.2

Bifurcations and routes to chaos and stochasticity 185

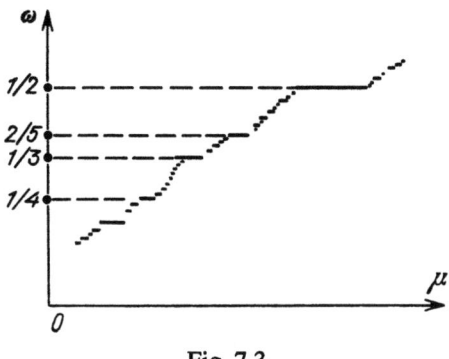

Fig. 7.3

Now, let (2.1) continuously depend on μ, so that

$$\bar{\theta} = f(\theta,\mu) \pmod{2\pi}, \qquad (2.4)$$

and let $\omega(\mu)$ be its Poincaré rotation number. It is natural to assume the general case to be where the values $\omega(\mu)$ of the form p/q are associated with generic maps. Any equality $\omega(\mu) = p/q$ then persists for any interval of μ values; therefore, being continuous, the function $\omega(\mu)$ is constant for any μ such that $\omega(\mu) = p/q$. Some idea of this function is given by the graph in Fig. 7.3, associated with a continuous function. Hence, $\omega(\mu)$ assumes all values in between, and, in any interval of variation, there are infinitely many bifurcations, with a countable number associated with those of cycles of fixed points. Thus, the bifurcation picture is not so simple as μ varies. It is manifest even in a point mapping of the form (2.4), occurring on the secant circle Σ of the two-dimensional torus, i.e., all the above is directly related to bifurcations of this coil. The rotation numbers $\omega(\mu)$ incommensurate with one are associated with quasi-periodic coils; on the contrary, for $\omega(\mu) = p/q$, there are, generally speaking, r stable and r unstable periodic motions on the torus. When traversing each of them, the secant circle is crossed q times, while the torus is traversed p times; in particular, the latter can be two-dimensional and stable in n-dimensional space with r=1,2,3,... stable periodic motions of $\Gamma^{n,1}$-type, and r of saddle $\Gamma^{n-1,2}$-type.

Above, we considered bifurcations of the coil of the two-dimensional integral torus, generated by variation of the Poincaré rotation number. In particular, they could occur on the torus generated from a periodic motion if the parameters vary, leading to a pass across the bifurcating surface N_ϕ of codimension one. As the torus is generated, the rotation number of the phase trajectories on it is $\phi/2\pi$, and can vary afterwards. Meanwhile, the generation is isolated, occurring when $\mu=\mu^*$, and, for μ close to μ^* but different from it, neither separation from nor merger of the tori with the periodic motion occurs. However, in special cases, and, in particular, for Hamiltonian systems, the integral tori can be generated from periodic motions in quite a different, "continuous", manner [61], related to the properties of Hamiltonian systems and, first of all, to the fact that there are periodic motions on them only with two complex roots of the form $e^{\pm i\phi}$ necessarily. As the parameters vary, the following is possible only:

(a) The roots are continuously displaced along the unit circle, remaining complex;

(b) the roots merge into two coincident roots equal to 1, then become real, one greater and the other less than 1;

(c) the roots merge into two coincident roots equal to -1, then become real, one greater and the other less than -1.

In the first case, the point mapping in the vicinity of the fixed point is representable in the form

$$\bar{\rho} = \rho + \rho^2(...), \quad \bar{\phi} = \phi + 2\pi\omega(\rho,\mu) + \rho(...), \tag{2.5}$$

and, replacing ρ by $\epsilon\rho$, in the form

$$\bar{\rho} = \rho + \epsilon\rho^2(...), \quad \bar{\phi} = \phi + 2\pi\omega(\epsilon\rho,\mu) + \epsilon\rho(...), \tag{2.6}$$

where ϵ is a small parameter introduced artificially, while μ the parameter of the system in question. For $\epsilon=0$, the system has a fixed centre-type point whose neighbourhood is filled with closed invariant curves ρ=const. A point mapping of the circle into itself is defined on each curve, with the Poincaré rotation number $\omega(\epsilon\rho,\mu)$. Let $\epsilon=0$, and the parameter μ vary. For $\omega'_\mu\omega'_\rho < 0$, as μ increases, new invariant curves ρ=const emerge from the fixed point $\rho=0$ with new values of the rotation number; on the contrary, for $\omega'_\mu\omega'_\rho > 0$, the available bifurcations merge (Fig. 7.4) [61]. This occurs for $\epsilon=0$. For small $\epsilon>0$, with ϵ possibly smaller the smaller the neighbourhood of the fixed point, some of the invariant curves vary little, remaining closed, while those associated with $\omega=p/q$ change substantially, and become synchronisms. The conclusions about preserving closed curves are based on the KAM theory created by Kolmogorov, Arnold and Moser. Generation from the resonance level, associated with a rational rotation number of the form p/q, of a synchronism made up of 2r cycles of fixed points of multiplicity q follows from a simplified argument in the Poincaré-Birkhoff theorem on area-preserving mappings of the annulus into itself. Theoretically, r can be arbitrary; however, r=1 and sometimes r=2 are used most often. Thus, we arrive at the conclusion that, in a neighbourhood of a fixed centre-point of an area-preserving mapping, bifurcation series occur as μ varies in accordance with variation of ϕ and dependence of ω on ρ. Will such series be observed if the bifurcation surface N_ϕ of the general dynamical system is crossed? No, they will not. They will only occur when crossing through the values $\phi=\pi/2$, $\phi=2\pi/3$ (and, certainly, through $\phi=0,\pi$) and if the analogue of the Lyapunov quantity changes sign (the latter determining stability of the fixed point in this critical case).

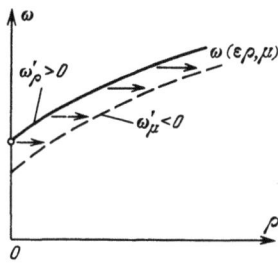

Fig. 7.4

Before considering the other possibilities, note that the above general argument about generation of closed curves and synchronisms from a fixed point can be repeated also for new fixed centre-type points generated by synchronisms, etc., and we arrive at an infinite hierarchy of nested structures, manifesting the extraordinarily complicated structure of a neighbourhood of a fixed centre-type point. The consideration of the hierarchy's properties for concrete systems was computerized in [61, 63–67]. We will discuss the relevant results below when considering a parametrically excited non-linear rotator.

As the complex roots $e^{i\phi}$ and $e^{-i\phi}$ coincide and equal +1 or -1, and afterwards become two distinct real ones (the root greater in absolute value being greater than one; the other, less), the fixed centre-type point turns into a fixed saddle, and simultaneously two centre-type points both of multiplicity one, or two, making up a cycle of double fixed points, separate from the saddle (Fig. 7.5). This is one possibility. The other is the merger of two fixed saddles (of multiplicity one in the case of two coincident roots +1, while of multiplicity two in the case of two coincident roots -1) with the fixed point in question as the two coincident roots are generated (Fig. 7.6). Such are the possible bifurcations of the fixed point, associated with the last two of the above displacements of the roots of the characteristic equation. Note that, in reverse displacement, the corresponding reversal of the bifurcations describes those of a fixed saddle.

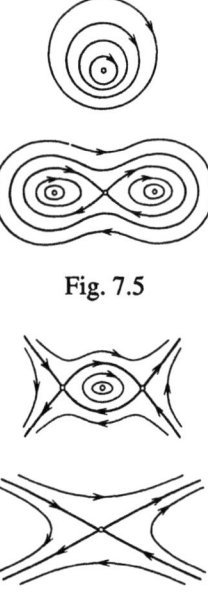

Fig. 7.5

Fig. 7.6

2. *Series of bifurcations and nested structures.* Considering a one-to-one mapping from a circle onto itself leads to the concepts of the Poincaré rotation number and bifurcation series, both are due to the special dependence of the rotation number on the parameter. Similarly, the study of a smooth, single-valued, but not one-to-one, mapping

$$x = f(x,\mu) \qquad (2.7)$$

from a straight line to a straight line made the discovery of a doubling bifurcation series possible. The series was made widely known by Feigenbaum, who discovered the remarkable universal behaviour now called after him. The series itself was observed many times before in studying concrete systems; its existence immediately follows from the remarkable coexistence of fixed points of a transformation of a straight line into a straight line, which was discovered by A. N. Sharkovsky as early as 1967 [368]. According to this law, fixed points of different multiplicities can only coexist in the order

$$3 < 5 < 7 < ... < 3 \cdot 2 < 5 \cdot 2 < ... < 3 \cdot 2^2 < 5 \cdot 2^2 < ... < ... < 2^3 < 2^2 < 2 < 1, \quad (2.8)$$

which means that, if a fixed point of one of the above multiplicities is there, then there exist fixed points of all subsequent multiplicities. In particular, if there is a point of multiplicity three, then all points of multiplicities $2, 2^2, 2^3, ..., 7 \cdot 2^3, 5 \cdot 2^3, 3 \cdot 2^3, ..., 7 \cdot 2^2$, $5 \cdot 2^2, 3 \cdot 2^2, ..., 7, 5, 3$ must precede it in the indicated sequence, i.e., first, points of multiplicity two go, then of multiplicities $2^2, 2^3$, etc., and, finally, of m=7,5,3.

Feigenbaum universality involves the consecutive values $\mu_1, \mu_2, \mu_3, ...$ of the parameter μ for which fixed points of multiplicities $2, 2^2, 2^3, ...$ appear, and consists in the existence of

$$\lim_{i \to \infty} \frac{\mu_i - \mu_{i+1}}{\mu_{i+1} - \mu_{i+2}} = 4.6692.... \quad (2.9)$$

The limit is the same for all general transformations of a straight line to itself, i.e., there is a certain universal constant (i.e., 4.6692...). This universal behaviour enables us, in particular, to give the limit value

$$\mu_\infty = \lim_{i \to \infty} \mu_i \quad (2.10)$$

approximately, after which the infinite fixed point multiplicity doubling bifurcation series is completed.

The fact is justified by a great number of numerical experiments performed both by Feigenbaum and many other researchers. Meanwhile, there are no complete theoretical foundations of Feigenbaum universality, though it is clear how to carry it out, and only computational difficulties are an obstacle. A survey of the works on Feigenbaum universality was published in 1984 [115]. Universality and its study are related there to the method of group renormalization, familiar in the theory of second kind phase transitions in statistical mechanics. However, universality can also be given a somewhat different aspect, apparently relating the theory of doubling bifurcations with the general property of the phase portrait of a dynamical system to possess nested structures [198]. That the structures are nested is discovered not only by increasing multiplicity of the fixed point of the bifurcation series but also by homoclinic structures, the location of synchronisms in a Hamiltonian system [65] or the often noted Cantor structure of a strange attractor.

If δ is a region of the phase space without singularities of the vector field, then a certain part $\bar{\delta}$ is homeomorphic to δ. In turn, we can indicate a still smaller part $\bar{\bar{\delta}}$ homeomorphic to $\bar{\delta}$, and, therefore, to δ, etc., up to infinity, i.e., a phase portrait without singularities is the same as the portrait of arbitrarily small parts. Similarly, if δ is a neighbourhood of an equilibrium state or a fixed point, then the phase portrait in δ is the same as in other arbitrarily small neighbourhoods. Thus, in both cases, we can say that the phase portrait admits infinite series of nested identical (homeomorphic) phase portraits. This statement seems to be trivial. However, suppose we ask the same question regarding a somewhat more complicated phase portrait; e.g., regarding the figure eight represented in Fig. 7.7. To construct the same series of nested phase portraits of the same type, we can embed a figure eight in one or both figure eight's loops, then embed it in each of the new figures eight,etc.,up to infinity. This is very similar to placing *matryoshkas*[4] into each other infinitely many times. Opening one, we invariably have "the same" *matryoshka* as before. If we subtract one from ∞, then we have the same infinity. The *matryoshka* decreases in size, but remains homeomorphic to the original.

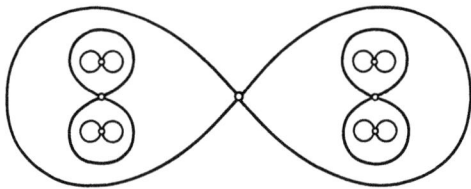

Fig. 7.7

We now give a formal, and somewhat more general, definition of a nested structure, and clarify conditions for its existence. Let η be a homeomorphism mapping a neighbourhood δ into a neighbourhood $\eta\delta$ inside δ. Then $\eta^2\delta$ is a neighbourhood inside $\eta\delta$, $\eta^3\delta$ inside $\eta^2\delta$, etc. (Fig. 7.8). Let a point mapping T be defined in δ. In δ, T is equivalent to the mapping T^m (where m is an integer) in $\eta\delta$ if

$$\eta^{-1}T^m\eta = T. \qquad (2.11)$$

---

[4]A series of nested wooden dolls, typical for old Russia; now, a popular souvenir (tr.).

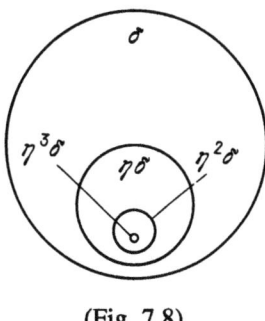

(Fig. 7.8)

In fact, according to (2.11), we can pass from a point $x \in \delta$ to the point Tx, mapping x to a point $y \in \eta x \in \eta \delta$, applying the transformation T^m to the latter, and returning back into δ via the transformation η^{-1} (Fig. 7.9), which just means that T and T^m are equivalent in δ, $\eta\delta$, respectively.

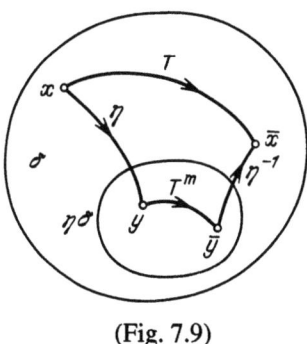

(Fig. 7.9)

The above examples of nested structures and *matryoshkas* were associated with m=1. The new and formal definition is somewhat more general. For structures with m>1 to be nested, there must be a homeomorphism η mapping δ to $\eta\delta$ inside, so that relation (2.11) holds.

We now eliminate the indeterminacy connected with the choice of η, and show that η can be regarded as an ordinary similitude; e.g., $x \to ax$, where $0 < a < 1$. Indeed, η is a homeomorphism mapping a spherical neighbourhood δ onto its part $\eta\delta$; therefore, η is equivalent to a similitude, i.e., there exists a homeomorphism ξ such that

$$\eta = \xi^{-1} \zeta \xi, \qquad (2.12)$$

where ζ is a similitude. Substituting this in (2.11), we have $\xi^{-1}\zeta^{-1}\xi\, T^m\, \xi^{-1}\zeta\xi = T$, or $\zeta^{-1}\xi\, T^m\, \xi^{-1}\zeta = \xi T \xi^{-1}$, or, finally,

$$\zeta^{-1} R^m \zeta = R, \qquad (2.13)$$

where $R = \xi T \xi^{-1}$ is a mapping homeomorphic to T. Thus, for R equivalent to T to admit nested structures, condition (2.13) should hold, where ζ is a similitude, viz.,

$$\zeta x = ax \quad (0<a<1), \qquad (2.14)$$

which means that when $0<a<1$ ranges over all possible values, the solution of equation (2.13) for R yields, up to equivalence, all mappings admitting nested structures with given m.

A solution R^* of (2.13) can be regarded as a fixed point of the mapping

$$\overline{R} = \mathcal{L} R = \zeta^{-1} R^m \zeta \qquad (2.15)$$

defined in the space of all possible admissible mappings R.

Thus, the question of nested structures is reduced to that of fixed points of the mapping $\mathcal{L}$ defined in the mapping space.

Let $\mathcal{F}$ be the bifurcation surface of R for bifurcations associated with generation of a fixed point of multiplicity m, so that, after intersection,

$$R^m x^* = x^* \qquad (2.16)$$

for a certain point $x^* \in \delta$. For the transformation $\underline{R} = \mathcal{L}^{-1} R$, we have $R = \zeta^{-1} \underline{R} \zeta$, or $R^m = \zeta^{-1} \underline{R}^{m^2} \zeta$ (by (2.15); therefore, according to (2.16),

$$\underline{R}^{m^2} \zeta x^* = \zeta x^*. \qquad (2.17)$$

Hence, ζx^* is a fixed point of $\underline{R}$, of multiplicity m^2, $\zeta^2 x^*$ is a fixed point of the mapping $\underline{\underline{R}} = \mathcal{L}^{-1} \underline{R} = \mathcal{L}^{-2} R$, of multiplicity m^3, etc. The following important conclusion is valid. If $\mathcal{F}$ is a bifurcation surface on which a fixed point of multiplicity m is generated, then the surface $\mathcal{L}^{-1} \mathcal{F}$ is a bifurcation surface on which a fixed point of multiplicity m^2 is generated, $\mathcal{L}^{-2} \mathcal{F}$ that on which is a fixed point of multiplicity m^3 is generated, etc. Therefore, the series of bifurcations for which fixed points of multiplicities m, m^2, m^3, etc. are generated consecutively, is associated with successive intersections of the surfaces $\mathcal{F}, \mathcal{L}^{-1}\mathcal{F}, \mathcal{L}^{-2}\mathcal{F}, \ldots$. The behaviour of the sequence

$$\mathcal{F}, \mathcal{L}^{-1}\mathcal{F}, \mathcal{L}^{-2}\mathcal{F}, \ldots \qquad (2.18)$$

essentially depends on the kind of fixed point of the mapping $\mathcal{L}^{-1}$, its invariant manifolds S^+, S^-, and their position relative to $\mathcal{F}$. We can distinguish the case of $\mathcal{F}$ intersecting the one-dimensional invariant manifold S⁻, where S⁺ is of codimension one (Fig. 7.10), and where (2.18) has the invariant surface S⁺ as a its limit. This does not mean that (2.18) has no other limit points, but S⁺ is made up of the limit points of (2.18), which tend to S⁺ in an ordered way, surface after surface (Fig. 7.10). This occurs because $\mathcal{L}^{-1}$ is expanding along S⁺, and contractive along S⁻.

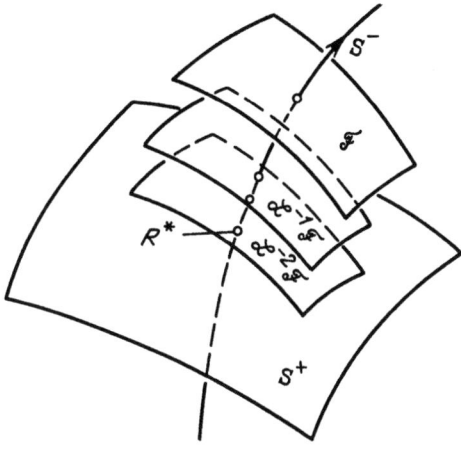

Fig. 7.10

The above case is also of interest, since the limit set of bifurcating surfaces contains a manifold of codimension one, which makes the intersection of curves with S⁺ sufficiently natural and frequent. The curves are associated with change in one scalar parameter of the dynamical system, i.e., the bifurcation series in question in parameter space is linked with a surface of codimension one. It is now fairly obvious that the Feigenbaum law holds for the sequence of parameter bifurcation values associated with intersection by (2.18). It is clear also that the Feigenbaum constant equals the unique root exceeding one, of the characteristic equation of the fixed point of $\mathcal{L}$.

Such is a possible explanation for a doubling bifurcation series to appear. The described picture did not contain complete proof either for a one-dimensional or, much more so, a multidimensional mapping, though the pattern is well supported by numerical computations of a fixed point of $\mathcal{L}$, possible approximate determination of the number a, and of an eigenvalue greater, and of several others, less than one. That there is intersection of the curve S⁻ and the surface $\mathcal{F}$, and its nature, was not made clear.

In addition to doubling series, we can expect tripling [197] and quadrupling bifurcation series as well as those associated with other nested structures.

In conclusion, Fig. 7.11 represents a nested structure for a mapping from a straight line into a straight line when m=2, a multiplicity doubling bifurcation of a fixed point. The graphs of the mappings T, T^2 are shown. The graph of T in the square D_1 is similar to that of T^2 in the square D_2. A fixed point x_2^* of T^2 of multiplicity one is associated with a point x_1^* of T of multiplicity one; however, for T, x_2^* is of multiplicity two.

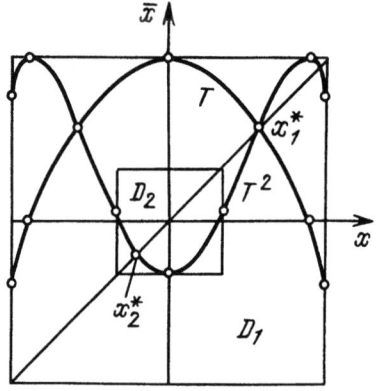

Fig. 7.11

3. *Bifurcation series for tangency of the invariant manifolds* S^+, S^-. The bifurcation series below were discovered in [117–119, 137, 262], and arise when the integral manifolds of saddle equilibria or saddle periodic motions approach each other or touch. The tangency of S^+, S^- leads to homoclinic structures or to changes both of the original invariant manifolds and of the new ones arising in the structures. The bifurcation series consist of the pairwise generation of periodic motions of different types, e.g., $\Gamma^{2,2}$-, $\Gamma^{3,1}$- types, and of subsequent transformation of the periodic motion $\Gamma^{3,1}$ of the type of period-doubling bifurcation series. A sort of double series of bifurcations of pair generation and subsequent doubling of one motion in each pair then arises. Generation "out of nothing" of pairs of periodic motions was actually described in Chapt.6 in Situations 2, 3 and 6. The consideration should only be somewhat extended from the point of view of the bifurcations occurring there.

We now retrace our steps in Situation 2 to the "negative", where contact of the manifolds S^+ and S^- of a fixed saddle occurs, but now introducing the parameter ε so that

Bifurcations and routes to chaos and stochasticity 195

there is no tangency for $\varepsilon > 0$, there is for $\varepsilon = 0$, and there is intersection for $\varepsilon < 0$. The corresponding "negatives" for mappings T^m, $m = 1, 2, \ldots$, and the mapping L, which is also equal to a certain power of T, associated with these cases, are in Fig. 6.11b. T^m transforms the region A into the region $\bar{A}$; L, the region B into $\bar{B}$. The "positive" associated with these mappings is represented in Fig. 6.13, where the mapping $\tilde{L}$ transforms the region B_1 into the region $\bar{B}_1$. $\tilde{L}$ is two-valued, and $\bar{B}_1$ accordingly consists of two regions $\bar{B}_{11}$, $\bar{B}_{12}$; $\tilde{T}^m$ transforms $\bar{B}_1$ into the region $\bar{\bar{B}}_1^m$. If $\bar{\bar{B}}_1^m$ is inside B_1, then this implies the existence of two fixed points of $T^m L$. For sufficiently large m, depending on the value of ε, $\bar{\bar{B}}_1^m$ is either outside B_1 or inside it. $\bar{\bar{B}}_1^m$ enters B_1 via the boundary, on which two regular branches of $\tilde{L}$ are generated, and the Lipschitz constant of $\tilde{L}$ tends to infinity. For the sake of being specific, let T, L be represented in the form

$$T: \bar{u} = \lambda u \ (0 < \lambda < 1), \ \bar{v} = v \ v \ (1 < v < \infty),$$
$$L: \bar{u} = -a \ (v-1) + bu + 1, \ \bar{v} = \varepsilon + c \ (v-1)^2 + du. \quad (2.19)$$

For $\varepsilon = 0$, the point $(0,1)$ on the invariant curve S^- is sent by L to $(1,0)$ on S^+, whereas the curve $\bar{S}$ is sent to $\bar{S}^-$, which touches S^+ at $(1,0)$. When $\varepsilon > 0$, S^- and S^+ do not meet; they do intersect when $\varepsilon < 0$ (Fig. 7.12). The auxiliary mappings are written as

$$\tilde{T}: \bar{u} = \lambda u, \ v = v^{-1} \bar{v},$$
$$\tilde{L}: \bar{u} = -\frac{a}{\sqrt{c}} \sqrt{\bar{v} - du - \varepsilon}, \ v = 1 + \frac{1}{\sqrt{c}} \sqrt{\bar{v} - du - \varepsilon} \quad (2.20)$$

by (2.19). It follows that the boundary b of the region B_1 is determined by the equation

$$\bar{v} - du - \varepsilon = 0. \quad (2.21)$$

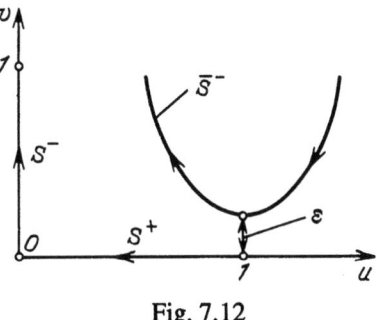

Fig. 7.12

We now return to the interrupted argument. As ε decreases, the region $\bar{\bar{B}}_1^m$ gets into B_1 via b. Meanwhile, two fixed points are generated "out of nothing", and, as we know, should be of different types. As b is then made still more distant from $\overline{B}_1^m$, both branches of the auxiliary mapping $\widetilde{T^m L}$ become contractive, while both points become saddles of the same type. This means that one point undergoes some bifurcations, due to which the types of both become the same. We draw the reader's attention to the narrowing of $\overline{B}_1^m$ as m increases, and, according as $v\lambda<1$ or $v\lambda>1$, approaching the point O ($\bar{u}=\bar{v}=0$) differently. If $v\lambda<1$, the approach proceeds along the v-axis; if $v\lambda>1$, along the ū-axis. The fixed points of $T^m L$ consequently arise only for small $\varepsilon>0$ or only for small $\varepsilon<0$.

Thus, as b is made more distant from $\overline{B}_1^m$, the arising fixed points undergo bifurcations, and then become of the same type. Meanwhile, none can vanish; nor can there appear new fixed points of multiplicity one, since the Jacobian of $T^m L$ is zero only on the curve b. This is a sufficient hint for understanding what occurs with one of the fixed points of multiplicity one: It changes its type, and separates a fixed point of multiplicity two and of the same type.

Let us see how this occurs, considering (2.20) as an example. $T^m L$ can be written in the form

$$\bar{u} = \lambda^m [-a(v-1)+bu+1], \quad \bar{v} = v^m [\varepsilon + c(v-1)^2 + du]. \tag{2.22}$$

We find from the equation that the coordinates u^*, v^* of the fixed points are determined by

$$c(v^*-1)^2 + \left[\frac{ad\lambda^m}{1-b\lambda^m} - v^{-m}\right](v^*-1) - v^{-m} + \frac{d\lambda^m}{1-b\lambda^m} + \varepsilon = 0,$$

$$u^* = -\frac{a\lambda^m}{1-b\lambda^m}(v^*-1) + \frac{\lambda^m}{1-b\lambda^m}. \tag{2.23}$$

We also write the characteristic equation of the fixed points:

$$\chi(z) = z^2 - [b\lambda^m + 2cv^m(v^*-1)]z + \lambda^m v^m [ad - 2bc(v^*-1)] = z^2 + \tau z + \mu = 0. \tag{2.24}$$

The critical value $\varepsilon = \varepsilon^*$ when two fixed points are generated is, according to (2.23), found from

$$\left[\frac{ad\lambda^m}{1-b\lambda^m} - v^{-m}\right]^2 - 4c\left[-v^{-m} + \frac{d\lambda^m}{1-b\lambda^m} + \varepsilon^*\right] = 0. \tag{2.25}$$

When $\varepsilon = \varepsilon^*$, one root of (2.24) is one, so that

$$1 + \tau + \mu = 0, \tag{2.26}$$

and τ, μ are, by (2.24), determined by

$$\tau = -b\lambda^m - 2Cv^m(v^*-1),$$

Bifurcations and routes to chaos and stochasticity 197

$$\mu = (\lambda v)^m [ad + 2bc (v^* - 1)]. \quad (2.27)$$

The partitioning of the parameter (τ, μ)-plane for (2.24) into the regions $D(s)$ is illustrated by Fig. 7.13, associated with s roots inside the unit circle, and 2-s outside. The partition boundaries are N_{+1}, N_{-1}, N_ϕ. When $\varepsilon = \varepsilon^*$, the point (τ, μ) is on the boundary of the bifurcation manifold N_{+1}. For large m, according to which of the inequalities $v\lambda < 1$, $v\lambda > 1$ holds, the point is either near the τ-axis or, on the contrary, quite far from it. For the sake of being specific, let $v\lambda < 1$. The point M is associated with a fixed point when the point is generated. As ε decreases further ($\varepsilon < \varepsilon^*$), two fixed points appear, associated with the corresponding values of τ, v. This makes M bifurcate into two points M_1, M_2. As ε decreases from ε^*, after the bifurcation, one of the points on the (τ, μ)-plane shifts to the left; the other, to the right. First, they fall into the regions $D(1)$, $D(2)$, and are therefore associated with fixed points of $O^{1,1}$- or $O^{2,0}$- type. As ε decreases still further, the point shifting to the right attains the boundary of the bifurcation manifold N_{-1}, and crosses it. Meanwhile, the fixed point of $O^{2,0}$-type turns into one of $O^{1,1}$-type; simultaneously, two fixed points of multiplicity two and of $O^{2,0}$-type are generated.

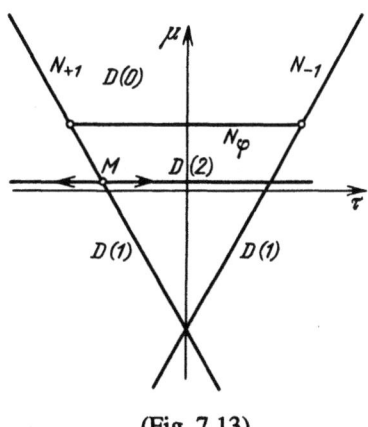

(Fig. 7.13)

The above example can be made still clearer if we note that all formulas simplify if the coefficient d is zero. Moreover, the mapping $T^m L$ is, according to (2.22), reduced to the familiar one-dimensional mapping

$$\bar{v} = v^m \varepsilon + c (v-1)^2 v^m, \quad (2.28)$$

whose graph is in Fig. 7.14. As ε decreases, the parabola descends. As $\varepsilon = \varepsilon^*$, a critical fixed point arises, which subsequently separates into two fixed points v_1^*, v_2^*. At first, one of these is stable, and the other unstable. The unstable one remains unstable also if ε decreases further; the stable one becomes unstable under the bifurcation associated with N_{-1}. Under this last bifurcation, a fixed point of multiplicity two is generated, undergoing the same bifurcation, etc., until the stable point vanishes, and there only remain infinitely many unstable ones.

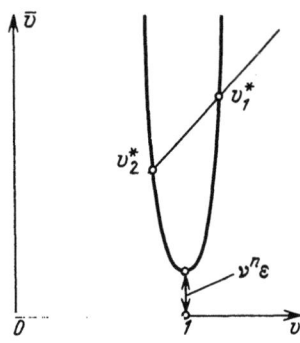

Fig. 7.14

Above, we considered bifurcation series of fixed points for $T^m L$ and all possible $m > m^*$. Similarly, bifurcation series for mappings of more general form,

$$T^{m_1}LT^{m_2}L\ldots T^{m_s}L, \qquad (2.29)$$

can be considered, and even of sequences of mappings,

$$\ldots T^{m_1}LT^{m_2}L\ldots T^{m_s}L\ldots, \qquad (2.30)$$

infinite in both directions. The key to considering (2.30) is the "positive" of T^nLT^m, i.e., the auxiliary mapping $\widetilde{T^nLT^m}$.

The subsequent argument is of quite general nature. For the sake of being specific, we illustrate by the mappings T, L determined by (2.19), in which case

$$T^nLT^m : \begin{array}{l} \bar{u} = \lambda^n\left[-a\,(v^m v - 1) + b\lambda^m u + 1\right] \\ \bar{v} = v^n\left[\varepsilon + c(v^m v - 1)^2 + d\lambda^m u\right] \end{array},$$

$$\widetilde{T^nLT^m} : \begin{array}{l} \bar{u} = \lambda^n\left[-\dfrac{a}{\sqrt{c}}\sqrt{v^{-n}\bar{v} - d\lambda^m u - \varepsilon} + b\lambda^m u + 1\right] \\ v = v^{-m}\left[\dfrac{1}{\sqrt{c}}\sqrt{v^{-n}\bar{v} - d\lambda^m u - \varepsilon} + 1\right] \end{array}. \qquad (2.31)$$

The auxiliary mapping $\widetilde{T^nLT^m}$ transforms its domain of existence $\tilde{G}$ into a certain region $\bar{G}$. The greater the integers m, n, the smaller $\bar{G}$. One of the domain

Bifurcations and routes to chaos and stochasticity

boundaries is the line

$$v^{-n}\bar{v} - d\lambda^m u - \varepsilon = 0 \qquad (2.32)$$

on which $\widetilde{T^n L T^m}$ ramifies into two regular branches, so that $\tilde{G}$ can be determined by the inequalities

$$0 \leq u \leq a, \; 0 \leq \bar{v} \leq b, \; v^{-n}\bar{v} - d\lambda^m u - \varepsilon \geq 0. \qquad (2.33)$$

According to (2.33), $\overline{\tilde{G}}$ is not larger than the region determined by

$$\left|\bar{u} - \lambda^n\right| \leq K_1 \lambda^n \max\left(|\varepsilon|^{1/2}, \lambda^{m/2}, v^{-n/2}\right),$$

$$\left|v - v^m\right| \leq K_1 v^m \max\left(|\varepsilon|^{1/2}, \lambda^{m/2}, v^{-n/2}\right), \qquad (2.34)$$

for certain K_1, and lies inside $\tilde{G}$ if the conditions

$$v^{-n-m} - d\lambda^{n+m} - \varepsilon > K_2 \max\left(\lambda^{n+m}, v^{-n-m}\right) \max\left(|\varepsilon|^{1/2}, \lambda^{n/2}, v^{-m/2}\right) \qquad (2.35)$$

hold.

On the contrary, $\overline{\tilde{G}}$ must lie inside $\tilde{G}$ if the condition

$$v^{-n-m} - d\lambda^{n+m} - \varepsilon < K_2 \max\left(\lambda^{n+m}, v^{-n-m}\right) \max\left(|\varepsilon|^{1/2}, \lambda^{n/2}, v^{-m/2}\right) \qquad (2.36)$$

holds. The ranges of m, n, determined by inequalities (2.35), (2.36), are separated by a very narrow slit

$$-K_2 \max\left(\lambda^{n+m}, v^{-n-m}\right) \max\left(|\varepsilon|^{1/2}, \lambda^{n/2}, v^{-m/2}\right) < v^{-n-m} - d\lambda^{n+m} - \varepsilon$$

$$< K_2 \max\left(\lambda^{n+m}, v^{-n-m}\right) \max\left(|\varepsilon|^{1/2}, \lambda^{n/2}, v^{-m/2}\right). \qquad (2.37)$$

Applying the theorem of Sect.6, we arrive at the conclusion that the sequence of mappings

$$\ldots T^{n_1} L T^{m_1} T^{n_2} L T^{m_2} T^{n_3} L T^{m_3} \ldots \qquad (2.38)$$

is realizable if all pairs n_s, m_s satisfy (2.35), and non-realizable if at least one pair satisfies (2.36).

It should be borne in mind that each mapping L can be treated either as the restriction L_1 associated with one regular branch of the two-valued auxiliary mapping $\tilde{L}$ or as the restriction L_2 associated with the other regular branch. For large positive ε, none of (2.35) can hold for $n_s, m_s > N$. On the contrary, for large negative ε, there is an infinite set, of continuum power, of possible sequences of the form (2.38).

We now clarify conditions for $\widetilde{T^n L T^m}$ to be contractive. It immediately follows from (2.31) that

$$\delta \bar{u} = \frac{\partial \bar{u}}{\partial u} \delta u + \frac{\partial \bar{u}}{\partial v} \delta v, \quad \delta \bar{v} = \frac{\partial \bar{v}}{\partial u} \delta u + \frac{\partial \bar{v}}{\partial v} \delta v,$$

where

$$\frac{\partial \bar{u}}{\partial u} = b\lambda^{n+m} - \frac{a}{2} D\lambda^{n+m}, \quad \frac{\partial \bar{u}}{\partial v} = \frac{a\lambda^n v^{-n}}{2} D, \quad \frac{\partial \bar{v}}{\partial u} = dv^{-n}\lambda^m D, \quad \frac{\partial \bar{v}}{\partial v} = -\frac{1}{2} v^{-m-n} D,$$

$$D = \frac{1}{\sqrt{c}} \frac{1}{\sqrt{v^{-n}\bar{v} - d\lambda^m u - \varepsilon}}.$$

It suffices that the inequalities

$$\left|\frac{\partial \bar{u}}{\partial u}\right| + \left|\frac{\partial \bar{v}}{\partial u}\right| < q, \quad \left|\frac{\partial \bar{u}}{\partial v}\right| + \left|\frac{\partial \bar{v}}{\partial v}\right| < q < 1, \qquad (2.39)$$

or

$$\left(v^{-m-n} - d\lambda^{m+n} - \varepsilon\right)^{1/2} > K_3 \max\left(\lambda^{m+n}, \lambda^n v^{-n}, v^{-m}\lambda^m, v^{-m-n}\right), \quad b\lambda^{n+m} < q, \qquad (2.40)$$

hold, providing for mappings (2.38) being saddle. As ε decreases, (2.40) is necessarily valid if (2.35) holds.

We can now summarize. As ε decreases, countably many bifurcations arise, generating a pair of fixed points (2.31) and pairs of sequences (2.38) of continuum power. Moreover, a multiplicity doubling bifurcation series of infinite length is generated from a point in each pair formed. Something similar also occurs with (2.38). After the doubling series are complete, only fixed saddle points remain; this is necessarily true if inequalities (2.40) hold.

All the above describes bifurcations only one in a neighbourhood of a homoclinic structure; there can be very many of them in the structure. Any fuller study of the question is sufficiently intricate and complicated; therefore, we confine ourselves to the above remark.

A similar situation also occurs as a saddle focus loop is crossed, i.e., in forming and making vanish the doubly asymptotic closed phase trajectory from the focus and back. The loop is the structurally unstable intersection of the integral manifolds S^-, S^+ of the saddle focus.

That there are bifurcation series was already noted in considering Situation 6 in Chapt.6. A more detailed investigation can be performed according to the above.

3. Bifurcations and the stochastic attractor in a Lorenz system

We have already said that the Lorenz differential equations arose as a three-mode discrete truncation in the problem of thermal convection between two horizontal planes. It was shown in Chapt.1 that the equations with parameter b=1 are basic in describing fluid convectional circulation in a closed circular contour. That there are aperiodic stable-state motions was established in 1963; however, a sufficiently complete study was made only in 1976-1978 in [46, 68, 69, 276–278, 280, 539, 551, 552, 679], involving two approaches: proceeding from the "Smale horseshoe", and from Poincaré's homoclinic structures.

The homoclinic structure discovered in a Lorenz system is comparatively very simple. It has nothing of that unfathomable complexity about which H. Poincaré wrote, and in this sense it is unique. This simplicity is due to the three-dimensional phase portrait's structure, reflected by the discontinuity line of the point mapping on the cutting plane, and the possibility to separate variables; we already discussed the matter briefly. Below, we give a more detailed treatment of the Lorenz system

$$\dot{x} = -\sigma x + \sigma y, \; \dot{y} = rx - y - xz, \; \dot{z} = -bz + xy, \qquad (3.1)$$

and analyze restructuring and bifurcations of its phase portrait for increasing r. The parameters σ, b and r are assumed to be positive. The most complete numerical studies are related to the case σ=10, b=8/3 and r <150; however, other cases also have been considered.

We will now study the Lorenz system (3.1) in steps.

1. Replacing the variables x, y, z by -x, -y, z does not change the form of equations (3.1); therefore, the phase portrait is symmetric about the z-axis.

2. All phase trajectories of (3.1) lie inside the sphere

$$V = x^2 + y^2 + (z-\sigma-r)^2 = K^2 (\sigma+r)^2 \qquad (3.2)$$

with centre at the point x=y=0, z= σ+r, and with radius K (σ+r). In fact, according to (3.1), the time derivative of the function $V = x^2+y^2+ (z-\sigma-r)^2$ equals

$$\dot{V} = -2\sigma x^2 - 2y^2 - 2b\left(z - \frac{\sigma+r}{2}\right)^2 + b\frac{(\sigma+r)^2}{2}.$$

Therefore, the above statement holds if (3.2) is outside the ellipsoid determined by the equation $\dot{V}=0$, which necessarily occurs for $K^2 = \frac{1}{4} + \frac{b}{4}\max(\frac{1}{\sigma}, 1)$.

3. In positive time, the phase volume contracts everywhere, since

$$\text{div}(\dot{x}, \dot{y}, \dot{z}) = -1 - b - \sigma < 0, \tag{3.3}$$

provided (3.1) holds.

4. The point O (x=y=z=0) is an equilibrium state for all b, σ, r. The characteristic equation of the state is

$$(\lambda + b)[\lambda^2 + (\sigma + 1)\lambda + \sigma(1 - r)] = 0. \tag{3.4}$$

It follows that, for r <1, an equilibrium state of $O^{3,0}$-type occurs; for r >1, of $O^{2,1}$-type; i.e., a stable and a saddle node, respectively, while bifurcation occurs in crossing the boundary of N_0.

5. For r >1, in addition to the above equilibrium state O (x=y=z=0), there are two other equilibrium states O_1, O_2, with coordinates $x = y = \pm\sqrt{b(r-1)}$ and z=r-1, arising as O loses stability for r=1, and lying symmetrically about the z-axis. By symmetry, the equilibrium types are the same, and determined by the roots of the characteristic equation

$$\lambda^3 + (\sigma + b + 1)\lambda^2 + b(\sigma + r)\lambda + 2b\sigma(r-1) = 0. \tag{3.5}$$

As r increases from r=1, O_1, O_2 become consecutively stable nodes of $O_1^{3,0}$ - and $O_2^{3,0}$ - types, then stable foci, and, for

$$r > \frac{\sigma + b + 3}{\sigma - b - 1} \sigma, \tag{3.6}$$

saddle foci $O_1^{1,2}$ and $O_2^{1,2}$.

6. O_1, O_2 change their stability type when the boundary of N_ω is crossed; therefore, either unstable saddle periodic motions merge with the equilibrium states, or stable periodic motions are generated. What exactly happens depends on the sign of the Lyapunov quantity. Its calculation shows that merger with an unstable periodic motion occurs. For b=8/3 and σ=10, the equilibrium states lose stability, and, simultaneously, unstable saddle periodic motions $\Gamma_1^{2,2}$, $\Gamma_2^{2,2}$ merge with the latter according to (3.6) for r =24.74.

To understand where the periodic motions Γ_1, Γ_2 come from, we decrease the parameter r from the bifurcation value r =24.74, and numerically track Γ_1, Γ_2. For r =13.92, Γ_1, Γ_2 turn into loops of invariant curves S_1^-, S_2^- of the saddle equilibrium state

$O^{2,1}$ (Fig. 7.15).

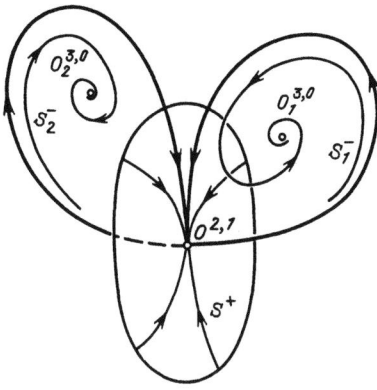

Fig. 7.15

7. We can now describe the phase portraits for $0 < r < 1$ and $1 < r < 13.92$. In the former case, the equilibrium state is globally stable, and all phase trajectories tend to it. When r increases and passes through the bifurcation value r=1, the equilibrium state O loses stability, and two new stable equilibria O_1, O_2 appear. All phase trajectories tend to one of the latter. The separating boundary is the invariant surface of the state O (Fig. 7.16). As r increases beginning with r=1, the two invariant curves S_1^-, S_2^- which emanate from O should be tracked. If r =13.92, they turn into homoclinic loops, and appear on the two-dimensional invariant surface S^+ of O (Fig. 7.15).

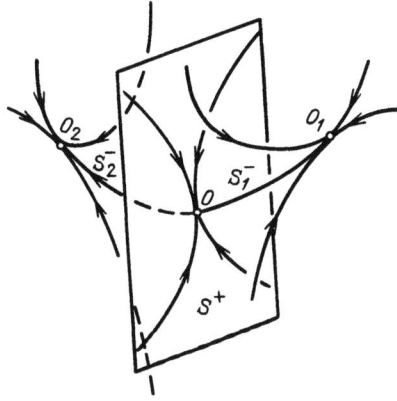

Fig. 7.16

8. The tracking of further changes in the phase portrait as r increases can be

done conveniently by using the cutting plane Σ with equation z=r−1 (Fig. 7.17).

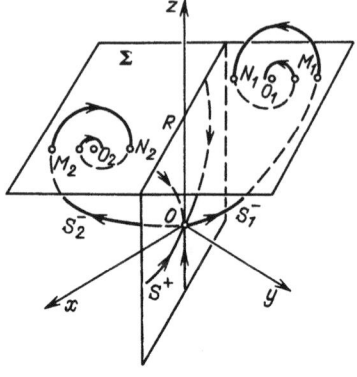

Fig. 7.17

For r =13.92, the phase portrait of the point mapping on the cutting plane z=r−1 is depicted in Fig. 7.18, with the line R as the intersection of the integral surface S^+ of O with Σ; O_1 and O_2 are equilibrium states appearing on z=r−1. M_1, N_1 (resp. M_2, N_2) are the two consecutive points of intersection of the invariant curve S_1^- (resp. S_2^-) with Σ. The phase portrait of the point mapping on Σ for r somewhat less than the bifurcation value r =13.92 can be seen in Fig. 7.17. M_1, N_1, ... and M_2, N_2, ... are consecutive points of intersection of S_1^- and S_2^- with the cutting plane. The curve R breaks the latter into two parts, each of which is a region of attraction of one of the stable states O_1, O_2 which are stable fixed points on Σ.

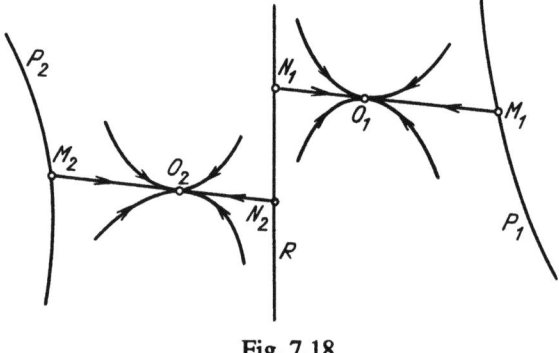

Fig. 7.18

R separates the regions of attraction of the fixed points O_1, O_2; the mapping T of Σ is discontinuous. Nearby points placed on opposite sides of R are sent into far ones.

Bifurcations and routes to chaos and stochasticity 205

That the transformation of Σ is discontinuous is due to an indefinite increase in duration of motion of the phase point as it approaches R until the next intersection of Σ. Starting at M near R, the phase trajectory moves near the surface S^+ towards O, then along S_1^- or S_2^- according to which side of R the point M is located on. Moving near S_1^- or S_2^-, the trajectory again intersects Σ at a point $\overline{M}$ close to M_1 or M_2, respectively. As M approaches R on one of its sides, the point coincides with M_1 or M_2 in the limit. Thus, the mappings T on each side of R can be continuously extended to the curve. However, one extension maps R to the point M_1; the other, to M_2 (Fig. 7.17).

The general forms of the phase space and cutting plane Σ for parameter values r somewhat less than 13.92 are given in the diagram. Recall that the other parameters σ and b are, for the sake of being specific, assumed to be fixed, viz., $\sigma=10$, b=8/3. As r increases up to r =13.92, the second intersection points, N_1, N_2, of the integral curves S_1^-, S_2^- with Σ find themselves in the discontinuity line R; this is associated with the appearance of two loops S_1^-, S_2^- for the equilibrium state O, represented in Fig. 7.15. Recall that both are on the integral surface S^+ of O, the point mapping on Σ being of the form schematically represented in Fig. 7.18 (in the sense that a separate fragment is given, and not the global picture, which is sufficiently complicated as a whole), where the distinct mappings T_1, T_2 are defined on opposite sides of R by symmetry. The curves P_1, P_2 are transformed by the mappings into R, and R, in turn, into M_1, M_2, which are again transformed into the points N_1, N_2 of R. The regions enclosed by the curves P_1, R, and P_2, R contract to fixed stable points O_1, O_2, respectively. The above means that any interior point of the regions under successive transformations asymptotically tends either to O_1 or O_2, respectively.

9. As r increases further, and becomes greater that 13.92, an unstable saddle periodic motion Γ_1 (resp. Γ_2) is generated from each closed loop of the integral curve S_1^- (resp. S_2^-). This bifurcation into periodic motions arising from saddle equilibrium loops was considered in Chapt.5. It is associated on the cutting surface with the separation of fixed points of multiplicity two from N_1, N_2, which we denote by the same letters Γ_1, Γ_2 as the associated periodic motions. The points N_1, N_2, where S_1^-, S_2^- meet the cutting plane Σ as r increases, shift and cross the line R for r=13.92. Immediately after r exceeds 13.92, the phase portrait on Σ is as in Fig. 7.19. The general form of the basic elements of the

three-dimensional phase portrait is in Fig. 7.20. New are the periodic motions Γ_1, Γ_2 as well as their two-dimensional integral manifolds S_1^-, S_1^+ and S_2^-, S_2^+. The intersections of the motions and their manifolds with the cutting plane are in Fig. 7.19, and also denoted by Γ_1, Γ_2, S_1^-, S_2^-, S_1^+, S_2^+. If r slightly exceeds 13.92, the points Γ_1, Γ_2, N_1 and N_2 are close to the line R (as r→13.92, they tend to R). Numerical computation[5] shows that the relative position of N_1, N_2 and S_1^+, S_2^+, associated with fixed points Γ_1, Γ_2, is exactly as in Fig. 7.19. As r increases further, their position changes, taking the form of Fig. 7.21 for r =24.06. Meanwhile, N_1, N_2 are on S_1^+, S_2^+, respectively.

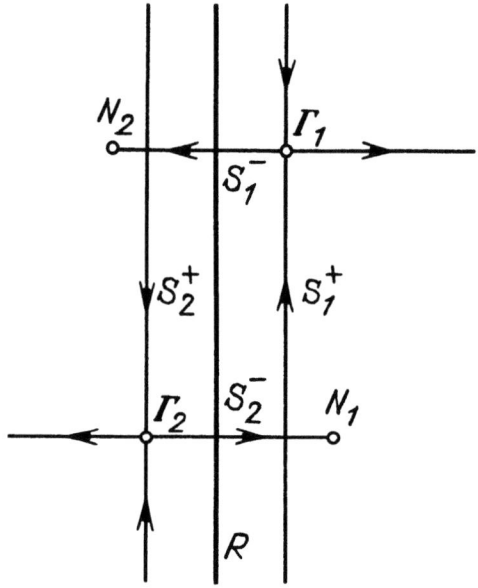

Fig. 7.19

---

[5]If r slightly exceeds 13.92, this location is consistent with theoretical foundations.

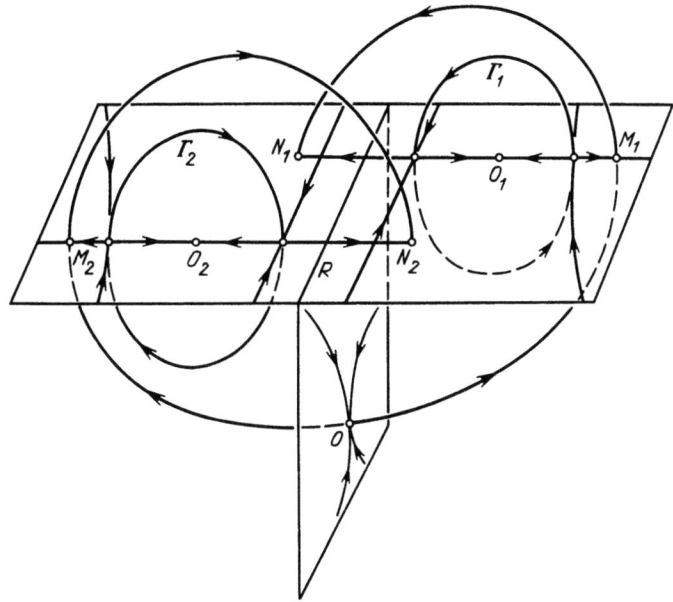

Fig. 7.20

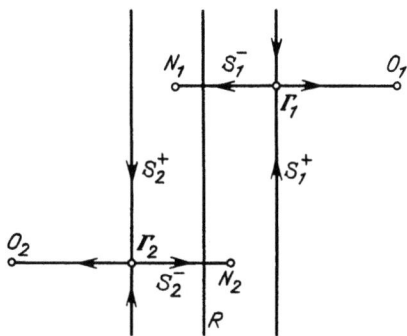

Fig. 7.21

Consider the phase portraits on Σ ($z=r-1$) represented in Figs. 7.19, 7.21, and also the phase portrait in Fig. 7.22 associated with the limiting case of bifurcation. There are two fixed saddle points on all the portraits, as well as corresponding integral manifolds S^+, S^-.

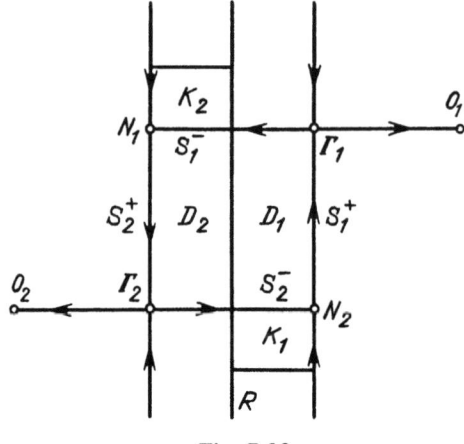

Fig. 7.22

The manifolds are intersecting in Fig. 7.19, and form a homoclinic structure. There is no intersection in Fig. 7.21; Fig. 7.22 is the limiting case.

Thus, a homoclinic structure in the Lorenz equations arises for $r = 13.92$, which means that there is an infinite set J of all possible saddle motions including periodic ones. However, for $13.92 \leq r < 24.06$, the saddle motions do not form an attractor, at least since the invariant curves S_1^-, S_2^- are outside J (they are outside the strip in between S_1^+, S_2^+, which encloses J) (Fig. 7.19).

As r increases further, subsequent changes make sure that the invariant manifolds S_1^-, S_2^- are included in the set of invariant saddle motions J; this occurs for r ≥ 24.06 (Figs. 7.21, 7.22). The nearest subsequent change of the phase portrait occurs for $r = 24.74$ when the periodic motions Γ_1, Γ_2 merge with the equilibrium states O_1, O_2, respectively. This is related to the merger of the fixed points Γ_1, Γ_2 with O_1, O_2 on the cutting plane Σ. Thus, when $r = 24.06$, a stochastic attractor arises, already considered in Sect.2, Chapt.6 (Situation 4). Moreover, we will now clarify the form of the limiting set J if $r = 24.06$, and prove that transformation of Σ into itself can be reduced to that of a straight line into a straight line, as was briefly mentioned in Chapt.4. We focus our attention on how all points of the phase space (of course, with the exception of unstable equilibria) are "absorbed" by the stochastic attractor, and on how the latter arises "out of nothing" as the parameter increases. The hard appearance of the stochastic attractor is easiest tracked down by means of the associated mapping of a straight line into a straight

Bifurcations and routes to chaos and stochasticity 209

line.

10. The phase portrait of the point mapping T on the cutting plane for r =24.06 is in Fig. 7.22. Under T, the rectangle $\Gamma_1 N_1 \Gamma_2 N_2$ is transformed into itself, so that the region D_1 is sent into $\overline{D}_1$, and D_2 into $\overline{D}_2$ (Fig. 7.23). The regions K_1, K_2 (Fig. 7.22) outside $\Gamma_1 N_1 \Gamma_2 N_2$ get inside the rectangle. This "absorption" of phase points occurs with each application of T. The rectangle is associated with a certain region G in the three-dimensional phase space. G is transformed into itself, so that its volume V decreases to $Ve^{-1-\sigma-b}$ per unit time, while the vacant space in G is filled with outside phase volume. This finally leads to sending all points of the phase space into G (of course, except the unstable equilibrium points O_1, O_2).

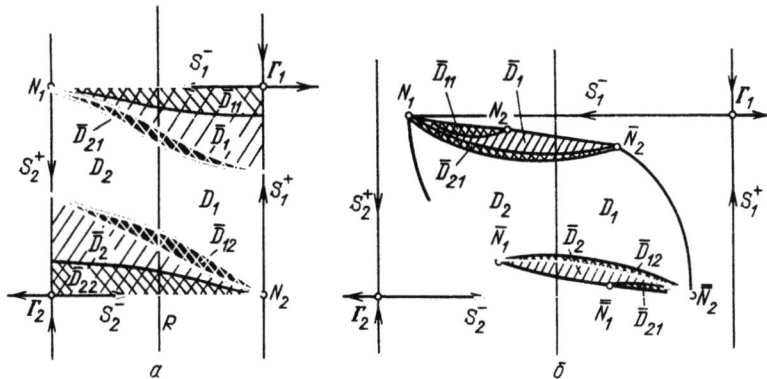

Fig. 7.23

Still, what happens inside G itself? We retrace our steps to the cutting plane. T transforms D_1, D_2 into two regions $\overline{D}_1$, $\overline{D}_2$ which are, in turn, transformed into $\overline{D}_{11}$, $\overline{D}_{22}$ and $\overline{D}_{21}$, $\overline{D}_{22}$ belonging to $\overline{D}_1$, $\overline{D}_2$. Each region D_{ij} is transformed into two regions $\overline{D}_{ij1}$, $\overline{D}_{ij2}$ in $\overline{D}_{j1}$, $\overline{D}_{j2}$, respectively. The process can be extended indefinitely, with the region

$$\overline{D}_{i_1 i_2 \ldots i_n} \tag{3.7}$$

being transformed into two regions

$$\overline{D}_{i_1 i_2 \ldots i_n 1} \text{ and } \overline{D}_{i_1 i_2 \ldots i_n 2}, \tag{3.8}$$

the former lying in $\overline{D}_{i_2 i_3 \ldots i_n 1}$, and the latter in $\overline{D}_{i_2 i_3 \ldots i_n 2}$.

It should also be added that, each time T is applied, contraction along S^+ and extension along S^- occur. The limiting set of the nested regions (their number increasing like 2^n) is just the stochastic attractor, which consists of a continuum of limiting curves.

Each can be ascribed an infinite sequence of subscripts $i_1, i_2, i_3, \ldots$, so that the curve $\omega_{i_1 i_2 i_3 \ldots}$ is the limit of the nested regions

$$\overline{D}_{i_1} \supset \overline{D}_{i_2 i_1} \supset \overline{D}_{i_3 i_2 i_1} \supset \overline{D}_{i_4 i_3 i_2 i_1} \supset \ldots . \tag{3.9}$$

11. We now reduce the point mapping T of the cutting plane Σ to a one-dimensional point mapping. The reduction is understood in the sense that if variables u, v are conveniently selected on Σ, then T is written as

$$\overline{u} = f(u), \quad \overline{v} = g(u,v) \tag{3.10}$$

with respect to u, v, where $g'_v(u,v) < q < 1$. The notation means that the coordinate lines u=const are transformed into each other, and the iterations of v tend to a certain sequence of values $v_1, v_2, \ldots$ determined only by the sequence of values $u_1, u_2, \ldots$. The required u-curves are constructed one after another as follows: T is discontinuous on R, and separates into two mappings T_1, T_2 coinciding with T on opposite sides of the discontinuity line R.

As the first series of u-curves, we take $T_1^{-s_1} R$, $T_2^{-s_1} R$ (s_1=1,2,...); as the second, $T_1^{-s_2} T_2^{-s_1} R$, $T_2^{-s_2} T_1^{-s_1} R$ (s_2=1,2,...); as the third, $T_2^{-s_3} T_1^{-s_2} T_1^{-s_1} R$, $T_1^{-s_3} T_2^{-s_2} T_1^{-s_1} R$ (s_3=1,2,...), etc. The curves do not intersect, and densely cover the region D. Accordingly, the choice of u-curves is unique. On the contrary, the v-curves can be selected rather arbitrarily. Transforming D into a rectangle, so that the u-curves are straight lines, as v-curves we can select straight lines orthogonal to u-curves. T with respect to the variables so selected can be written as (3.10), since, under T, u-curves are sent into each other, and contraction occurs with respect to the variable v.

We now consider the mapping from a straight line into a straight line, determined by the first relation in (3.10). For r =24.06, the mapping is given in Fig. 7.24. On its graph, N_1, N_2 are associated with the opposite sides of the line of discontinuity R; Γ_1, Γ_2, with points denoted in the same way on the cutting plane. The graph is symmetric about the origin of coordinates; N_1, Γ_1 and N_2, Γ_2, respectively, have the same coordinates $\overline{u}$ for r =24.06. The slightest decrease in r makes Γ_1 lower than N_2, and the strange attractor vanishes, since all points of the interval (+1,−1) turn out to be outside. On the contrary, for r =24.06 or slightly greater (Fig. 7.25), the point Γ_1 is higher than N_2, the interval J is transformed into itself, and the points of (+1,–1) are attracted to J, which manifests the hard appearance of the stochastic attractor as the parameter r passes through the critical value $r^*=24.06$, which dependens on the parameters σ, b involved in the Lorenz equation. We already pointed out that, for $\sigma=10$, $b=8/3$, r^* was equal to 24.06. The dependence of r^* on σ for $b=8/3$ is represented in Fig. 7.26.

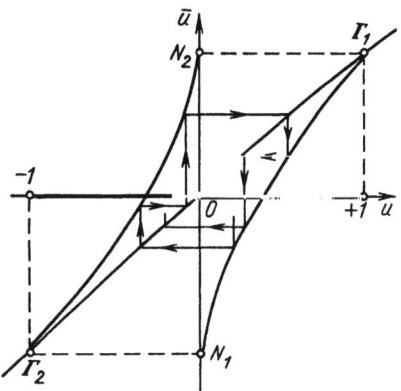

Fig. 7.24

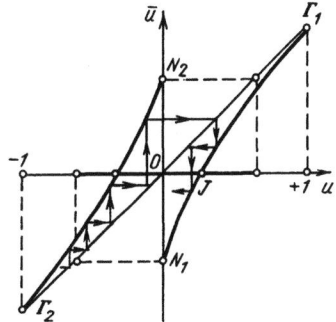

Fig. 7.25

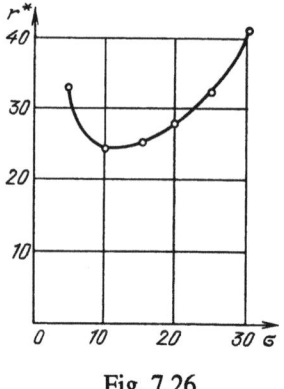

Fig. 7.26

A little retracing our steps to Fig. 7.23a,b, we see that the diagrams are schematic, and not associated with real scales. The regions $\overline{D}_1$, $\overline{D}_2$ are actually very thin, while the size of each of $\overline{D}_{11}$, $\overline{D}_{12}$, $\overline{D}_{21}$, $\overline{D}_{22}$ is characteristically different from zero only in the third significant figure. This means that successive transformations of Σ fall, after a few steps, into the vicinity of the curve. Thus, for the parameter values considered below, after one transformation, the circle $x^2+y^2 \leq 10$ turns into a strip not wider than 0.5, and, after two or three, into a strip not wider than 10^{-3}. These very thin strips are represented in Fig. 7.27 by the lines J_1, J_2. It is natural that the fixed points O_1, O_2, Γ_1 and Γ_2 are on the lines. Neglecting the width of the strips and regarding them as J_1, J_2, we thereby reduce the mapping T of the cutting plane Σ into itself to that of a curve into a curve. Introducing the coordinate u on this "curve" (J_1, J_2), we obtain a one-dimensional mapping

$$\overline{u} = f(u) \qquad (3.11)$$

which approximately describes the asymptotic behaviour of the mapping of Σ and, up to the approximations mode, coincides with the mapping from a straight line into a straight line to which the mapping of the attractor J was reduced, and which has the same description.

Bifurcations and routes to chaos and stochasticity 213

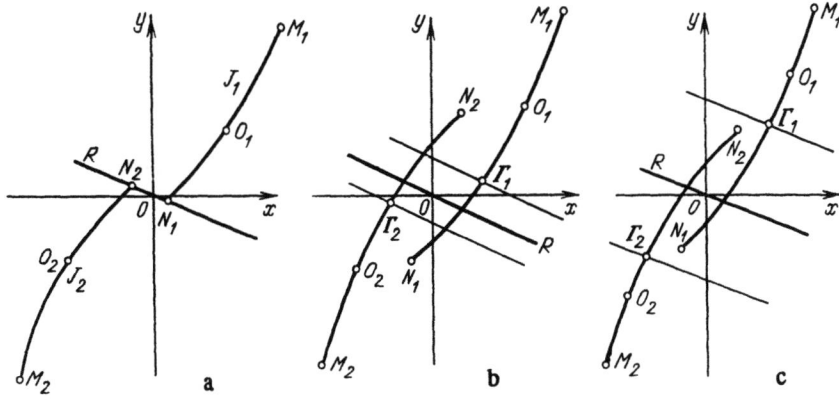

Fig. 7.27

The numerically determined curves J_1, J_2 are shown in Fig. 7.27 for a series of parameter values r. As r varies, (3.11) varies, too. As r increases considerably, the mapping ceases to be expanding everywhere. Moreover, points with a horizontal tangent appear on the graph, as in the case of the repeatedly mentioned transformation $\bar{x} = ax(1-x)$. This also leads to stable fixed points (of certain multiplicity) and stable periodic motions in the Lorenz system. In particular, for r =100, such a motion was determined numerically [68]. It turned out to exist in the following range of variation of r : [99.98, 100.06].

By means of the transformation (3.11) from a straight line into a straight line, the behaviour of the phase trajectories of the Lorenz equation can be represented as a series of point mappings from a straight line into a straight line, as shown in Fig. 7.28. Fig. 7.28a is associated with the stable equilibrium state O, 0< r <1; Fig. 7.28b, with two stable equilibrium states O_1, O_2; Fig. 7.28c, with generation of unstable periodic motions Γ_1, Γ_2, and a discontinuity jump; Fig. 7.28d, with a stochastic attractor; Fig. 7.28e, with Γ_1, Γ_2 merging with O_1, O_2; Fig. 7.28f, with horizontal tangents to the graph, and, accordingly, stable fixed points of certain multiplicity. We see that stochasticity in a Lorenz system is in this interpretation similar to that in an unstable impact oscillator with negative friction, which was considered in Chapt.3.

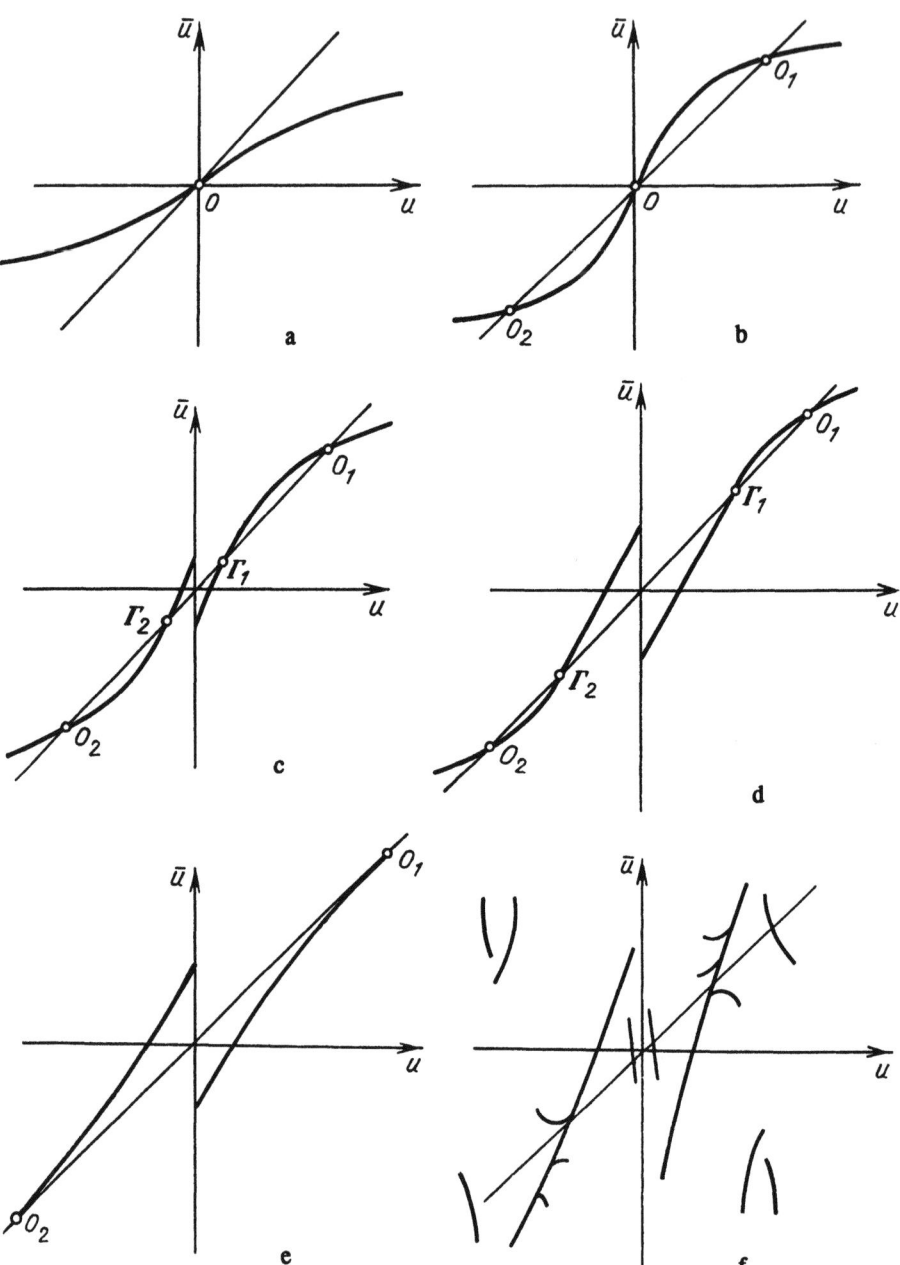

Fig. 7.28

4. Bifurcations and the phase portrait of a parametrically excited oscillator or rotator

Here, we consider equations of the form

$$\dot\phi = \omega, \quad \dot\omega = M - h\omega + (\nu + \mu \sin t) \sin \phi, \qquad (4.1)$$

where ϕ is an angle variable and M, h, ν, μ, are four parameters; e.g., such differential equations describe the motion of an unbalanced rotor or pendulum on a vibrating basis, the motion of a charged particle in the sine wave field, etc. In the system described by (4.1), both the phenomenon of synchronization and motion chaotization are possible. For M=h=0, (4.1) is a non-autonomous Hamiltonian system which becomes the familiar Mathieu equation on replacing $\sin \phi$ with ϕ, and describes linear parametric excitation and resonance. For $\nu=0$, (4.1) is directly related to the problem of synchronization of vibrators on a common basis, and was investigated in [60], where the discovery was made of synchronisms of various types and of aperiodic Poisson-stable, or chaotic, motions according to the classification of today. In more detail, the case $\nu=0$ was studied in [62, 65, 67]; M=h=0, M=ν=0, in [63, 64, 66].

A non-autonomous system (4.1) can be written as an autonomous system of order three

$$\dot\phi = \omega, \quad \dot\theta = 1, \quad \dot\omega = M - h\omega - (\nu + \mu \sin \theta) \sin \phi, \qquad (4.2)$$

using a new angle variable θ; the phase space is three-dimensional and cylindric with respect to ϕ, θ. The two-dimensional cylinder $\theta=0$ is a cutting surface for (4.2), on which the phase trajectories generate a mapping T, and to which the study of motions and of the phase portrait of the three-dimensional dynamical system (4.2) can be reduced.

The differential equations (4.1) involve four parameters, M, h, ν, and μ. It is then expedient to distinguish the following four cases of the dependence of the phase portrait on them: the case M=h=0 completed with subsequently clarifying the dependence on h, and the case ν=0 completed with considering small and large h. The role of M, h, ν, and μ in forming the phase portrait and its bifurcations differs in accordance with their physical meaning, viz.: h is dissipation; μ, parametric external action; M, constant external force; and ν, restoring elastic force.

1. Case M=h=0.

For convenience, we write out (4.1) separately as

$$\dot{\theta} = 1, \ddot{\phi}+(\nu+\mu \sin \theta) \sin \phi = 0, \qquad (4.3)$$

and computerize the study of the point mapping T on the cutting cylinder $\theta=0$ by using special programs [283], including those for numerical search of fixed points, roots of their characteristic equations, and their invariant curves S^+, S^-. The key to the investigation of the phase portrait and its dependence on parameters is to distinguish the synchronisms Γ_{pq} and to construct their invariant manifolds S^+, S^-.

Γ_{pq} is a periodic solution $\phi = \phi(t)$ such that the relation

$$\phi(t+2\pi p) = \phi(t)+2\pi p \qquad (4.4)$$

holds, where p, q are integers (p=1,2,..., q=0,±1,±2,...), and p≠0 cannot be decreased. The period of the periodic motion is $2\pi p$, while $2\pi q$ denotes the total change in the angle ϕ when the closed phase curve associated with Γ_{pq} is traversed by the phase point. Therefore, the synchronism with q=0 is related to oscillatory motion; that with q≠0, to rotation. On the cutting surface $\theta=0$, Γ_{pq} is associated with a cycle of fixed points of multiplicity p, also denoted by Γ_{pq}. Since the system in question is Hamiltonian, under motion of the phase points, phase volume is preserved, while the mapping T on $\theta=0$ is area-preserving. Therefore, the characteristic equation of any fixed point of the point mapping from $\theta=0$ into itself is of the form

$$\rho^2+A\rho+1 = 0, \qquad (4.5)$$

with either two real roots ρ_1, ρ_2 so that $\rho_1\rho_2=1$ or two complex conjugate roots $\rho_1=\rho e^{i\phi}$ and $\rho_2=\rho e^{-i\phi}$. In the former case, for $|\rho_{1,2}|\neq 1$, we obtain a fixed saddle point; in the latter, a centre-type point. Bifurcations of a fixed point are only possible for $\rho_1=\rho_2=1$ or $\rho_1=\rho_2=-1$, i.e., for A=−2 and A=2. Schematically, all possible simplest basic bifurcations are represented in Fig. 7.29a,b. For each bifurcation, either one fixed point turns into three or three fixed points into one. The difference between the cases $\rho_1=\rho_2=1$ and $\rho_1=\rho_2=-1$ is in replacement of the fixed points Γ_{pq} by fixed points Γ_{2p2q} of multiplicity twice as large, which is associated with period doubling of the corresponding periodic motions.

Bifurcations and routes to chaos and stochasticity

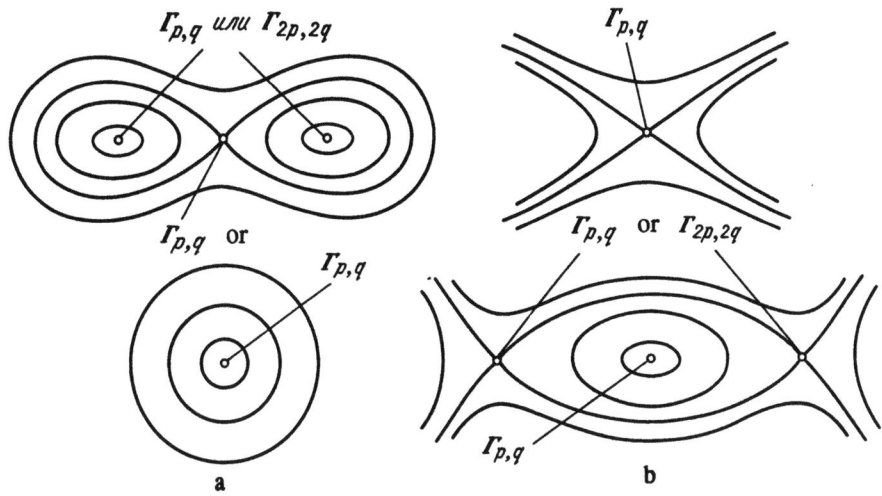

Fig. 7.29

To describe the synchronisms Γ_{pq} of (4.3), we begin with the parameter values $\mu=\nu=0$, for which the differential equations (4.3) are integrable, and

$$\phi(t) = \phi+\omega t, \quad (4.6)$$

where ϕ, ω are the initial values of $\phi(t)$, $\dot{\phi}(t)$ for $t=0$. Γ_{pq} is associated with the initial conditions, provided that

$$\phi(t+2\pi p) = \phi+(2\pi p+t)\omega = \phi+\omega t+2\pi q, \quad (4.7)$$

i.e., $\omega=q/p$ and ϕ be arbitrary. On the cutting cylinder $\theta=0$, Γ_{pq} are associated with fixed points with $\omega=q/p$ and arbitrary ϕ. The fixed points make up a circle (Fig. 7.30). The phase trajectory emanating from any point $(\phi, \omega=q/p)$ intersects the cylinder at the points

$$\left(\phi+\frac{2\pi q}{p}, \omega\right), \left(\phi+2\pi q\frac{2}{p}, \omega\right), ..., \left(\phi+2\pi q\frac{p-1}{p}, \omega\right),$$

in the above order, and, finally, at the point (ϕ, ω) again, traversing the cylinder q times in the positive direction for $q>0$, and in the negative direction for $q<0$. The value $\omega=0$ (for $q=0$) is associated with equilibrium states.

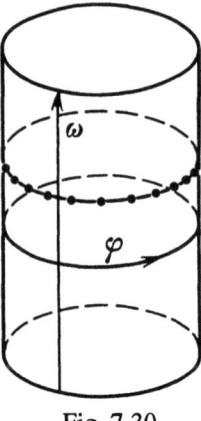

Fig. 7.30

We note that Γ_{pq} are everywhere dense on the cylinder, and therefore in the phase space. We may assume that the most important are those with small p, q. At any rate, it is only those which can be preserved when dissipation forces appear in the case of h≠0. A general proof of this fact is in [142].

From the trivial case $v=\mu=0$, we turn to the simple case $\mu=0$, i.e.,

$$\ddot{\phi}+v \sin \phi = 0. \tag{4.8}$$

The equation is also integrable, with the first integral

$$\frac{\omega^2}{2} - v \cos \phi = H. \tag{4.9}$$

The phase portrait of the point mapping on the cylinder $\theta=0$, which coincides in form with that of the autonomous conservative rotator (4.8), is represented in Fig. 7.31. O_1, O_2 are fixed center-type and saddle points, while S closed curves representing the coincident invariant curves S^+, S^- of O_2. The point O_1 is associated with $H=-v$; O_2, with $H=v$. The remaining phase curves are associated with values $H>-v$, closed for $-v<H<v$, encircle O_1, and represent periodic oscillatory motions; for $H>v$, they are closed, encircle the cylinder and represent periodic rotatory motions. The latter's period τ depends both on v and H. As $H\to -v$, it tends to $2\pi/\sqrt{v}$; as $H\to v$, it increases indefinitely; as $H\to\infty$, it tends to zero. The qualitative form of the graph of the dependence of τ on H is as in Fig. 7.32. As $v\to 0$, the phase portrait in Fig. 7.31 turns into that in Fig. 7.30. Meanwhile, the region of oscillatory motions, enclosed by the curves S, is transformed into the line $\omega=0$, so that each oscillatory motion of (4.8) contracts to the equilibrium state $\phi=\dot\phi=0$. Each rotatory motion Γ_{pq} (q≠0) of (4.8) arises from the corresponding rotatory motion Γ_{pq} of (4.8) for $v=0$, continuously varying as the parameter v varies. On the contrary, oscillatory synchronisms Γ_{p0} separate from O_1 ($\phi=\omega=0$) as v increases, Γ_{p0} of (4.8) being separated from O_1 if $v = q^2/p^2$, where q is the number of oscillations during the period $2\pi p$. Γ_{pq} intersect the cylinder $\theta=0$ along closed curves, and, therefore, form in the original three-dimensional space two-dimensional toroidal surfaces with periodic coil and Poincaré rotation number q/p. The other closed curves on the cutting surface $\theta=0$ are also associated with toroidal integral manifolds, but having quasiperiodic coil.

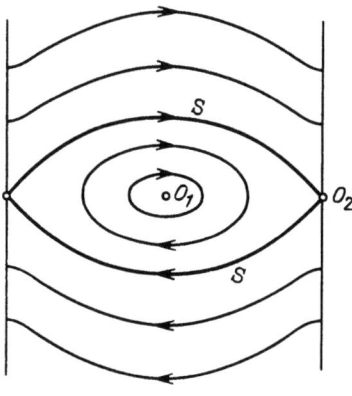

Fig. 7.31

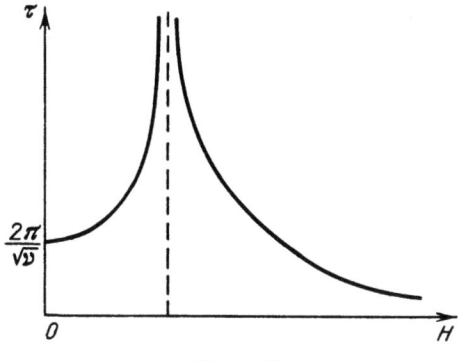

Fig. 7.32

We now turn to the case $\mu\neq 0$, starting from $\mu=0$, by continuously changing the parameter μ. Observe what occurs with Γ_{pq}. As μ varies continuously from zero, r synchronisms Γ_{pq} related to 2r cycles of fixed points of multiplicity p, with r of center-type and r saddles, arise in a neighbourhood of each curve on the cutting cylinder θ=0 associated with Γ_{pq}. Each fixed saddle point has its own invariant curves S^+, S^-. All this forms a somewhat simplified phase portrait (Fig. 7.33; r=1, p=4), since, as a rule, S^+, S^- actually meet, forming a homoclinic structure called a *stochastic synchronism* earlier. Neighbourhoods of fixed center-type points are also represented schematically in Fig. 7.33; we discuss them in the sequel.

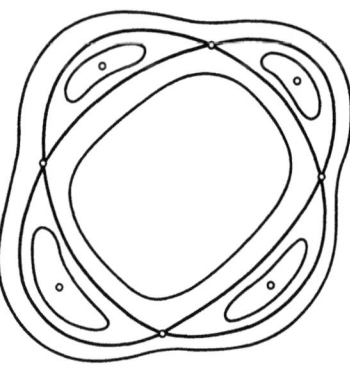

Fig. 7.33

The actual construction of phase portraits arising from each Γ_{pq} as μ varies, or, equivalently, from the corresponding resonance level of equation (4.8), requires computerization. Some computed portraits for ν=0, μ=0.185 are given in Fig. 7.34. Cycles of fixed points Γ_{pq} of center- and saddle-types, 2r in number, arising together from the corresponding resonance level, are denoted by Γ(p,q,r). We stress that the "phase portrait" in the diagram only represents the fixed points and invariant curves. The computations showed that, for odd p, q, r=1; if p, q are not simultaneously both even or both odd, then r=1 or r=2, according to which of two certain regions contains the values of ν, μ.

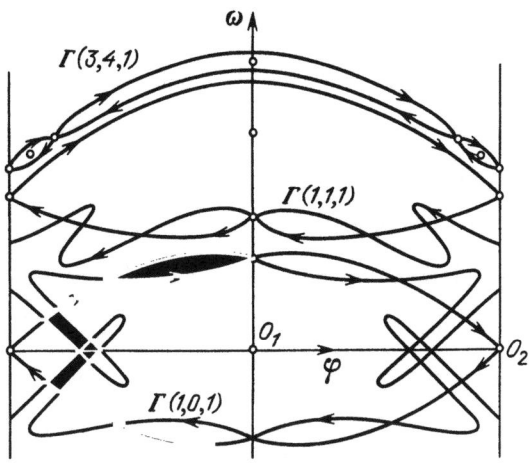

Fig. 7.34

Thus, when μ increases from zero, each resonance level of (4.8) generates the corresponding synchronism Γ(p,q,r) as the collection of 2r cycles of fixed points associated with individual $Γ_{pq}$, one half of center- and the other of saddle-type. There are countably many Γ(p,q,r), and they arise from an everywhere dense set of resonance levels of $Γ_{pq}$ in (4.8). However, the complexity of the phase portrait of (4.3) is not covered by this for μ≠0. The regions arising inside each synchronism near its center-type points, in turn, generate countably many synchronisms, each of which thereby generates new synchronisms, etc. Thus, a most complicated infinite hierarchy of nested structures arises. Precise laws can be determined numerically; however, any full theoretical analysis of their order of appearance has not yet been given. What is known was determined numerically in concrete systems on the basis of the above argument (Sect.2) related to synchronism generation from a fixed center-type point as the parameter changes. It is just this argument that leads to an understanding the general picture of nested structures, and that forms a basis for numerical investigation.

Other useful information is supplied by a general theorem about finitely many synchronisms with dissipation. Numerical computation enables us to understand how and which synchronisms vanish. In the remaining synchronisms, fixed center-type points become stable foci, saddles preserve their own type, and the invariant manifolds of the synchronisms may form complicated homoclinic structures.

In the above hierarchy, bifurcations of each synchronism can occur as parameters vary; these bifurcations were described above, and are represented in Fig. 7.29. In particular, they can lead to the period doubling of the "stable" synchronisms associated with center-type fixed points, simultaneously generating yet new saddle-type unstable synchronisms, and leading to homoclinic structures and chaotization of motions. Concrete numerical investigations show that the process starts with low-rank synchronisms, attains high-rank ones, and covers the whole nested structure hierarchy. By the *rank* of a synchronism [61, 65], we mean its subscript. Those generated from resonance levels are of rank zero; from fixed center-type points of rank-zero synchronisms, of rank one; from center-type fixed points of rank-one synchronisms, of rank two, etc. The region in which all synchronisms turn out to be saddle, because the period-doubling bifurcation series with synchronisms of all ranks are complete, is represented in Fig. 7.35. This does not exclude the possibility that there are points which are not saddle ones because of bifurcations to contacts of the invariant curves S^+, S^-, not always discovered in numerical computation. In Fig. 7.36, 2×10^3 iterations for $\mu=2$, $\nu=0$ are represented. How S^+, S^- of only a few saddle fixed points intersect is shown in Fig. 7.37.

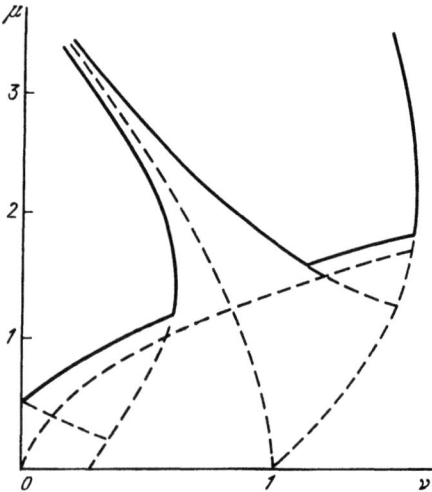

Fig. 7.35

Bifurcations and routes to chaos and stochasticity 223

Fig. 7.36

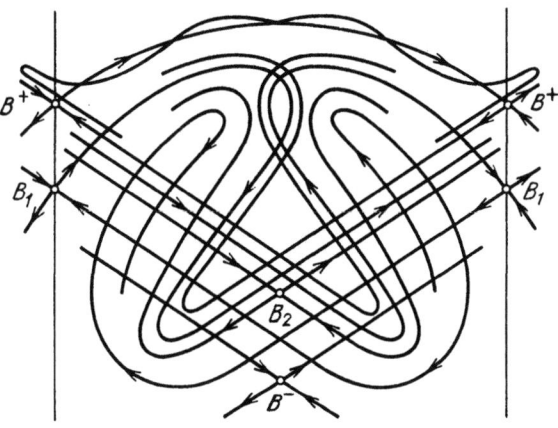

Fig. 7.37

We already noted that the whole magnificent picture of the phase portrait of a hierarchy of nested structures, whose examples are in Fig. 7.38, vanishes in damping. The vanishing starts with high-rank synchronisms, so that arbitrarily small damping leaves only finitely many of them. Any remaining center-type synchronism is turned by damping into a stable focus; however, a saddle remains a saddle. Therefore, after center-type synchronisms are turned into period-doubling synchronisms and only saddle-type ones remain, one can assume that sufficiently small damping does not destroy chaotic behaviour of motion.

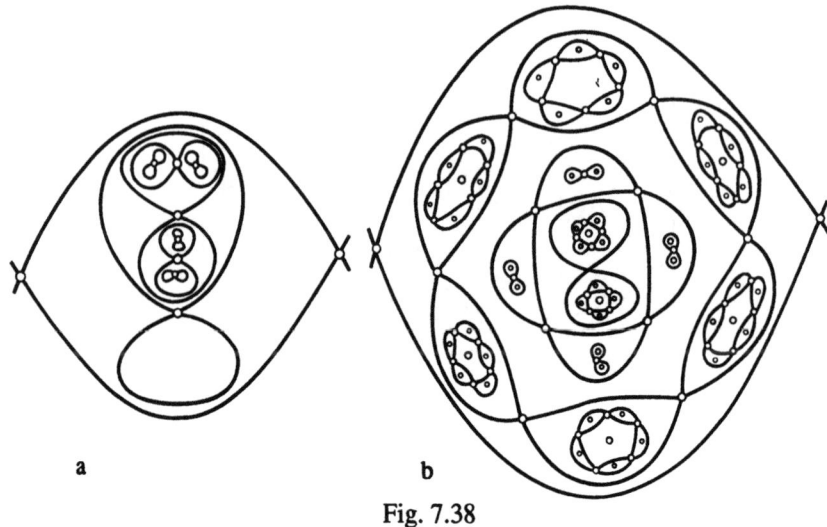

Fig. 7.38

2. *Case ν=0.* Equation (4.1) takes the form

$$\dot{\phi} = \omega, \ \dot{\theta} = 1, \ \dot{\omega} = M - h\omega - \mu \sin\theta \sin\phi. \tag{4.10}$$

The subcase h ≫M should be distinguished; it has the remarkable property of contraction of the whole three-dimensional phase (ϕ, θ, ω)-space into a very small neighbourhood of a certain two-dimensional surface, leaving as trace on the cutting cylinder $\theta=0$ a curve J as shown in Fig. 7.39. The consideration of the behaviour of phase trajectories is thereby approximately reduced to the point mapping from a circle into itself (topologically, J is a circle).

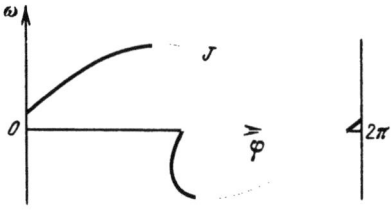

Fig. 7.39

For μ=0, equations (4.10) are integrable, so that

$$\phi = C_1 + C_2 e^{-ht} + \frac{M}{h}t, \ \omega = -C_2 h\, e^{-ht} + \frac{M}{h}. \tag{4.11}$$

The phase portrait of the point mapping on $\theta=0$ is very simple, and shown in Fig. 7.40, using (4.11). All phase trajectories asymptotically tend to the circle $\omega=M/h$ at exponential rate as t→+∞. The mapping

$$\bar{\phi} = \phi + 2\pi \frac{M}{h} \pmod{2\pi} \tag{4.12}$$

is defined on J. We now see how (4.12) varies as the parameter μ increases. For the sake of being specific, let M=0.1, h=1. For μ = 0.0, 0.5, 0.7, 0.94, 1.1, 1.32, 1.9, 2.2, 3.5, 4.3, respectively, the point mappings from J into J look as in Fig. 7.41. For small μ < 0.92, either quasi-periodic motions or a high-order synchronism does occur. For $\mu \approx 0.92$, the graph of the mapping touches the bisector, and intersects the latter first at two, and then at four points. This means that at first one and then two stable periodic motions of period 2π appear. As μ increases further, they lose stability; however, stable periodic motions of period 4π occur instead. For $\mu=1.44$, stable fixed points cannot be determined numerically, while the motions are chaotic (naturally, in a small neighbourhood of a certain toroidal surface intersecting the cutting cylinder $\theta=0$ along the closed curve J). As μ increases further, new intersections of the graph with the bisector appear, i.e., again, but for a very narrow band of parameters, stable periodic motions of periods 2π, 4π, etc., develop. Chaos then occurs. As μ increases still further, the picture repeats itself, but, each time, the domains of existence of the motions get smaller and smaller on the straight line of μ. For large μ, the domains practically vanish, while the motions become chaotic. It is interesting to note that, although with growth of μ, stable periodic motions of period 2π develop again and again, they all are of different types. First, there arise the synchronisms Γ_{10}; for the second time of their development, Γ_{11}; for the third, Γ_{12}.

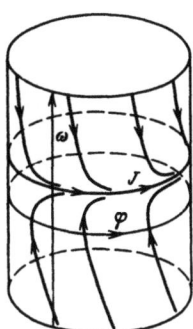

Fig. 7.40

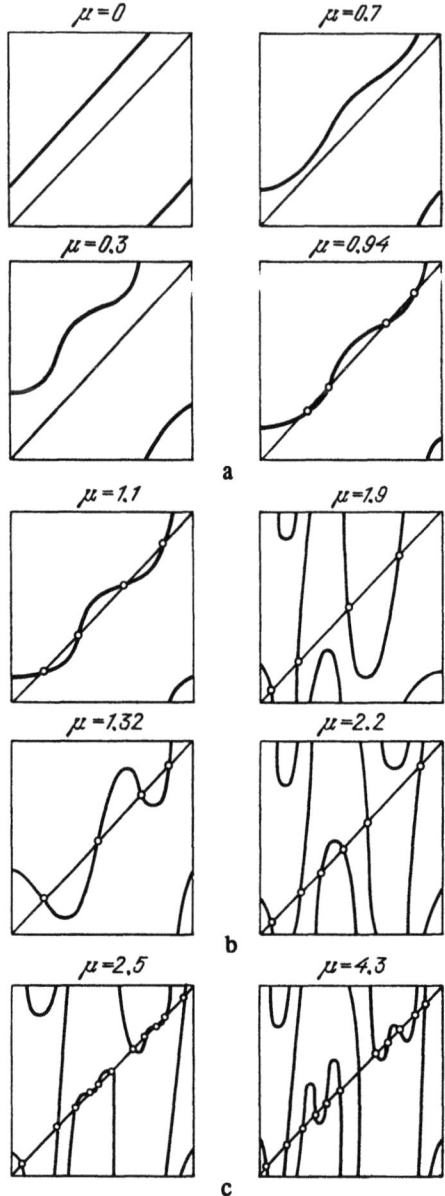

Fig. 7.41

This is what immediately follows from the graphs in Fig. 7.41, obtained by numerical computation. Meanwhile, this description is still somewhat coarse. It could easily be made exact if an asymptotically stable two-dimensional torus were attached to the system (4.10), and if J were the intersection of the torus with θ=0. This is not so for all μ. As the inverse mapping from the circle into itself becomes two-valued, J cannot be

Bifurcations and routes to chaos and stochasticity 227

the intersection of the two-dimensional integral torus with the cutting cylinder θ=0. The multi-valuedness arises because J is the intersection with θ=0 of a thin region between two tori, where motions depend on small transverse deviations from J and not only on the u-coordinate along J. As a matter of fact, we have approximately replaced the point mapping from a thin annulus into itself by one on the circle J. Mappings from an annulus into itself were considered in Sect.2, Chapt.6 (Situation 5). It follows that strong transversal contractibility of the mapping, i.e., in radial direction, does not exclude the absence of an invariant smooth curve or the chaotic behaviour of motions in the annulus. The sufficient conditions formulated there for chaotic behaviour only make use of contractions along the annulus, and of the way the coordinate (we call it u) is transformed along the annulus. This information yields the mapping from the circle into itself in Fig. 7.41. To make the motion chaotic, under these conditions, the point mapping from a circle into itself should have parts at which the mapping is expansive, and the inverse mapping should be two-valued.

Above, we discussed the numerical investigation of equation (4.10) for M=0.1, h=1. However, similar investigations show that the same results are obtained for other values of parameters M, h as soon as h »M. If the latter condition is not met and h<M, reduction to the point mapping from a circle into itself is not possible, and it is necessary to study the point mapping of a two-dimensional cylinder into itself. Generally, the phase portrait turns out to be as follows: For small μ, stable rotational synchronisms arise, whose regions of attraction are bounded by the separatrices S^+, S^- of fixed saddle points. As μ increases, they increase in number; meanwhile, separatrices of fixed saddle points, associated with different synchronisms, intersect, which leads to more complicated regions of attraction of stable synchronisms. As μ increases further, new intersections of separatrices and cycle-containing homoclinic structures appear. The phase points tend non-monotonically to stable synchronisms in quite a complicated way: The phase point now approaches a synchronism, now moves farther from it, and tends to the synchronism, only getting in a sufficiently small neighbourhood of it. Accordingly, regions of attraction of stable synchronisms are very complicated and thin. As μ increases further, period-doubling bifurcations of stable synchronisms occur, together with the appearance of new saddle-type synchronisms, which lead to still greater chaotization of

motions and to thinner regions attracting stable synchronisms. For negligibly small perturbations, the phase point walks across the surface of the cutting cylinder without getting into small neighbourhoods of stable synchronisms.

A fuller description of the phase portraits and their changes for M=0.1, h=0.05 and $0<\mu\leq1.8$ is given below. For $\mu=0$, the phase portrait is the same as in Fig. 7.40. The first essential bifurcation as μ increases occurs for $\mu=0.11$. Meanwhile, a complicated fixed saddle node arises, splitting as μ increases further into a stable fixed node A^+ and an unstable saddle B^+. As μ increases in the interval $0.12<\mu<1.43$, A^+ becomes a stable focus associated with a Γ_{11}-type synchronism. Note that, during the onset of a synchronism for $\mu=0$, its Poincaré rotation number is two; it is actually one for this stable synchronism.

The general form of the phase portrait for $\mu=0.25$ is shown in Fig. 7.42. All phase points either tend to the synchronism J or to the synchronism A^+.

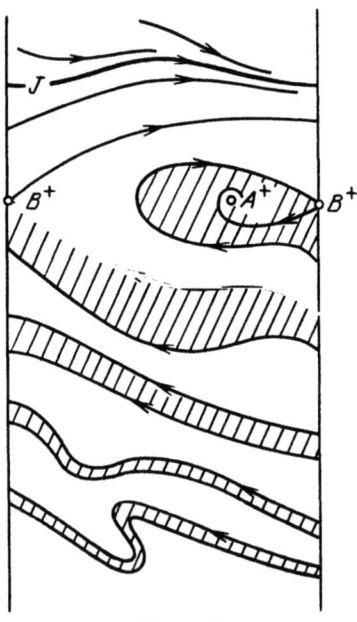

Fig. 7.42

For $\mu=0.301$, a new complicated fixed saddle node arises. Similarly to the preceding point, it decomposes into a stable node A^+ and a saddle B^-, associated with a Γ_{1-1}-type synchronism.

The phase portrait for µ=0.35 is represented in Fig. 7.43. The regions of attraction for different synchronisms are shaded (also in Fig. 7.42).

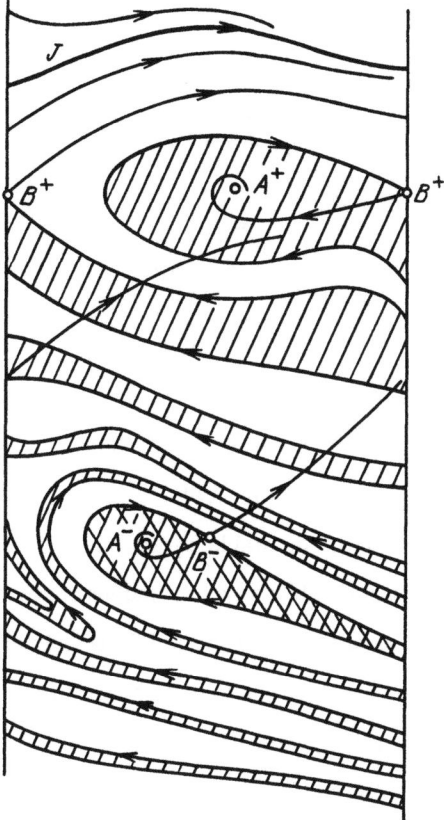

Fig. 7.43

Despite the complicated shape, the general structure of the phase portrait is very simple: The cylinder θ=0 decomposes into three regions of attraction, A^+, A^- and J; this is associated with decomposition of the original three-dimensional phase space into three regions of attraction of stable synchronisms. When µ increases further, contacts, and then intersections, of the separatrices S^+, S^- arise, not only of different but also of the same synchronisms. As a consequence, homoclinic cycle-containing structures appear. The phase portrait becomes extremely complicated, since there are infinitely many saddle-type synchronisms. The regions of attraction of stable synchronisms get complicated, and very thin. The phase point may move chaotically for a long time before it starts tending to one of the stable synchronisms A^+, A^- or J. The simplified phase portrait of the point mapping on the cutting cylinder θ=0 for µ=1.15 is given in Fig. 7.44. The invariant separatrices of the fixed points B^+, B^- form a rather complicated mutual intersection network. Only the first points of intersection of S^+ and S^- of the saddle-type synchronisms B^+, B^- are given in Fig. 7.44.

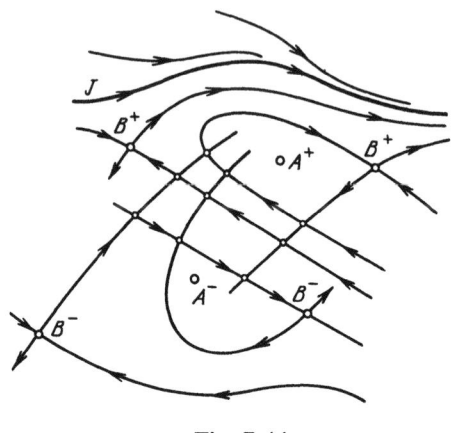

Fig. 7.44

As μ increases further, the stable synchronism J first vanishes, and the synchronisms A^+, A^- then change stability via bifurcations of N_{+1}-type.

J vanishes as the separatrices of the saddle B^+ interchange. This bifurcation occurs for $\mu=1.205$; the phase portrait assumes the form in Fig. 7.45. Both bifurcations of N_{+1}-type of A^+, A^- occur as the parameter varies from 1.39 to 1.40; each leads to two new stable synchronisms of the same type. For $\mu=1.5$, the corresponding portrait is in Fig. 7.46. As μ increases still further, period-doubling bifurcations occur and generate new stable synchronisms from A^+, A^-. The newcomers are of periods twice as large, and of Γ_{22^-}, Γ_{2-2^-} types. We can assume that, as μ increases further, the stable period-doubling synchronisms vanish; however, it will then be impossible to assert that stable synchronisms are totally absent, since they can appear due to more and more contacts of the invariant curves S^+, S^- of the infinitely many saddle-type synchronisms. Meanwhile, the regions of attraction of the synchronisms are so thin that any attempt to determine them directly or numerically is usually of no avail.

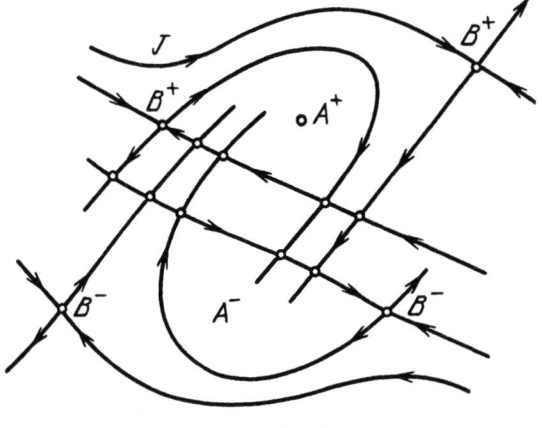

Fig. 7.45

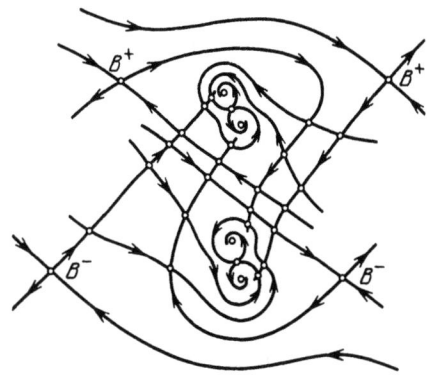

Fig. 7.46

5. On the appearance of chaos and stochasticity in dissipative dynamical systems

By summarizing the reasons for chaos and stochastic behaviour in dissipative dynamical systems, in Chapt.9, we will substantially complete the above-mentioned examples of Lorenz systems and parametrically excited rotators with descriptions and analysis of concrete systems admitting stochastic and chaotic oscillations.

The common reason for chaotic and stochastic behaviour of motions of dynamical systems is stability loss and exponential divergence of adjacent phase trajectories, accompanied by total boundedness of the trajectory and a certain global contraction. The simplest model of exponential instability together with global contraction is exponential divergence of trajectories and subsequent collapse. This combination can be treated both physically and in terms of phase space transformations, reflecting the real motion of the physical system. In the phase space, it is associated with a region G whose phase trajectories diverge at an exponential rate, and then leave it. The motion changes, and makes the phase points return to the original region G. The return is just the phase of deviation collapse. Everything is then repeated. Such is the simplest model; it can be formalized by two consecutive transformations T, S of the phase space as in Fig. 7.47. The excitation due to instability, energy pumping, and more or less rapid development of the process is associated with the phase corresponding to T; return to the original state on account of some restructuring (processes occurring in the system and changes in their nature), with the phase corresponding to S. The simplest example of the phases is the above-mentioned unstable damping impact oscillator, where amplification of oscillations is replaced by their decrease because of an impact. For motions of the Lorenz system, something similar is the build-up near unstable circulations, which results in the transition to another symmetric circulation, the build-up nearby, and the transition to the prior one, etc. Pendular motion on a vibrating basis, where replacement means change in the difference between the oscillation and the point of suspension phases, is also somewhat similar. Similar general argument can be developed for the systems described in Chapt.9. Certainly, these are only qualitative general ideas of mechanisms for making motions stochastic and chaotic.

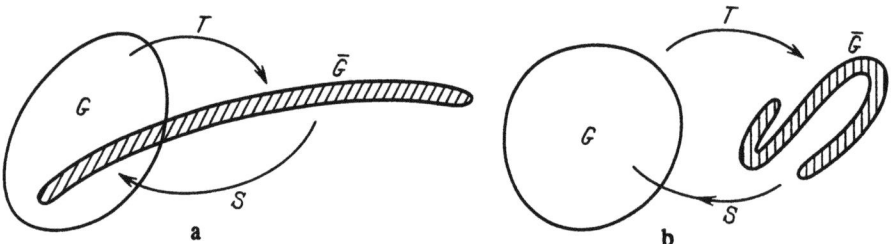

Fig. 7.47

The ideas are insufficient, already because periodic self-excited oscillations are also due to the conflict between two tendencies: instability and the system build-up, and build-up restriction. Meanwhile, the reasons for instability and build-up can be the same.

The difference lies in the nature of the restriction: in the former case, slow-down and restriction of amplification; in the latter, collapse. But, what is meant by restriction and collapse? What are the models for more exact characterization of the differences and their mechanisms?

Further progress requires studying the point mappings T, R which describe the divergence and collapse, respectively. For many cases, this was done in Chapt.6 by considering Situations 1–10. Now, we continue in a more general form.

More explicitly, increase and collapse can be seen in the transformation

$$x_{n+1} = qx_n - [qx_n], \qquad (5.1)$$

where q>1 and [z] is the entier of z.

The amplification phase is

$$y_{n+1} = qx_n \quad (q>1), \qquad (5.2)$$

while collapse the transition from y_{n+1} to x_{n+1}, viz.,

$$x_{n+1} = y_{n+1} - [y_{n+1}], \qquad (5.3)$$

after which x_{n+1} again becomes less than one.

We now compare (5.1) with

$$x_{n+1} = q(x_n) x_n, \qquad (5.4)$$

where $q(x_n)$ is a function decreasing with respect to x_n, so that, for certain a, b (a<b), q(a)>1 and q(b)<1. Given the last two inequalities, for certain $a<x^*<b$, $g(x^*)=1$. For $x_n<x^*$, the subsequent value x_{n+1} is greater than x_n; however, as x_n tends to x^*, the difference $x_{n+1}-x_n$ decreases. Moreover, for $x_n>x^*$, the value is negative.

The above examples can be regarded as simplest models of mechanisms for smooth restriction of the amplification and its collapse. Still, why do unpredictability and randomness occur for (5.1), and not for (5.2)? The difference is in the dependence of very distant values x_n on the original value x_0, or, generally, on the correlation between x_n and x_m for n much greater than m. In the second case, in order to find x_n up to a number ε, it is necessary to specify x_0 with accuracy ε/K, where K is a certain large, but possibly constant, number for all n. In the first case, the matter is different: To find x_n with accuracy ε, it is necessary to know x_0 with accuracy at least ε/q^n. As n increases, the required accuracy increases indefinitely at an exponential rate, making x_n unpredictable for large n; in the second case, all x_n are predictable with accuracy ε if only the accuracy of specifying x_0 is not less than ε/K. The fact that (5.1) loses accuracy as n increases indefinitely can be seen also from the fact that the values x_n are determined by stil more and more distant decimal places of the original value x_0. This is especially well demonstrated for q=10, since x_n is then the number obtained by discarding the first n places in the decimal notation of x_0.

Dependence of subsequent values x_n on the preceding ones is general for stochastic motions, and follows from the so-called *symbolic notation* of motions of deterministic dynamical systems, based on theorems on the unique existence of some or other sequence of point mappings (see Chapt.6). Thus, for the point mapping T from a straight line into a straight line, represented in Fig. 7.48, the inverse is contractive and two-valued. We take the latter's regular branches as the auxiliary mappings $\tilde{T}_1$, $\tilde{T}_2$. Defined only on [0,1/2] or [1/2,1], respectively, the restrictions T_1, T_2 of the original mapping T associated with the regular branches $\tilde{T}_1$, $\tilde{T}_2$ make up T.

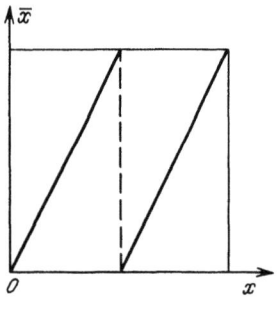

Fig. 7.48

Take an arbitrary point $x_0 \in [0,1]$. Its iterates $x_1 = Tx_0$, $x_2 = Tx_1$, ..., $x_{s+1} = Tx_s$, ... are formed either by the restriction T_1 if the corresponding value x_s is in [0,1/2], or T_2 if $x_s \in [1/2,1]$. Each point x_0 is associated with a sequence of numbers

$$i_0, i_1, i_2, ..., i_s, ..., \tag{5.5}$$

each of which is either 1 or 2. By theorems on sequences of point mappings, for any sequence (5.5), there is a point x_0 associated with the sequence. The point is unique; therefore, the correspondence between all x_0 in the segment [0,1] and all possible (5.5) is one-to-one, which follows from T being expansive. In fact, if there were two points, x_0, x_0', associated with one sequence, then, for any n,

$$|x_n - x'_n| = 2|x_{n-1} - x'_{n-1}| = \ldots = 2^n |x - x'_0| < 1,$$

which is impossible.

Thus, all x_0 in [0,1] and all possible (5.5) are in one-to-one correspondence, while the sequences can be regarded as certain generalized binary notation of x_0. For the mapping in Fig. 7.48, the sequence is the usual binary notation of x_0. If the diagram is replaced by Fig. 7.49, then bijectivity is preserved; however, (5.5) is no longer the usual binary notation of x_0.

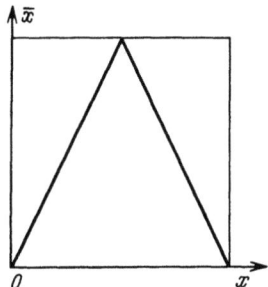

Fig. 7.49

Let us see how T looks if we make use of the notation of x_0 as in (5.5). It is immediately clear that the value $x_1 = Tx_0$ is associated with the sequence $i_1, i_2, i_3, \ldots$; $x_2 = Tx_1$, with the sequence $i_2, i_3, i_4, \ldots$, etc. Therefore,

$$T(i_s, i_{s+1}, \ldots) = (i_{s+1}, i_{s+2}, \ldots), \tag{5.6}$$

which is the symbolic notation of T. Note that the notation is the same for both Fig. 7.48 and Fig. 7.49; it is also identical with the notation of any T from a multi-dimensional region G into itself if only the mapping has a contractive two-valued inverse mapping $\tilde{T}$ which decomposes into two regular branches $\tilde{T}_1$, $\tilde{T}_2$. The above can be carried over to the case of a multi-valued contractive inverse mapping, with (5.6) preserved and with the only difference that (5.5) will consist of a large set of numbers or symbols $\omega_1, \omega_2, \ldots, \omega_m$, and not of just ones and twos.

The above can easily be made to hold for an arbitrary point mapping T with multi-valued contractive auxiliary mapping $\tilde{T}$. However, in the case of a saddle mapping T, it is now necessary to regard (5.5) as infinite in both directions, of the form

$$\ldots, i_{-2}, i_{-1}, i_0, i_1, i_2, \ldots \, . \tag{5.7}$$
$\uparrow$

It is again related to the point x_0 in a one-to-one way, and the symbol i_0 is distinguished. Under T, i_0 is replaced by i_1. As the transformation is applied s times, i_0 is carried into the symbol i_s. The infinite sequence (5.7) itself remains unaltered. By writing T in the form (5.6), or, in a more general case, as

$$T(..., i_{-1}, \underset{\uparrow}{i_0}, i_1, ...) = (..., i_{-1}, i_0, \underset{\uparrow}{i_1}, i_2, ...), \qquad (5.8)$$

where the arrows indicate the distinguished symbols, we can explain why, after some time, a subsequent motion of the phase point is determined by more and more distant decimal places of the decimal notation of the components of the initial position. The set of points associated with all possible (5.7) with given $i_{-s}, ..., i_0, i_1, ..., \underset{\uparrow}{i_s}$ makes up a region contracting to a point as $s \to \infty$.

It follows that specifying a few first decimal places can determine only finitely many symbols to the right and left of the distinguished one. Meanwhile, for sufficiently large m, the position of the point $T^m x_0$ is determined by these symbols in the notation of x_m, and depends on more distant decimal places of x_0 when m »s.

Above, we discussed the ideal combination of amplification, divergence and subsequent vanishing. However, there is both contraction and expansion. A characteristic example is the mapping from a straight line into a straight line, given in Fig. 7.50. The mapping is expansive everywhere except small regions near the extrema M_1, M_2, ... Chaotization of motion is now different in nature, being due to insufficient stability of existing periodic motions, and not instability itself. Instability necessarily takes place if the stable periodic motions occur in a finite region of the phase space and are represented by sufficiently long phase curves. Meanwhile, even very weak perturbations lead to random jumps of the phase point from some parts of the periodic motions to others. From the viewpoint of physics and computer simulation, chaotic motions are observed here as well as in the first case of global instability.

Bifurcations and routes to chaos and stochasticity 237

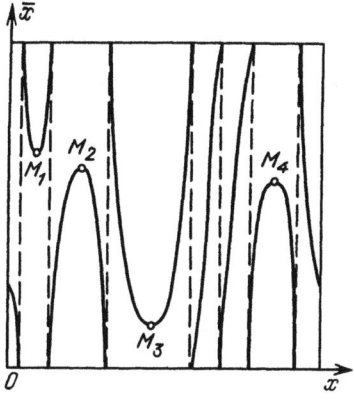

Fig. 7.50

For individual parameter values, such mappings (as in Fig. 7.50) may really be locally unstable, though the instability is, so to speak, ephemeral (a structurally unstable process), and can vanish because of arbitrarily small parameter changes. Nevertheless, instability occurs for individual parameter values. We illustrate by the above-mentioned mapping $\bar{x} = ax(1-x)$ for a=4. The change $x = \frac{1}{2}(1 - \cos 2\pi\theta)$ reduces it to $\bar{\theta} = 2\theta - 2[\theta]$, which coincides with (5.1) for q=2 [9, 143, 534].

We now discuss how divergence goes along with collapse in the phase space. How amplification and its boundedness go side by side is well known; we do not dwell on it here. Of interest are only amplification and collapse, which lead to chaotic and stochastic behaviour.

Stochastic and chaotic motions are based on homoclinic structures which generate instability accompanied by local divergence and global contraction. Meanwhile, the transition from stability to instability requires the vanishing of stable equilibrium states and of stable periodic motions or sufficiently large increase in the periods; more exactly, of the lengths of the corresponding phase curves. Stable periodic motions and equilibrium states can lose stability or vanish only in several specific ways. In this sense, we can speak of different routes to chaos and stochastic behaviour, as described in Sect.1 of the present chapter. Let us list them briefly.

1. Either a stable equilibrium state or a periodic motion merges with the corresponding unstable motion, and then both states vanish.

2. Either an equilibrium state or a periodic motion loses stability,

simultaneously generating a stable periodic motion or a stable two-dimensional toroidal manifold with periodic or quasi-periodic coil, respectively.

3. A stable periodic motion either contracts to a point, generating a stable equilibrium, or merges with an equilibrium, generating a doubly asymptotic curve (being the intersection of its integral manifolds S^+, S^-).

4. A periodic motion loses stability, simultaneously generating a stable periodic motion with period twice as large. The stability loss can be repeated many times, forming an infinite period-doubling bifurcation series.

As a result of the above bifurcations, either a stable equilibrium state or a periodic motion either vanishes or leaves a two-dimensional stable integral torus in its place, which, in turn, can lose stability or vanish in several different ways. Here, we can contrast the case of transformations of a torus as a whole and that of transformations of the periodic motions on the torus, which also eventually lead to torus destruction. As a whole, the two-dimensional stable torus can merge with an unstable two-dimensional one, and vanish; it can become unstable in order to generate a stable "double" two-dimensional torus (only for phase space dimension exceeding three); finally, it can become unstable with simultaneous appearance of a stable three-dimensional torus, which, in turn, can undergo similar changes. Meanwhile, we should recall that, as the dimension of the stable torus increases, chaotic behaviour on the torus may become possible, and need not be accompanied by torus destruction.

This brief description should be completed with bifurcations of periodic motions for resonance frequency ratio 1:3 and 1:4, as well as with resonant bifurcations on the torus.

All possible bifurcations can occur for the periodic motions on the torus. Still, we should draw a distinction between bifurcations on the torus, and outside.

These are possible ways for the vanishing of stable motions, a necessary basis both for the chaotic and the stochastic behaviour of motions of dynamical systems. As such, they do not lead to chaotic motions, but are necessary for chaotization. Moreover, in the regions with neither stable equilibrium states nor periodic motions, chaotization can arise also without the above "prerequisites", and is not due to the replacement of a simple attractor by chaotic motions. The latter's hard appearance can be observed in the region of attraction of an equilibrium state or a periodic motion.

Chaotic motions are generated by a homoclinic structure, leading to the divergence and subsequently "covering" of the phase trajectories. There are very many types of homoclinic structures; as a rule, it is extremely complicated to describe them in full. However, we can confine ourselves to a simplified version. A homoclinic structure includes infinitely many fixed saddles of all possible multiplicities, and their invariant manifolds S^+, S^- meeting each other in a most intricate manner. (S^+, S^- can intersect; and distinct S^+, S^+, as well as distinct S^-, S^-, are pairwise disjoint.) This most complicated intersection network involves considerable interdependence; in particular, the intersection of S^+ with S^- of one or more fixed saddles implies the existence of infinitely many other fixed saddles of greater multiplicity, and the intersection of their S^+, S^-. The interdependence also consists in the fact that other intersections follow from those of S^+, S^-, and, in turn, generate subsequent intersections, etc. It is natural that there arises a hierarchy of fixed saddles of increasing multiplicity, and of still finer and more complicated intersections of the invariant manifolds S^+, S^-. Apparently, similarity to nested structures can be detected. It is to the hierarchy that the inevitability, naturalness and generality of the hierarchy homoclinic structures are related. To a considerable extent, the hierarchy is characterized by its upper level, i.e., by the fixed points of least multiplicities and "first" intersections of their S^+, S^-. Examples are the simplified representations in Figs. 7.44–7.46.

Comparatively simple homoclinic structures arise in transforming a straight line into a straight line, and in the system of Lorenz equations.

After chaos arose, the homoclinic structure generating it can be remodelled, which implies a "chaos-chaos" type restructuring. There is a non-manipulable set of such restructurings. However, as far as a simplified description goes, they can be treated easily, and consist either in the appearance of new intersections of S^+, S^-, or in the reverse process (existing saddles and intersections disappear). Restructuring of the upper level leads to infinitely many reorganizations of the lower levels or is brought about by them. The restructuring is accompanied by the merging or vanishing of infinitely many bifurcations of fixed points and period-doubling bifurcations; we already discussed the matter in Sect.2. Meanwhile, stable fixed points are also generated and vanish. If the periods of the generated and existing stable periodic motions are large enough, then this does not lead to destruction of the chaotic motions of the dynamical system.

CHAPTER 8

QUANTITATIVE CHARACTERISTICS OF STOCHASTIC AND CHAOTIC MOTIONS. SOME UNIVERSAL PROPERTIES IN ORDER-CHAOS AND INVERSE TRANSITIONS

To quantitatively describe stochastic and chaotic motions, such concepts as *probability distribution, correlation function, spectral density, Lyapunov exponents, dimension,* and *entropy* of a stochastic system are used. Some (e.g., spectral density) can easily be obtained from an experiment; others can only be determined by numerical analysis. In a number of cases, the above quantities can be calculated analytically.

1. Statistical characteristics

In Chapt.4, we have said that one basic objection to the existence of stochastic solutions of differential equations was the uniqueness theorem. In fact, in posing the Cauchy problem, the solution should be unique, completely determined by the initial conditions, and therefore completely predictable. Still, how can unpredictability arise? It turns out, that, in investigating stochastic solutions, the Cauchy problem is irrelevant. The problem is never related to the conditions of a (physical or numerical) experiment, for the initial conditions cannot be given absolutely correctly in principle. Therefore, it is appropriate to state the problem in the statistical language. Let a probability distribution close to the delta-distribution be specified at an initial moment. If the distribution is not at wide at subsequent moments, then from the start we can regard it as the delta-function, and consider the Cauchy problem. Thus, then the solution is regular and predictable. Otherwise, when the original probability distribution spreads and gets finite in spread even if the initial distribution tends to the delta-distribution, the solution is stochastic (or chaotic if spreading occurs because of small random perturbations). It immediately follows from the problem so posed that stochastic motions of dynamical systems are a special kind of random processes, it thus being convenient to use the usual statistical characteristics for the description of motion, such as probability distribution, correlation function, spectral density, etc. These processes are of a particular kind: in the absence of random perturbations, the transition probability is always the delta-function; therefore, the correlation function can be expressed in terms of a one-dimensional probability distribution[6].

For dynamical systems described by an exponentially unstable discontinuous mapping from a line segment into itself, the existence of a limit probability density has been proved rigorously [553]. We now qualitatively clarify why this probability density exists. Consider the expansive mapping $\bar{u} = Tu$, selecting u_0 as the initial point. Then $u_1 = Tu_0$, $u_2 = Tu_1$, ... Assuming that u_0 can be specified absolutely exactly, this sequence is determined uniquely, and no statistical description can be spoken of, since there is neither randomness nor indeterminacy. Indeterminacy can be introduced either because of specifying a certain initial probability distribution for the value u_0 or because of some uncontrollable noise destroying the exact transformation T. Since the latter is expansive, even for an arbitrarily narrow initial probability distribution and arbitrarily small noise, randomness should arise in the sequence of points u_0, u_1, u_2, ..., and the statistical description does not depend on the type of the initial distribution and the noise. Therefore, in this formulation, stochasticity of a dynamical system is due to arbitrarily small fluctuations, and is remarkable in that it does not depend on them; the stochasticity is determined by the dynamical system itself.

Generally speaking, the calculation of the limiting probability density is a difficult problem for dynamical systems, and must be done on a computer. In certain most simple cases, it can also be performed analytically. Apparently, S. Ulam and J. von Neumann [670] were the first to calculate the density, by considering the quadratic mapping $\bar{x} = 1 - \mu x^2$, and by showing that, for $\mu = 2$, the limiting probability density exists

---

[6]In this stochastic motions of deterministic dynamical systems differ from "really" random ones; e.g., due to a random external force (reported by R. L. Stratonovich).

and equals $w(x)=1/(\pi\sqrt{1-x^2})$. The work was based on the possibility to reduce the above mapping to the linear one $\bar{y}=2y$ on $0 \leq y \leq 1$ by the transition $x=-\cos(\pi y)$. Since the probability distribution $w(y)$ is uniform for a linear mapping, the formula $w(x)=1/(\pi\sqrt{1-x^2})$ is obtained immediately after transition to the variable x. Below, two other examples of calculating the limiting probability distribution are given. One is the linear damping impact oscillator with negative friction, described by equation (3.4), Chapt.3, for which the point mapping of the variable $y = \dot{x}|_{x=0_+}$ can be calculated easily, and is of form (3.5), Chapt.3. For the phase space of the oscillator in question to have a limiting stochastic set J (a strange attractor), the quantities p, q should satisfy the condition $a(q-1) < p < a$.

The stationary distribution $w(y)$ for the variable y described by $\bar{y}=f(y)$ should satisfy the equation

$$w(y) = \int \delta(y - f(y')) w(y') \, dy'. \tag{1.1}$$

Using properties of the delta-function, we then obtain the linear functional equation

$$w(y) = \sum_{i=1}^{n} w(y_i'(y)) \, Dy_i'(y)/Dy, \tag{1.2}$$

where $y_i'(y)$ is the *i*th root of the equation $f(y')=y$, and $Dy_i'(y)/Dy$ is the Jacobian of the transformation from y to y'. Solving exactly or approximately equation (1.2), we can find $w(y)$, and then calculate the probability distribution in the whole region of the strange attractor. For the system in question, simpler tricks can be used, involving certain relations between the parameters, [227, 228].

All phase trajectories in the (x,y)-space, intersecting the y-axis in the interval $I_0=[y_1,y_0]$, where $y_1=a$, $y_0=qa$, undergo a jump. Assume that

$$1-q^{-1} = q^{-k}(1-q^{-m}), \quad p = qa(1-q^{-k}), \tag{1.3}$$

where k, m are any two integers. The conditions mean that, under a jump, one elementary interval I_0 is transformed to m intervals $I_{k+m-1}=[y_{k+m},y_{k+m-1}], \ldots, I_k=[y_{k+m},y_k]$, where $y_k=y_0 q^{-k}$, i. e., the points $y_0, y_1, \ldots, y_{k+m}$ form a Markov partition of $[y_{k+m},y_0]$ [325]. If relation (1.3) holds, the stationary probability distribution $w(y)$ can be calculated analytically [227, 228]. Simultaneously, the stationary distribution $\tilde{w}(y)$ for the values $y|_{x=0_-}$ can also be found. The results can be written in the form

$$\tilde{w}(y) = \begin{cases} Cq^{\alpha} & (\alpha = 0,1,\ldots,k-1), \\ C\dfrac{1-q^{\alpha+1-k-m}}{q-1} & (\alpha = k-1,k,\ldots,k+m-1), \end{cases}$$

$$w(y) = \begin{cases} C(1-\delta_{\alpha_0})q^{\alpha} & (\alpha = 0,1,\ldots,k), \\ C\dfrac{1-q^{\alpha-k-m}}{1-q^{-1}} & (\alpha = k,k+1,\ldots,k+m-1), \end{cases} \qquad (1.4)$$

where α is the subscript of the interval related to the corresponding value of y (i. e., $y \in I_{\alpha}$); δ_{mn} is the Kronecker delta.

The graphs of $w(y)$, $\tilde{w}(y)$ constructed on the basis of (1.4) for m=3, k=6 (q=1.171, p/a=0.717) are given in Fig. 8.1.

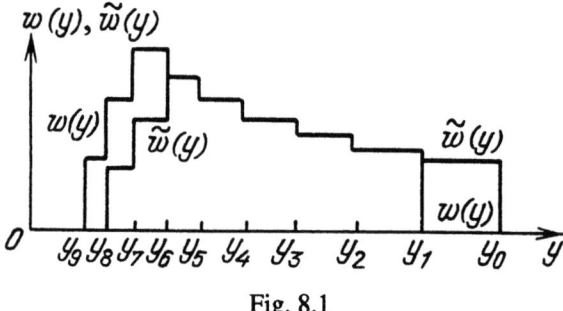

Fig. 8.1

The constant C is determined by a normalization condition, and equals

$$C = a^{-1}\left[1-k+q(k-mq^{-(k+m)})\right]^{-1}.$$

Note that the probability distribution found is completely determined by the dynamical properties of the system in question; therefore, the system is a stochastic oscillation generator.

As another example, consider the piecewise linear mapping [125]

$$\bar{u} = \begin{cases} -\mu+(1+\mu)u & \text{for } u>0, \\ \mu+(1+\mu)u & \text{for } u<0. \end{cases} \qquad (1.5)$$

Its graph is given in Fig. 8.2. The mapping is of the same general form as a one-dimensional mapping for the Lorenz system, and in many cases it is obtainable from the latter by conveniently choosing the variable u.

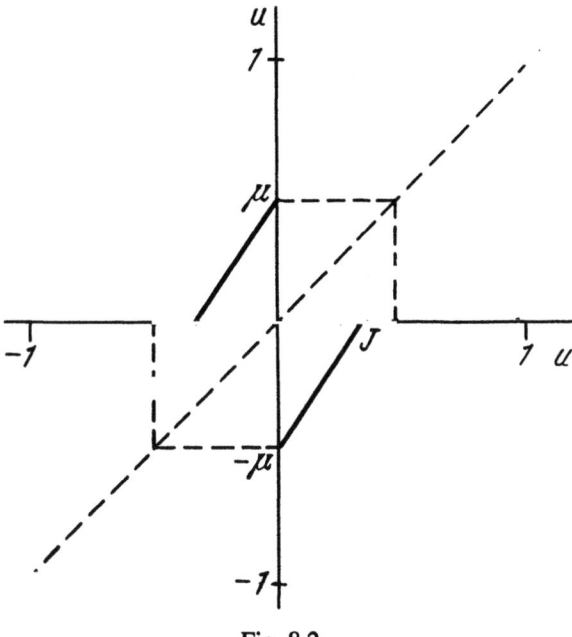

Fig. 8.2

It is easy to see that the mapping (1.5) for $0<\mu<1$ transforms the interval $[-1,1]$ into itself, and is expansive with coefficient $[1+\mu]$. It can be seen from the diagram that the limiting set J fills $[-\mu,\mu]$.

Let $w_0(u)$ be an arbitrary smooth probability density on $[-1,1]$. Under (1.5), it varies according to the recursive relation

$$w_{n+1}(u) = \frac{1}{1+\mu}\left[w_n\left(\frac{u+\mu}{1+\mu}\right) + w_n\left(\frac{u-\mu}{1+\mu}\right)\right] \quad (-\mu<u<\mu) \tag{1.6}$$

(see equation (7.45) in [100]). We assume that all values of u, $(u+\mu)/(1+\mu)$, $(u-\mu)/(1+\mu)$ differ from zero, i.e., from the discontinuity point of point mapping (1.5). The points $(u+\mu)/(1+\mu)$, $(u-\mu)/(1+\mu)$ are inverse images of u, i.e., $(u\pm\mu)/(1+\mu) = T^{-1}u$.

Consider the consecutive points $u_0=\mu$, $u_1=Tu_0$, $u_2=Tu_1$, $u_3=Tu_2$, ... Let, for a certain integer m, $T^m u_0=0$. Add the points $u_{m+1}=-\mu$, $u_{m+2}=Tu_{m+1}$, ..., $u_{2m}=Tu_{2m-1}=0$ to $u_0, u_1, u_2, ..., u_m = 0$. The set $M=\{u_0, u_1, ..., u_{2m}\}$ has the properties that it contains the point of discontinuity $u=0$, and each u_s is sent into $u_{s+1} \neq 0$ by the transformation T if $s \neq m-1$, $s \neq 2m-1$, and into the point $u=0$ if $s=m-1$ or $s=2m-1$.

Consider an arbitrary interval $\Delta_i=[u_i,u_j]$ not containing other points of M; it is transformed by T into the interval $\Delta=[u_{i+1},u_{j+1}]$, which is $1+\mu$ times as long as the original interval. Therefore, Δ can certainly contain other points of M. Thus, $u_0, u_1, ..., u_{2m}$ determine a partition of $[-\mu,\mu]$ into intervals $[u_i,u_j]$, so that each is transformed into one or more of the same intervals. The partition is said to be *Markov* with respect to T [8, 325]. Under the Markov partition, none of the points $u \notin M$ can be sent into a point $u_k \in M$ by the inverse transformation T^{-1}; therefore, the point of discontinuity of T is never involved in (1.6) for such initial points u. Consequently, (1.6) holds for all n and above u. Differentiating (1.6) with respect to u, we have

$$w'_{n+1}(u) = \frac{1}{(1+\mu)^2}\left[w'_n\left(\frac{u-\mu}{1+\mu}\right)+w'_n\left(\frac{u+\mu}{1+\mu}\right)\right]. \quad (1.7)$$

Hence,

$$\sup_u |w'_{n+1}(u)| < \frac{2}{(1+\mu)^2} \sup_u |w'_n(u)|, \quad (1.8)$$

and the limiting probability density is constant in each region of smoothness, or on each $[u_i,u_j]$ of the Markov partition, provided $2/(1+\mu)^2 < 1$, i.e.,

$$\mu > \sqrt{2} - 1. \quad (1.9)$$

This piecewise constant density can be easily found by (1.6), since this equation is reducible to a system of linear equations determining limiting probability densities $w^*_1, w^*_2, ..., w^*_{2m}$ on the Markov partition intervals. In fact, the points $u_0, u_1, ..., u_{2m}$ form 2m intervals of the Markov partition $\Delta_1, \Delta_2, ..., \Delta_{2m}$ on which the probability densities for the *n*th iteration are equal to $w_1^{(n)}, w_2^{(n)}, ..., w_{2m}^{(n)}$, respectively. By (1.6), they are transformed into certain new constant densities $w_1^{(n+1)}, w_2^{(n+1)}, ..., w_{2m}^{(n+1)}$, related to the old formulas by

$$w_i^{(n+1)} = \sum_{j=1}^{2m} p_{ji} w_j^{(n)} \quad (i = 1, 2, ..., 2m), \quad (1.10)$$

where

$$p_{ji} = \begin{cases} 1/(1+\mu) & \text{for } \Delta_i \subset T\Delta_j, \\ 0 & \text{for } \Delta_i \cap T\Delta_j = \emptyset. \end{cases}$$

The graphs of $w^*(u)$, calculated for a number of values μ corresponding to Markov partitions, are given in Fig. 8.3. Note that numerical computations confirmed the periodicity of Markov systems for $\mu < \sqrt{2}-1$, and ergodicity for $\mu > \sqrt{2}-1$ [125].

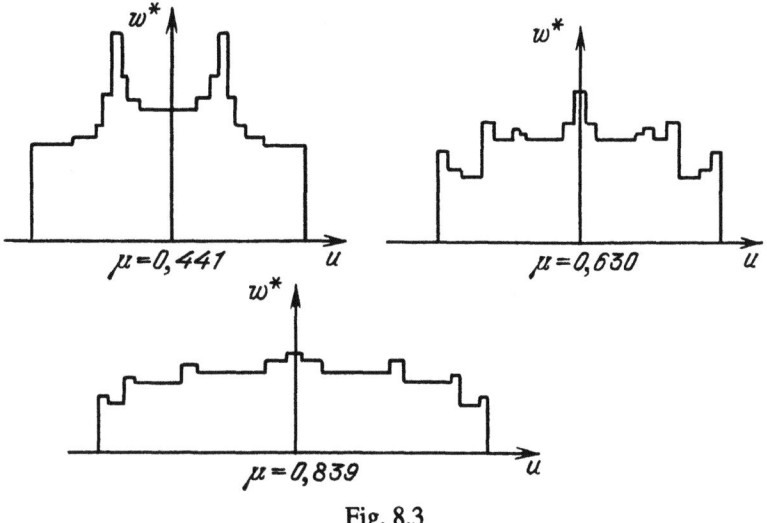

Fig. 8.3

We now retrace our steps in order to analyze the above assumption that

$$T^m\mu = 0 \tag{1.11}$$

for a certain integer m. Solving the equation for the parameter μ with different m, we find μ such that there is a Markov partition. For small m, μ can be determined analytically. We show that the solution set of equations (1.11) for all possible m is everywhere dense in the interval [0,1]. In fact, we derive from (1.5) that

$$T^m\mu = \varepsilon_m\mu + \varepsilon_{m-1}\mu(1+\mu) + \ldots + \varepsilon_0\mu(1+\mu)^m, \tag{1.12}$$

where $\varepsilon_s = -\mathrm{sign}(T^{s-1}\mu)$, $\varepsilon_0 = 1$. Assume that there are no roots in an interval $[\underline{\mu}, \overline{\mu}]$. All the quantities $\varepsilon_1, \varepsilon_2, \ldots, \varepsilon_m$ are then identical, and $\left|\varepsilon_m\mu + \varepsilon_{m-1}\mu(1+\mu) + \ldots + \varepsilon_0\mu(1+\mu)^m\right| \leq \mu$ for all $\mu \in [\underline{\mu}, \overline{\mu}]$. The inequality means that

$$\left|\varepsilon_0 + \varepsilon_1\frac{1}{1+\mu} + \ldots + \varepsilon_m\frac{1}{(1+\mu)^m}\right| \leq \frac{1}{(1+\mu)^m} \tag{1.13}$$

for any m. It follows that the power series $\varepsilon_0 + \varepsilon_1\frac{1}{1+\mu} + \varepsilon_2\frac{1}{(1+\mu)^2} + \ldots$ converges for $\mu > 0$, and is identically zero on $[\underline{\mu}, \overline{\mu}]$; therefore, all its coefficients $\varepsilon_0, \varepsilon_1, \ldots$ should be zero, which is impossible. Hence, there are roots of (1.11) on any $[\underline{\mu}, \overline{\mu}]$, and the Markov partitions of $[-\mu, \mu]$, related to (1.5), are everywhere dense relative to parameter μ.

Other examples of calculating the stationary probability distribution for systems described by one-dimensional and two-dimensional mappings (of the form (3.10),

Chapt.7) are given in [481, 532, 536].

We already noted above that knowledge of the limiting probability distribution $w(x)$ for the dynamical system described by the vector x enables us to calculate in principle the correlation function of the process $x(t)$, and the spectral density related to this function by Khintchine's theorem [362]. The function decreases for stochastic motions in positive time. In some cases, this decrease is exponential [99, 343, 422]. However, it is often of power character [97].

The spectral densities $S(\omega)$ are simplest in order to quantitatively describe stochastic motions of a dynamical system. Measuring $S(\omega)$ experimentally or on a computer, we can rather precisely detect the onset of chaotic behaviour, and also the bifurcations occurring as the parameters change smoothly. Thus, in the case of periodic motion, the spectrum consists of a number of equidistant lines at frequencies ω, 2ω, 3ω, ... (e.g., as in Fig. 8.4a, where the experimental velocity spectra in thermal convection in a layer of water are represented [469]; R is the Rayleigh number; R_c, the critical value of R for which convection occurs, $f_2=\omega/2\pi$). Under period-doubling bifurcations, additional lines appear at the frequencies $\omega/2$, $3\omega/2$, $5\omega/2$, ... (Fig. 8.4b), $\omega/4$, $3\omega/4$, $5\omega/4$, ... (Fig. 8.4c). In the transition to a quasiperiodic motion, another incommensurate frequency ω_2 appears in addition to the fundamental one ω_1, as well as linear combinations of them (e.g., see Fig. 8.5, where the experimental temperature power spectra are represented for thermal convection in mercury [443], $f=\omega/2\pi$). In chaotic behaviour, along with a discrete spectrum, background first appears to increase gradually; finally, the spectrum becomes continuous (Fig. 8.4 d,e). Changes in the spectrum are governed by certain laws which will be partly discussed in the sequel. Many concrete examples of spectra in transition to chaos and in chaotic and stochastic oscillation regimes are given in the next chapter.

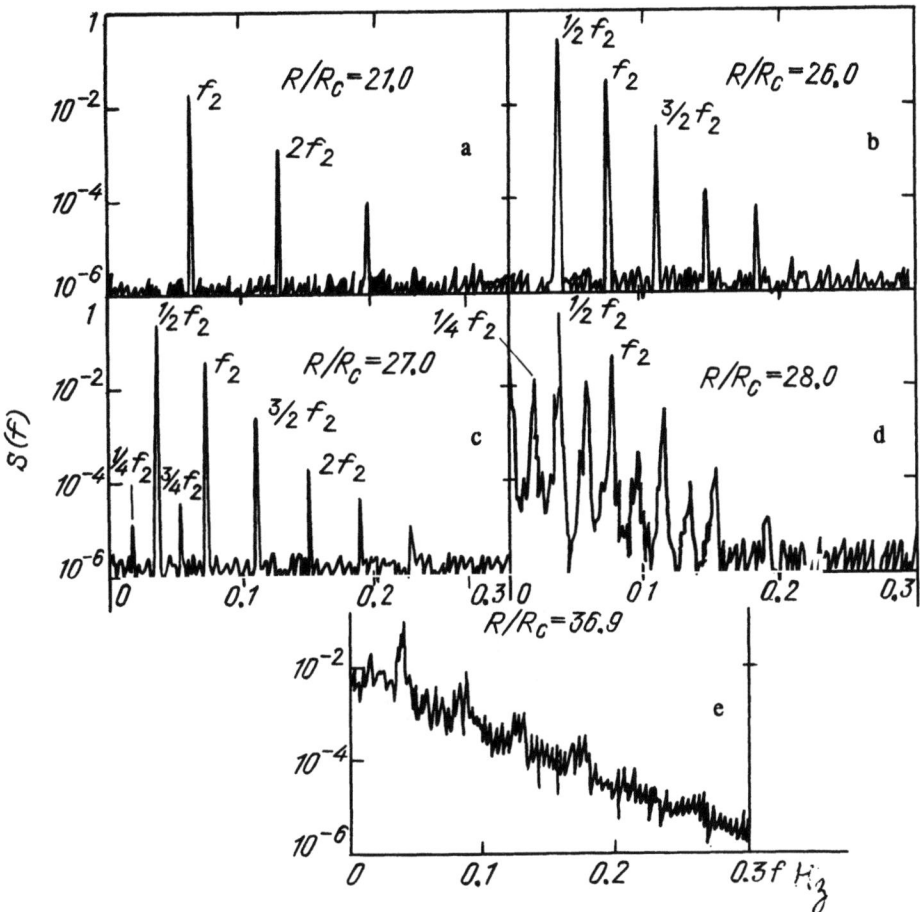

Fig. 8.4

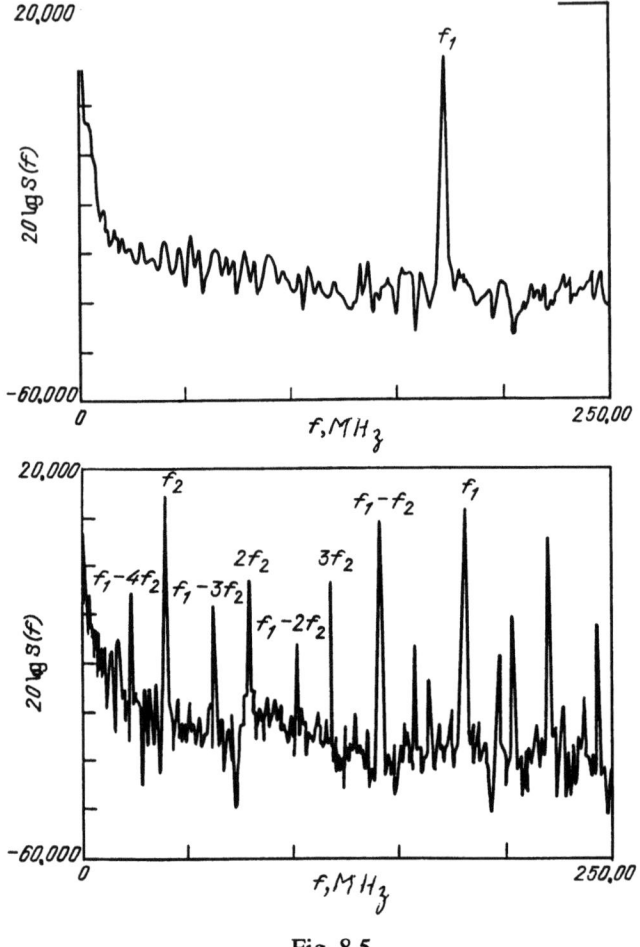

Fig. 8.5

In conclusion, we consider the effect of random perturbations on the invariant limiting probability density. Assume that the mapping T sends points x to $\bar{x}$, while the random perturbation replaces $\bar{x}$ by a point y with probability density $f(\bar{x},y)$. The consecutive probability densities $w_s(x)$ and $w_{s+1}(y)$ are related by

$$w_{s+1}(y) = \sum \int w_s(T^{-1}x) \left| \frac{dT^{-1}x}{dx} \right| f(x, y) \, dx ,$$

where summation is done for all values of the inverse mapping T^{-1}, and integration with respect to the domain of T, transformed by T into itself.

According to this formula, the limiting probability density $w^*(x)$ depends both on the point mapping T and the noise probability density $f(x, y)$. However, according to the above, the correlation can be different, according as whether we are in the case of a stochasticity transformation, or amplification, or generation. Below, we give the data of mathematical simulation carried out by I. S. Gelfer with the mappings associated with stochasticity amplifiers and generators, and described by the formulas $\bar{x} = ax(1-x)$ and $\bar{x} = \mu \operatorname{sign} x + (1+\mu)x$.

The probability density f(x, y) selected was

$$f(x, y) = \begin{cases} \gamma e^{-\upsilon(x-y)^2} & \text{for } |x-y| < h, \\ 0 & \text{for } |x-y| \geq h, \end{cases}$$

where υ, h are given and γ is determined by a normalization condition.

The mathematical simulation consisted in successive computations by the formula for $w_{s+1}(y)$ of the probability densities $w_1(x)$, $w_2(x)$, $w_3(x)$, ... for the above transformations T and functions f(x,y), and in subsequently reducing the data obtained. Meanwhile, the difference of the probability densities calculated from the limiting one $w^*(x)$ was estimated by the quantity

$$\rho_s = \int |w_s(x) - w^*(x)| \, dx.$$

T	α	κ
μ = 0,618	0,03	0,57
μ = 0,839	0,09	0,5
a = 3,92	0,32	0,29
a = 3,98	0,2	0,3

Table 8.1

It was discovered that $\rho_s \approx \alpha e^{-\kappa s}$. The values α, κ are given in Table 8.1 for above a and μ, respectively. For the same values of T, a, μ, the deviations $\rho = \int |w^*(x) - \tilde{w}^*(x)| \, dx$ of $w^*(x)$ and $\tilde{w}^*(x)$ in the absence and in the presence of random perturbations are given in Table 8.2; the random perturbation in the table characterizes the perturbation dispersion D.

μ=0,618		μ=0,839		a=3,92		a=3,98	
D	ρ	D	ρ	D	ρ	D	ρ
0,2·10⁻⁶	0,0036	0,3·10⁻⁶	0,003	10⁻⁷	0,13	10⁻⁷	0,12
0,2·10⁻⁴	0,027	0,8·10⁻⁵	0,013	0,5·10⁻⁵	0,22	0,5·10⁻⁵	0,20
0,5·10⁻²	0,11	0,2·10⁻²	0,089	0,5·10⁻⁴	0,32	0,5·10⁻⁴	0,29

Table 8.2

It can be seen how different noise levels are, which cause considerable changes in the limiting probability density. The differences can be greater, since "intrinsic computer perturbations" are inevitable in machine computation.

2. Lyapunov exponents. Dimension and entropy of a stochastic attractor

We have already repeatedly said that all trajectories that form a stochastic attractor are Lyapunov unstable motions. Therefore, the trajectories should have at least one positive Lyapunov exponent. That there is a positive exponent is one basic criterion to decide whether a motion is stochastic.

The Lyapunov exponents are determined as follows [101, 240, 297]. Let a dynamical system be described by the equation

$$\dot{x} = F(x), \qquad (2.1)$$

where x is a vector with components $x_1, x_2, ..., x_n$. Consider two adjacent trajectories $x(t)$, $x_1(t)$ emanating from points x_0, x_{10}, supposing that $y(t)=x_1(t)-x(t)$, $(y(0)=x_{10}-x_0)$. If the trajectories are nearby, then the evolution of the vector $y(t)$ can be given by the linearized equation

$$\dot{y} = \left.\frac{\partial F}{\partial x}\right|_{x=x(t)} y, \qquad (2.2)$$

where $\partial F/\partial x$ is the matrix with entries $\partial F_i/\partial x_j$. The system of equations (2.2) is known to have n fundamental particular solutions $y_i(t)=e_i(t)$ such that

$$\lambda_i(x_0) = \lim_{t \to \infty} \frac{1}{t} \ln \frac{|e_i(t)|}{|e_i(0)|} \qquad (2.3)$$

exists for each i. In general, λ_i are different. The quantities $\lambda_i(x_0)$ are called *Lyapunov exponents* [240]. It is easy to see that, for any initial vector $y(0)$ in general position, the quantity

$$\lambda(x_0) = \lim_{t \to \infty} \frac{1}{t} \ln \frac{|y(t)|}{|y(0)|}$$

equals $\lambda_1(x_0)$, or the maximal Lyapunov exponent.

Direct computation by (2.3) for systems with exponentially unstable trajectories is practically impossible, because, even for very small $|y(0)|$, $|y(t)|$ increases indefinitely as t increases, which leads to overflow of the computer and to substantial errors. To overcome the difficulties, G. Benettin et al. [397–400] offered the following algorithm for computing Lyapunov exponents: First of all, we compute λ_1, for which we select a certain trajectory x(t) on the interval $0 \le t \le T$, a solution of (2.1), densely covering the whole attractor. We solve (2.2) together with (2.1) in the vicinity of x(t) for a certain initial condition y_0 such that $|y_0|=1$. After some given time τ, the vector y(t) takes a value $y_1 = y(\tau)$. Denote the length of the vector y_1 by d_1. Furthermore, extend the solution of (2.2), proceeding from another initial condition $y_{10} = y_1/d_1$. Let the value of y(t) at the moment $t = 2\tau$ be y_2. Denote the length of the vector y_2 by d_2. As a result of repeatedly applying the above procedure (Fig. 8.6), we obtain a sequence of the numbers d_i, where i=1, 2, ..., m=T/τ. The maximal exponent λ_1 equals

$$\lambda_1 = \lim_{m \to \infty} \frac{1}{m\tau} \sum_{i=1}^{m} \ln d_i. \qquad (2.4)$$

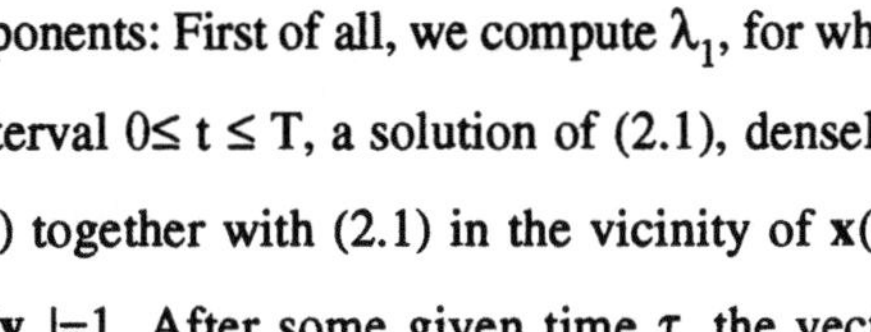

Fig. 8.6

Note that, for the trajectories on the attractor, λ_i do not depend either on the initial point x_0 on the selected trajectory (which follows from the random choice of the initial moment) or the trajectory itself.

To calculate the remaining Lyapunov exponents, [397–400, 647] suggest the use of a similar procedure but with Gram-Schmidt orthogonalization. To clarify, consider the calculation of the next Lyapunov exponent $\lambda_2 \le \lambda_1$. Denote the vectors y_i, y_i/d_i found in calculating λ_1 by $w_i^{(1)}$, v_i^1, respectively ($v_i^{(1)} = w_i^{(1)}/d_i$). The initial vector for (2.2) is $v_0^{(2)}$ orthogonal to $v_0^{(1)}$, i.e., satisfying the condition $(v_0^{(2)} v_0^{(1)}) = 0$. After time τ, $v_0^{(2)}$ is transformed to $w_1^{(2)}$. Make up a linear combination of $w_1^{(2)}$, $v_0^{(1)}$ so that it is orthogonal to $v_1^{(1)}$, e.g., put $u_1^{(2)} = w_1^{(2)} + \beta v_1^{(1)}$, where β is an undetermined multiplier, and require that $(u_1^{(2)} v_1^{(1)}) = 0$. Hence, $\beta = -(w_1^{(2)} v_1^{(1)})$. As the initial vector for the second step we take

$v_1^{(2)} = u_1^{(2)}/d_1^{(2)}$, where $d_1^{(2)} = |u_1^{(2)}|$. Proceeding similarly at each ith step, we find all $d_i^{(2)}$. The Lyapunov exponent λ_2 is determined by the formula

$$\lambda_2 = \lim_{m \to \infty} \frac{1}{m\tau} \sum_{i=1}^{m} \ln d_i^{(2)}. \qquad (2.5)$$

To calculate the jth one, orthogonalization at each ith step should be carried out relative to the vectors $v_i^{(1)}, v_i^{(2)}, \ldots, v_i^{(j-1)}$, for which we transform the vector

$$u_i^{(j)} = w_i^{(j)} + \beta_1 v_i^{(1)} + \beta_2 v_i^{(2)} + \ldots + \beta_{j-1} v_i^{(j-1)},$$

and require that $(u_i^{(j)} v_i^{(1)}) = 0$, $(u_i^{(j)} v_i^{(2)}) = 0$, ..., $(u_i^{(j)} v_i^{(j-1)}) = 0$. Hence, $\beta_k = -(w_i^{(j)} v_i^{(k)})$ and

$$u_i^{(j)} = w_i^{(j)} - \sum_{k=1}^{j-1} (w_i^{(j)} v_i^{(k)}) v_i^{(k)}.$$

The Lyapunov exponent λ_j is found by the relation

$$\lambda_j = \lim_{m \to \infty} \frac{1}{m\tau} \sum_{i=1}^{m} \ln d_i^{(j)}, \qquad (2.6)$$

where $d_i^{(j)} = |u_i^{(j)}|$. Thus, we can calculate all n Lyapunov exponents consecutively or simultaneously[7].

We now define the entropy and dimension of a stochastic attractor. We first introduce the concept of *topological entropy* [380] of a dynamical system described by differential equations as follows: Assume that we can distinguish between points of the phase space at a distance greater than a certain quantity $\varepsilon > 0$. Consider the flow of trajectories emanating from a neighbourhood of radius ε, of the initial point, i.e., initially indistinguishable. Denote the number of distinguishable trajectories at a moment t by $N(\varepsilon, t)$. Topological entropy is

$$h = \lim_{\varepsilon \to 0} \lim_{t \to \infty} \ln(N(\varepsilon, t))/t.$$

It characterizes the rate of divergence of adjacent phase trajectories. If the divergence is not sufficiently rapid (e.g., as a power function), then h=0; otherwise, h>0.

Such quantitative characterictics of stochastic motions as *dimension* or *metric*

---

[7] Note that the above method is based on Oseledets' theorem for generalized Lyapunov exponents [297].

entropy to be described in the sequel are, strictly speaking, only related to generators of stochastic oscillations. The notions can to some extent also be used for stochasticity amplifiers with negligibly small chaotic behaviour threshold if the small random perturbations are fixed.

We have already indicated that there must be a limiting probability density w(x) for the process x(t) describing stochastic generator oscillations. Cover the attractor with $N(\varepsilon)$ small cubes of edges $\varepsilon_i \leq \varepsilon$. Denote the *i*th cube by ε_i. The probability that the representing point will be in ε_i is

$$p_i = \int_{\varepsilon_i} w(x)dx .$$

According to C. Shannon [642], the system's entropy is

$$H(\varepsilon) = -\inf_{\varepsilon_i} \sum_{i=1}^{N(\varepsilon)} p_i \log_2 p_i . \quad (2.7)$$

It characterizes the indeterminacy that the representing point be in ε_i. If the cover is refined, $H(\varepsilon)$ increases indefinitely. However, the limit

$$\sigma = \lim_{\varepsilon \to 0} H(\varepsilon)/\log_2 \varepsilon^{-1} \quad (2.8)$$

may exist, called the *information dimension* of the attractor [440, 442, 473, 474, 476, 477, 502, 610, 689]. In the case where (2.8) does not exist, the concept of the *greatest lower bound*, $\underline{\sigma} = \lim_{\varepsilon \to 0} \inf(H(\varepsilon)/\log_2 \varepsilon^{-1})$, and of the *least upper bound* of information dimension, $\overline{\sigma} = \lim_{\varepsilon \to 0} \sup(H(\varepsilon)/\log_2 \varepsilon^{-1})$, are introduced.

The entropy $H(\varepsilon)$ attains its maximum value if $p_1 = p_2 = ... = 1/N(\varepsilon)$, the latter being equal to $\log_2 N(\varepsilon)$. Then σ attains its maximum

$$d = \lim_{\varepsilon \to 0} \log_2 N(\varepsilon)/\log_2 \varepsilon^{-1}, \quad (2.9)$$

called the *fractal dimension,* or *capacity,* of the attractor [442, 570, 604, 688, 689]. If (2.9) does not exist, then the concepts of greatest *lower capacity* $\underline{d}$, least *upper capacity* $\overline{d}$ are introduced similarly.

If the attractor is a point, line or two-dimensional surface, then d is 0, 1 or 2, respectively. However, if the attractor is complicated, then d can be a fraction. Thus, for the one-dimensional Cantor set depicted in Fig. 8.7, $\varepsilon = (1/3)^n$, $N(\varepsilon) = 2^n$; therefore,

$$d = \lim_{n \to \infty} \log_2 2^n / \log_2 3^n = \log_2 2 / \log_2 3 = 0.6309 \cdots .$$

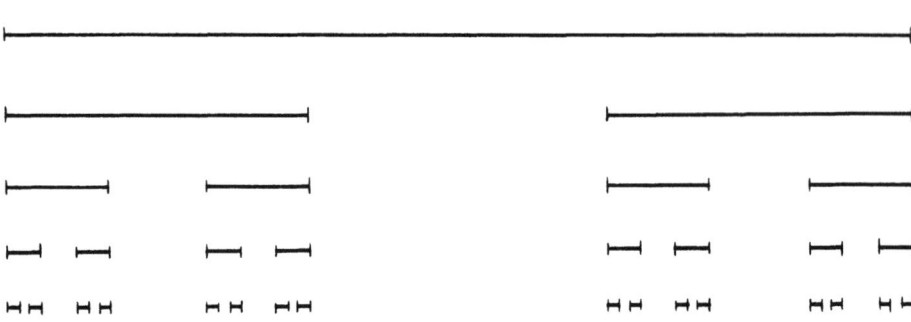

Fig. 8.7

We stress that the measure of the Cantor set on the straight line is zero, whereas the capacity is non-zero, and a fraction.

The information and fractal dimensions defined above are not unique. The Hausdorff dimension is also widely known [493]. It is defined as follows: Let a set of points be given in an n-dimensional space, Cover its n-dimensional cubes of edges $\varepsilon_i \leq \varepsilon$, and determine the quantity $l_D = \lim_{\varepsilon \to 0} l_2(\varepsilon)$ where

$$l_D(\varepsilon) = \inf \sum_i \varepsilon_i^D.$$

Hausdorff showed that there is a critical value of D with $l_D = 0$ greater than it, and $l_D = \infty$. The critical value $D = d_H$ for which the quantity l_D called *Hausdorff measure* is finite is called the *Hausdorff dimension*. It is easy to show that $d_H \leq d$. In fact, if we select cubes with the same edge ε, then $\tilde{l}_d(\varepsilon) = N(\varepsilon)\varepsilon^d \geq l_d(\varepsilon)$ (due to the operation inf). As $\varepsilon \to 0$, by (2.9), $\tilde{l}_d(\varepsilon) \to 1$; therefore, $l_d \leq 1$. Hence, $d \geq d_H$ (in the general case, $\bar{d} \geq \underline{d} \geq d_H$).

Calculating the Hausdorff dimension and even attractor capacity by formula (2.9) directly as the number of degrees of freedom of a dynamical system increases is very complicated [480]. Meanwhile, the calculation of the so-called *correlation dimension* ν [473, 474, 476, 477, 479, 502, 610] turns out to be considerably simpler; it coincides with the averaged *pointwise dimension* $\rho_i = \lim_{\varepsilon \to 0} \ln p_i / \ln \varepsilon_i$. The quantity ν is determined by means of the correlation integral

$$C(\varepsilon) = \lim_{N \to \infty} \frac{1}{N^2} \sum_{i,j=1}^{N} \vartheta(\varepsilon - |x_i - x_j|), \qquad (2.10)$$

where $\vartheta(z)$ is the Heaviside function, x_i the vector describing the position of the representing point in the phase space at moment $t_0 + i\tau$, τ a certain given time interval, and

N the number of samples. The quantity $C(\varepsilon)$ determines the relative number of pairs of points distant from each other at most at ε. For small ε, $C(\varepsilon) \sim \varepsilon^{\nu}$. Hence, $\nu = \lim_{\varepsilon \to 0}[\ln C(\varepsilon)/\ln(\varepsilon)]$. Paper [474] pays attention to the fact that the dimensions σ, d, and ν are particular instances of the so-called *generalized dimension* $d_q = \lim_{\varepsilon \to 0}[I_q(\varepsilon)/\ln \varepsilon^{-1}]$, where

$$I_q(\varepsilon) = (1/(1-q))\ln \sum_{i=1}^{N(\varepsilon)} p_i^q$$

is the order-q Rényi entropy [612]. It is easy to see that $d = d_q|_{q=0^+}$, $\sigma = \lim_{q \to 1} d_q$, $\upsilon = d_q|_{q=2}$. Since, for $q \geq 0$, $I_q \leq I_{q'}$ when $q > q'$, we have $\nu \leq \sigma \leq d$. Thus, the value ν can be used to give a lower estimate to the fractal dimension of an attractor.

To determine ν, it suffices to only know one realization of the process $x(t)$ from which the quantity can be computed numerically; therefore ν is very convenient for investigating the results of experimental study. Moreover, it is shown that ν can be also computed in the case where small noise affects the dynamical system, somewhat distorting its behaviour [459, 475, 479]. Meanwhile, it turns out that, for small ε, comparable with noise variance, $C(\varepsilon) \sim \varepsilon^n$, where n is the phase space dimension; for ε exceeding the variance, $C(\varepsilon) \sim \varepsilon^{\nu}$, where $\nu < n$. The characteristic dependence of $\ln C(\varepsilon)$ on $\ln \varepsilon$ is demonstrated in Fig. 8.8.

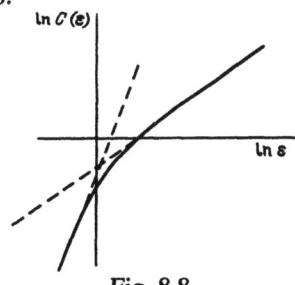

Fig. 8.8

It is easy to see that, for homogeneous attractors, i.e., such that, all pointwise dimensions ρ_i are equal, all dimensions d_q coincide. For inhomogeneous attractors, d_q can be substantially different. To characterize the dimension spectrum, it is convenient to introduce the function $f(\alpha)$ connected with d_q by the relation $d_q = 1/q^{-1}(\alpha_q q - f(\alpha_q))$ [698], where α_q is determined by the condition $df(\alpha)/d\alpha|_{\alpha=\alpha_q} = q$. The approximate form of $f(\alpha)$ is illustrated in Fig. 8.9.

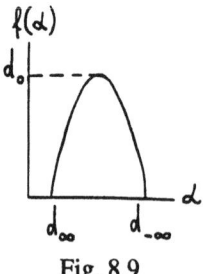

Fig. 8.9

In addition to the above, dimensions calculated in terms of the Lyapunov exponents $\lambda_1, \lambda_2, ..., \lambda_n$ are often made use of.

The Kaplan-Yorke formula [540, 554] determining the so-called *Lyapunov dimension* is

$$d_L = j + \sum_{i=1}^{j} \lambda_i / |\lambda_{j+1}|, \qquad (2.11)$$

where all λ_i are in descending order, $\lambda_1 \geq \lambda_2 \geq ... \geq \lambda_n$, and j is determined by the conditions $\lambda_1+\lambda_2+...+\lambda_j \geq 0$, $\lambda_1+\lambda_2+...+\lambda_j+\lambda_{j+1} < 0$. In a three-dimensional autonomous dissipative system, only one Lyapunov exponent can be positive, since another is zero and the sum of all the three should be negative. Therefore, for such a three-dimensional system, (2.11) is of the form

$$d_L = 2 + \lambda_1 / |\lambda_3|. \qquad (2.12)$$

Accordingly, $|\lambda_3| > \lambda_1$, $2 < d_L < 3$. If the contraction rate for the phase volume at all points of the space is the same, equal to div $\dot{x} = -\alpha$, then $\lambda_1 - |\lambda_3| = -\alpha$. To use (2.12), it then suffices to know only one positive Lyapunov exponent λ_1.

In [688, 689], another formula to estimate the dimension of an attractor was offered, viz.,

$$D_L = k + \sum_{i=1}^{k} \lambda_i / |\lambda_n|. \qquad (2.11')$$

It also involves the Lyapunov exponents, and k is the number of non-negative ones. It was shown that $D_L \leq \underline{d} \leq \overline{d} \leq d_L$. It is easy to see that, for three-dimensional systems or those described by two-dimensional point mappings, formula (2.11') coincides with (2.12); therefore, $D_L = \underline{d} = \overline{d} = d_L$, which is also corroborated by the data of [628], in which the quantities d_L and d are numerically compared by considering the Hénon [500], $x_{n+1} = y_n + 1 - ax_n^2$, $y_{n+1} = bx_n$, the Kaplan-Yorke [540], $x_{n+1} = 2x_n \{\text{mod} 1\}$, $y_{n+1} = \alpha y_n + \cos 4\pi x_n$, and Zaslavski [690],

$$x_{n+1} = [x_n + v(1+\mu y_n) + \varepsilon v\mu \cos 2\pi x_n]\{\text{mod } 1\},$$

$$y_{n+1} = e^{-\Gamma}(y_n + \varepsilon \cos 2\pi x_n), \text{ where } \mu = (1-\exp(-\Gamma))/\Gamma,$$

mappings, the latter taking dissipation into account, and also the Rabinovich-Fabrikant equations [315] (see (4.31), Chapt.9). The results are given in Table 8.3.

System	d_L	d
Hénon		
$a = 1{,}2, b = 0{,}3$	$1{,}200 \pm 0{,}001$	$1{,}202 \pm 0{,}003$
$a = 1{,}4, b = 0{,}3$	$1{,}264 \pm 0{,}002$	$1{,}261 \pm 0{,}003$
Kaplan-Yorke		
$\alpha = 0{,}2$	$1{,}4306766$	$1{,}4316 \pm 0{,}0016$
Zaslavsky		
$\Gamma = 3{,}0, \varepsilon = 0{,}3, v = 400/3$	$1{,}387 \pm 0{,}001$	$1{,}380 \pm 0{,}007$
Rabinovich-Fabrikant	$2{,}317 \pm 0{,}001$	$2{,}318 \pm 0{,}002$

Table 8.3

Mori's formula determines the quantity

$$d_M = k + \sum_{i=1}^{m} \lambda_i^+ \Big/ \sum_{i=1}^{l} |\lambda_i^-| \qquad (2.13)$$

[586], where k is the number of non-negative Lyapunov exponents and m, l are the numbers of positive and negative ones, respectively. In the general case, $d_M \leq D_L \leq d_L$. However, for three-dimensional systems, $d_M = d_L$; for infinite-dimensional systems, $d_M = k$, thus coinciding with the number of instability directions [439].

The knowledge of the dimension of an attractor is quite essential in many physical problems. First of all, it enables us to estimate the minimum number of dynamical variables by which the motion on the attractor can be described in principle. This is especially important for continuous systems. As a rule, the Hausdorff dimension of an attractor for the latter is finite [51, 165, 489, 571]. As is known, any compact set with finite Hausdorff dimension d_H can be projected into a hyperplane of dimension not exceeding $2d_H+1$, using a one-to-one mapping [571]. Besides, in certain cases, a Hausdorff type dimension is directly involved in expressions for measurable physical manifestations. Apparently, it was L.Richardson who first discovered the above in his attempt to measure the irregular coastline of England [614]. Replacing the coastline by a broken line with each segment ε long, he found that the whole length $L_\varepsilon=N\varepsilon$ increased indefinitely as ε decreased. However, it turned out that the quantity $\lambda=N\varepsilon^d$, where $d>1$, then remained unaltered [57, 570]. Thus, we can write $L_\varepsilon=\lambda\varepsilon^{1-d}$. If we take two parts of the coastline approximated by line segments $L_{1\varepsilon}$, $L_{2\varepsilon}$, then their ratio is λ_1/λ_2, which is independent of ε. Therefore, different coastline parts can be compared using λ, and λ has the meaning of Hausdorff measure. The quantity d is considered in the sense of the Hausdorff dimension. For the coastline of England, $d=1.24$; for Australia, $d=1.13$. Another example is the problem of determining the efficient conductance of a three-dimensional anisotropic medium, considered by A. M. Dykhne, with chaotic directions of maximum and minimum conductances, so that, on the average, the medium is isotropic. Finally, knowledge of a dimension often helps to identify the type of motion in the system, and the transitions from one to another.

The metric entropy K [205], introduced by A. N. Kolmogorov [202, 203] and by Ya. G. Sinai [324], is defined by

$$K=\lim_{\varepsilon\to 0}\lim_{\tau\to 0}\lim_{N\to\infty}\left\{-\frac{1}{N\tau}\sum_{i_1,\dots,i_N} p(i_1,\dots,i_N)\ln p(i_1,\dots,i_N)\right\}, \qquad (2.14)$$

where $p(i_1, i_2, \dots, i_N)$ is the probability that the point $x(t=0)$ in the represented phase trajectory on the attractor will belong to the cube ε_{i_1}; $x(t=\tau)$, to ε_{i_2}, ..., etc.; $x(t=(N-1)\tau)$, to ε_{i_N}. Metric entropy is a particular case of the order -q Rényi entropy

$$K_q=\lim_{\varepsilon\to 0}\lim_{\tau\to 0}\lim_{N\to\infty}\left\{\frac{1}{N\tau(1-q)}\ln\sum_{i_1,\dots,i_N} p^q(i_1,\dots,i_N)\right\}. \qquad (2.15)$$

We can show that (2.14) is the limit of expression (2.15) as $q\to 1$. Note that K never exceeds the topological entropy [382, 426].

To calculate the metric entropy, it is convenient to use a formula relating it to Lyapunov exponents. Ya. B. Pesin [301] established that on the attractor K is the sum of all m positive Lyapunov exponents

$$K=\sum_{i=1}^{m}\lambda_i^+. \qquad (2.16)$$

If the dimension of the system is great or infinite, then finding the numerical values of Lyapunov exponents, as well as computing σ, d and K directly, is complicated. Therefore, of interest is a comparatively simple computational procedure enabling us to estimate the exponents, dimension of an attractor, and metric entropy if we only possess the time series of one phase space coordinate. Such a procedure was suggested by N. Packard [600] and F. Takens [657] and its use is especially convenient when there are experimental results for continuous systems, where to know the whole infinite-dimensional vector x(t) is just impossible [681].

Let $x_1, x_2, ..., x_n$ be consecutive values of one coordinate x(t) of the phase space of the system with time intervals τ, or $x_i = x(i\,\tau)$. A new dynamical system of dimension m can be constructed from them by taking $y_i^{(m)} = \{x_i, x_{i+1}, ..., x_{i+m-1}\}$ as the ith value of the vector $y^{(m)}$ which describes the point position in the new phase space. Takens' theorem can be formulated as follows: For almost any observable time series of x(t) and time delay τ, the attractor of the system constructed has the same properties (e.g., the same dimension and spectrum of Lyapunov exponents) as the original one if only $m \geq 2d_H + 1$, where d_H is the Hausdorff dimension of the original attractor. It is a consequence of Mañé's theorem [571].

Note that $x(t), \dot{x}(t), \ddot{x}(t), ..., x^{(m-1)}(t)$ can be taken instead of $x_i, x_{i+1}, ..., x_{i+m-1}$.

In some way or other, we can calculate the attractor's dimension and the entropy of the system constructed for each m. First, as m increases, the dimension also increases, and entropy changes somehow; both then attain certain constant values, which can be assumed to be the original dimension and entropy.

When using this procedure, the estimates of the dimension of the attractor and of the metric entropy become especially convenient if we employ the correlation exponent ν and the order-2 Rényi entropy [478]. To calculate ν by specifying the time series of one phase space coordinate x(t), dynamical systems of different dimensions m are constructed, and the correlation integrals

$$C_m(\varepsilon) = \lim_{N \to \infty} \sum_{i,j=1}^{N} \frac{1}{N^2}\, \vartheta\!\left(\varepsilon - \left|y_i^{(m)} - y_j^{(m)}\right|\right) \qquad (2.17)$$

are calculated.

The quantity m should be increased until the slope of the dependence $\ln C_m(\varepsilon)$ on $\ln \varepsilon$ is constant.

To estimate the entropy K_2, somewhat different correlation integrals

$$C_m^{(s)}(\varepsilon) = \lim_{N \to \infty} \sum_{i,j=1}^{N} \frac{1}{N^2} \prod_{k=0}^{s-1} \vartheta\left(\varepsilon - \left|y_{i+k}^{(m)} - y_{j+k}^{(m)}\right|\right) \quad (2.18)$$

should be found for sufficiently large m. It is easy to see that

$$C_m^{(s)}(\varepsilon) = \sum_{i_1, i_2, \ldots, i_s} p^2(i_1, i_2, \ldots, i_s),$$

where $p(i_1, i_2, \ldots, i_s)$ are the probabilities involved in expression (2.15), with the consequence that

$$C_m^{(s)}(\varepsilon) = \exp(-s\tau K_2^{(s)}(\varepsilon)), \quad (2.19)$$

where

$$K_2^{(s)}(\varepsilon) = -\frac{1}{s\tau} \ln \sum_{i_1, i_2, \ldots, i_s} p^2(i_1, i_2, \ldots, i_s).$$

Writing a similar expression for $C_m^{(s+1)}(\varepsilon)$ and taking into account that $K_2^{(s)}(\varepsilon) = K_2^{(s+1)}(\varepsilon) \approx K_2$ for sufficiently large s, we get

$$K_2 \approx K_{2s} = \frac{1}{\tau} \ln \frac{C_m^{(s)}(\varepsilon)}{C_m^{(s+1)}(\varepsilon)}. \quad (2.20)$$

Numerical computations [478, 479] show that, as s increases, the quantity K_{2s} first decreases and then tends to a constant identifiable with K_2 (see Fig. 8.9 for the dependence of K_{2s} on s for the system (9.3), (9.21), Chapt.9, when $\tau=23$ [478]). It follows from (2.14), (2.15) that $K_2 \leq K$; therefore, the positivity of K_2 can be a reliable criterion of stochastic behaviour of motion.

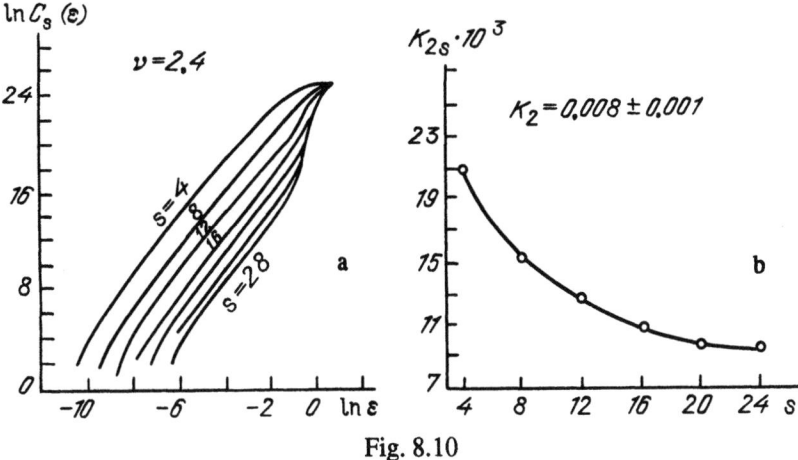

Fig. 8.10

Another method for estimating the dimension of an attractor by the Packard-Takens procedure (generalized to the case of several measurable variables) was offered in [238]. Meanwhile, it was suggested that the number of variables uniquely determining the system's state on the attractor should be used as the dimension n called by the authors the *embedding dimension*. Let the values $y_i^{(m)} = \{x_i, x_{i+1}, ..., x_{i+m-1}\}$ be selected as the phase variables of the dynamical system constructed by the procedure. The value n is determined as follows: If m≥n, then the variable x_{i+m} is a function of the prior m variables, i.e., $x_{i+m} = f(y_i^{(m)})$; otherwise, x_{i+m} is not. It is suggested in [238] that the following functional dependence criterion should be used: In the space of vectors $y_i^{(m)}$, we take a sphere with centre at the point $y_0^{(m)}$ and of radius ε, and consider all phase trajectory points inside if i is sufficiently large. If the variable x_{i+m} for the points turns out to be at a distance from x_m, not exceeding a value d(ε), decreasing as ε decreases, then we assume that x_{i+m} depends on the variables $y_i^{(m)}$. If there is no dependence, d(ε) should not depend on ε. Numerical computations [238] showed that the embedding dimension n is, as expected, always greater than the fractal dimension d. Thus, for the limit cycle, n=2; for the two-dimensional torus, n=3.

Other procedures for estimating the embedding dimension n, and recipes for constructing the phase space from experimental data, are suggested in [696, 698, 699]. They also enable us to substantially filter the noise added to the dynamical process.

In [681, 700], algorithms are described for calculating the spectrum of

Lyapunov exponents on the basis of the time series and the Packard-Takens procedure.

3. Synchronization threshold as a quantitative characteristic of chaotic motions

It was shown in a number of papers [28, 76, 78, 151, 214, 215, 223, 428] that, under the action of a harmonic external force on a self-oscillatory system in the chaotic regime, transition from chaotic to periodic oscillations is possible, with a period which is a multiple of that of the force, i.e., the synchronization phenomenon can occur. It is essential that, for whatever change in force frequency, the transition arises only beginning with a certain critical value of the force amplitude (e.g., see Fig. 8.11 for the synchronization regions in a backward wave tube [78] under the action of an external harmonic signal). Accordingly, it makes sense to define the concept of *synchronization threshold* as the minimum amplitude of the harmonic action, with synchronization occurring at some or other frequency. That synchronization threshold is non-zero for chaotic self-oscillatory systems is related to the tendency to order (i.e., to synchronize) the motion, occurring against the background of the reverse tendency to chaos. Synchronization takes place when the former tendency starts prevailing. If the system under synchronization functions in the periodic regime, then the synchronization threshold is zero [89, 90]. Therefore, a non-zero threshold may serve as a criterion of chaos in dynamical systems.

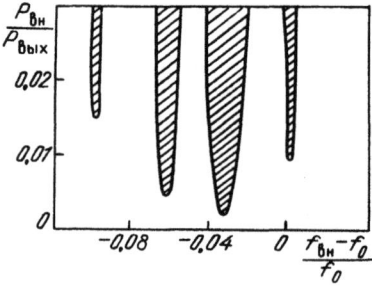

Fig. 8.11

It is natural to assume that a synchronization threshold is connected with such a quantitative characteristic of stochastic motions as Kolmogorov's *metric entropy*. The conjecture was checked and corroborated by numerical computation for a number of concrete systems [214, 215, 224]. As it turned out, the correlation between the synchronization threshold B_{thr} and the entropy K is sufficiently well approximated by the formula

$$B_{thr} = CK^\chi, \qquad (3.1)$$

where C is a quantity depending on the scales of dynamical variables and on time, while χ is approximately the same for all systems investigated, i.e., $\chi \approx 0.33 \pm 0.10$[8].

For three-dimensional systems, entropy was identified with a positive Lyapunov exponent λ^+ calculated by the Benettin algorithm. The dependence (3.1) for $\chi=0.33$, and the points found numerically for all investigated three-dimensional systems are in Fig. 8.12 (where v_0 denotes the external action frequency associated with the synchronization threshold, $C_1 = Cv_0^\chi$). The constant C was determined from minimum mean square deviation of B_{thr}/C from K^χ. With the accuracy given by errors of the numerical experiment, mostly by error in the synchronization threshold determination, the points found, with rare exceptions, correspond to (3.1).

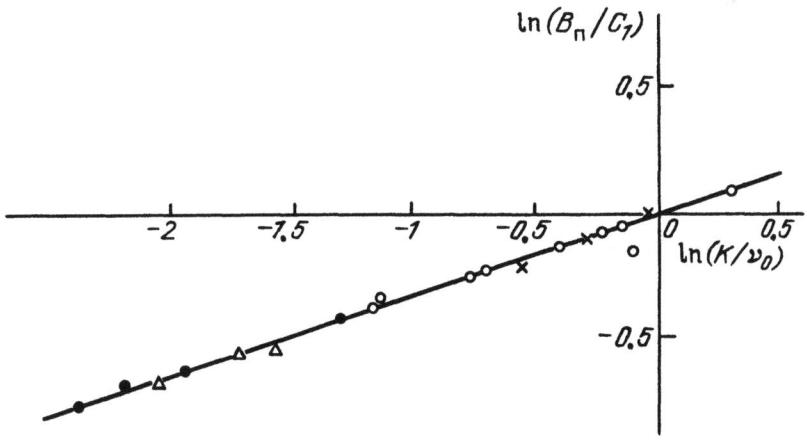

Fig. 8.12

To see that formula (3.1) involves entropy, and not the largest Lyapunov exponent, synchronization of chaotic oscillations in the Mackey-Glass system was

---

[8]Note that the synchronization threshold should apparently be determined by the onset of complicated behaviour of the system, when the synchronization region is left behind; e.g., from occurrence of period-doubling bifurcations.

investigated [224] (see equations (9.3), (9.21), Chapt.9). The system with a delay equation is remarkable in that the maximum Lyapunov exponent decreases as τ increases, and the entropy remains approximately the same, from a certain delay time value τ onwards. It was obtained that, in this range of τ, the synchronization threshold is practically unchanged, which indicates its relation to entropy.

Concrete forms of the systems under investigation and more detailed results are considered in the next chapter.

4. Certain universal laws in order-chaos transitions, and analogy with phase transitions

At present, the literature often contains statements regarding an analogy between transitions of dynamical systems from motions of one type to those of another (e.g., from a state of equilibrium to a periodic motion, from a regular to a chaotic motion, from some chaotic regimes to others, etc.) and the second-kind phase transitions [56, 106 127, 232, 241, 298, 309, 327, 338, 339, 355, 356] known in statistical physics. In fact, formally, the phenomena have much in common, which is, in particular, manifested by the power dependence of certain quantities serving as order (or disorder) parameters on the difference between a bifurcation parameter and its critical value. (In statistical physics, the role of a bifurcation parameter is played by temperature.) Especially important is the fact that the power index called *critical* is universal for as much as a class of systems performing a phase transition.

The idea is due to in H. Haken [486], who paid attention to the analogy between phase transitions and transitions through an excitation threshold in generators with soft excitation. The critical index for the dependence of the oscillation amplitude near the threshold on the excess over the excitation threshold is 1/2. The analogy was subsequently developed by Yu. L. Klimontovich [191, 192], who considered the effect of noise on the generator near the threshold, or in the critical region. In contrast to phase transitions in equilibrium systems considered in statistical physics, the ones related to bifurcations of dynamical systems are usually called *non-equilibrium phase transitions* [355, 487].

Although, as indicated above, the analogy between order-order, order-chaos, chaos-chaos, and phase transitions is formal in many respects, it is quite useful, and enables us to successfully use well-developed methods in the theory of equilibrium phase transitions while analyzing the phenomena; e.g., scaling and renormalization group methods [106, 115, 237, 241, 298, 327, 513].

First of all, we dwell on the transition from order to chaos, accompanied by an infinite period-doubling bifurcation sequence in accordance with the Feigenbaum law [444, 445, 447, 448]. The transition is typical for systems whose motion is exactly or approximately described by a one-dimensional point mapping with differentiable maximum. If the mapping is written as

$$f(x) = 1 - \mu |x|^z, \qquad (4.1)$$

where $z > 1$, μ is a bifurcation parameter, and without loss of generality, the maximum can be assumed to be at the point $x=0$, then the generalized Feigenbaum law can be written in the form

$$\mu_\infty - \mu_n \underset{n \to \infty}{\sim} \delta^{-n}, \qquad (4.2)$$

where μ_n is the value of μ such that the nth period-doubling bifurcation occurs, $\mu_\infty = \lim_{n \to \infty} \mu_n$, δ is a constant depending on the index z in expression (4.1) [430, 494, 513, 514, 516, 541][9].

The values of δ for different z are given in Table 8.4 [516]. It can be seen that the quantity δ increases monotonically as z increases. The *Feigenbaum constant* is usually the value $\delta = 4.6692...$ associated with $z=2$. In the sequel, if not stated otherwise, by δ, we just mean that value.

---

[9]A more complicated form of a mapping near its maximum is considered in paper [541], in which two other universal constants are introduced.

z	δ	a	z	δ	a
1,0	2	∞	5,0	8,345	1,556
1,5	3,8004	3,8889	6,0	9,31	1,468
2,0	4,6692	2,5029	7,0	10,18	1,405
2,5	5,4127	2,1368	8,0	10,98	1,35
3,0	6,0847	1,9277	9,0	11,72	1,32
3,5	6,7053	1,7895	10,0	12,48	1,29
4,0	7,2851	1,6903	11,0	13,15	1,27

Table 8.4

The Feigenbaum law can also be written in a different form, viz.,

$$\omega_n \underset{n\to\infty}{\sim} (\mu_\infty - \mu_n)^\nu, \qquad (4.2^*)$$

where $\omega_n = 2\pi/T_n = 2\pi/(T \cdot 2^n)$, $\nu = \ln 2/\ln\delta = 0.4498...$ [663, 664]. Indeed, it follows from (4.2*) that, for large n,

$$\ln(\mu_\infty - \mu_n)^\nu = -\nu n \ln\delta + \text{const} = -n\ln 2 + \text{const} = \ln\omega_n - \ln\omega_0 + \text{const}.$$

Clearing logarithms, we obtain (4.2*). Feigenbaum universality in the form (4.2) is similar to the law of variation for the order parameter ω_n and the second-kind phase transition with critical index ν. Note that

$$\left.\frac{d\omega_n}{d(\mu_\infty - \mu_n)}\right|_{\mu_n\to\mu_\infty} \sim \left.\frac{\nu}{(\mu_\infty - \mu_n)^{1-\nu}}\right|_{\mu_n\to\mu_\infty} \to \infty.$$

When $\mu > \mu_\infty$, either the positive Lyapunov exponent λ, or the topological entropy h [303], or the threshold for synchronization B_{thr} can be taken as a disorder parameter. Though for systems described by a one-dimensional point mapping differentiable at its maximum, there almost everywhere exist stable limit cycles [576] for μ greater than μ_∞, computations in, e.g., [518, 643, 663, 664] show that, due to inevitable small noise, as a rule, the cycles are not manifest, and, near μ_∞,

$$\lambda \sim (\mu - \mu_\infty)^\nu, \qquad (4.3)$$

where ν is the same critical index as in (4.2*). If there is a small external noise of level σ^2, then the Lyapunov exponent λ can become positive for $\mu = \mu_c < \mu_\infty$ [424, 427, 651]. It is shown in [651] that, for the quadratic mapping, $\mu_\infty - \mu_c \sim \sigma^\rho$, where

$\rho = \ln\delta/\ln(2\beta) = 0.8217...$, and

$$\beta = \frac{\sqrt{2}\, a^2}{\sqrt{1+a^2}} = 3.287\,...\,10 \qquad (4.4)$$

is a universal constant related to the second Feigenbaum constant, a=2.5029... . Note that the values of a as well as of δ depend on the exponent z in expression (4.1). The above value of a is associated with z=2. The values of a for other z are given in Table 8.4. It can be seen that a decreases monotonically from ∞ to 1 as z increases from 1 to ∞.

The qualitative dependence of λ on μ for $\mu \sim \mu_\infty$ and the noise-free quadratic mapping (solid line) and for small noise (dashed line) are in Fig. 8.13.

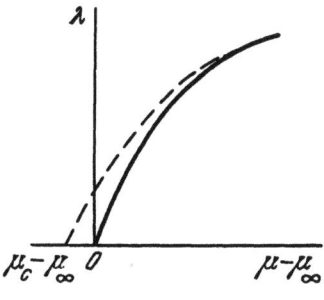

Fig. 8.13

If the mapping is not quadratic in the vicinity of the maximum, then the external noise can stabilize the motion, and decrease the Lyapunov exponent; e.g., it is shown in [575] for the mapping

$$\bar{x} = \begin{cases} \left[(x-0.125)^{1/3} + 0.50607357\right]e^{-x} + b & \text{when } x \leq 0.3, \\ 0.121205692[10x\exp(-10x/3)]^{19} + b & \text{when } x \geq 0.3, \end{cases}$$

whose graph is in Fig. 8.14a, with b=0.023288 (associated with the Markov partition), that if white noise uniformly distributed in $[-\sigma,\sigma]$ is added, then the Lyapunov exponent decreases as σ increases (Fig. 8.14b), whereas sharp peaks related to a close-to-periodic motion appear in the spectrum. This reaction to noise is apparently because the noise sends the representing point to the contracting parts more often than to the expansive ones. Note that the above mapping is different near the maximum x=0.3 from the quadratic mapping in having a discontinuous second derivative.

---

[10] A more exact value is $\beta=3.2375...$.

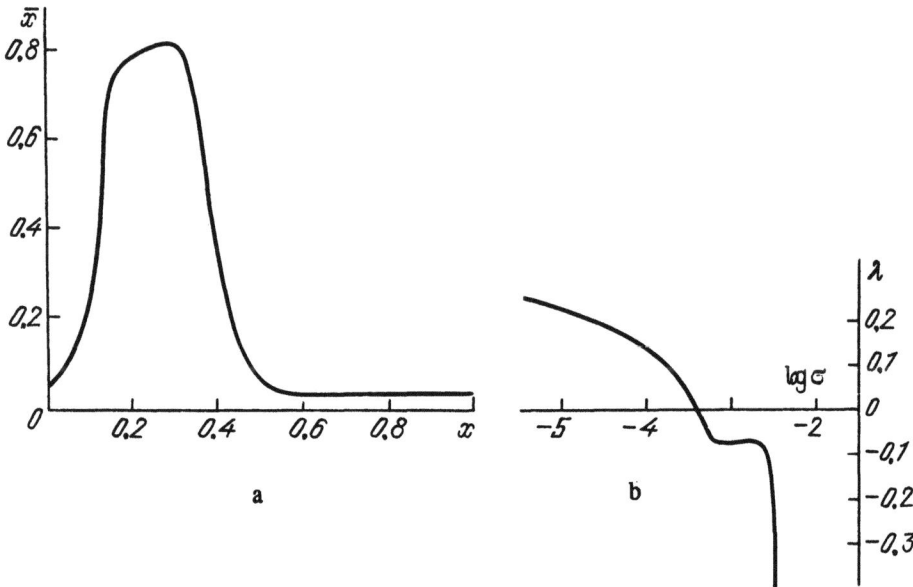

Fig. 8.14

The results of computing the dependence of λ on μ numerically in a wide range of variation of μ, obtained for the quadratic mapping $x_{n+1}=2x_n(\mu+x_n)$ by the formula

$$\lambda = \lim_{N\to\infty} \frac{1}{N}\sum_{i=1}^{N} \ln|f'(x_i)|, \qquad (4.5)$$

where $f(x)=2x(\mu+x)$, are given in Fig. 8.15 and taken from [643]. The regions of negativity of λ are associated with the domains of existence of stable periodic solutions of small period. Near the value $\mu=\mu_\infty=-0.78497...$, the dependence of λ on μ is determined by (4.3), and is illustrated in Fig. 8.13.

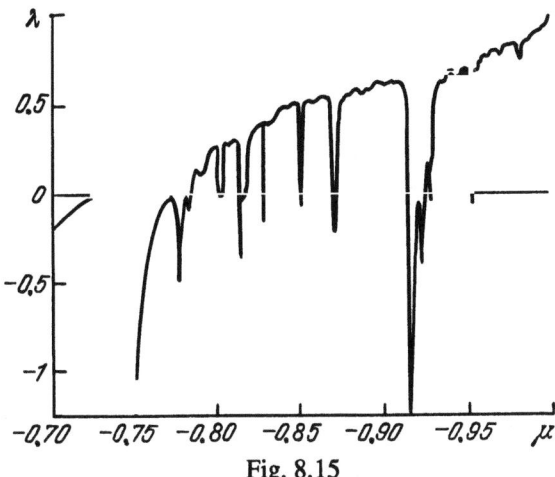

Fig. 8.15

Since, as shown in the preceding section, the synchronization threshold B_{thr} is connected with the Lyapunov exponent λ by $B_{thr} \sim \lambda^{\chi}$, we derive from (4.3) that

$$B_{thr} \sim (\mu - \mu_{\infty})^{\alpha} \tag{4.6}$$

near μ_{∞}, where $\alpha = \chi \nu$.

Other disorder parameters after transition to chaos can be the intensity N of the continuous spectrum and the width $\Delta \omega$ of spectrum peaks. The dependence

$$N(\mu) \sim (\mu - \mu_{\infty})^{\zeta}, \tag{4.7}$$

where $\zeta = 2\ln\beta / \ln\delta = 1.5184...$, is obtained in [519, 680]. We now derive it. The intensity changes as the parameter μ increases and attractor's individual layers merge in the region of reverse doubling bifurcations [344, 421, 425, 448, 564]. The diagram of direct and reverse doubling bifurcations is given in Fig. 8.16 [564]. Consider the mapping $\bar{x} = f(x)$. Let $x_m = f^m(x_0)$. The quantity x_m is determined both by periodic transitions from layer to layer and by chaotic displacement within the layers themselves. Therefore, x_m can be represented in the form

$$x_m = \sum_j A_j \exp(i\omega_j m) + n(m), \tag{4.8}$$

where n(m) is the noise term. Accordingly, the power spectrum for the process x_m is

$$S(\omega,\mu) = \sum_j |A_j(\mu)|^2 \delta(\omega - \omega_j) + |n(\omega,\mu)|^2 \tag{4.9}$$

(where μ is a bifurcation parameter), i.e., the superposition of discrete lines[11] and

---

[11] Actually, they are of finite width (see below).

background noise. The intensity of the continuous spectrum is

$$N(\mu) = \int_0^{2\pi} |n(\omega,\mu)|^2 d\omega = \zeta(0,\mu), \qquad (4.10)$$

where
$$\zeta(k,\mu) = \lim_{N \to \infty} \frac{1}{N} \sum_{m=1}^{N} n(m+k,\mu) n(m,\mu)$$

is the noise's correlation function. Since the noise is due to random walk within each layer, it is obvious that $\zeta(0,\mu) = W^2(\mu)$, where $W(\mu)$ is the layer's average width for given μ. According to the similarity law established by Feigenbaum, in the transition through a parameter's bifurcation value, half of the layers decreases a times; the other half, a^2 times. Therefore, the mean-square width of one layer is

$$W_{m+1} = \sqrt{\frac{1}{2a^2}\left(1+\frac{1}{a^2}\right)} W_m \equiv \frac{1}{\beta} W_m \qquad (4.11)$$

for $\mu = \bar{\mu}_{m+1}$, the value associated with the $(m+1)$th reverse doubling bifurcation, where β is determined by expression (4.4). It follows from the above and (4.10) that

$$N(\bar{\mu}_{n+1}) = N(\bar{\mu}_n)\beta^{-2}. \qquad (4.12)$$

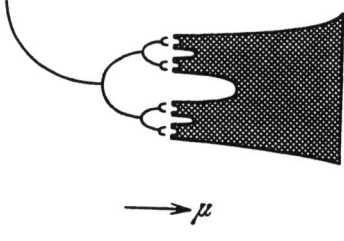

Fig. 8.16

Formula (4.7) is a consequence of (4.12), since $\bar{\mu}_n - \mu_\infty = \text{const}\,\delta^{-n}$ for large n. In fact, taking the logarithms of (4.7), we obtain the equalities

$$\ln N(\bar{\mu}_n) = 2(\ln\beta/\ln\delta)\ln(\bar{\mu}_n - \mu_\infty) + C = -2n\ln\beta + C_1,$$
$$\ln N(\bar{\mu}_{n+1}) = -2(n+1)\ln\beta + C_1,$$

which are equivalent to (4.12). We have already said that, after transition to chaos due to infinite period-doubling bifurcation sequence, the spectrum consists of narrow lines owing to alternating attractor layers and background noise. The latter are due to random walks in each layer. As the transition point is made more distant, both the width $\Delta\omega$ of the narrow spectral lines and the background noise level increase. It is shown in [305, 438] that $\Delta\omega \sim (\mu-\mu_\infty)^\rho$, where $\rho=\ln 2\beta^2/\ln\delta=1.9745...$ according to [438], and $\rho=2.426...$ according to [305].

Other universal characteristics of Feigenbaum transition are the ratio of intensities of developing subharmonics in the nth and $(n+1)$th period-doubling bifurcations, and change in the shape of the continuous spectrum for reverse period-doubling bifurcations. It was obtained in [446] that, as $n \to \infty$, $S_{n+1}(2k) \approx S_n(k)$, $S_{n+1}(2k+1) = \gamma^{-2} S_n(k+1/2)$, where $\gamma = 2\beta = 4a^2/\sqrt{2(1+a^2)} \approx 6.57$, and $S_n(k)$ is the intensity of the kth harmonic of fundamental frequency $\omega_n = 2\pi/T_n$ far from the nth doubling bifurcation. These relations mean that the envelope of the spectrum of the subharmonics generated in the $(n+1)$th bifurcation (far from the bifurcation point) should be lower than that in the nth bifurcation by $20 \log\gamma \approx 16.35$ dB. However, as shown in [593], Feigenbaum was in error when calculating γ. Actually, $\gamma = \sqrt{2}\beta = 4.5785...$, which is associated with the difference of $13.214...$dB between the envelopes. The result has been repeatedly corroborated both in physical and numerical experiments (e.g., see [535, 658]). After transition to chaos, the continuous spectrum is governed by similar laws. It was shown in [680] that the power spectrum in the $(n+1)$th reverse bifurcation ($S_{n+1}(\omega)$) is related to that in the nth bifurcation ($S_n(\omega)$) via

$$S_{n+1}(\omega) = \frac{1}{2a^4} (1+a^2 - 2a\cos\omega T) S_n(2\omega). \quad (4.13)$$

Meanwhile, $S_0(\omega)$ is assumed to be constant (i.e., corresponds to white noise) from $\omega=0$ to $2\pi/T$. Formula (4.13) is consistent with direct numerical computations of the power spectrum for the quadratic mapping, performed in [680]. The comparison results are in Fig. 8.17, where the curves marked by 1 are associated with the numerical data; those marked by 2, with the theory without an account of the discrete components of the spectrum (for convenience, the latter are shifted below).

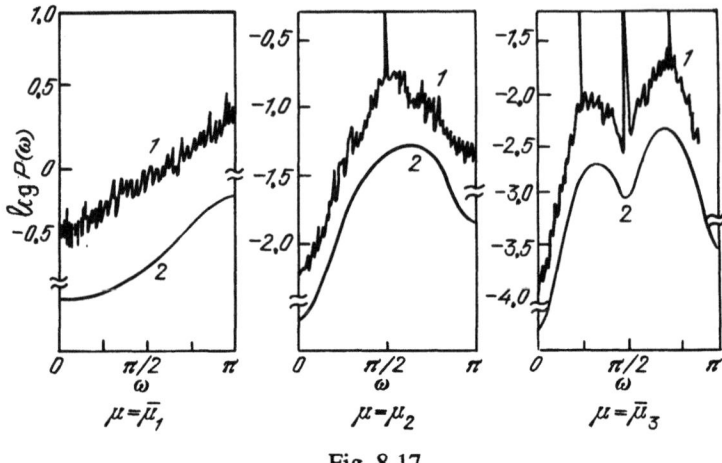

Fig. 8.17

Similar universal laws hold for period-doubling bifurcations in systems described by two- and n-dimensional point mappings depending on one parameter [420, 693]. Thus, it was shown in [693] by considering the two-dimensional mapping

$$x_{n+1} = 2\mu (x_n + x_n^2) - y_n, \quad y_{n+1} = Bx_n \quad (0 \leq B \leq 1), \tag{4.14}$$

that μ for which period-doubling bifurcations occur were governed by Feigenbaum universality; however, with another constant depending on the coupling coefficient B, viz.,

$$\lim_{n \to \infty} \frac{\mu_n(B) - \mu_{n-1}(B)}{\mu_{n+1}(B) - \mu_n(B)} = \lim_{n \to \infty} \delta_n\left(B_e^{(n)}\right), \tag{4.15}$$

where $B_e^{(n)} = B2^n$. The dependence $\delta_n\left(B_e^{(n)}\right)$ found on a computer for sufficiently large, but finite, n is given in Fig. 8.18. For $B = B_e^{(n)} = 0$, $\delta_n\left(B_e^{(n)}\right) \underset{n \to \infty}{\longrightarrow} 4.6692...$; for $B = B_e^{(n)} = 1$, $\delta_n\left(B_e^{(n)}\right) \underset{n \to \infty}{\longrightarrow} 8.7210...$.

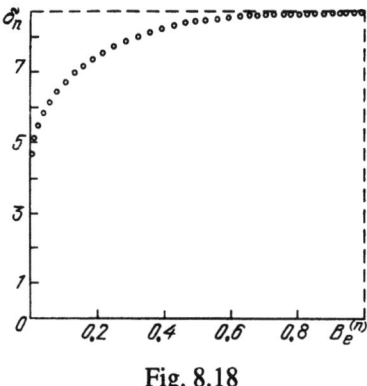

Fig. 8.18

Although, for any B<1, $\lim_{n \to \infty} B_e^{(n)} = 0$, and, therefore, $\lim_{n \to \infty} \delta_n(B_e^{(n)}) = 4.6692...$, for practical computations and with the parameter B close to one, we can obtain a constant substantially different from Feigenbaum's. Note that B has the same meaning as a dissipation parameter, and B=1 is associated with no dissipation, i.e., with a Hamiltonian system (the product of the multipliers of the fixed points of the mapping being one). Therefore, the obtained result can be treated as follows: In dissipative systems described by a two-dimensional point mapping, a cascade of period-doubling bifurcations near a condensation point occurs similarly to in the systems described by a one-dimensional mapping. However, in Hamiltonian systems with a two-dimensional mapping, the transition constants are different, viz., $\delta = 8.7210...$, a =4.018... [237].

It was already noted that, besides infinitely many period-doubling bifurcations, an infinite sequence of period-tripling bifurcations is possible in principle. This sequence should also be governed by a number of universal laws [130, 516]. For systems described by a one-dimensional point mapping of form (4.1), the laws are similar to Feigenbaum universality; however, with other constants. The dependence of δ, a on the exponent z in expression (4.1) for a period-tripling bifurcation sequence is demonstrated in Table 8.5 [516].

z	δ	a	z	δ	a	z	δ	a
1,5	73,12	30,10	1,8	56,64	17,75	2,1	55,50	8,21
1,6	63,71	20,80	1,9	55,55	10,72	2,2	56,10	7,38
1,7	59,02	15,81	2,0	55,26	9,28	3,0	66,99	4,37

Table 8.5

It is interesting to note that, as z increases, the constant a decreases monotonically, similarly to the period-doubling case; however, δ first decreases, attains a minimum for z=2, and then increases. This behaviour of δ substantially differs from the case of a sequence of period-doubling bifurcations. Besides, in tripling bifurcations, δ and a are much greater than in doublings, which practically does not enable us to observe more than one bifurcation.

Certain universal laws were obtained in [210] under small external periodic force acting on the system described by a one-dimensional parabolic point mapping. It is shown that, as the force increases, the bifurcation parameter μ_n associated with the nth period-doubling bifurcation also increases monotonically. (In the case of non-resonant force, the values of μ_n found are associated with quasi-period-doubling bifurcations of the torus.) Note that extending the results to the region of chaos may enable us to account for a synchronization threshold and for the latter's connection with a positive Lyapunov exponent.

In the transition to chaos, accompanied by so-called *intermittency*, a number of universal laws also arise similarly to those in second-kind phase transitions. For systems described by a one-dimensional point mapping, these were discovered in papers [229, 304, 435, 503, 504, 572, 573, 577, 609]. The mapping can be represented as in Fig. 8.19 in the vicinity of transition through intermittency. In the transition, a stable fixed point of the mapping, M, fuses with the unstable one, M', and vanishes. The bifurcation is often said to be *tangential*. Let the bifurcation parameter ε be zero when the mapping touches the bisector. For sufficiently small ε and small $|x_1|$, x_2, the part marked by I can be approximated by

$$\bar{x} = x + ax^z + \varepsilon, \qquad (4.16)$$

where z is an even number (Fig. 8.19). When the representing point moves in the part marked by I, the motion for sufficiently small ε>0 is close to periodic (a "laminar" phase). The long parts of almost regular behaviour alternate with short irregular peaks ("turbulent" phases), which are associated with transitions of the point to the part marked by II, and back. Hence the term "intermittency". It was shown in [504, 515] by the renormalization method that the laminar phase duration τ for small ε>0 was proportional to $\varepsilon^{-(1-1/z)}$. The same was obtained in [503] by replacing the difference equation (4.16) by a differential equation, and by subsequent integration.

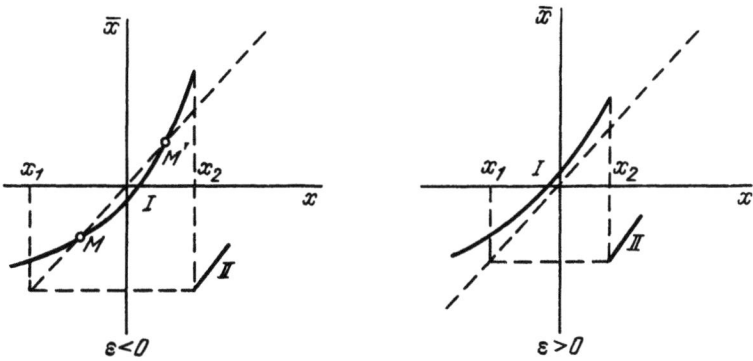

Fig. 8.19

The above papers and [304, 435] consider the effect of external noise on the behaviour, and show that, for ε=0, the average laminar phase duration τ is proportional to $g^{-2(z-1)/(z+1)}$, where g^2 is the noise level. To obtain the result, the authors turn from the Langevin equation to the corresponding Fokker-Planck equation. However, the latter was solved with boundary conditions not related to Fig. 8.19. The corresponding conditions were obtained in [229], giving the results to be discussed in the sequel.

External noise can be described by an additional term in equation (4.16), viz.,

$$\bar{x} = \varepsilon + x + ax^z + g\xi, \qquad (4.17)$$

with the factor g selected so that $\langle\xi\rangle^2 = 1$. Assume that the noise ξ is uncorrelated, of zero mean value, i.e., $\langle\xi\rangle = 0$, $\langle\xi_n\xi_{n'}\rangle = \delta_{nn'}$, where $\delta_{nn'}$ is the Kronecker delta and n, n' are iteration subscripts.

For sufficiently small ε, g and $ax_{1,2}^z$, equation (4.17) can be replaced by a differential equation, by dividing throughout by certain small Δt. The time scale can be selected so that Δt =1. The corresponding equation is

$$\frac{dx}{dt} = \varepsilon + ax^z + g\xi(t), \tag{4.18}$$

where $\langle\xi(t)\rangle = 0$, $\langle\xi(t)\,\xi(t')\rangle = \delta(t-t')$.

If the parts of the mapping, on which the representing points are exchanged in between the curves marked by I, II, are small (so that $\varepsilon + ax_2^z$, $k(\varepsilon + ax_2^z) \ll x_2 - x_1$, where k is the slope of the part marked by II, approximable by a line segment), then (4.18) can be associated with the Fokker-Planck equation

$$\frac{\partial w}{\partial t} = -\frac{\partial}{\partial x}\left[(\varepsilon + ax^z)w\right] + \frac{g^2}{2}\frac{\partial^2 w}{\partial x^2} \tag{4.19}$$

for the probability density $w(t,x)$ [342].

First, we consider the steady-state solution of (4.19). To obtain it, it suffices to specify one boundary condition and the normalization condition. Since all points attaining the boundary x_2 leave the interval in question,

$$w(x_2) = 0. \tag{4.20}$$

The steady-state solution of (4.19) with boundary condition (4.20) is of the form

$$w(x) = \frac{2G_0}{g^2} \exp\left[\frac{2}{g^2}\left(\varepsilon x + \frac{ax^{z+1}}{z+1}\right)\right] \int_{x_1}^{x_2} \exp\left[-\frac{2}{g^2}\left(\varepsilon y + \frac{ay^{z+1}}{z+1}\right)\right] dy \tag{4.21}$$

for flow $x \in [x_1, x_2]$, where G_0 is the probability flow $G = (\varepsilon + ax^z)w - \frac{g^2}{2}\frac{dw}{dx}$ in $[x_1, x_2]$. For $x < x_1$, the flow is zero; therefore,

$$w(x) = \frac{2G_0}{g^2} \exp\left[\frac{2}{g^2}\left(\varepsilon x + \frac{ax^{z+1}}{z+1}\right)\right] \int_{x_1}^{x_2} \exp\left[-\frac{2}{g^2}\left(\varepsilon y + \frac{ay^{z+1}}{z+1}\right)\right] dy. \tag{4.22}$$

Note that, in shifting to the left of the point x_1, the function $w(x)$ decreases rapidly; and the lesser the noise level g^2, the faster the decrease. For $g=0$, $w(x)=0$ when $x < x_1$.

The constant G_0 is determined by the normalization condition when integrating with respect to all values of x from $-\infty$ to x_2, i.e.,

$$G_0^{-1} = \frac{2}{g^2}\left\{\int_{x_1}^{x_2}\exp\left[\frac{2}{g^2}\left(\varepsilon x + \frac{ax^{z+1}}{z+1}\right)\right]\int_x^{x_2}\exp\left[-\frac{2}{g^2}\left(\varepsilon y + \frac{ay^{z+1}}{z+1}\right)\right] dy\, dx\right.$$
$$\left. + \int_{-\infty}^{x_1}\exp\left[\frac{2}{g^2}\left(\varepsilon x + \frac{ax^{z+1}}{z+1}\right)\right] dx \int_{x_1}^{x_2}\exp\left[-\frac{2}{g^2}\left(\varepsilon y + \frac{ay^{z+1}}{z+1}\right)\right] dy\right\}. \tag{4.23}$$

In the simplest case where $g=0$, it immediately follows from (4.19) that

$$w(x) = \begin{cases} G_0/(\varepsilon + ax^z) & \text{for } x \in [x_1, x_2], \\ 0 & \text{for } x < x_1. \end{cases} \quad (4.24)$$

The quantity G_0 in formula (4.24) can be calculated explicitly for sufficiently small ε when $\varepsilon \ll ax_{1,2}^z$. Meanwhile,

$$G_0 = \frac{z}{2\pi} \sin\left(\frac{\pi}{z}\right) (a\varepsilon^{z-1})^{1/z}. \quad (4.25)$$

It can be seen from (4.24) that, for $z=2$, the noise-free probability distribution is of Lorenz curve shape with maximum at the point $x=0$ and with width $\sqrt{\varepsilon/a}$.

If the noise level is small ($g^2 \ll \varepsilon$), the shape of the probability distribution described by expressions (4.21), (4.22) is close to (4.24); however, the maximum is somewhat shifted towards negative x. Numerical computations have shown that, for $g^2 \gtrsim \varepsilon$, the function $w(x)$ depends very weakly on ε. The graphs of $w(x)$, constructed for the computation results and for different values of the parameters, are given in Fig. 8.20.

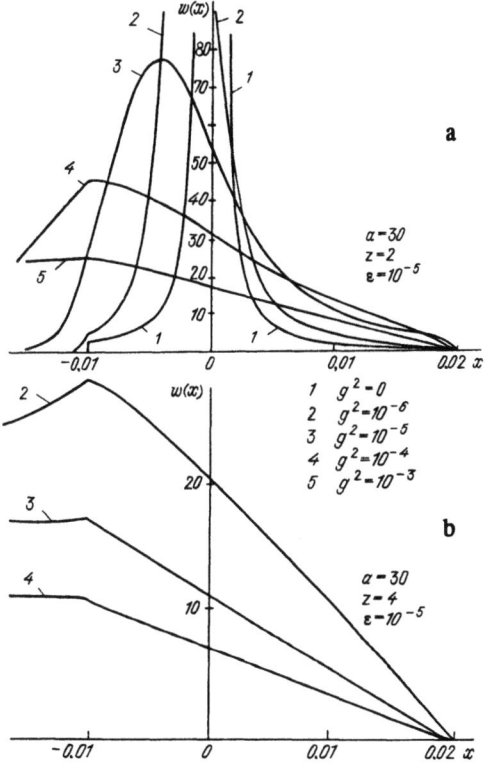

Fig. 8.20

We now calculate the average duration of the laminar phase τ. According to the results of Chapt.XVIII in [342],

$$\tau = G_0^{-1}, \qquad (4.26)$$

where G_0 is the probability flow on $[x_1,x_2]$, determined by the expression (4.23). For $g=0$ and $\varepsilon \ll ax_{1,2}^z$, we derive from (4.26), (4.25) that $\tau \sim \varepsilon^{-(1-1/z)}$, which coincides with the results of [503, 504, 515]. In the other extreme case where $\varepsilon=0$, but $g\neq 0$, we have

$$\tau = 2^{(z-1)/(z+1)} ((z+1)/a)^{2/(z+1)} g^{-2(z-1)/(z+1)} B, \qquad (4.27)$$

with

$$B = \int_{u_1}^{u_2} \exp(u^{z+1}) \int_{u}^{u_2} \exp(-v^{z+1}) \, dv \, du + \int_{-\infty}^{u_1} \exp(u^{z+1}) \, du \int_{u_1}^{u_2} \exp(-v^{z+1}) \, dv,$$

$$u_{1,2} = \left[\frac{2a}{(z+1)g^2}\right]^{1/(z+1)} x_{1,2}.$$

For $g^2 \ll a|x_{1,2}|^{z+1}$, we can put $u_1=-\infty$, $u_2=\infty$, and the value B will not depend on g. It follows from (4.27) that $\tau \sim g^{-2(z-1)/(z+1)}$. The obtained dependence of τ on g coincides with the results in [435, 503, 504, 515].

The dependences of τ on g^2, calculated by formulas (4.26), (4.23), are given in Fig. 8.21 for different values of the other parameters. As well as w(x), the values of τ for $g^2 \gtrsim \varepsilon$ are practically independent of ε, which is manifested by the data in Table 8.6.

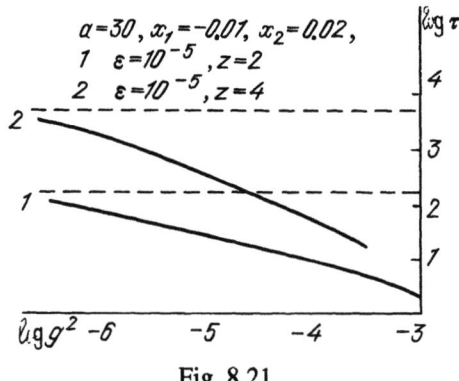

Fig. 8.21

g^2	ε	τ
10^{-5}	0	25,32
	10^{-6}	25,30
	10^{-5}	25,08
	10^{-4}	23,11
10^{-6}	0	54,55
	10^{-7}	54,50
	10^{-5}	51,12
10^{-4}	0	9,30
	10^{-5}	9,28
10^{-3}	0	2,208
	10^{-5}	2,207

Table 8.6

We now calculate the Lyapunov exponent λ for the mapping in question, using (4.5) as the original formula. Meanwhile, the next term in the expansion of the mapping on the part marked by I, involving an odd power of z, may be of importance. Taking it into account, we specify the mapping on the part marked by I in the form

$$\bar{x} = \varepsilon + x + ax^z + bx^{z+1} + g\xi, \quad (4.28)$$

and on the part marked by II as

$$\bar{x} = x_1 + k(x - x_2). \quad (4.29)$$

Substituting (4.28), (4.29) in (4.5), we obtain

$$\lambda = \lim_{N \to \infty} \frac{1}{N+Q} \left[\sum_{i=1}^{N} \ln\left(1 + azx_i^{z-1} + b(z+1)x_i^z\right) + Q \ln |k| \right], \quad (4.30)$$

where Q is the number of transitions of the representing point from the part marked by I to the part marked by II in N+Q iterations. It is obvious that, for large N, $Q = N/\tau \ll N$. Since the process is ergodic on the part marked by I, expression (4.30) can be written as

$$\lambda = \int_{-\infty}^{x_2} w(x) \ln\left(1 + azx^{z-1} + b(z+1)x^z\right) dx + \frac{1}{\tau} \ln |k|, \quad (4.31)$$

where w(x) is the stationary probability distribution whose shape is close to the above, the term bx^{z+1} being small. The latter can be seen in Fig. 8.22, where w(x) is given for $g^2=10^{-6}$, b=30 (other parameters being the same as in Fig. 8.20a). The dependence almost exactly coincides with the curve marked by 2 in the diagram.

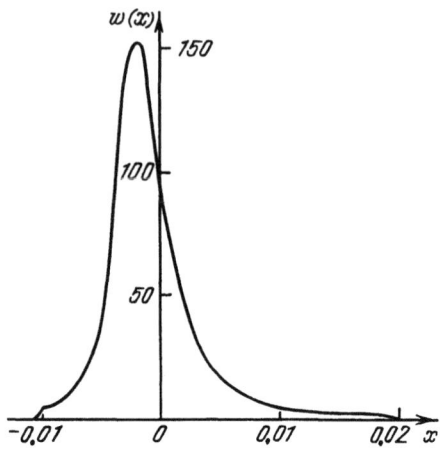

Fig. 8.22

For g=0, $\varepsilon \ll ax_{1,2}^z$, we can put $w(x) = G_0/(\varepsilon+ax^z+bx^{z+1})$, where G_0 is approximately determined by expression (4.25), in which case the integral in (4.31) can be calculated easily, provided $az|x_{1,2}|^{z-1}+b(z+1)x_{1,2}^z \ll 1$. As a result, we obtain

$$\lambda = \frac{z \sin(\pi/z)}{2\pi} \left(a\varepsilon^{z-1}\right)^{1/z} \left\{ z \ln\frac{x_2}{|x_1|} + \frac{b}{a}(x_2-x_1)+\ln|k| \right\}. \qquad (4.32)$$

Hence $\lambda>0$, i.e., the motion is stochastic if the expression in the brackets is positive. Otherwise, after the transition, a periodic motion with large period develops, and the behaviour of the system can be chaotic in the sense of Chapt.3. Note that the attractor in the phase space, associated with the chaotic behaviour, was termed a *quasi-attractor* in [45, 381], though it should have been called a *quasi-strange attractor*, to be more precise.

In the case of a stochastic motion, after transition, the Lyapunov exponent λ smoothly increases according to $\lambda \sim \varepsilon^{1-1/z}$, in spite of the transition in question being hard. The dependence of λ on ε, free of noise (the solid line), and with noise of level $g^2=10^{-6}$ (the dashed line), found from formula (4.31) using a computer, is shown in Fig. 8.23. In contrast to Feigenbaum transition, the noise stabilizes the system under transition if |k| is not very large, since the noise sends the representing point to the contractive part of the mapping[12]. For sufficiently large |k|, the decrease in τ owing to noise is the most important, which leads to increase of the Lyapunov exponent as the noise level increases, i.e., to destabilization of motion.

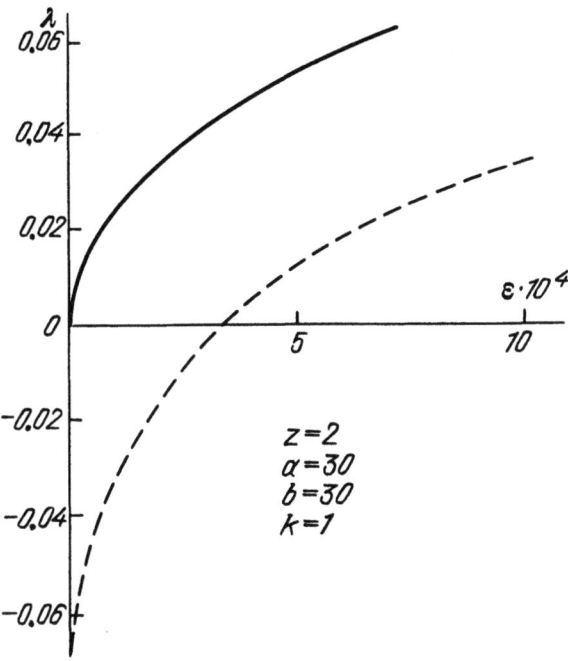

Fig. 8.23

It should be noted that expression (4.31) determines the Lyapunov exponent in quite an approximate way, since the basic contribution to the integral is given by values of x far from zero, for which the error arising in the transition from the difference to the differential equation is greater; therefore, the calculation of w(x) is less exact.

Another type of transition to chaos through intermittency is considered in [218, 606], where the mapping is of the form illustrated by Fig. 8.24. During the transition, the stable fixed point M fuses with two unstable points M', M", and becomes unstable. Owing to the parts marked by II, II', the representing point always returns to the region near M. For small x_0, the part marked by I can be approximated by the formula

$$\bar{x} = (1+\varepsilon+ax^z)x, \qquad (4.33)$$

where z is an even number. The onset of transition is associated with ε=0. For sufficiently small ε, the representing point is near M for a long time, i.e., the motion is close to

---
[12]This effect is not taken into account in [304], since $x_{1,2}=\pm\infty$.

periodic (a laminar phase). Parts of the motion alternate with short "turbulent" peaks associated with transitions in the regions marked by II, II', and back.

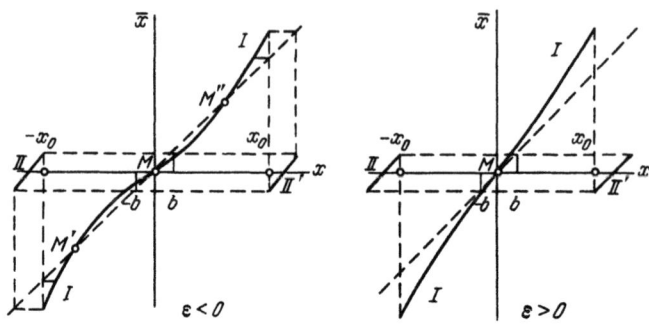

Fig. 8.24

Taking external noise into account, the mapping (4.33) is of the form

$$\bar{x} = (1+\varepsilon+ax^2)x+g\xi, \qquad (4.34)$$

where $\langle\xi\rangle=0$, $\langle\xi^2\rangle=1$. For $|\varepsilon|$, g, $ax_0^2 \ll 1$, equation (4.34) can be replaced by the differential equation

$$\frac{dx}{dt} = (\varepsilon+ax^2)x+g\xi(t), \qquad (4.35)$$

where $\langle\xi(t)\rangle=0$, $\langle\xi(t)\,\xi(t')\rangle = \delta(t-t')$.

For simplicity, we consider the case of a totally symmetric mapping, as in Fig. 8.24, and assume that the regions in which the representing points in between the parts marked by I, II and II' are exchanged are small. The corresponding smallness conditions can easily be written similarly to the above. Meanwhile, exchange parts near the points Fx_0, because they are small, can be simply neglected, while the part $[-b,b]$ (Fig. 8.24) near the point $x=0$ should be necessarily taken into account, since the fixed point M is in it. (As b tends to zero, the probability distribution at $x=0$ acquires a singularity.) In the regions $b \le |x| \le x_0$, (4.35) can be associated with the source-free Fokker-Planck equation

$$\frac{dw}{dt} = -\frac{\partial}{\partial x}[(\varepsilon+ax^2)\,xw]+\frac{g^2}{2}\frac{\partial^2 w}{\partial x^2}. \qquad (4.36)$$

The boundary conditions at the points $x=\pm x_0$ are, according to (4.20), of the form

$$w(\pm x_0) = 0. \qquad (4.37)$$

In the region $-b \le x \le b$, a source of calculable form should be added to (4.36).

However, for the sake of simplicity, since b is small, the source can be regarded as uniform and equal to an unknown quantity R. R is determined by the condition that the density and probability flow at the points x=0, x=±b are continuous.

The steady-state solution of (4.36), satisfying (4.37), is

$$w(x) = \frac{2G_0}{g^2} \exp\left[\frac{2x^2}{g^2}\left(\frac{\varepsilon}{2} + \frac{ax^z}{z+2}\right)\right] \int_x^{x_0} \exp\left[-\frac{2y^2}{g^2}\left(\frac{\varepsilon}{2} + \frac{ay^z}{z+2}\right)\right] dy. \quad (4.38)$$

(By symmetry, it suffices to consider only the region of positive x.)

The condition $dw/dx|_{x=0} = 0$, by the symmetry of the probability density for positive and negative x, and the expression for the probability flow at x=0, imply $R = G_0/b$. Taking into account the relation of R to G_0, the solution of (4.36) (with constant R in $0 \le x \le b$), satisfying the condition for continuity at x=b, is of the form

$$w(x) = \frac{2G_0}{g^2} \exp\left[\frac{2x^2}{g^2}\left(\frac{\varepsilon}{2} + \frac{ax^z}{z+2}\right)\right] \left\{ \int_b^{x_0} \exp\left[-\frac{2y^2}{g^2}\left(\frac{\varepsilon}{2} + \frac{ay^z}{z+2}\right)\right] dy \right.$$
$$\left. + \frac{1}{b}\int_x^b y \exp\left[-\frac{2y^2}{g^2}\left(\frac{\varepsilon}{2} + \frac{ay^z}{z+2}\right)\right] dy \right\}. \quad (4.39)$$

We find from the normalization condition that

$$G_0^{-1} = \frac{4}{g^2}\left\{\int_0^{x_0}\exp\left[\frac{2x^2}{g^2}\left(\frac{\varepsilon}{2}+\frac{ax^z}{z+2}\right)\right]\int_x^{x_0}\exp\left[-\frac{2y^2}{g^2}\left(\frac{\varepsilon}{2}+\frac{ay^z}{z+2}\right)\right] dy\, dx \right.$$
$$\left. + \frac{1}{b}\int_0^b\exp\left[\frac{2x^2}{g^2}\left(\frac{\varepsilon}{2}+\frac{ax^z}{z+2}\right)\right]\int_x^b(y-b)\exp\left[-\frac{2y^2}{g^2}\left(\frac{\varepsilon}{2}+\frac{ay^z}{z+2}\right)\right]dy\,dx\right\}. \quad (4.40)$$

In the limiting case g=0, we derive directly from (4.36) that

$$w(x) = \begin{cases} G_0/[(\varepsilon+ax^z)|x|], & b \le |x| \le x_0, \\ G_0/[(\varepsilon+ax^z)b], & 0 \le |x| \le b, \end{cases} \quad (4.41)$$

where

$$G_0^{-1} = \frac{2}{\varepsilon}\left[\ln\frac{x_0}{b} - \frac{1}{z}\ln\frac{\varepsilon+ax_0^z}{\varepsilon+ab^z} + \frac{1}{b}\left(\frac{\varepsilon}{a}\right)^{1/z} I_z\right], \quad (4.42)$$

$$I_z = \int_0^{b(a/\varepsilon)^{1/z}} \frac{dy}{1+y^z}.$$

For z=2, $I_2 = \tan^{-1}\left(\sqrt{a/\varepsilon}\, b\right)$; for z=4,

$$I_4 = \frac{1}{4\sqrt{2}}\left[\ln\frac{1+\sqrt{2}\,b(a/\varepsilon)^{1/4}+b^2\sqrt{a/\varepsilon}}{1-\sqrt{2}\,b(a/\varepsilon)^{1/4}+b^2\sqrt{a/\varepsilon}} + 2\tan^{-1}\frac{\sqrt{2}\,b(a/\varepsilon)^{1/4}}{1-b^2\sqrt{a/\varepsilon}}\right].$$

As $\varepsilon \to 0$, $I_2 \to \pi/2$, $I_4 \to \sqrt{2}\pi/4$.

The graphs of the function $w(x)$ are given in Fig. 8.25 for $a=30$, $x_0=0.01$, $\varepsilon=10^{-5}$, $b=(k/2)(\varepsilon+ax_0^z)x_0=3\cdot 10^{-4}$ and a number of values of the parameters z, g^2. It follows from numerical computations that, for $g^2 \gtrsim g_c^2(z)$, where $g_c^2 \sim 10^{-7}$ for $z=2$ and 10^{-8} for $z=4$, the probability distribution practically becomes independent of g^2. Besides, for $g^2 \gtrsim 10^{-7}$, it is almost identical both for $z=2$ and $z=4$ (cf. the curve marked by 3 in Fig. 8.25a and that marked by 2 in Fig. 8.25b).

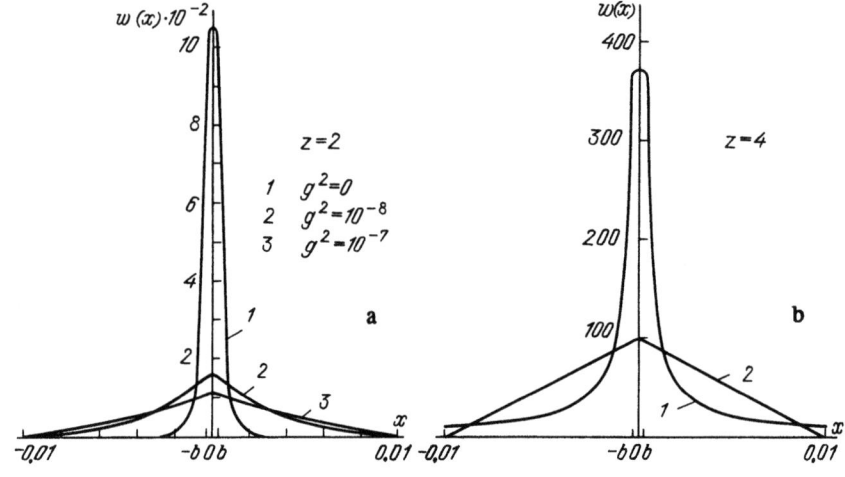

Fig. 8.25

Since b is small in regard to x_0, to calculate the average laminar phase duration, we can still assume that the probability flow is constant and equal to G_0. Taking into account that the representing point lies with probability $1/2$ in the region where $x>0$, we have

$$\tau = G_0^{-1}/2. \tag{4.43}$$

For $g^2=0$ and sufficiently small ε, we derive from the above and (4.42) that

$$\tau = I_z/(ba^{1/z}\varepsilon^{(z-1)/z}). \tag{4.44}$$

Hence, $\tau \sim \varepsilon^{-(1-1/z)}$ for small ε, similarly to the case above. In particular,

$\tau \sim \varepsilon^{-1/2}$ for z=2, which coincides with the results in [606].

We now estimate the dependence of τ on g^2 in the limiting case $\varepsilon=0$. Putting $u = (2a/(z+2)g^2)^{1/(z+2)} x$, we obtain

$$\tau = 2 \left(\frac{z+2}{2a} \right)^{2/(z+2)} g^{-2z/(z+2)} B, \qquad (4.45)$$

where

$$B = \int_0^{u_0} \exp(u^{z+2}) \int_u^{u_0} \exp(-v^{z+2}) \, dv \, du + \frac{1}{b_0} \int_0^{b_0} \exp(u^{z+2}) \int_u^{b_0} (v-b_0) \exp(-v^{z+2}) \, dv \, du,$$

$$u_0 = \left[\frac{2a}{(z+2)g^2} \right]^{1/(z+2)} x_0, \quad b_0 = \left[\frac{2a}{(z+2)g^2} \right]^{1/(z+2)} b.$$

For $g^2 \ll ab^{z+2}$, we can put $u_0 = \infty$ and $b_0 = \infty$. Then,

$$B \approx \frac{1}{b_0} \int_0^\infty \exp(u^{z+2}) \int_u^\infty v \exp(-v^{z+2}) \, dv \, du \sim g^{2/(z+2)}.$$

Therefore, $\tau \sim g^{-2(z-1)/(z+2)}$, i.e., the dependence of τ on g^2 in the limiting case is also the same. (For z=2, this also coincides with the results in [606], which, however, were not obtained quite correctly.)

Numerical computations by formulas (4.40), (4.43) for a=30, b=3·10^{-4}, x_0=0.01 show that, for $g^2 \gtrsim 10^{-8}$, the dependence of τ on g^2 is practically independent of ε; for $g^2 \gtrsim 10^{-6}$, it is also independent of z. Its shape is illustrated in Fig. 8.26, where the asymptotic dependence of τ on g^2 for $\varepsilon=0$, z=2 is shown by the dashed line.

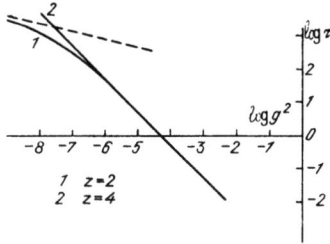

Fig. 8.26

The Lyapunov exponent for the mapping in question can be calculated by the formula

$$\lambda = 2 \int_0^{x_0} w(x) \ln \left(1+\varepsilon+a(z+1)x^z \right) dx + \frac{1}{\tau} \ln |k| \qquad (4.46)$$

similarly to (4.31), where k is the slope of the parts marked by II and by II'. For g=0, $\varepsilon+a(z+1)x_0^z \ll 1$, we get

$$\lambda = \frac{1}{\tau}\left[1+z+\ln\frac{x_0(\varepsilon+ax_0^z)}{b(\varepsilon+ab^z)} - \frac{z}{b}\left(\frac{\varepsilon}{a}\right)^{1/z} I_z + \ln |k|\right], \qquad (4.47)$$

taking into account (4.41), (4.43). It can be seen that the Lyapunov exponent can be positive even for sufficiently small $|k|$, since the part marked by I is expanding for $\varepsilon>0$. For sufficiently small ε, it follows from (4.47), (4.44) that λ increases in proportion to $\varepsilon^{1-1/z}$. The dependence of λ on ε for different g^2 (a), and on g^2 for $\varepsilon=0$ (b) is shown in Fig. 8.27. Note that, in contrast to the above case, external noise destabilizes the system's behaviour, and substantially increases the Lyapunov exponent; the difference is also due to the fact that the mapping in the part marked by I has no contracting regions for $\varepsilon>0$.

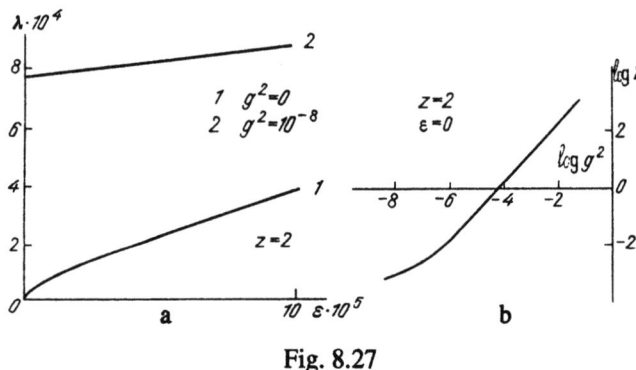

Fig. 8.27

It follows from these and the above results that transition to stochastic behaviour through intermittency is similar to a second-kind phase transition, where we can consider either λ or τ^{-1} as disorder parameters. For both parameters, the critical indices obtained are identical, and depend only on the power index z, or on the nature of the mapping near the point of contact or of inflection.

Note that hard appearance of the stochastic regime, not accompanied by intermittency (as in the Lorenz system for small values of r [686]), is similar to a first-kind phase transition, since the Lyapunov exponent λ and entropy K increase in a jump-like manner with hysteresis characteristic of hard transitions (Fig. 8.28).

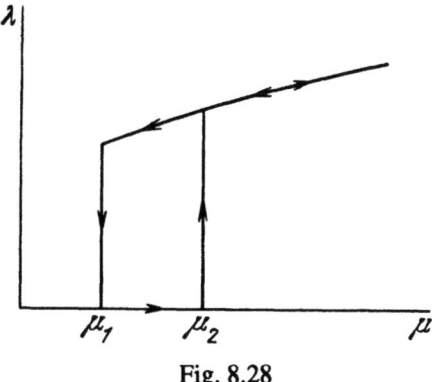

Fig. 8.28

Transitions to chaos, accompanied by both the appearance and destruction of the torus, also have to be characterized by a number of universal laws; however, at present, they have not been studied to the full. Certain laws occurring during torus destruction are, along with the above, made manifest in [24, 50, 167, 390, 416, 417, 463, 615]. Something is also known for the case where torus quasi-period-doubling bifurcations occur before its destruction [37, 389, 537, 538]. Thus, it is shown in [537, 538] that there are finitely many torus-doubling bifurcations as the parameter A varies. These are determined by the coupling parameter ε, by considering the system described by the two-dimensional point mapping

$$x_{n+1} = 1 - Ax_n^2 + \varepsilon \sin 2\pi y_n, \quad y_{n+1} = y_n + C + \varepsilon x_n, \tag{4.48}$$

where C is a constant playing the role of Poincaré rotation number. This can be seen in a diagram demonstrating the domains of existence of the torus with quasi-period $2^n \times T$ on the parameter (A,ε)–plane. One such diagram for $C=(\sqrt{5}-1)/2$ is given in Fig. 8.29, where A_∞ is the critical value of A, signalling the end of the period-doubling bifurcation cascade if $\varepsilon=0$. Dependence of the torus-doubling number l on ε in the transition to chaos can then be investigated (Fig. 8.30). It is seen that the maximum or minimum values of ε, associated with given l, are approximately in the straight line with the equation

$$\log 2^l = \text{const} - \frac{1}{3} \log \varepsilon. \tag{4.49}$$

chaos

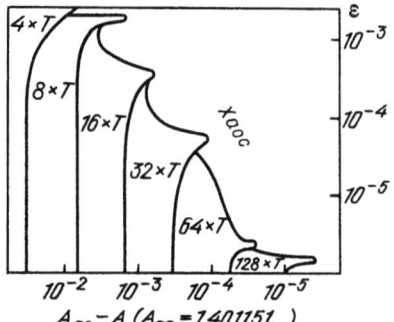

Fig. 8.29

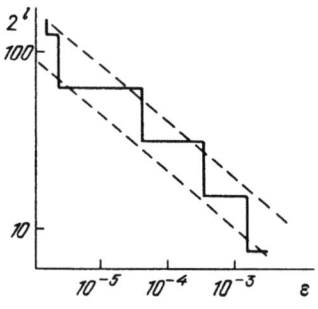

Fig. 8.30

A number of laws governing replacement of resonances on a two-dimensional torus are derived in [449, 597, 611, 644] by considering the two-dimensional mapping

$$\theta_{n+1} = \theta_n + \Omega + br_n - (K/2\pi)\sin 2\pi\theta_n, \quad r_{n+1} = br_n - (K/2\pi)\sin 2\pi\theta_n. \qquad (4.50)$$

The two-dimensional torus associated with (4.50) is specified by the Poincaré rotation number $W = \lim_{n\to\infty}(\theta_n - \theta_0)/n$. Under a resonance in the torus, $W = p/q$, where p, q are integers, if and only if there are θ_0, r_0 such that $\theta_q = \theta_0 + p$, $r_q = r_0$. Each irrational value $W = \overline{W}$ can be represented as the limit of a sequence of rational numbers $W_i = p_i/q_i$, or $\overline{W} = \lim_{i\to\infty} W_i$. In the above papers, the value $\overline{W} = (\sqrt{5}-1)/2$ (golden section) is considered. Meanwhile, $W_i = F_i/F_{i+1}$, where F_i are the so-called *Fibonacci numbers* (satisfying the recurrence relation $F_{i+1} = F_i + F_{i-1}$ with initial conditions $F_0 = 0$, $F_1 = 1$). It is shown in [644] for $b=0$ that the sequence of differences of numbers $\Omega_i(K)$ associated with $W_i = W(K, \Omega_i(K))$ is a geometric progression for sufficiently large i, whose common ratio

$$\delta(K) = \lim_{i\to\infty} \frac{\Omega_{i-1}(K) - \Omega_i(K)}{\Omega_i(K) - \Omega_{i+1}(K)}, \qquad (4.51)$$

generally speaking, depends on K. For $0 \leq K < 1$, $\delta(K) = -\overline{W}^{-2} = -2.61803...$; for $K=1$, $\delta(K) = -\overline{W}^{-y} = -2.83360...$, where $y = 2.16443...$. It is easy to see that formula (4.51) is similar to the Feigenbaum law with constant $\delta(K)$. In [644], another constant similar to the Feigenbaum constant a is determined, viz., $\alpha(K) = \lim_{i \to \infty} \alpha_i(K)$, where $\alpha_i(K) = d_{i-1}/d_i$, $d_i = \theta_{q_{i-1}} - \theta_0 - p_{i-1}$ for $\Omega = \Omega_i(K)$, equal to $-\overline{W}^{-1} = -1.61803...$ for $0 \leq K < 1$, and to $-\overline{W}^{-x} = -1.28857...$ for $K=1$ ($x = 0.52687...$).

In the other papers quoted, (4.50) is considered for $0 < b < 1$. It turns out that property (4.51) also holds in this case, with the difference that $\delta(K) = -\overline{W}^{-2}$ for $0 \leq K < K_c < 1$, and $\delta(K) = -\overline{W}^{-y}$ for $K = K_c$. For $b = 0.5$, $K_c = 0.9788...$.

It is shown in [531] by considering (4.50) with $b=0$ that, for the critical value $K=1$, the ranges of Ω, where $W = p/q$, make up the Cantor set with fractal dimension $d = 0.87$.

CHAPTER 9

EXAMPLES OF MECHANICAL, PHYSICAL, CHEMICAL, AND BIOLOGICAL SYSTEMS WITH CHAOTIC AND STOCHASTIC MOTIONS

At present, a large number of real mechanical, physical, chemical, and biological systems and their models are known that display chaotic processes. Since, as a rule, an analytic study of such systems is impossible, the conclusion that there are chaotic or stochastic motions is made by analyzing the results of numerical or physical experiments. There are already a great many similar results. We discuss some of them, focussing our attention on numerical experiments.

1. Non-linear impact negative-friction oscillator and other systems with discontinuous characteristics

As noted above, one of the simplest and most representative examples of stochastic generators is a negative-friction oscillator whose vibrations are bounded due to impacts decreasing the oscillator velocity by p at moments t_s, where $x=0$, $\dot{x} \geq a > 0$ [150, 280]. An example of a negative-friction oscillator is the Froude pendulum [15, 216, 345]. Assuming friction to be weakly non-linear, we write the oscillator's equation as

$$\ddot{x} - 2\delta(1-\alpha\dot{x}^2)\dot{x} + \omega_0^2 x = -p\sum_s \delta(t-t_s). \qquad (1.1)$$

We analyzed the possible solutions for $\alpha=0$ in Sect.3, Chapt.3; therefore, below, we consider the case $\alpha \neq 0$, where the point mapping $\bar{y}=f(y)$, $y=\dot{x}$, cannot be determined analytically (at least, exactly). However, its behaviour can easily be imagined. In relation to the parameters, the mapping can be one of those illustrated by Fig. 9.1. [150]. In (a), for any initial conditions, periodic oscillations are excited, corresponding to the stable fixed point M_2. In (b), in relation to the initial conditions, either periodic or stochastic oscillations are excited, i.e., there are two attractors, the limit cycle and the strange attractor, in the phase space. Finally, in (c), (d), for any initial conditions, only stochastic oscillations are possible. The ranges of y, associated with steady-state stochastic oscillations, are given by solid lines, and denoted by the letter J. The transition from (a), (b) to (c), (d) occurs when the stable and unstable limit cycles merge (the unstable fixed point M_1 being associated with the unstable limit cycle). According as whether merging takes place below the point A or above it, the transitions turn out to be different. In the former case, there was no strange attractor before the transition, and only a non-attracting homoclinic structure existed. The stable and unstable limit cycles merge just in the structure's region which becomes attracting, and forms a stochastic attractor. Hence, the appearance of stochastic behaviour is afterwards accompanied by intermittency. The above is well illustrated by the graph of the point mapping in Fig. 9.1d, associated with the transition to stochasticity described. As indicated in the preceding chapter, when the transition point is getting more distant, the duration of the "laminar" phases decreases, while that of the "turbulent" phases increases; as a result, the "laminar" phases vanish.

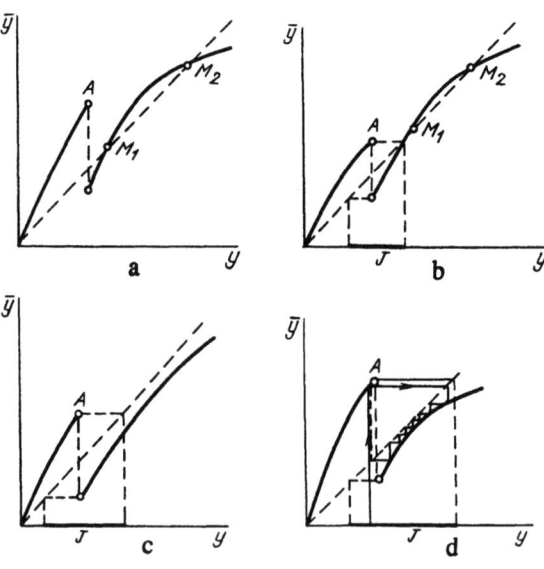

Fig. 9.1

Examples of mechanical, physical, chemical, and biological systems   295

In the latter case where the stable and unstable fixed points merge above A, the strange attractor also existed prior to the transition, along with a stable limit cycle. Since the stable cycle merges with the unstable one outside the attractor's region, there is no intermittency. As the parameter changes in the reverse direction, hysteresis characteristic of hard transitions is observed.

The motion of a solid in a plane under the action of a spring with one end displacing at constant velocity was investigated in [184]. The motion equation is

$$m\ddot{x}+k(x-vt)=F_{fr}(\dot{x}).$$

It is assumed that, for $\dot{x} \neq 0$, $F_{fr}(\dot{x}) = -F_0 \operatorname{sign} \dot{x}$ and that, for $\dot{x}=0$, the force of friction does not in absolute value exceed a quantity $F(\tau)$ depending on the duration (τ) of contact between the body and the plane. For $\tau=0$, $F=F_0$, while, as $\tau \to \infty$, F tends to a certain finite value F_1. Consequently, the dependence $F(\tau)$ is given by

$$F(\tau)=F_1-(F_1-F_0)e^{-\delta\tau}.$$

It was derived analytically that, for certain values of F_1, δ and v, stochastic self-oscillations develop; they can be described by an expansive piecewise linear mapping from a line segment into itself.

2. Tunnel-diode generators

Other sufficiently simple examples of stochastic self-oscillatory systems are tunnel-diode generators of different types [190, 324, 607, 682]. We first consider the one with circuit diagram as in Fig. 9.2 [190, 314, 607]. Assuming that amplifier response is linear, the equations of the generator are:

$$L\dot{I}+RI+\int \frac{I}{C} dt = -u+\frac{MS}{C}I, \quad C_1 \dot{u}+i(u)=I, \qquad (2.1)$$

where i(u) is the current in the tunnel diode D, C_1 the spurious capacitance, and S the slope of the amplifier response. The voltage-current characteristic is qualitatively of the form as in Fig. 9.3a.

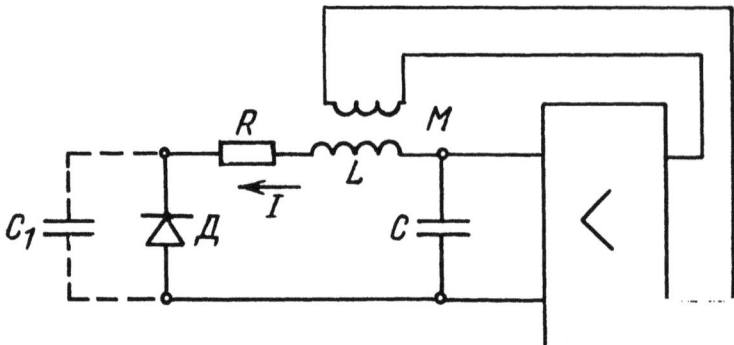

Fig. 9.2

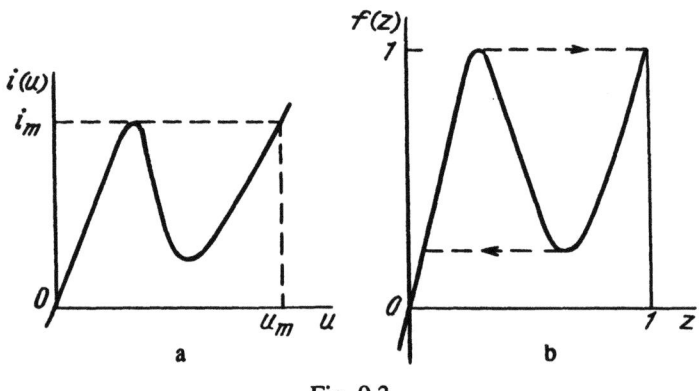

Fig. 9.3

Equations (2.1) can conveniently be written in dimensionless variables as

$$\dot{x} = 2\delta x + y - gz, \quad \dot{y} = -x, \quad \varepsilon \dot{z} = x - f(z), \qquad (2.2)$$

where

$$x = I/i_m, \quad y = -\frac{1}{\sqrt{LC}}\int x\,dt, \quad z = u/u_m, \quad \tau = t/\sqrt{LC}, \quad 2\delta = \frac{MS - RC}{\sqrt{LC}}, \quad g = \frac{u_m}{i_m}\sqrt{\frac{C}{L}},$$

$$f(z) = \frac{i(u_m z)}{i_m}, \quad \varepsilon = g\frac{C_1}{C} \ll 1,$$

and the dot denotes differentiation with respect to dimensionless time τ. The function f(z) is represented in Fig. 9.3b. For sufficiently small ε, motion in the phase space consists of slow uncoilings relative to the equilibrium state x=0, y=0 on the surface of small z, associated with the left-hand branch of f(z), and of sudden jumps up at the right-hand branch, and vice versa. If the jump is reverse, the quantity y turns out to be small, i.e., it decreases rapidly. In this respect, the generator is similar to the one considered in the preceding section.

The form of the point mapping for the system (2.2) if the idealized tunnel diode characteristic is approximated (Fig. 9.4) is shown in Fig. 9.5a [607]. The time series of the process y(t), obtained experimentally [190, 314, 607] for g=2.4, ε=4.8·10^{-5}, is given in Fig. 9.5b. The series is very much similar to the oscillograph recording of the damping impact negative-friction pendulum oscillations in Fig. 3.15.

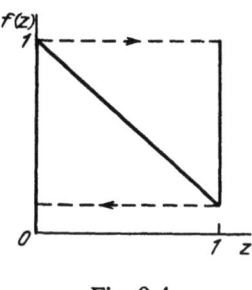

Fig. 9.4

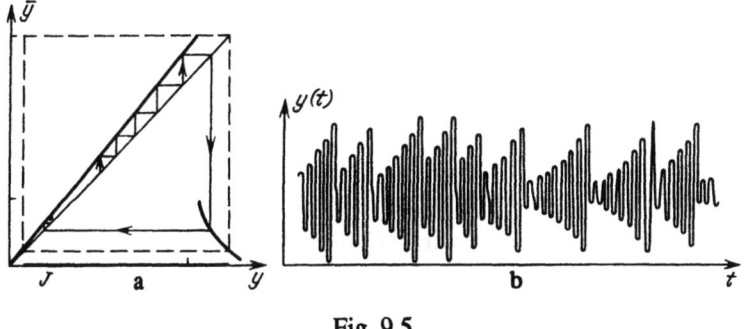

Fig. 9.5

The circuit diagram of a two-tunnel-diode generator is in [682] (Fig. 9.6). Numerical solution of the equations of the generator showed that, in a certain parameter range, the oscillations turn out to be stochastic, the corresponding point mapping discontinuous and exponentially unstable (Fig. 9.7a), while the vibration spectrum is continuous (Fig. 9.7b). It is interesting to note that the point mapping in Fig. 9.7a fully coincides with the one in Fig. 3.14b.

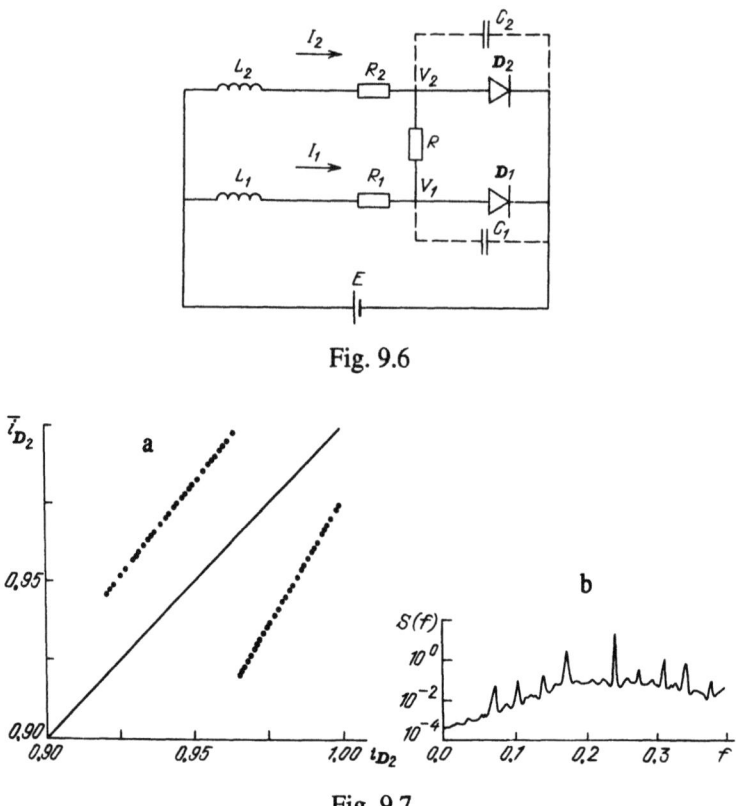

Fig. 9.6

Fig. 9.7

3. Non-linear oscillators with periodic external force

The third coordinate of the three-dimensional phase space of oscillatory systems with one degree of freedom under the action of external forces is time. Chaotic oscillations are possible even under the action of periodic external force. Similar oscillations were observed both in numerical and physical experiments.

First of all, we consider the non-linear oscillator described by the Duffing equation under the action of a harmonic external force whose average value is, generally speaking, non-zero, viz.,

$$\ddot{x} + \alpha\dot{x} + kx + \gamma x^3 = B_0 + B\cos\omega t. \tag{3.1}$$

The equation has been studied in many papers. Various types of elastic characteristics were considered, such as "soft" ($k>0$, $\gamma<0$) [153, 419, 436, 517], "hard" ($k>0$, $\gamma>0$) [208], with zero ($k=0$, $\gamma>0$) [394, 495, 496, 542] and negative ($k<0$, $\gamma>0$)

[386, 506, 508, 585] linear stiffness. Note that, by translating the origin, the equations for k<0 and $B_0 \neq 0$ lead to

$$\ddot{y}+\alpha\dot{y}+ky+\beta y^2+\gamma y^3 = B\cos\omega t \qquad (3.2)$$

with k >0, i.e., to that of a non-linear oscillator with asymmetric elastic characteristic and symmetric external effect.

Both for "soft" and "hard" elastic characteristics, chaotic oscillations occur for sufficiently large amplitude of the external force in the range of frequencies ω (the bistable region) where the corresponding amplitude-frequency curve is not single-valued. Numerical experiments show that the oscillations developed are due to a sequence of period-doubling bifurcations. The regions of these bifurcations and chaotic oscillations were determined in [517] on an analog computer for k=1, γ =-4, α =0.4, B=0.115, B_0=0. The critical value of ω with the onset of chaos is ω_∞=0.5567. Decreasing ω further again leads to periodic oscillations, with amplitude corresponding to the lower branch of the resonance curve.

Simulating equation (3.1) numerically for k=1, γ<0, B_0=0, B=1 [153] shows that, for the parameter values considered in [517] (α=0.4, γ =-0.0529), period-doubling bifurcations and the transition to chaos could not be observed. These phenomena only occur for larger $|\gamma|$ ($|\gamma|>|\gamma_c|$ =0.058), which is equivalent to a larger amplitude of the external force; e.g., for γ =-0.061 (α=0.4), the first doubling bifurcation was observed when ω_1=0.5904; the second, for ω_2=0.5898; the third, for ω_3=0.5897; etc. The doubling bifurcations cascade terminated for ω_∞=0.5896, and chaos then started.

In the region of chaos, there are narrow windows with oscillations of period three, five, and seven times that of the external force. The windows are characteristic of systems described by a one-dimensional point mapping that is differentiable near the maximum [576]. As frequency gets less than ω_{thr}=0.585, the solution tends to infinity.

The results are in Fig. 9.8. The solid lines represent periodic regimes; oblique shading, the region of chaos; the horizontal lines inside the region of chaos, the period-tripling regime; the vertical ones, the period-quintupling regime. The dependence of the region of bifurcations and of chaotic oscillations on the parameter γ for several α is given in Fig. 9.9; it can be seen that, as γ increases, the region with complicated behaviour shifts towards higher frequencies. The dashed line denotes the boundary where a homoclinic structure is generated, and α=0.4; it was constructed on the basis of V. K. Melnikov's criterion [250, 507, 580] by the formula $\gamma_{cr}=(2\alpha^2/9)(\sinh x /x)^2$, where $x=\pi\omega/\sqrt{2}$. That the boundary does not coincide with the boundary for chaos means there is a range of values γ, ω with a homoclinic structure; however, the latter is not attracting. The transition processes must be chaotic in the range.

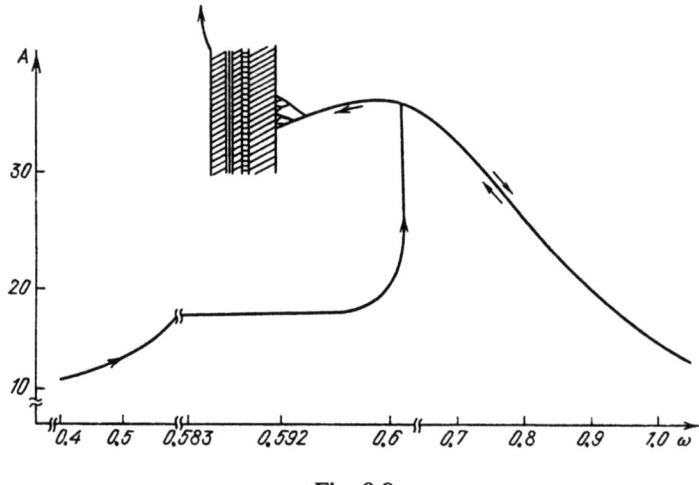

Fig. 9.8

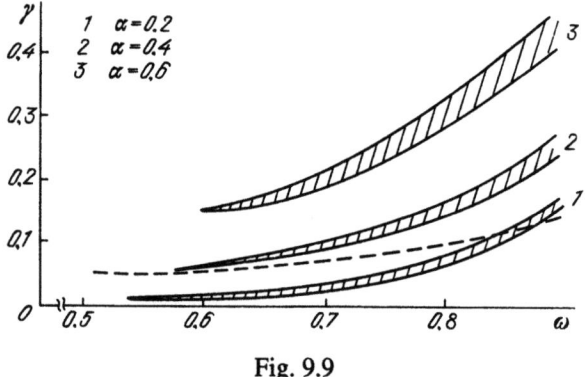

Fig. 9.9

If the elastic characteristic is hard, then chaos arises for considerably larger external force amplitude. Thus, for α=0.2, k=1, γ=1, B_0=0, chaos was observed when B=50 in the frequency range 1.5≲ω≲2 [208]. We illustrate this by the point distribution on the cutting plane t=0[mod 2π/ω] when ω=1.9 (Fig. 9.10). The regions marked by 1, 2, 3 were distinguished, for which the relative frequencies p for the representing point to get there under different initial conditions were determined. It turns out that, as the number of iterations increases, the frequencies tend to limiting values which determine the limiting probability that the representing point will be there (for the region marked by 1, p≈0.05; by 2, p≈0.26; by 3, p≈0.10). However, it remains unclear whether the probability characterizes the internal dynamics of the system or external noise.

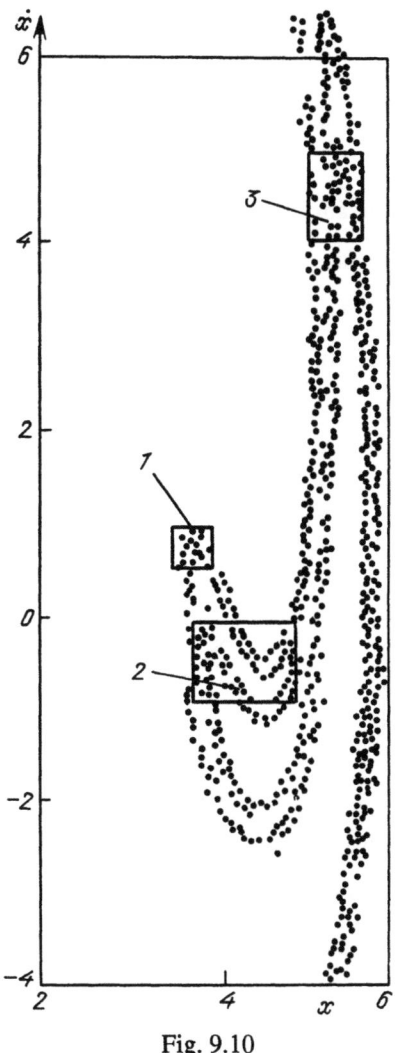

Fig. 9.10

For an oscillator with assymetric elastic characteristic (or asymmetric external action), chaotic solutions arise in a wide range of external force amplitudes and frequencies. Thus, for $\alpha=0.05$, $k=0$, $\gamma=1$, $B_0=0.04$, $\omega=1$, chaos was observed when

B>0.12 [496, 542]. The point mapping on the cutting plane t=0[mod $2\pi/\omega$] when B=0.14 is shown in Fig. 9.11.

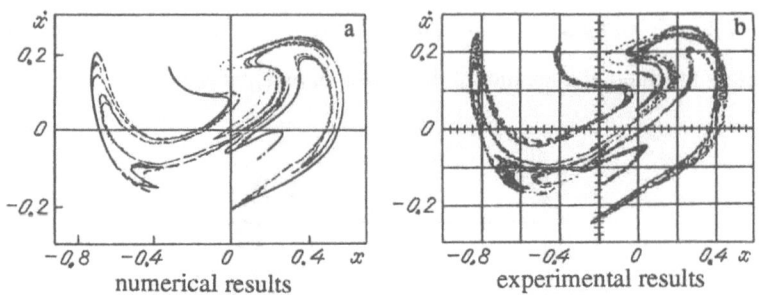

numerical results   experimental results

Fig. 9.11

The structure of the attractor on the cutting plane t=0[mod 2π] for (3.1) with k=−1, γ=1, ω=1, B_0=0 and for different α and B was studied in [508]. The fractal dimension d in relation to the damping factor α was estimated in [585] for a similar case (γ=−k=1/2, ω=0.833, B=0.16, B_0=0). It turns out that, for small α, d is close to three; and d decreases up to two for increasing α (Fig. 9.12).

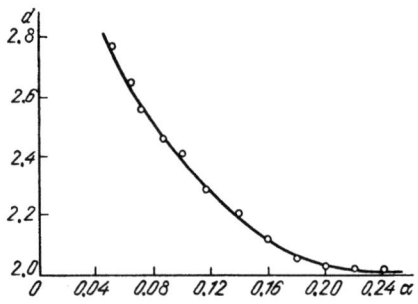

Fig. 9.12

The transition to chaos in the oscillator described by (3.1) for k<0, γ>0, B_0=0 was observed experimentally in [386]. The scales were selected so that k=−1, γ=4, α=0.154. When the frequency of the external force is ω=1.253, the transition to chaos for increasing amplitude of the external force B occurs via a sequence of period-doubling bifurcations. From certain B onwards, a chaotic attractor on the (x,ẋ)-phase plane has been observed to enclose (depending on the initial conditions) one of the singular points (±1/2, 0) of the autonomous oscillator. Meanwhile, the vibration spectrum becomes continuous, but strongly truncated in the low-frequency region (Fig. 9.13a). As B increases, infrequent random jumps occur from the attractor to another one enclosing the other (symmetrically located) singular point. Because of the jumps, the spectrum substantially expands into the low-frequency region (Fig. 9.13b). As B increases further, each attractor starts enclosing both singular points. Random jumps from attractor to attractor also occur, leading to a low-frequency part which varies according to 1/f (Fig. 9.13c). The associated time series is given in Fig. 9.13d. In the opinion of [386], the phenomenon can serve as a model for *flicker noise*. A more general picture of flicker-like noise appearance because of jumps between different attractors is given in [385][13].

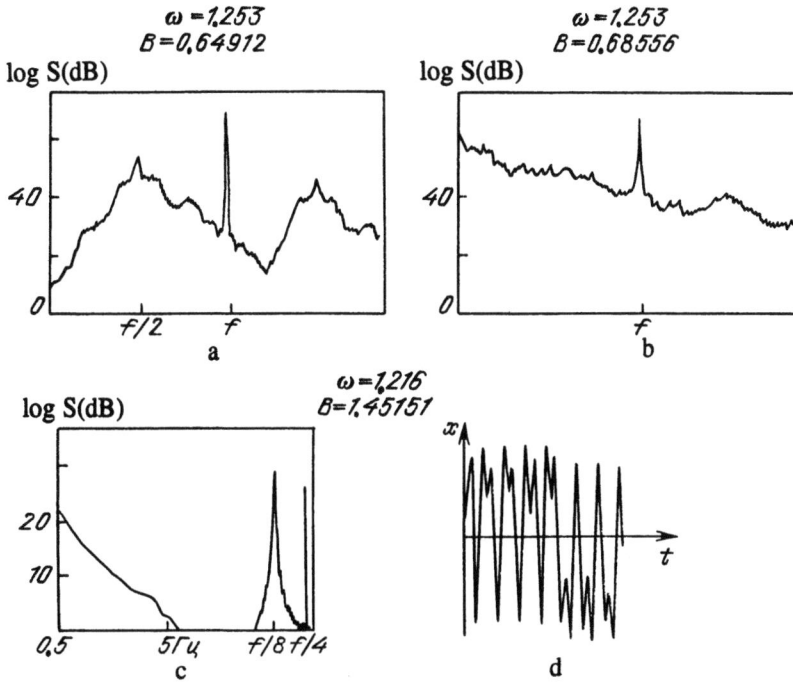

Fig. 9.13

Experimental results for the non-linear oscillatory circuit specified by the equation

$$LC\ddot{V} + RC\dot{V} + V = V_0 \sin(2\pi ft),$$

where $C = C_0(1+V/0.6)^{1/2}$, $C_0 = 300$pF, $L = 10$mH and $R = 28\Omega$, are given in [529, 562, 603,

---

[13]A totally different attempt to account for flicker noise on the basis of dynamical chaos was made in [94].

658]. As V_0 increases, a sequence of period-doubling bifurcations arises, and then chaos sets in. The region of chaos is not continuous; it has windows with oscillations of different periods. The data in Table 9.1 [658] with boundary values of V_0 corresponding to the regions of periodic regimes (i.e., windows) manifest just that. The sequence of periodic solutions strongly resembles the order in which multiplicities of fixed points alternate for the one-dimensional mapping (established by A. N. Sharkovsky [368]). The transition to chaos from different periodicity windows occurs differently: either via a sequence of period-doubling bifurcations [562, 658], via intermittency [603], or is hard with typical hysteresis [529].

Period	V_0		Period	V_0	
2	0,639		5	2,353	Window
4	1,567	Bifurcation	10	2,363	
8	1,785	values	7	2,693	Window
16	1,836		14	2,696	
32	1,853		3	3,081	
Chaos	1,856	Chaos onset	6	3,338	Extensive
12	1,901	Windows	12	3,711	Window
24	1,902		24	3,821	
6	2,073	Windows	9	4,145	Window
12	2,074		18	4,154	

Table 9.1

Similar and more complicated solutions are also obtained for the equation of the forced physical pendulum [139, 140, 401, 408, 432, 482, 543],

$$\ddot{\phi}+\alpha\dot{\phi}+\sin\phi = B_0+B\cos\omega t, \tag{3.3}$$

and for the equation

$$\ddot{\phi}+\alpha\dot{\phi}+\sin\phi+\beta\phi = B_0+B\cos\omega t, \tag{3.3*}$$

[548]. Within certain approximations, they describe the phase of the wave function in a Josephson junction injected into an a.c.-circuit [139, 140, 401, 408, 482, 548] and a phase

Examples of mechanical, physical, chemical, and biological systems 305

synchronization system [82–87, 432][14]. We illustrate by the bifurcation diagram on the parameter (B,ω)-plane, (Fig. 9.14) the evolution of the phase portrait, and the vibration spectrum (Fig. 9.15g-j) as B increases for fixed $\omega=0.67$, obtained experimentally in [432] for the phase synchronization system described by (3.3). It can be seen that, first, for the symmetric system (3.3), the phase portrait ($B_0=0$) can be assymetric (Fig. 9.15b-f); second, the transition to chaos occurs via period-doubling bifurcations; as B increases further, inverse bifurcations are observed (Fig. 9.15d,g). For still larger B, islands in the periodic motions appear; e.g., in Fig. 9.15e,i, the phase portrait and spectrum of a period-tripling motion are represented.

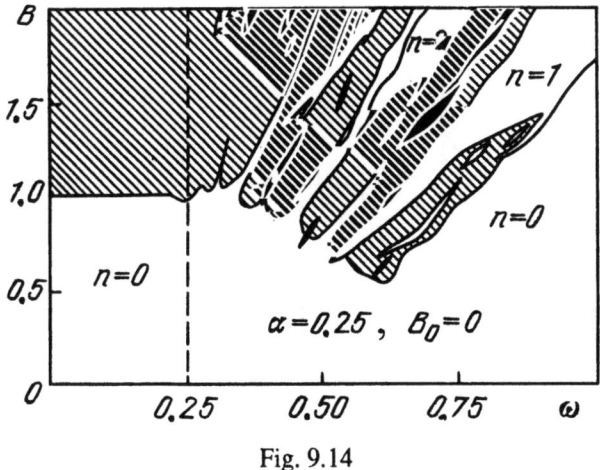

Fig. 9.14

---

[14]Note that, apparently, [82–87] are the first to indicate the possibility of chaotic regimes in phase synchronization systems, and to show the role of homoclinic structures.

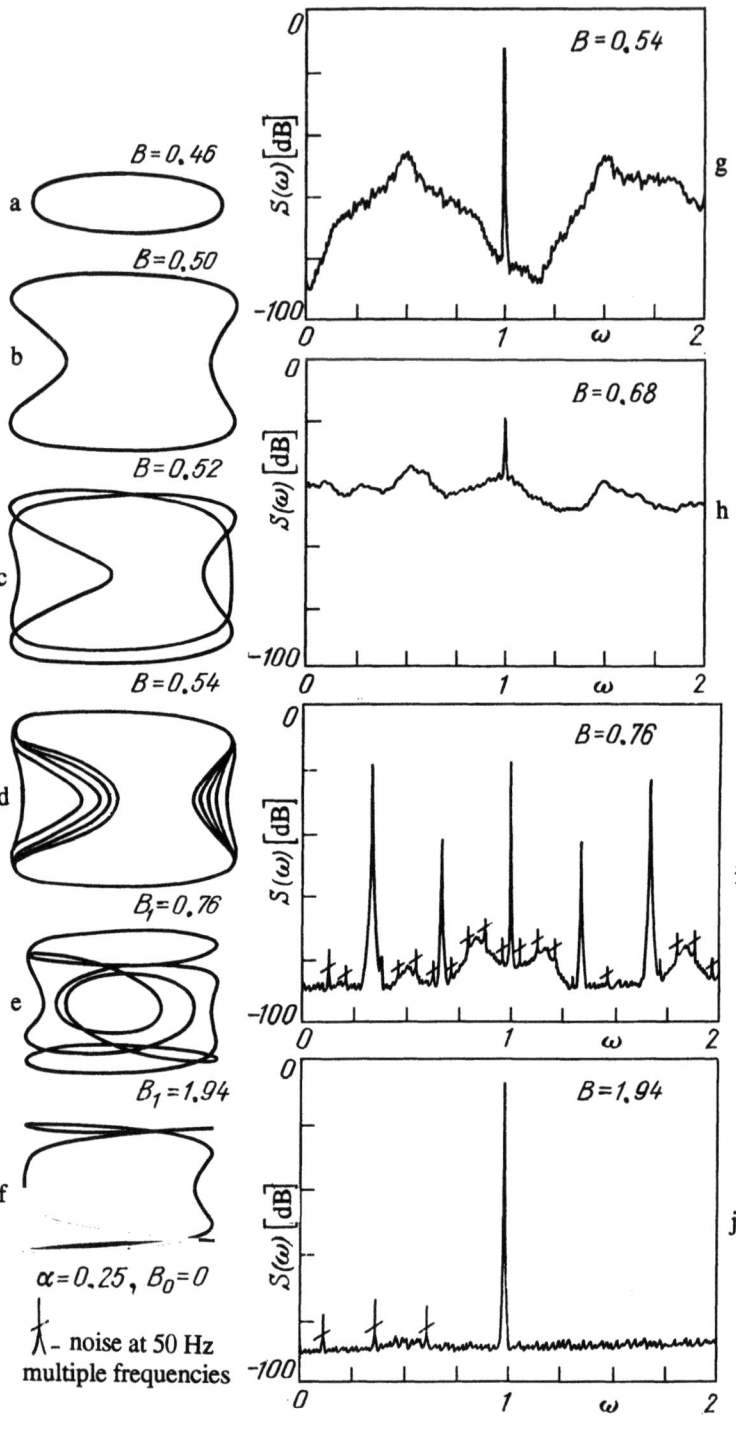

Fig. 9.15

Examples of mechanical, physical, chemical, and biological systems 307

In the opinion of [432], the chaos is due to random transitions of the representing point from a region near one unstable cycle into a nearby region of another such cycle.

The numerical calculations in [543] for (3.3) and $B_0=0$, $\alpha=0.2$, $\omega=0.6$ are somewhat more detailed. Thus, Table 9.2 supplies different types of solution arising as the parameter B varies with sufficiently small step. Due to rapid change, they naturally cannot be noticed in the experiment, nor can they be reflected in Fig. 9.14. The most interesting difference is that, as B changes very slightly from 0.61278 to 0.61279, transition from the periodic motion to the chaotic one occurs, and is not accompanied by period doublings. The authors account for the jump as being due to change in symmetry of the solution. In turn, the chaotic regime is rather rapidly replaced with the period-7 regime.

B	Solution type	B	Solution type
0,61278	Period-1	0,7607—0,77	2 x 2 sequence
0,61279	chaotic	0,77—0,80	chaotic
0,62	chaotic	0,81	perios-2
0,626	period-7	0,82	period-16
0,63465	period-14	0,83	period-2
0,64	chaotic	0,84—0,89	chaotic
0,65	period-7	0,90	period-3
0,66	chaotic	0,91	period-3
0,67—0,71	period-1	0,92	period-6
0,72	period-2	0,93—0,96	chaotic
0,725	period-4	0,97	chaotic
0,73	chaotic	0,98	chaotic
0,7309	period-5	0,99	period-3
0,74—0,76	chaotic	1,0	chaotic

Table 9.2

In contrast to (3.1), equations of the form (3.3), (3.3*) admit solutions associated with regular and irregular pendulum rotation in some or other sense. The solutions were obtained numerically in [543], and observed experimentally in [432].

In the bifurcation diagram (Fig. 9.14), these are characterized by the number n of rotations in the period of the external force. In the region of chaos, n and the sense of rotation can vary from period to period.

Period-doubling bifurcations and the transition to chaos under the action of an external harmonic force were also discovered in such simple systems as an impact oscillator and a piecewise linear oscillator. The equations of motion can be written as

$$\ddot{x}+(2h/\eta)\dot{x}+(1/4\eta^2)x = (1/\eta^2)\sin t, \quad x>0$$

and

$$\ddot{x}+2h\dot{x}+k(x)x = \sin(2\pi\tau+\phi),$$

where $k(x)=k_1$ for $x>0$, $k(x)=k_2$ for $x<0$. They are of interest, since they admit an exact analytic solution, in principle. The former equation was studied in [659] for $\eta=4.5$; the latter, in [639] for $k_1=(2\pi/1.1)^2$ and $k_2=(2\pi/4.5)^2$. As bifurcation parameter, h was chosen. Six period-doubling bifurcations were studied in [639]. The corresponding values of h_n and $\delta_n=(h_n-h_{n-1})/(h_{n+1}-h_n)$ are given in Table 9.3.

n	1	2	3	4	5	6
h_n^{-1}	0,5889	0,6479	0,65546	0,657038	0,6573748	0,65744680
δ_n			7,80	4,79	4,69 -	4,68

Table 9.3

Oscillation chaotization also occurs under a periodic force acting on non-linear oscillators specified by more complicated equations [496, 497], or on a system of coupled non-linear oscillators [123, 211, 583]. Some authors have attempted to give an account of certain heart diseases as due to chaotization; e.g., arrhythmic contractions of the cardiac muscle [466, 530], observed experimentally [467]. However, it should be noted that, if healthy, the heart must beat in the chaotic regime [701]. Transition to a periodic regime means a disease.

The paper [497] including numerical simulations of the equations

$$\ddot{x}+k_1\dot{x}+(1/8)(3y^2+x^2)x = B\cos t,$$

$$\ddot{y}+(k_2/8)(3x^2+y^2)y = B_0 \qquad (3.4)$$

for $k_1=k_2=0.5$, $B_0=0.03$ and varying parameter B is especially interesting in that, apparently for the first time, quasi-period-doubling bifurcations of the torus were observed in it. Fig. 9.16 giving the evolution of the invariant curve on the projection of the cutting plane t=0 [mod 2π] attests just to this fact. It can be seen in the diagram that the authors observed at least two doubling bifurcations (Fig. 9.16b,c), after which the chaotic attractor (d) arose.

Until now, we have only considered external actions (external forces) involved additively in the motion equations. Under parametric external actions on non-linear oscillators, no less interesting effects were observed. Possible bifurcations and chaotic regime sets for such oscillators are described in Sect.4, Chapt.7 (mostly, on the basis of [60, 62–67]). Here, we dwell on other results, and focus our attention on the quality of behaviour of the system and on the spectral characteristics of oscillations.

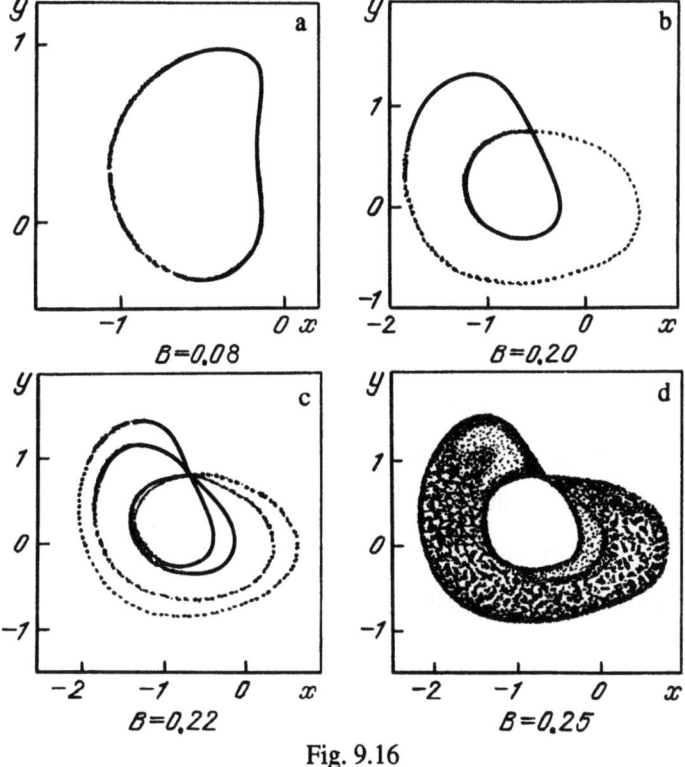

Fig. 9.16

The equation of the physical pendulum with vibrating suspension axis is

$$I\ddot{\phi}+H\dot{\phi}+mb(g+av_0^2 \sin v_0 t)\sin\phi = 0, \qquad (3.5)$$

where I is the moment of inertia, m the mass of the pendulum, b the distance from the mass centre to the axis, $H\dot{\phi}$ the moment of forces of friction, g the acceleration due to gravity, a the amplitude, and v_0 the frequency of oscillation of the suspension axis. By varying the time scale and shifting the reference point, (3.5) can be reduced to

$$\ddot{x}+\alpha\dot{x}+(1+2q\cos v_0 t)\sin x = 0. \qquad (3.6)$$

In the fundamental parametric resonance region ($v_0=2$), the equation was given a detailed study in [546, 547, 559, 560, 581]. For small α, limit of stability of the equilibrium state for (3.6) is determined by the condition $q=q_0 \approx \alpha$; for $q>q_0$, oscillations with period $T=2\pi$ develop. They remain stable up to a certain value $q=q_1$ depending on α. Thus, $q_1=0.713\pm0.012$ when $\alpha=0.2$. Meanwhile, rotatory pendular motion of two different kinds becomes possible, with period equal to that of the suspension axis π: clockwise and counterclockwise. As q increases, each motion is subject to period-doubling bifurcations. The bifurcation values of the parameter of q are given in Table 9.4.

T/π	q
1	0,7925±0,0025
2	0,984±0,010
4	1,01975±0,00025
8	1,03075±0,00025
16	1,03275±0,00025

Table 9.4

From the cascade of doubling bifurcations, two chaotic attractors are formed in the phase space. They are associated with irregular pendular motion in each direction (depending on the initial conditions). When q=1.045, the sense of rotation starts changing. The average frequency of sense change increases with increasing q. Meanwhile, chaotic attractors are placed nearer and nearer to unstable manifolds passing through the points $x=\pm\pi$, $\dot{x}=0$ on the cutting surface $t=0$ [mod π] (Fig. 9.17). For q=1.58, the inverse transition from chaotic to periodic motions occurs, associated with stable rotations in both directions with period π. As the coefficient of friction α decreases, the picture does not change qualitatively; however, all bifurcation values of q somewhat decrease.

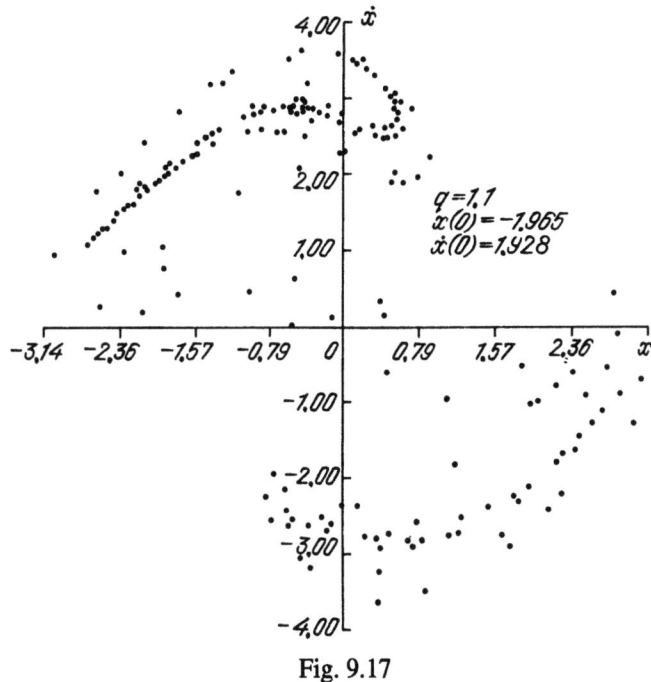

Fig. 9.17

Outside the regions of parametric resonance, the behaviour of the solution of (3.6) was studied in [91, 93, 109, 171, 423, 671] and other papers. It is known from the theory of the Mathieu equation that, for sufficiently large frequency and amplitude of the oscillation of the suspension axis ($v_0 \gg 1$, $q \gg q_1$), the upper equilibrium position of the pendulum becomes stable; see [91, 93, 171]. (Meanwhile, the lower equilibrium position also remains stable.) The limit of stability for the upper equilibrium position can be conveniently estimated analytically if the damping α is neglected, and (3.6) written as

$$\frac{d^2\xi}{d\tau^2} = (\varepsilon^2 + 2\varepsilon c \cos \tau)\, \xi \tag{3.7}$$

for small $\xi = x - \pi$, where $\tau = v_0 t$, $\varepsilon = 1/v_0$ is a small parameter, $c = q/v_0$. As we will see below, the limit of stability of c is of order one.

The solution of equation (3.7) for time τ less than, or of order, one is sought for as a power series in ε, viz.,

$$\xi = y_0 + \varepsilon y_1 + \varepsilon^2 y_2 + \dots \, . \tag{3.8}$$

Substituting (3.8) in (3.7), equating the terms involving equal powers of ε, we find $y_0, y_1, y_2, \dots$. Restricting to infinitesimals of order two, we see that

$$\xi = C_1 \left[\tau + 2c\varepsilon (2\sin\tau - \tau\cos\tau) + \varepsilon^2 \left(\frac{1}{6}\tau^3 - \frac{3}{2}c^2\sin 2\tau - \frac{1}{3}c^2\tau^3 + \frac{1}{2}c^2\tau\cos 2\tau \right) \right]$$

$$+ C_2 \left[1 - 2c\varepsilon\cos\tau + \varepsilon^2 \left(\frac{1}{2}\tau^2 + c^2\tau^2 + \frac{1}{2}c^2\cos 2\tau \right) \right],$$

$$\dot\xi = C_1 \left[1 + 2c\varepsilon (\cos\tau + \tau\sin\tau) + \varepsilon^2 \left(\frac{1}{2}\tau^2 - \frac{5}{2}c^2\cos 2\tau - c^2\tau^2 - c^2\tau\sin 2\tau \right) \right]$$

$$+ C_2 \left[2c\varepsilon\sin\tau + \varepsilon^2 (\tau - 2c^2\tau - c^2\sin 2\tau) \right], \tag{3.9}$$

where C_1, C_2 are two arbitrary constants determined by the initial conditions.

To find the limit of stability by Floquet theory [359], it suffices to have two fundamental particular solutions of (3.7) for $\tau = 2\pi$. Denote them by $\xi_1(\tau)$, $\xi_2(\tau)$. If $\xi_1(\tau)$ satisfies the initial conditions $\xi_1(0)=1$, $\dot\xi_1(0)=0$, and $\xi_2(\tau)$ the initial conditions $\xi_2(0)=0$, $\dot\xi_2(0)=1$, then the stability condition for the equilibrium state is

$$[\xi_1(2\pi) + \dot\xi_2(2\pi)]/2 \equiv \beta \leq 1 \tag{3.10}$$

[359]. Meanwhile, the multipliers are determined by the expression

$$\mu_{1,2} = \beta \pm \sqrt{\beta^2 - 1}. \tag{3.11}$$

It follows from (3.9) that $\xi_1(2\pi) = \dot\xi_2(2\pi) = 1 + 2\pi^2\varepsilon^2(1-2c^2)$. The stability condition $c^2 \geq 1/2$, or $q \geq q_1 = v_0/\sqrt{2}$, is found from the above and (3.10). Under this condition and for $\alpha=0$, we find that if the deviation from the upper equilibrium position is small, then the pendular oscillations are (with frequencies v_0 and $\Omega = \sqrt{2c^2-1}$) undamped quasiperiodic (Fig. 9.18). (The expression for Ω follows from (3.11).) If $\alpha \neq 0$, then the oscillations are damped, and the equilibrium position is asymptotically stable.

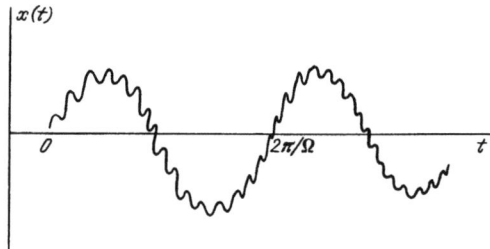

Fig. 9.18

As q increases, the "natural" frequency Ω of pendular oscillations relative to the upper equilibrium position increases; when Ω attains the value $v_0/2$, instability arises again (where the multipliers pass through the value -1). The upper stability limit q^* can be estimated if we put $c=q^*/v_0$, $\xi = A \sin(\tau/2) + B \sin(3\tau/2)$ in (3.7). For q^* and the ratio B/A, we then obtain $q^* = (\sqrt{117}-9)v_0^2/8$, $B/A = (\sqrt{117}-9)/18 \approx 0.1$. Numerical computations by T. S. Landa yielded the following results: For $\alpha=0.1$ and $v_0=9.5$, we have $q_1=6.9$, $q^*=21.5$; for $v_0=20$ and the same value of α, $q_1=14.15$, $q^*=91.70$. The quantities are close to the theoretical values for $\alpha=0$: $q_1=6.72$, $q^*=20.49$ for $v_0=9.5$; $q_1=14.14$, $q^*=90.83$ for $v_0=20$.

Stability of the lower equilibrium position is lost similarly. The ``natural'' oscillation frequency of the pendulum relative to the position can easily be shown to equal $\Omega=\sqrt{2c^2+1}$. As the frequency attains the value $v_0/2$, stability is lost. Since the effect of gravitational force is negligible for the values of q ($v_0 \gg 1$) associated with stability loss, the limit value q is approximately equal to the value q^* found above.

For $q>q^*$, depending on the initial conditions, the pendulum can swing with undamped oscillations of period $4\pi/v_0$ either relative to the lower or to the upper equilibrium position. Besides, the regular rotatory regimes are possible when the pendulum makes one complete rotation in one or the other direction in the period of oscillation of the suspension axis. The computerized projection of the phase portrait onto the $(x,\dot{x})$-plane for $\alpha=0.1$, $v_0=20$, $q=95.92$ is in Fig. 9.19a [225]. The shapes of the process $\dot{x}(t)$, its spectral density and associated limit cycle, with oscillations relative to the upper equilibrium position, are all given in Figs. 9.19b,c,d for the same parameter values. Note that the exteriors of the regions of attraction of the limit cycles are rather narrow. For comparatively small deviations from the cycles, the pendulum passes into the regime of rotation.

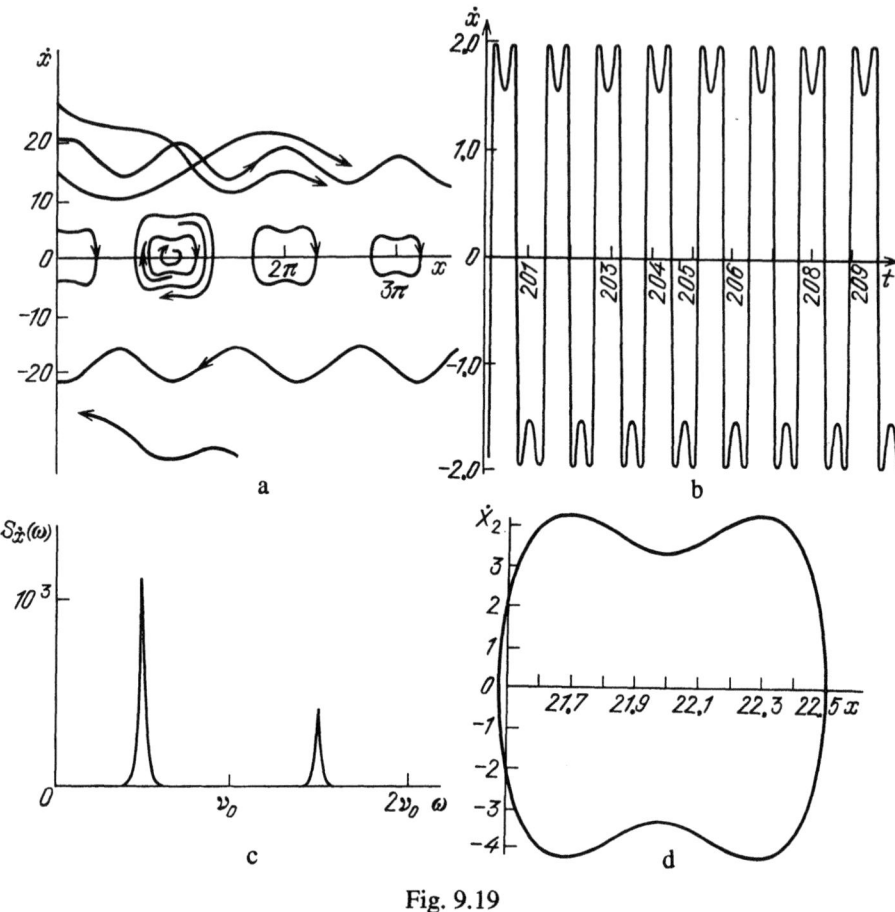

Fig. 9.19

As noted above, for $v_0 \gg 1$ and $q \gtrsim q^*$, the effect of the gravitational force is negligible; therefore, equation (3.6) can be written as

$$\ddot{x} + \alpha \dot{x} + 2q \cos v_0 t \, \sin x = 0. \tag{3.12}$$

Note that (3.12) follows from investigations of particle motion in a 2-wave field [109, 225, 671], of compass needle motions in an alternating magnetic field perpendicular to the compass axis [423], etc.

To estimate the amplitude of oscillations for small $q-q^*$, we put $\alpha=0$ in (3.12), expand $\sin x$ into a series in the vicinity of $x=0$ (or $x=\pi$), and restrict ourselves to the cubic term, viz.,

$$\ddot{x} + 2q \cos v_0 t \, (x - x^3/6) = 0.$$

If we substitute $x = A \sin(v_0 t/2)$ (amplitudes of the third and higher harmonics

Examples of mechanical, physical, chemical, and biological systems 315

near the instability limit being small), then $A^2=6(1-q^*/q)$. The dependence is consistent with numerical computation (the dots in Fig. 9.20). From a certain value $q=q_2$ onwards ($q_2 \approx 98.47$ for $v_0=20$), stable limit cycles of the upper and lower equilibrium positions merge with unstable ones, and vanish. As a result, regular modes of rotation only remain. The conditions

$$x(t+2\pi n/v_0) - 2\pi m = x(t), \quad \dot{x}(t+2\pi n/v_0) = \dot{x}(t),$$

where $n=1,2,...$, $m=\pm 1,\pm 2,...$, then hold (synchronisms of n,m-type; see Chapt.7). The above conditions mean that, first, the rotation velocity of the pendulum changes with period n times that of the external action; second, the pendulum makes m rotations in n external action periods. The domains of existence of regimes associated with different values of n, m were found in [65, 66]. Below, only those regimes of rotation are mentioned in which the pendulum makes one complete rotation during the oscillation period of the suspension axis, i.e., for m=n. For $q<q_2$, regimes associated with m=n=1 were observed. Meanwhile, the velocity spectrum is as in Fig. 9.21a. It is interesting to note that the frequency v_0 is not observed in the spectrum, which is related to practical equivalence of the upper and lower equilibrium positions. After the limit cycles associated with oscillatory modes vanish ($q>q_2$), the frequency $v_0/2$ appears in the rotatory velocity spectrum of $\dot{x}$ (Fig. 9.21b), i.e., m=n=2. This reveals a kind of memory of the system for oscillations with frequency $v_0/2$. The corresponding point mapping on the cutting plane t=0 [mod $2\pi/v_0$] and part of the projection of the phase portrait are in Figs. 9.22a, 9.23a. As q increases, a period-doubling bifurcation arises, i.e., the frequency $v_0/4$ appears in the spectrum. It then vanishes along with $v_0/2$, while the point mapping becomes as in Fig. 9.22b. This regime is associated with m=n=1, and exists for $124.44 < q < 173.40$. For q=173.40, frequencies $2v_0/3$, $4v_0/3$ appear in the velocity spectrum (Fig. 9.21c), i.e., the regime with m=n=3 arises. The point mapping associated with the latter regime, and the projection of the phase portrait are in Figs. 9.22c, 9.23b. The regime is stable for $173.40 < q < 189.72$. It then loses stability, and a period-doubling motion arises. The next period-doubling bifurcation occurs for q=190.10. The sequence of doubling bifurcations subsequently terminates, and, for q=190.70, the regime of rotation arises again with m=n=1 (Fig. 9.22d).

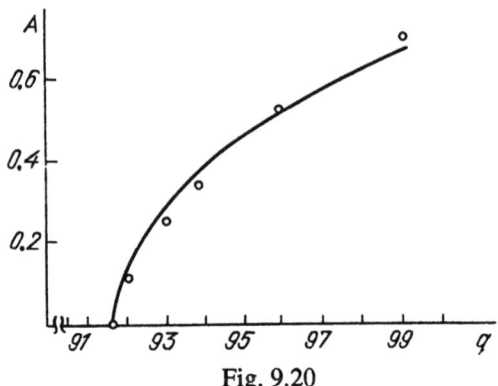

Fig. 9.20

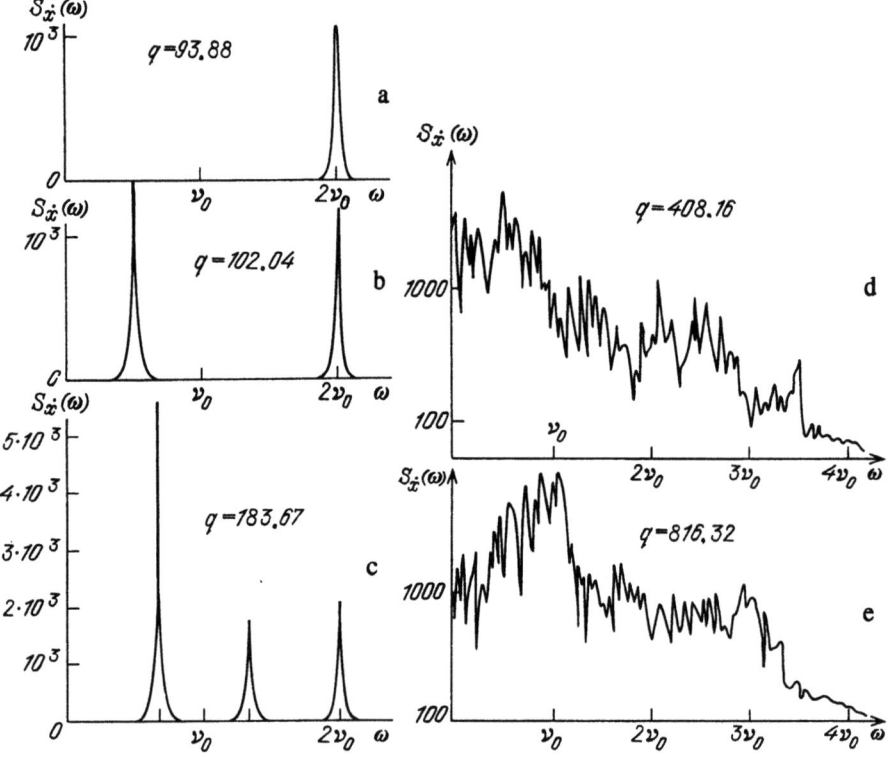

Fig. 9.21

Examples of mechanical, physical, chemical, and biological systems

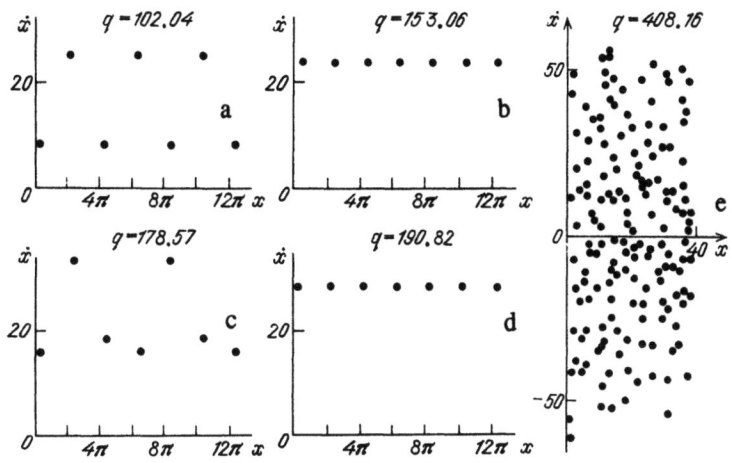

Fig. 9.22

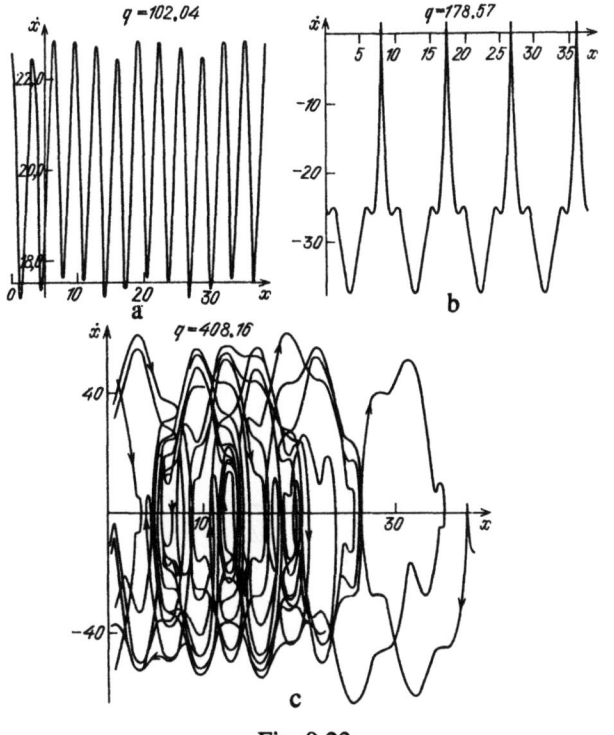

Fig. 9.23

For q≈204, the initial homoclinic structure (its existence can be seen by the chaotic behaviour of transient processes) becomes attracting. Besides the regular attractors associated with periodic rotation, another, chaotic, attractor is generated in the system. Its region of attraction increases as q increases, while those of regular attractors decrease. For q≈306, multipliers of cycles associated with regular rotations attain the value +1, and the cycles vanish. Only chaotic attractors remain. For $q>q_{cr}≈306$, regular modes of rotation were not observed [225]. (It follows from [65] that, for $\alpha=0$, the regimes must vanish when q>335.) By B. V. Chirikov's criterion for overlapping resonances [418], which holds if $\alpha=0$, chaos should set in when $q \sim \pi v_0^2 ≈ 1,257$, i.e., for q substantially larger than q_{cr}.

In the region of chaos, the power spectrum of the process $\dot{x}(t)$, the point mapping and the projection of the phase diagram are given in Figs. 9.21d,e, 9.22e and 9.23c, respectively. The dependence of $\dot{x}$ on t (Fig. 9.24a) and the correlation function of the random process $\dot{x}(t)$ (Fig. 9.24b) are illustrated for the same parameter values. It can be seen in Fig. 9.23c that, in the chaotic regime, pendular motion consists of regions of oscillations about equilibrium positions, rotatory motions in some sense, and jumps in between. Along with the unpredictability at any moment of x, $|\dot{x}|$, the sense of pendular rotation is also unpredictable, irrespective of the initial conditions.

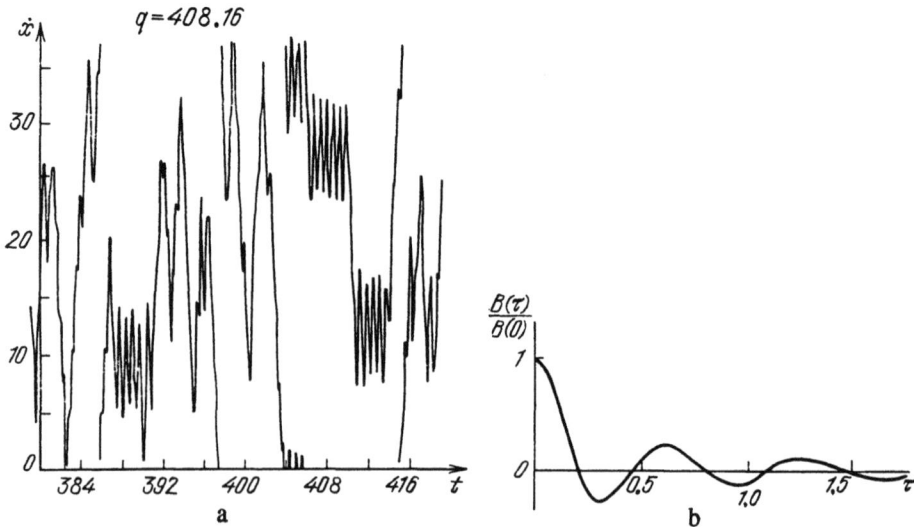

Fig. 9.24

Examples of mechanical, physical, chemical, and biological systems 319

Note that the point mapping obtained is the same in the case α=0.1 under consideration as when there is no dissipation (α=0), which can be determined from comparing Fig. 9.22e with Fig. 9.25a [671]. For sufficiently large α, the point mapping changes substantially (Fig. 9.25b).

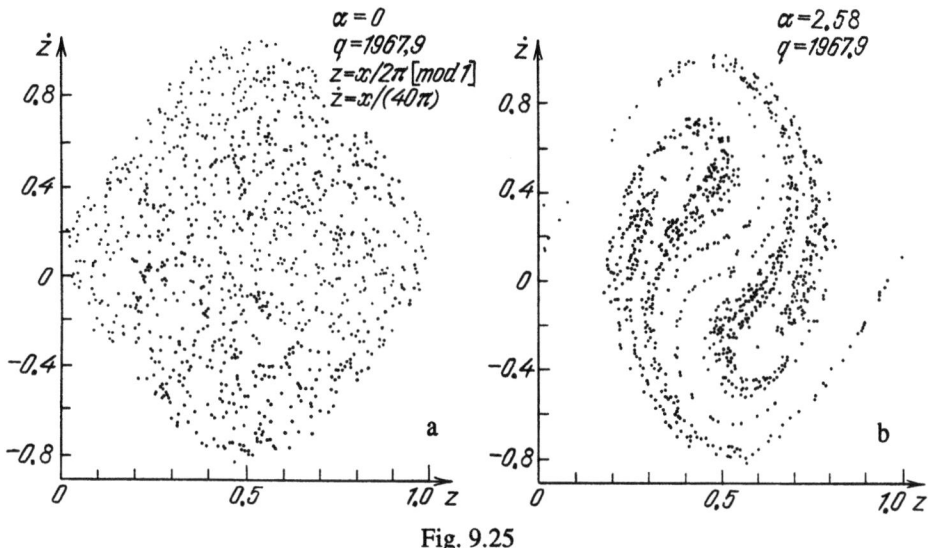

Fig. 9.25

The correlation exponent ν specifying the dimension of the attractor was calculated in the region of chaos for the time series of the process $\dot x(t)$. The results of computation for α=0.1, v_0=20 are given in Table 9.5. As expected, for increasing q, the dimension of the attractor increases.

q	306	408	510
ν	2,4	2,6	2,7

Table 9.5

Period-doubling bifurcations and transition to chaos were observed in [47, 161, 526, 528] for the equation

$$\ddot x+\alpha\dot x+(1+q\cos v_0 t)x+x^3 = 0. \qquad (3.13)$$

It was shown that, in the region of chaos, the point mapping can be reduced to a one-dimensional mapping that is differentiable at its maximum. For fixed q and varying α, transitions from one type of attractor to another were discovered [47]. One was observed for $v_0=2$, $q=50$ and α decreasing up to $\alpha^*=0.4576$. For $\alpha<\alpha^*$, the fractal dimension d of the attractor increases sharply, as do the average power N and spectral density $S(\omega)$ of oscillations (Fig. 9.26).

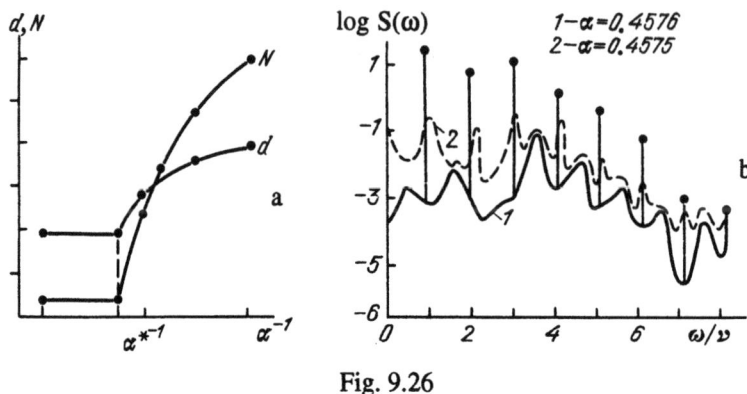

Fig. 9.26

The equations for the amplitude of the field strength in a gas laser in an external modulated light field [683] or in a constant field with modulated inversion in one-mode approximation [636] are reduced to those of the externally driven non-linear oscillator (both under parametric and power action); the same concerns the equations for field strength and difference between level populations in a solid-state laser in periodic modulation of loss [248, 249, 527]. The equations are

$$\dot{x} = g\epsilon x\,(n-1-\beta\cos\Omega t), \quad \dot{n} = \epsilon[\alpha-n(x+1)], \qquad (3.14)$$

where x is the dimensionless field strength, n the relative difference of level populations, α the population's relative difference in the absence of the field, specified by inversion, $g=\Delta\omega_p/\gamma \gg 1$ the product of the resonator strip width $\Delta\omega_p$ and the population difference's relaxation time γ^{-1}, $\epsilon=g^{-1}(\alpha-1)^{-1} \ll 1$, β the amplitude, and Ω the frequency of loss modulation. The evolution of oscillations and the phase portrait for typical solid-state laser parameters, for $\Omega=0.4$ and changing β are illustrated in Fig. 9.27 [249]. It can be seen that transition to chaos occurs via a sequence of period-doubling bifurcations. A similar transition was observed experimentally for a CO_2-laser in [387], and for a solid-state neodymium garnet laser in [357, 544].

Examples of mechanical, physical, chemical, and biological systems

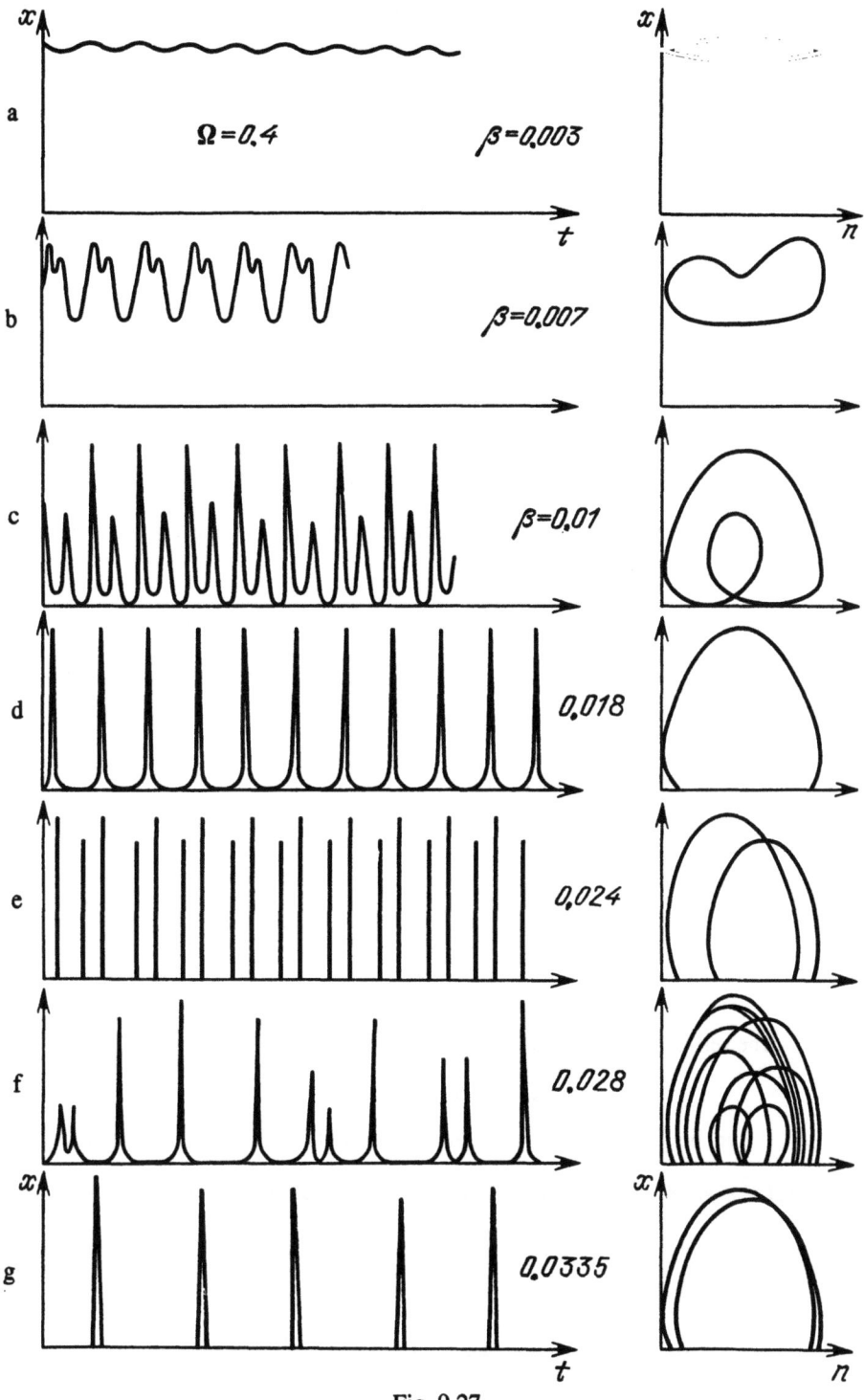

Fig. 9.27

4. Lorenz equations and other systems of order three

Essentially, the first dynamical system with numerically discovered and investigated stochastic self-oscillations were the Lorenz equations [563]. As mentioned above, its three-mode truncation describes convective motion in a liquid layer heated from below, of the form

$$\dot{X} = -\sigma(X-Y), \quad \dot{Y} = r X-Y-XZ, \quad \dot{Z} = -b Z+XY \qquad (4.1)$$

[217, 414].

The Lorenz equations belong to the class of self-oscillatory systems with inertial self-excitation [52, 53, 391]. In those with block diagram as in Fig. 9.28, generation appears due to the inertial behaviour of the feedback circuit, which leads to so-called *inertial intercoupling* between the dynamical variables. In the simplest case, the corresponding oscillation equations are

$$\ddot{x}+2\delta\dot{x}+\omega_0^2 x = -ky+f(x,\dot{x},y), \quad \dot{y}+\gamma y = ax+\phi(x,\dot{x},y), \qquad (4.2)$$

where δ is the coefficient of friction, γ the parameter of inertial behaviour, k and a are linear couplings, and f and ϕ non-linear functions not involving linear terms[15].

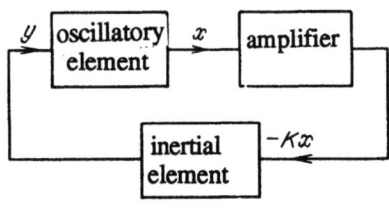

Fig. 9.28

The self-excitation condition for (4.2) is

$$ak \geq 2\delta (\omega_0^2+2\delta\gamma+\gamma^2). \qquad (4.3)$$

It can be seen that the oscillations can only be excited for $\gamma \leq \gamma_{cr}$, where $\gamma_{cr}=-\delta+\sqrt{\delta^2+ak/2\delta-\omega_0^2}$, or for sufficiently inertial behaviour of the circuit.

The system of Lorenz equations (4.1) can be reduced to (4.2) if the variable Y is eliminated, and

$$x = X-\sqrt{b(r-1)}, \quad y = \sigma\left[Z-\frac{1}{2\sigma}x^2-(r-1)\left(1-\frac{b}{2\sigma}\right)\right], \quad 2\delta = \sigma+1,$$

---

[15] A more general case is considered in [55].

$$\omega_0^2 = b\,(r-1), \quad k = \sqrt{b(r-1)}, \quad \gamma = b, \quad a = (2\sigma-b)\,\omega_0,$$

$$f(x,y) = -\left(\frac{3}{2}\omega_0 x + \frac{1}{2}x^2 + y\right)x, \quad \phi(x) = (\sigma-b/2)\,x^2. \tag{4.4}$$

The Lorenz equations admit a purely mechanical model suggested by S. I. Zlochevsky and consisting of two solid bodies, a lifting and the lifted axi-symmetric rotor (Fig. 9.29). The system has the property that, as the rotor moves relative to the lifting body, mass distribution in space remains unaltered. Therefore, the tensor for the system is constant. Such a system is called a *gyrostat*. It is assumed that its centre of mass (the point 0) is fixed, and that the ellipsoid of inertia is of the form of an ellipsoid of revolution. The rotor rotates with constant angular velocity Ω about its symmetry axis, which is fixed on the lifting body perpendicular to the symmetry axis of the gyrostat. The moment M of external force applied to the gyrostat depends on the angular velocity ω of the lifting body, which rotates about the fixed centre of mass.

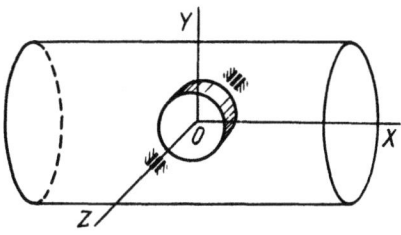

Fig. 9.29

The equations of motion of the gyrostat reduced to projections onto the coordinate axes are

$$I_X \dot\omega_X + h\omega_Y = M_X,$$
$$I\dot\omega_Y + (I_X - I)\omega_X\omega_Z - h\omega_X = M_Y,$$
$$I\dot\omega_Z + (I - I_X)\omega_Y\omega_X = M_Z.$$

Here ω_X, ω_Y, ω_Z are the projections of the instantaneous angular velocity ω onto the X-, Y-, Z- axes, I_X is the axial moment of inertia of the gyrostat, I the equatorial moment of inertia, $h = J\Omega$ the relative angular momentum of the rotor, J the axial moment of inertia of the rotor, Ω=const, the X-axis coincides with the axis of symmetry of the gyrostat, and the Z-axis is parallel to the rotating axis of the rotor.

If $M_X = M_Y = M_Z = 0$, $\omega \neq 0$, $\Omega \neq 0$, then the gyrostat is known [108] to make periodic oscillations or to be in permanent rotation. In the velocity space, the trajectories of motion of the representing point are on the energy ellipsoid of the lifting body, viz., on

$$I_X\omega_X^2 + I(\omega_Y^2 + \omega_Z^2) = 2T = \text{const.}$$

If M_X, M_Y, M_Z are different from zero, and equal to

$$M_X = (a+h)\omega_Y - a\omega_X, \quad M_Y = -c\omega_Y, \quad M_Z = -d\omega_Z,$$

then the equations of motion of the gyrostat can be written as

$$\dot\omega_X = \frac{a}{I_X}(\omega_Y - \omega_X),$$

$$\dot\omega_Y = \frac{h}{I_X}\omega_X - \frac{I_X - I}{I}\omega_X\omega_Z - \frac{c}{I}\omega_Y,$$

$$\dot\omega_Z = \frac{I_X - I}{I}\omega_X\omega_Y - \frac{d}{I}\omega_Z.$$

Putting

$$X = (I_X - I)\omega_X/c, \quad Y = (I_X - I)\omega_Y/c, \quad Z = (I_X - I)\omega_Z/c,$$

$$\tau = (c/I)t, \quad \sigma = aI/cI_X, \quad r = hI/cI_X, \quad b = d/c,$$

we obtain the Lorenz equations in standard form (4.1).

We have already said much about the properties of the well-studied Lorenz equations (see Sect.3, Chapt.7). Besides the above, we note [10, 11, 81, 98, 233, 302, 308, 321, 375, 384, 458, 460, 556, 567, 572, 578, 579, 625, 635, 648–650, 655, 662, 686]. Certain results of these papers substantially elaborating the above are given in the sequel.

In the region of stochastic behaviour, the vibration spectrum of the Lorenz system is continuous, and sufficiently wide (Fig. 9.30). This indicates mixing strongly [441]; an approximate computation of the spectrum is in [567]. The Lorenz attractor's capacity is close to two, viz., $d = 1.98 \pm 0.02$ for $b = 4$, $\sigma = 16$, $r = 40$ [578, 579]. (The Lyapunov dimension calculated by the Kaplan-Yorke formula is $d_L = 2.06$ [587].) The dependence of the maximal Lyapunov exponent on the parameter r for the above values of b, σ, on whose basis the Lyapunov dimension was found, is illustrated by Fig. 9.31 [686]. It is interesting to note that, in the region of r close to $r_{cr} \approx 33.45$, the dependence is of the form as in Fig. 8.29. The dot-and-dash line in Fig. 9.31 is associated with metastable chaos.

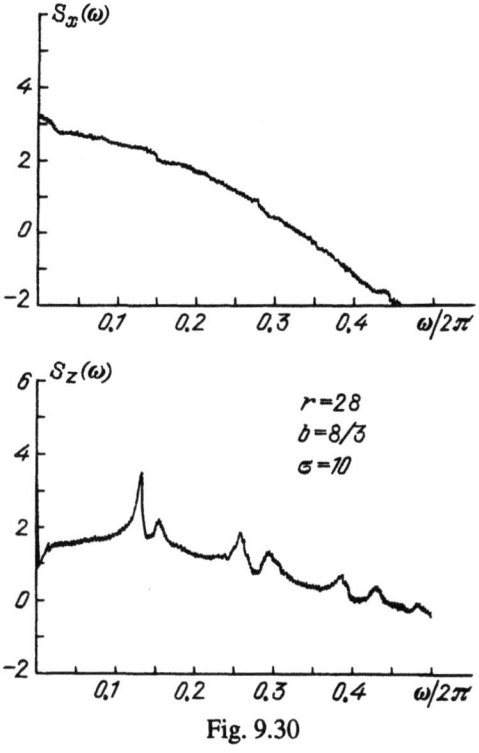

Fig. 9.30

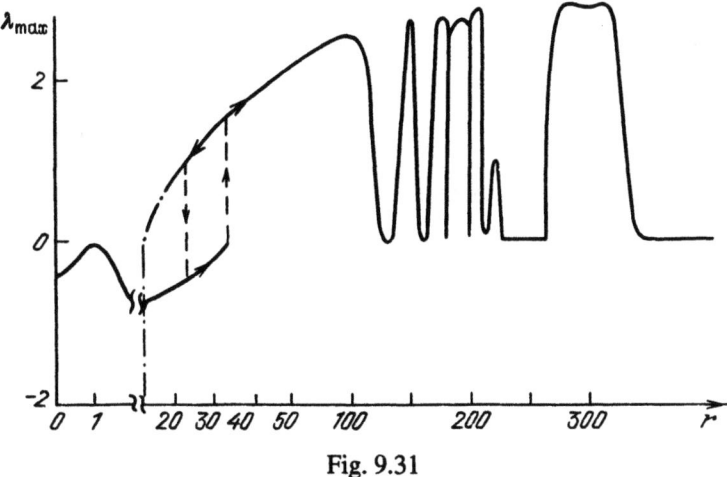

Fig. 9.31

We now dwell on the behaviour of the Lorenz system for large r. It is known that, for sufficiently large r ≥313, the solution is always periodic. The shape and amplitude of the limit cycle for these values of r can be calculated approximately in an analytic way [52, 53, 308, 375, 391, 648], for which it is convenient to change the variables as in [375, 648, 649]. Eliminate the quantity Y in (4.1), and introduce the new variables

$$\tau = \sqrt{\sigma(r-1)}\, t, \quad \xi = [2\sigma(r-1)]^{-1/2}X, \quad \eta = (Z - X^2/2\sigma)/(r-1).$$

The equations can then be written as

$$\ddot{\xi} + (\eta - 1)\xi + \xi^3 = -\mu\dot{\xi}, \quad \dot{\eta} = -\frac{\mu}{\sigma+1}\left[b\eta - (2\sigma - b)\xi^2\right], \qquad (4.5)$$

where $\mu = (\sigma+1)/\sqrt{\sigma(r-1)}$ is a small parameter. The first equation describes non-linear oscillator with frequency varying slowly in accordance with the other equation. The averaging method for systems of this class was worked out by V. M. Volosov [111].

If $\mu = 0$, equations (4.5) have the periodic solution

$$\xi = \xi(A, \eta, \psi), \quad \eta = \text{const}, \qquad (4.6)$$

where $\psi = \omega(A, \eta)\tau + \phi$, $\omega(A, \eta) = 2\pi/T(A, \eta)$, $T(A, \eta)$ is the oscillation period in the 0-th approximation with respect to μ, and A the maximum value of ξ. The solution should satisfy the system of the integrals of (4.5) for $\mu = 0$,

$$E = \frac{1}{2}\dot{\xi}^2 + \frac{1}{2}(\eta-1)\xi^2 + \frac{1}{4}\xi^4 \equiv H(\xi, \dot{\xi}, \eta), \qquad (4.7)$$

$$\pm \int_{\xi_0}^{\xi} \frac{d\xi}{\left[2E - (\eta-1)\xi^2 - \xi^4/2\right]^{1/2}} = \tau - \tau_0, \qquad (4.8)$$

where E, τ_0 are constants used in the sense of energy and zero time, and $H(\xi, \dot{\xi}, \eta)$ the Hamilton function.

Following Volosov's method, we can obtain equations for slow variables without knowing (4.6) explicitly, and only resorting to (4.7), (4.8). We select the quantity A as one of the variables, related to the energy E by $E = \frac{1}{2}(\eta-1)A^2 + \frac{1}{4}A^4$.

It follows from (4.7) that the energy's averaged equation is

$$\dot{E} = \frac{1}{T}\int_0^T \left(\frac{\partial H}{\partial \xi}\dot{\xi} + \frac{\partial H}{\partial \dot{\xi}}\ddot{\xi} + \frac{\partial H}{\partial \eta}\dot{\eta}\right) d\tau. \qquad (4.9)$$

Calculating the integrand and taking (4.5), (4.7) into account, we have

$$\frac{\partial H}{\partial \xi}\dot{\xi} + \frac{\partial H}{\partial \dot{\xi}}\ddot{\xi} + \frac{\partial H}{\partial \eta}\dot{\eta} = \left[(\eta-1)\xi + \xi^3\right]\dot{\xi} - \dot{\xi}\left[(\eta-1)\xi + \xi^3 + \mu\dot{\xi}\right]$$

$$-\frac{\xi^2\mu}{2(\sigma+1)}\left[b\eta-(2\sigma-b)\xi^2\right] = -\mu\left\{\dot\xi^2 + \frac{\xi^2}{2(\sigma+1)}\left[b\eta-(2\sigma-b)\xi^2\right]\right\}. \quad (4.10)$$

Since

$$d\tau = \frac{d\xi}{\left[2E-(\eta-1)\xi^2-\xi^4/2\right]^{1/2}}$$

as it follows from (4.8), and

$$\int_0^T F(\xi(\tau))\,d\tau = 4\int_0^A \frac{F(\xi)\,d\xi}{\left[2E-(\eta-1)\xi^2-\xi^4/2\right]^{1/2}},$$

we derive from (4.9), (4.10) that

$$\dot E = \mu(I_1-I_2), \quad (4.11)$$

where

$$I_1(A,\eta) = \frac{2}{(\sigma+1)T}\int_0^A \frac{\xi^2\left[(2\sigma-b)\xi^2-b\eta\right]\,d\xi}{\left[2E-(\eta-1)\xi^2-\xi^4/2\right]^{1/2}},$$

$$I_2(A,\eta) = \frac{4}{T}\int_0^A \left[2E-(\eta-1)\xi^2-\xi^4/2\right]^{1/2}d\xi,$$

$$T(A,\eta) = 4\int_0^A \frac{d\xi}{\left[2E-(\eta-1)\xi^2-\xi^4/2\right]^{1/2}}. \quad (4.12)$$

The averaged equation for the slow variable η is

$$\dot\eta = -\mu\left[b\eta-(2\sigma-b)I_3\right]/(\sigma+1), \quad (4.13)$$

where

$$I_3(A,\eta) = \frac{4}{T}\int_0^A \frac{\xi^2\,d\xi}{\left[2E-(\eta-1)\xi^2-\xi^4/2\right]^{1/2}}.$$

In the steady state, the solution of equations (4.11), (4.13) is determined by the system of transcendental equations

$$I_1(A,\eta) = I_2(A,\eta), \quad \eta = \frac{2\sigma-b}{b}I_3(A,\eta). \quad (4.14)$$

The integrals I_1, I_2, I_3, and T can be expressed in terms of the complete elliptic integrals $K(k)$, $E(k)$ of the first and second kind [133]. Substituting the expressions in (4.14) and taking into account the relation of the modulus k of the elliptic integral to the amplitude A, we obtain

$$3(2\sigma-b)E^2(k) - \left[(\sigma+1)(1-2k^2)+2(2\sigma-b)(2-k^2)\right]E(k)K(k)+(3\sigma+1-b)(1-k^2)K^2(k) = 0,$$

$$\left[1-4\frac{\sigma}{b}(1-k^2)+2\frac{2\sigma-b}{b}\frac{E(k)}{K(k)}\right]\tilde{A}^2 - 1 = 0,$$

$$\eta = \tilde{A}^2(1-2k^2)+1, \quad T = 4K(k)/\tilde{A}. \tag{4.15}$$

This can be used to determine the stationary values of k, A^2, η, and T, where $\tilde{A}^2 = A^2/2k^2$.

The equations can easily be solved numerically or graphically; e.g., for $\sigma=10$, $b=8/3$, we have $k^2 \approx 0.9$; $A^2 \approx 0.35$, $\eta \approx 0.84$, $T \approx 23.4$. Substituting the values of A, η in (4.8), taking into account the relation of A^2 to E, we determine the shape of the oscillations of the variable ξ, $\xi(\tau)=0.59\mathrm{cn}\,(0.44\tau, 0.95)$. The form of the function $\xi(\tau)$ and of the projection of the phase portrait on the $\eta=0.84$ plane are given in Fig. 9.32a,b, while the singular points are O ($\xi=\dot{\xi}=\eta=0$), O' ($\xi=-0.36$, $\dot{\xi}=0$, $\eta=0.87$), O'' ($\xi=0.36$, $\dot{\xi}=0$, $\eta=0.87$).

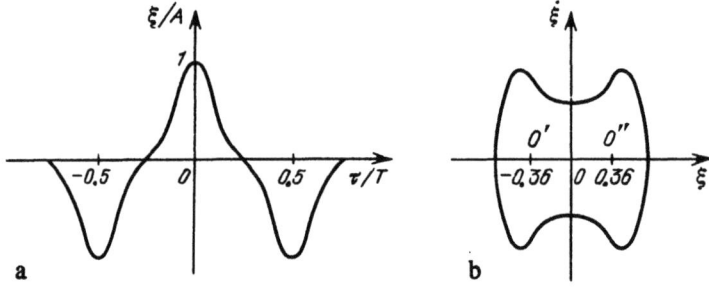

Fig. 9.32

We stress that, for increasing r, the amplitude of oscillation of the variable X increases as $\sqrt{r}$, while its period relative to the original time t decreases as $1/\sqrt{r}$.

As r decreases, the periodic solution found loses stability, and generates a solution with period twice as large. The latter solution again loses stability, etc., which means that an infinite sequence of period-doubling bifurcations arises. A region where the periodic solution is stable arises, and, in turn, is subject to period-doubling bifurcations; e.g., the following regions of such bifurcations were determined [460]: A sequence of bifurcations starts for r =233.5 and terminates for r =r_∞=210, or starts for r =148.4 and terminates for r_∞=145. In the region 148.4 < r < 166.07, the solution of the Lorenz equations is periodic. As r increases from the value r =r_{thr}=166.07, transition to chaos occurs through intermittency [572], the average duration of a "laminar" phase being sufficiently exactly governed by the law $\tau \sim (r-r_{thr})^{-1/2}$. It is just in considering this transition that the very concept of *intermittency* was established. Thus, in the region where r is large, the Lorenz equations behave similarly to the dynamical system described by a one-dimensional point mapping that is differentiable at its maximum. That transition to chaos through intermittency is possible can be well seen in Fig. 9.33, depicting the graph of the point mapping for $\sigma=16$, $b=4$, $r=140$ [662].

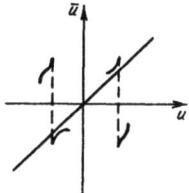

Fig. 9.33

Dependence of the behaviour of the solutions of the Lorenz equation on the parameters σ, b has not been studied in detail. It is shown in [655] that, for fixed $\sigma=10$, the whole range of variation of the parameter b may be conveniently separated into three qualitatively different regions, viz., the one marked by I, $0<b<b^*$, where $2<b^*<8/3$; by II, where $b^*<b<b_2=14.5$; and by III, where $b>b_2$. In the region marked by II, solutions are qualitatively the same as when $b=8/3$ (though, for $b>b_1=\sigma-1=9$, the singular points O_1, O_2 are always stable). In the region marked by I, the picture of bifurcations and the topology of the attractor are more complicated than in the region marked by II. Finally, in the region marked by III, on the contrary, everything is simple, and, therefore, uninteresting. The region marked by II was given a detailed study in [635]. Evolution of the solution as the parameter σ varies for fixed sufficiently large r, which enables us to resort to asymptotic methods, was studied analytically in [308].

Equations for slow amplitudes of field strength, polarization and level population difference in lasers and masers with one-mode approximation and no detuning of frequency of oscillations from the centre of the amplification curve are reduced to the Lorenz equations [134, 296, 308, 356, 592, 692]. However, as a rule, the real parameters are such that the steady-state solution is always stable, i.e., stochastic regimes do not arise[16]. If the detuning is non-zero, a system of equations of order five is obtained, easily reducible to the complex Lorenz equations studied in [457] in the form

$$\dot{x} = -\sigma(x-y), \quad \dot{y} = rx-y-xz, \quad \dot{z} = -bz+xy^*+x^*y, \qquad (4.16)$$

where σ, r, x, y are complex and b, z real numbers.

Another form of the complex Lorenz equations is studied in [533] (equivalent to six equations of order one):

$$\dot{x} = -\sigma(x-y), \quad \dot{y} = rx-y-x^*z, \quad \dot{z} = -bz+xy, \qquad (4.17)$$

where r, x, y and z are complex and σ, b real numbers. The authors point out that, in a certain approximation, (4.17) describe waves in the magnetic field of the Sun.

In contrast to the usual Lorenz system (4.1), the transition to chaos in systems (4.16), (4.17) can occur via generation and subsequent destruction of the torus. Thus, for (4.17), where $b=0.5$, $\sigma=1$, $r=2iD$, where D is a real bifurcation parameter, the bifurcation scheme is as follows [533]: For $0<D<1$, (4.17) has the unique stable solution $x=y=z=0$. In the region where $D>1$, the solution becomes unstable; however, there is a stable limit cycle existing until $D=2.07$; the pair cycle's complex conjugate multipliers lie on the unit circle, and a two-dimensional torus is generated. The latter's projection onto the (Re y, Re z)-plane for $D=3$ is in Fig. 9.34a. The projection of the Poincaré cutting surface Re $x = 0$ onto the (Re y, Re z)-plane is in Fig. 9.34d. For $D=3.47$, a three-dimensional torus is generated, whose projections onto the latter plane and of the corresponding Poincaré cutting surface are in Fig. 9.34b,e ($D=3.6$). Finally, for $D=4.0$, the torus breaks down, and a chaotic attractor arises, of the shape illustrated in Fig. 9.34c,f ($D=8$).

---

[16] It is reported in [545] that such single-mode operation is experimentally attained on a $^{15}NH_3$-laser pumped by CO_2-laser radiation, for which the steady-state solution becomes unstable, and stochastic behaviour arises.

Examples of mechanical, physical, chemical, and biological systems

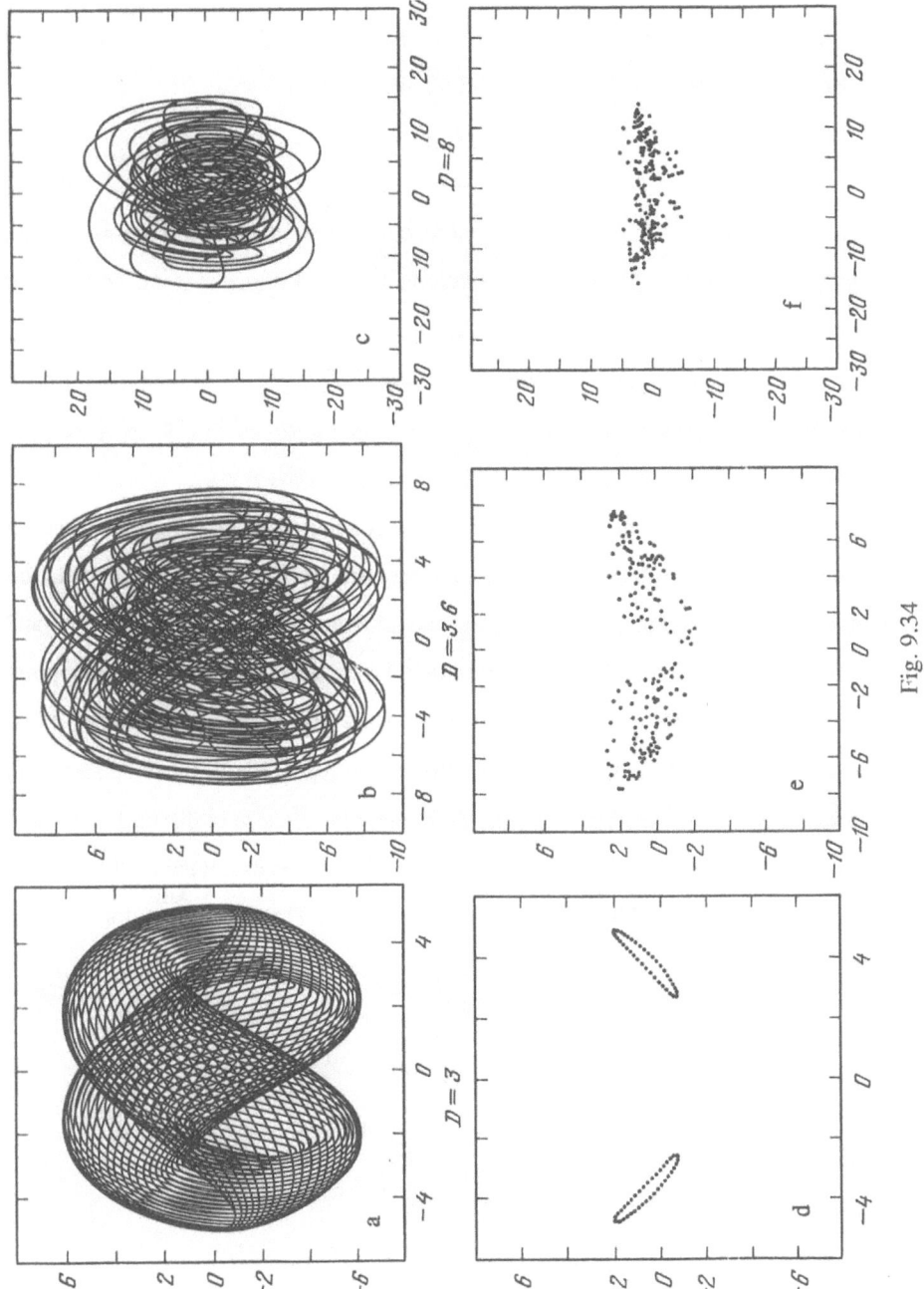

Fig. 9.34

As noted above, the Lorenz equations can easily be reduced to (4.2), where the non-linear functions f, ϕ are determined by expressions (4.4). For other, simpler, forms of f, ϕ, equations (4.2) were studied in [54, 220, 392]. The two cases

$$f(x,y) = -gxy, \quad \phi(x) = lx^2; \tag{4.18a}$$

$$f(x) = -nx^3, \quad \phi(x) = -lx^2 - hx^3 \tag{4.18b}$$

were considered. In the former, the physical mechanism for a rapid collapse in energy, necessary for chaos to appear, is the same as for the Lorenz system, and due to modulation of the oscillation frequency of the variable x by oscillations of the variable y. The same mechanism is called *parametric* in [54, 220, 392] for the latter reason. In contrast, the mechanism in the case of (4.18b) is called *forced* in [54, 392]. In both cases, the system (4.2) is dissipative, since div$\{\dot{x}, \ddot{x}, y\} = -(2\delta + \gamma) < 0$.

In [54, 220, 392], numerical simulation shows that there are parameter ranges in which equations (4.2), (4.18) have solutions that are irregular, with the correlation function sharply decreasing, spectrum continuous, Kolmogorov entropy positive, and point mapping reducible, at least approximately, to a one-dimensional mapping. Characteristic examples of time series of the process x(t), correlation function B(τ), power spectrum S(f), and point mapping (4.18a) are in Fig. 9.35a-d. It can be seen that, for the specified values of parameters, the point mapping is close to discontinuous, and strongly expansive initially. The dimension of the attractor, as estimated by the Kaplan-Yorke formula, is within 2.05–2.30 in the region of chaos. The typical attractor is in Fig. 9.35e.

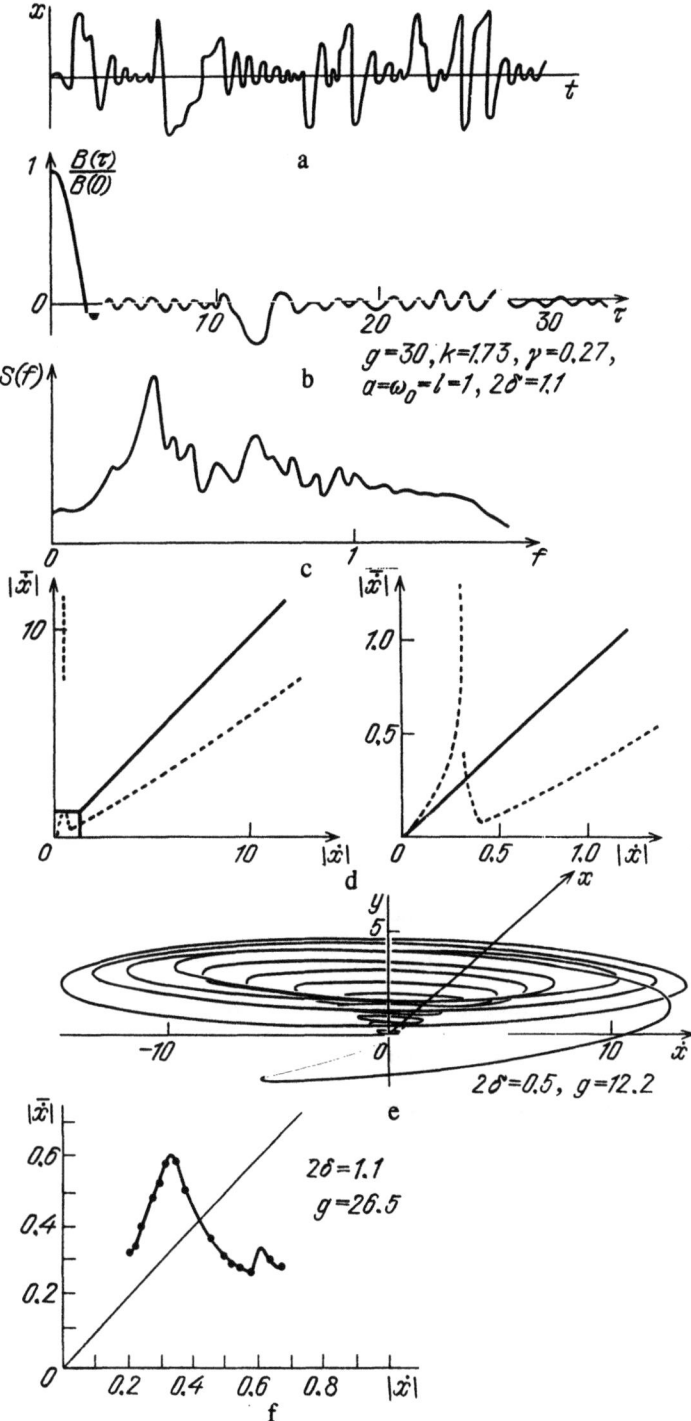

Fig. 9.35

The transition from periodic to chaotic oscillations as parameters vary can occur in Fig. 9.35a both via an infinite sequence of period-doubling bifurcations and in a hard way [54, 222, 392]; e.g., the former transition was observed for $2\delta=1.1$, $k=1.73$, $a=\omega_0=l=1$, $\gamma=0.27$ as the parameter g increased. The sequence of bifurcation values $g=\{15.66, 24.07, 25.50, 25.80, 25.86\}$, leading to the sequence of numbers $\delta_j = (g_{j+1}-g_j)/(g_{j+2}-g_{j+1}) = \{5.88, 4.767, 4.761\}$ was obtained. We may assume that the sequence converges to the Feigenbaum constant $\delta=4.6692...$. The point mapping near the boundary where transition to chaos occurs is shown in Fig. 9.35f. As g increases further, the maximum of the mapping increases sharply, while the falling portion becomes still steeper. As a result, the mapping is as in Fig. 9.35d.

Hard transition to chaos occurs, e.g., for $2\delta=1.1$, $a=\omega_0=l=1$, $\gamma=0.5$, $g=30$, and as the parameter k increases. For $k<k_1\approx1.91$, the phase space has only one attractor, a stable singular point at the origin. For $k=k_1$, another, strange, attractor arises, coexisting with the former until $k=k_2=2.06$, the value for which the stable singular point loses stability. The point mapping of general form for $k<k_2$ is shown in Fig. 9.36a. It has two fixed points O, A, the singular point at the origin, and the unstable limit cycle. When $k<k_1$, A is above B (Fig. 9.36b); therefore, for any initial conditions, the image point is at O; on the contrary, for $k_1<k<k_2$, A is below B (Fig. 9.36c), and the system has two attractors. For $k=k_2$, A and O fuse. There remains only one, strange, attractor. The mapping near the origin for $k\geq k_2$ is shown in Fig. 9.36d. Since B is a little above O, a phenomenon similar to intermittency arises when k is close to k_2. Once near the origin, the image point stays there for a sufficiently long time, which is associated with a large number of small oscillations with slowly increasing amplitude. After leaving the region, the point then returns to it again. The time series of characteristic form for x(t) and k=2 is in Fig. 9.37.

Examples of mechanical, physical, chemical, and biological systems

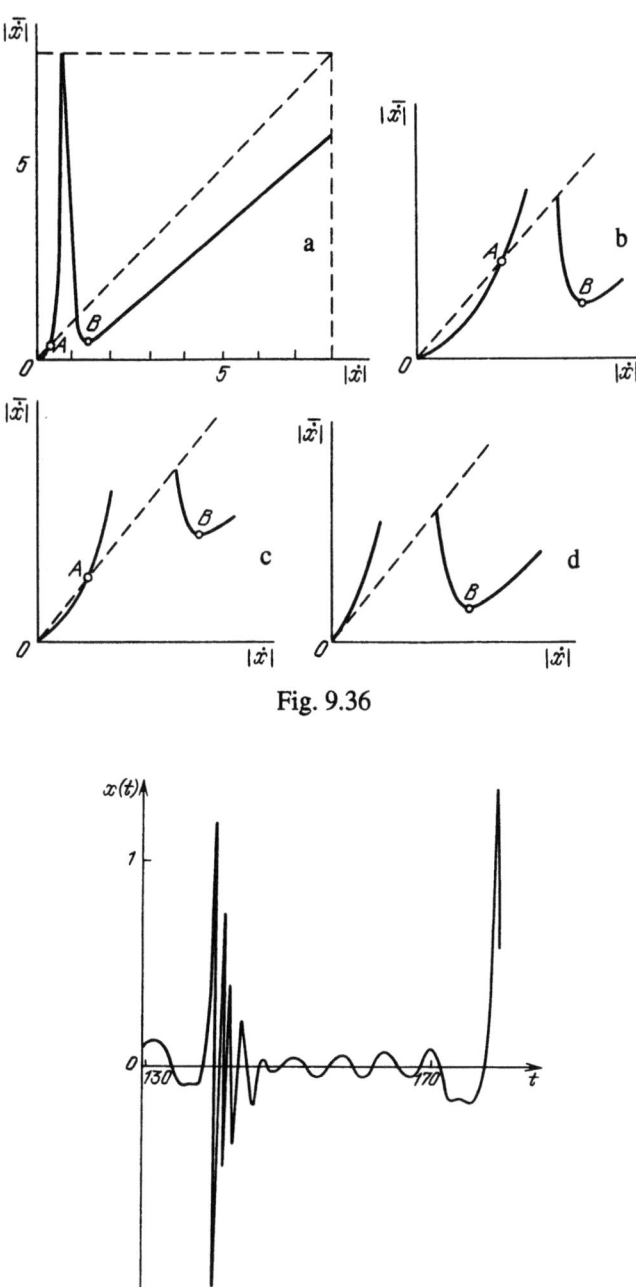

Fig. 9.36

Fig. 9.37

In (b), the types of attractor and point mapping substantially differ from Fig. 9.38a. The graph of the dependence of $\bar{x}$ on x, constructed on the basis of the point mapping on the cutting plane $\dot{x}=-0.13$ (Fig. 9.38b), is close to a parabola in shape (Fig.

9.38c). Transition from the periodic to the chaotic regime as parameters vary was only observed via period-doubling bifurcations. Compared with (a), the vibration spectrum in the chaotic regime is narrower, while the correlation function decreases more slowly (Fig. 9.39).

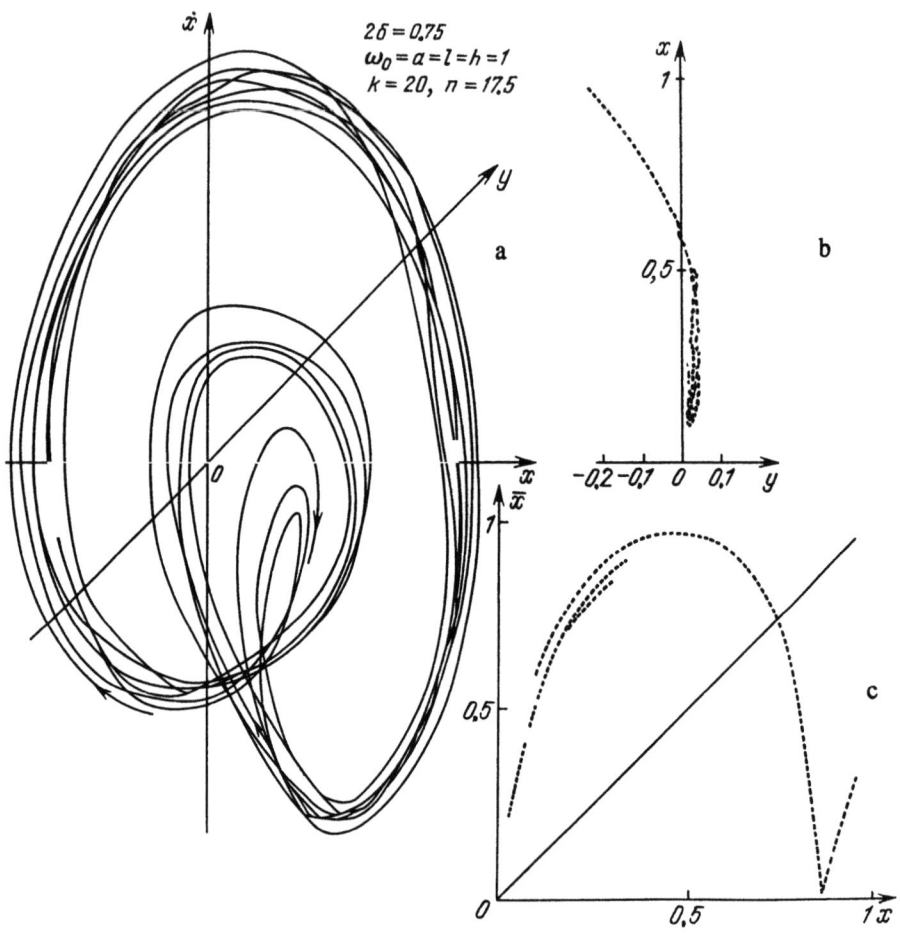

Fig. 9.38

Examples of mechanical, physical, chemical, and biological systems 337

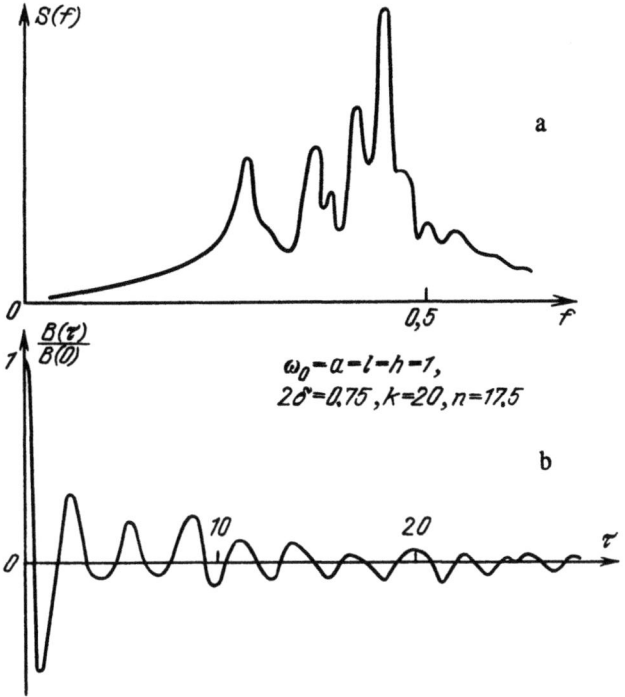

Fig. 9.39

A system of equations easily reducible to (4.2) was also considered in [148], and a numerical study conducted for

$$\omega_0=1,\ \delta=0.05,\ k=1,\ \gamma=0.5,\ a=\ln M-0.5,\ f(x,\dot x,y)=0,$$

$$\phi(x) = \{0.5(1-\exp[-x(x-2\sqrt{\ln M})])+x\sqrt{\ln M}\}(x-\sqrt{\ln M})-2x^2\sqrt{\ln M}$$

with varying parameter M. The system has three singular points $O_1=(x,\dot x,y)=(0,0,0)$, $O_2=(\ln M,0,-\ln M)$, $O_3=(2\ln M,0,-2\ln M)$. As in the Lorenz system, O_1, O_3 are symmetric about O_2. For M>1, O_2 is unstable; O_1, O_3 are stable for M<$M_0\approx1.88$, and oscillatorily unstable for M>M_0. For M=M_0, stable limit cycles are generated around O_1, O_3, losing stability as M increases further, and subject to period-doubling bifurcations which lead to chaos. As M changes, the regions of chaotic and periodic oscillations alternate.

Along with the Lorenz system, the system suggested by O. Rössler is widely known, viz.,

$$\dot X = -Y-Z,\ \dot Y = X+eY,\ \dot Z = f-\mu Z+XZ, \qquad (4.19)$$

where e, f, and μ are parameters [618]. In contrast to the Lorenz equations, (4.19), depending on the parameters, either does not have singular points at all or has the two singular point:

$$X_{1,2}=\frac{\mu}{2}\pm\sqrt{\frac{\mu^2}{4}-ef}, \quad Y_{1,2}=-Z_{1,2}=-\frac{X_{1,2}}{e}.$$

The singular points can only be seen to exist for $\mu \geq 2\sqrt{ef}$. The first is always aperiodically unstable, the second can be stable. For certain μ, at the second singular point, the oscillatory instability condition is fulfilled, and self-excited oscillations develop. Their nature depends on μ (if the other parameters are fixed).

Equations (4.19) were studied numerically in [425, 618] for e=f=0.2. The oscillatory instability condition for the second point then coincides with the condition for its generation. If μ is not very large, oscillations are periodic, and associated in the phase space with a one-revolution cycle (Fig. 9.41a). Starting with $\mu=3.5$, a sequence of period-doubling bifurcations arises (Fig. 9.41b,c,d). The critical value is $\mu_\infty=4.20$. A chaotic attractor with "laminated" structure then arises in the phase space (Fig. 9.41e,f,g). This structure vanishes when $\mu=4.60$ (Fig. 9.41h). Meanwhile, the intensity of the continuous spectrum increases substantially. The bifurcation diagram associated with Fig. 9.41 is in Fig. 9.42 [425]. A three-dimensional plot of the attractor for $\mu=5.7$ is given in Fig. 9.43 [618]. O. Rössler called an attractor of this kind *spiral* (see below).

Examples of mechanical, physical, chemical, and biological systems

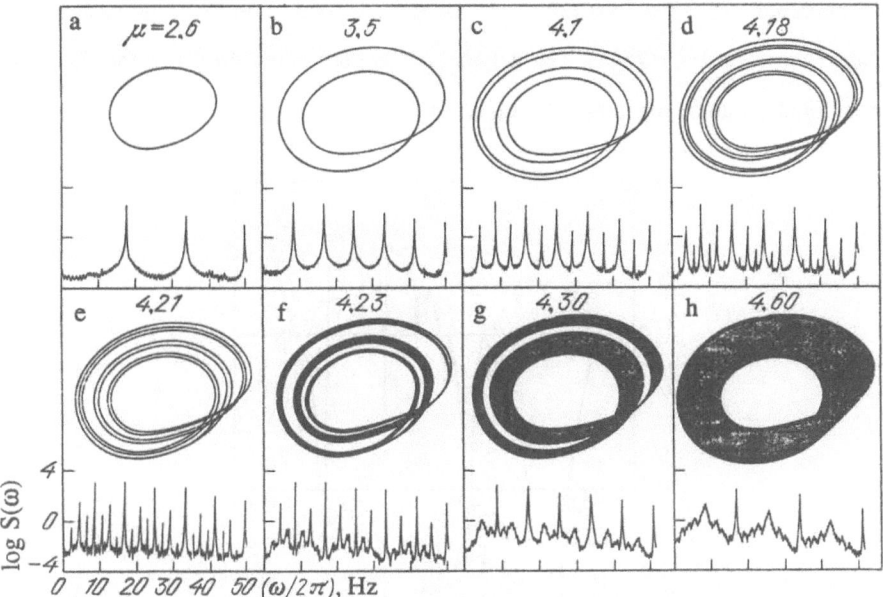

Fig. 9.40

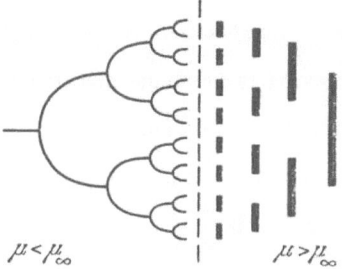

Fig. 9.41

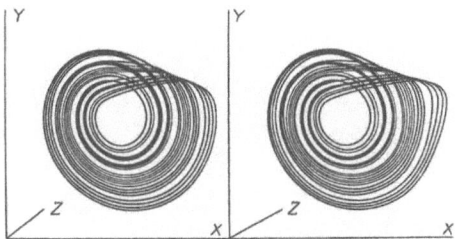

Fig. 9.42

For $\mu<\mu_\infty$, (4.19) has no positive Lyapunov exponents; for $\mu=\mu_\infty$, there is one. The maximal non-zero Lyapunov exponent λ depends non-monotonically on the parameter μ (Fig. 9.43) [618]. The dependence is very much similar to that for a one-dimensional quadratic mapping (cf. Fig. 8.14).

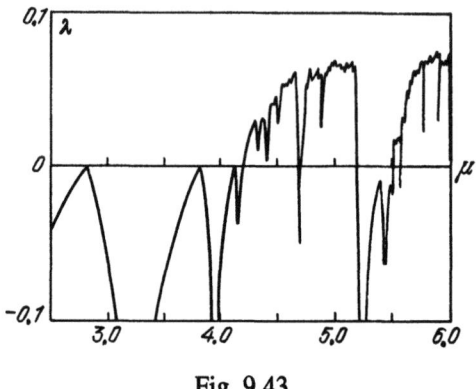

Fig. 9.43

For large μ, the attractor and power spectrum vary substantially. Thus, for $\mu=8.5$, $f=0.4$, and e varying from 0.15 to 0.30, the evolution of the attractor and the corresponding power spectra are in Fig. 9.44. It can be seen from the diagrams that, for e=0.3, discrete components of the spectrum vanish completely. The authors of [441] account for a spectrum of this kind as due to the appearance of a "whirlwind" (Fig. 9.45) instead of a "fold" (Fig. 9.42). The whirlwind induces a continuous transition of phase trajectories from a neighbourhood of a saddle-type cycle into a neighbourhood of a saddle-type equilibrium state, and back.

Examples of mechanical, physical, chemical, and biological systems   341

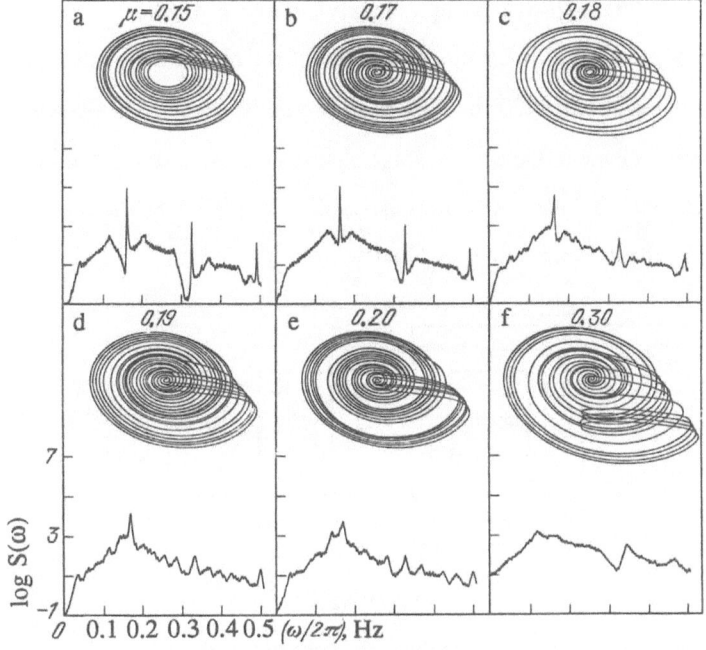

Fig. 9.44

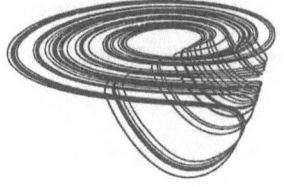

Fig. 9.45

Another system of order three, studied in detail, is

$$\dot{X} = \mu X+Y-XZ, \quad \dot{Y} = -X, \quad \dot{Z} = \gamma[\vartheta(X)X^2-Z], \tag{4.20}$$

where $\vartheta(X)$ is the Heaviside function, μ the excess over the oscillation threshold, and γ the inertial behaviour parameter. The system describes one of the circuit diagrams of so-called *generators with inertial non-linearity*. The term was first introduced by K. F. Teodorchik in 1945 [347]. It refers to self-oscillatory systems with self-excitation occurring on account of inertia-free positive feedback leading to negative resistance, whereas the restriction of oscillators on account of non-linear inertial intercoupling between the dynamical variables. In the simplest case, the equations of oscillators in a generator with inertial non-linearity are

$$\ddot{x}-(\mu-f(y))\dot{x}+\omega_0^2 x = \phi(x,y), \quad \dot{y}+\gamma y = \gamma F(x), \tag{4.21}$$

where $\phi(x,y)$, $F(x)$ are non-linear functions. One of the circuit diagrams is in Fig. 9.46 [173, 276], with equations (4.20) holding as shown in the sequel. The detector plays the role of the inertial element. Detector voltage changes the slope of the amplifier voltage-current characteristic, and thereby bounds the oscillation amplitude. The inertial behaviour is determined by the detector's time: $R_d C_d \equiv 1/\gamma$.

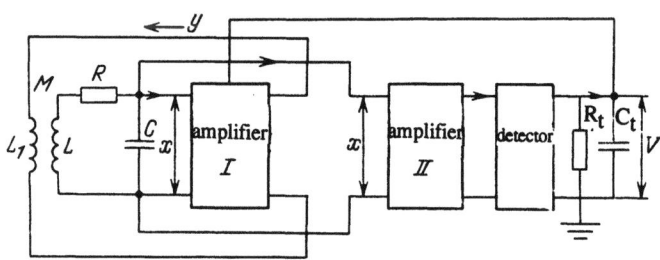

Fig. 9.46

The equations describing the diagram work in Fig. 9.46 are:

$$\ddot{x}+\omega_0^2 x = (\mu-bV)\dot{x}-b\dot{V}x, \quad \dot{V}+\gamma V = \gamma F(x), \tag{4.22}$$

where $\omega_0^2=1/LC$, $\mu=\omega_0^2 MS_0-R/L$, S_0 is the slope of the characteristic of the amplifier for $V=0$, and $F(x)$ the non-linear function determined by the response curve of the detector [216].

Substituting the expression for $\dot{V}$ from the second relation of (4.22) in the first, we obtain equations of form (4.21), where $y=bV$, $f(y)=y$, $\phi(x,y)=\gamma xy-\gamma bxF(x)$. Putting $x=X$, $\dot{x}-\mu x+xy=Y$, $y=Z$, $\omega_0=1$, $bF(x)=v(x)x^2$, we get (4.20).

The system described by (4.20) was numerically and experimentally studied in [22, 23, 27 –35, 44], where in a certain range of μ, γ, the oscillations were formed to be irregular, and depend on the initial conditions [27, 29, 31].

Similar results were also obtained in a physical experiment (Fig. 9.47). If one parameter (e.g., γ) is fixed, and the other (μ) increases, chaos appears via a sequence of period-doubling bifurcations in accordance with Feigenbaum universality. The data of Table 9.6 testify to this, specifying experimental bifurcation values of the parameter μ designed and obtained on a computer for $\gamma=0.3$ and corresponding $\delta_i=(\mu_{i-1}-\mu_{i-2})/(\mu_i-\mu_{i-1})$ [31]. When i=5, δ_i practically coincides with the Feigenbaum constant.

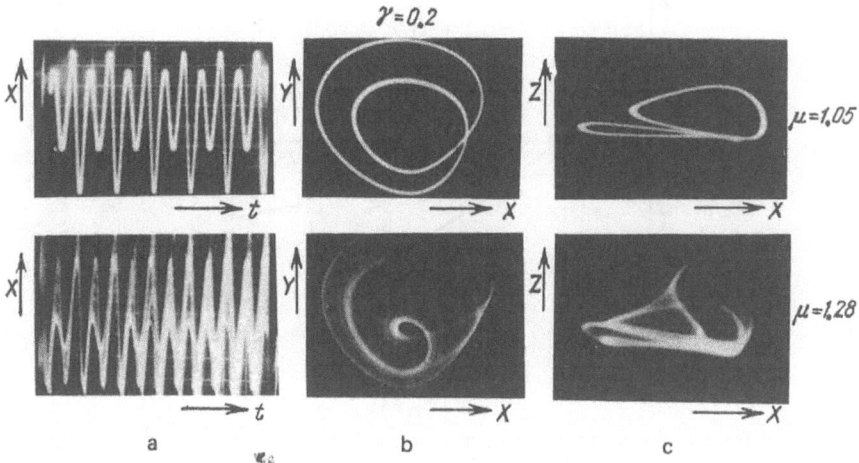

i	μ_i theor.	μ_i exp.	δ_i theor.
1	0,770	0,77	—
2	1,020	1,02	—
3	1,0713	1,07	4,873
4	1,08216	1,08	4,724
5	1,08449	—	4,66896

$\mu_{\infty \text{ theor.}} = 1,08512$, $\mu_{\infty \text{ exp.}} = 1,09$

Table 9.6

If μ deviates slightly from μ_∞ (viz., for $\mu'_1=1.10112$), two-revolution limit cycles Γ'_2, Γ''_2 are generated in the phase space: one stable, the other saddle-type [32], which can well be seen in the graph of Fig. 9.48 with the multipliers m of Γ_2, Γ'_2, Γ''_2 (Γ_2 being the two-revolution cycle which develops for the value $\mu=0.770$ associated with the first period-doubling bifurcation).

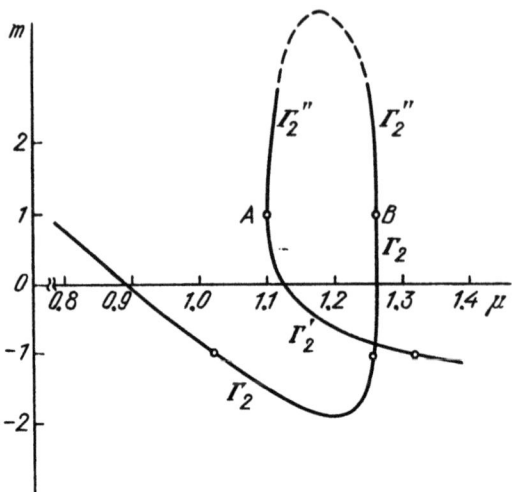

Fig. 9.48

It follows from the diagram that, in the region where $\mu'_1 < \mu < \mu'_2 = 1.31902$, there is at least one stable periodic solution associated with Γ'_2. If, not leaving the cycle, we decrease μ, then, for $\mu = \mu'_1$, numerical computations show transition to chaos through intermittency. The evolution of the spectrum is given in Fig. 9.49, where the frequency f_0 is associated with Γ'_2 and f_1 is the frequency characteristic of the chaotic attractor arising due to the Γ_2 doubling sequence. Note that there are no linear combinations of f_0, f_1 in the spectrum, typical for spectra in intermittency.

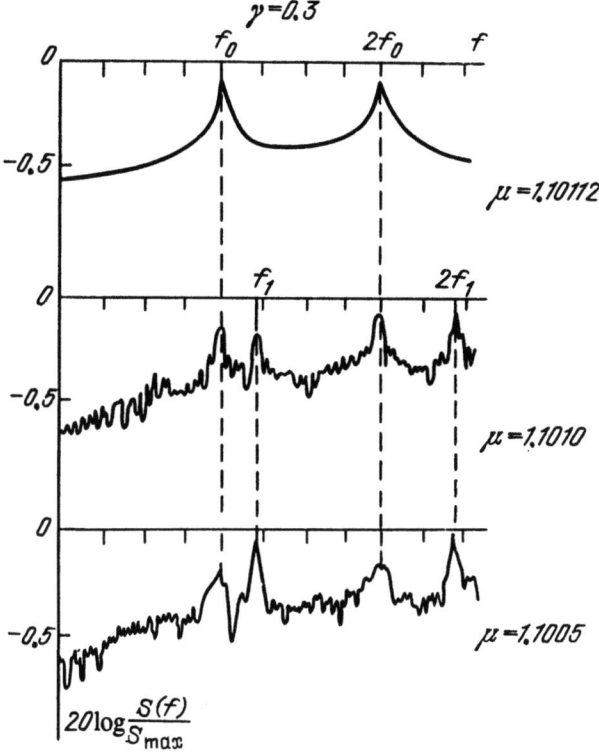

Fig. 9.49

Besides the above, intermittency of "chaos-chaos" type in (4.20) was discovered [22], and the phenomenon investigated for $\gamma=0.097$ and varying μ. If μ increases, a sequence of cycle-doubling bifurcations arises for the fundamental period T_0, and terminates when $\mu=\mu_1 \approx 2.39$. The chaotic attractor J_1 then appears in the phase space. For increasing μ, the region occupied by the attractor increases, too. On the other hand, at the point $\mu=2.31$, two three-revolution cycles are generated by trajectory condensation: one stable, and the other of saddle-type. The stable cycle is also subject to period-doubling bifurcations for increasing μ, which terminate at the point $\mu=\mu_2=2.35$, and lead to a chaotic attractor J_2. In the region where $2.40<\mu<\mu^*\approx 2.445$, both J_1 and J_2 exist independently, with regions of attraction divided by a separatrix surface. When $\mu=\mu^*$, the surface is destroyed, and the attractors merge into one attractor J_0. However, for $\mu \gtrsim \mu^*$, the representing point is for a long time in the region of J_1. The transition from J_1 into J_2 occurs irregularly as far as time is concerned, and is represented by brief peaks. The average time τ of the representing point remaining in the region of J_1 for small $\mu-\mu^*$ is governed by the same universality as in "periodic motion-chaos" intermittency, viz., $\tau = C(\mu-\mu^*)^{-1/2}$ where C=18. As $\mu-\mu^*$ increases, the oscillation's average energy increases, too, the spectrum becomes more uniform, shifts to the low-frequency region, and the autocorrelation time decreases (Fig. 9.50).

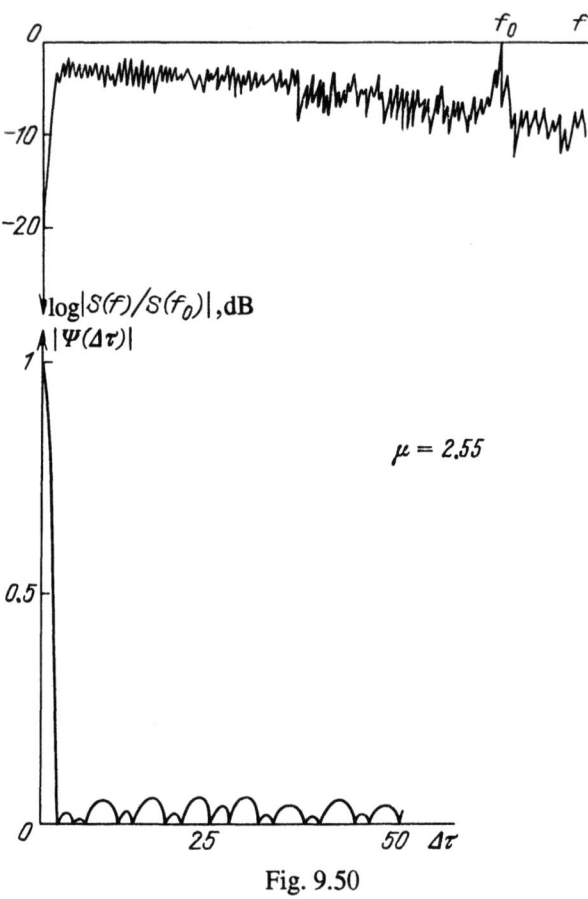

Fig. 9.50

In setting up equations (4.20), the amplifier response is assumed to be linear. If we take into account that the response is non-linear, i.e., replace S_0 by $S_0-S_1 x^2$ in (4.22), then (4.20) take the form

$$\dot{X} = \mu X+Y-XZ-dX^3, \quad \dot{Y} = -X, \quad \dot{Z} = \gamma[\vartheta(X)X^2-Z],$$

where $d=\omega_0^2 MS_1$. The numerical and physical experiments carried out in [33] show that the non-linear inertia-free dissipation characterized by the parameter d leads to the destruction of chaos and the start of a periodic motion, the destruction occurring via an infinite sequence of period-halving bifurcations.

Certain models of non-linear wave interaction lead to third-order systems. Thus, resonant interaction of three quasi-harmonic waves in a non-linear medium, only taking quadratic non-linearity into account, is described by the equations

$$\frac{dA_1}{dt} = \gamma_1 A_1 + mA_2 A_3 e^{-i\delta t}, \quad \frac{dA_{2,3}}{dt} = \gamma_{2,3} A_{2,3} - mA_1 A^*_{3,2} e^{i\delta t}, \quad (4.23)$$

where A_k are complex wave amplitudes, $\delta = \omega_1 - \omega_2 - \omega_3$ is the detuning, ω_k the kth wave frequency, and m the non-linear coupling [429]. In forming (4.23), it is assumed that the waves are sine waves with respect to spatial coordinates.

The case where one of the waves (viz., the first) is unstable, with $\gamma_1 > 0$, and the other two stable, with $\gamma_{2,3} < 0$, is considered in [676, 677]. Introducing dimensionless time, amplitudes and wave phases by the formulas

$$t = \gamma_1^{-1} \tau, \quad A_k = \gamma_1 m^{-1} a_k \exp\{i(\phi_k - \omega_k t)\},$$

and separating the real and imaginary parts, we obtain

$$\dot{a}_1 = a_1 + a_2 a_3 \cos \phi, \quad (4.24)$$

$$\dot{a}_{2,3} = -\Gamma_{2,3} a_{2,3} - a_1 a_{3,2} \cos \phi, \quad (4.25)$$

$$\dot{\phi} = \Delta + \left(\frac{a_1 a_3}{a_2} + \frac{a_1 a_2}{a_3} - \frac{a_2 a_3}{a_1} \right) \sin \phi, \quad (4.26)$$

where $\phi = \phi_1 - \phi_2 - \phi_3$, $\Delta = \delta/\gamma_1$, $\Gamma_{2,3} = |\gamma_{2,3}|/\gamma_1$, and the dot means differentiation with respect to τ. If $\Gamma_2 = \Gamma_3 = \Gamma$, then it follows from equations (4.25) that $d(a_2^2 - a_3^2)/d\tau = -2\Gamma(a_2^2 - a_3^2)$, or that the difference between the intensities of the second and third wave tends to zero in the course of time. Therefore, we can assume that $a_3 = a_2$ in the steady state. Equations (4.24)–(4.26) then take the form

$$\dot{a}_1 = a_1 + a_2^2 \cos \phi, \quad \dot{a}_2 = -\Gamma a_2 - a_1 a_2 \cos \phi, \quad \dot{\phi} = \Delta + \left(2a_1 - \frac{a_2^2}{a_1} \right) \sin \phi. \quad (4.27)$$

Furthermore, assuming that $X = a_1 \cos \phi$, $Y = a_1 \sin \phi$, $Z = a_2^2$, we obtain

$$\dot{X} = X - \Delta Y + Z - 2Y^2, \quad \dot{Y} = \Delta X + Y + 2XY, \quad \dot{Z} = -2\Gamma Z - 2XZ, \quad (4.28)$$

which was investigated in [676, 677] for $\Delta = 2$ and as Γ varied from 1 to 25. For smaller values, the condition that phase volume contracts $\left(\text{div}\{\dot{X}, \dot{Y}, \dot{Z}\} = 2(1-\Gamma) \right)$ is violated. This condition is necessary in order that the values X, Y, Z be finite. Hence, such Γ have no physical meaning. The regions of periodic and chaotic regimes were found, along with values Γ for which period-doubling and period-tripling bifurcations occur. It was shown that the point mapping for the system (4.28) was approximately one-dimensional.

Resonance interaction of three waves in a medium with cubic non-linearity was considered in [313, 315]. For complex wave amplitudes A_k, we have the equations

$$\frac{dA_1}{dt} = \gamma_1 A_1 + 2imA_1^* A_2 A_3 e^{-i\delta t} + imA_1 \left(|A_1|^2 + 2|A_2|^2 + 2|A_3|^2 \right),$$

$$\frac{dA_{2,3}}{dt} = \gamma_{2,3} A_{2,3} + imA_{2,3}^* A_1^2 e^{i\delta t} + imA_{2,3}^* \left(2|A_1|^2 + |A_{2,3}|^2 + 2|A_{3,2}|^2 \right), \quad (4.29)$$

where $\delta = 2\omega_1 - \omega_2 - \omega_3$ is the detuning. As before, we assume that $\gamma_1 > 0$, $\gamma_{2,3} < 0$. Substituting $t = \gamma_1^{-1}\tau$, $A_k = \sqrt{\gamma_1/2|m|}\, a_k \exp\{i(\phi_k - \omega_k t)\}$ in (4.29), we obtain

$$\dot{a}_1 = a_1 + a_1 a_2 a_3 \sin\phi, \quad \dot{a}_{2,3} = -\Gamma_{2,3} a_{2,3} - \frac{1}{2} a_1^2 a_{3,2} \sin\phi,$$

$$\dot{\phi} = \Delta + \left(2a_2 a_3 - \frac{a_3}{2a_2} a_1^2 - \frac{a_2}{2a_3} a_1^2 \right) \cos\phi - a_1^2 + \frac{1}{2}(a_2^2 + a_3^2),$$

where $\Gamma_{2,3} = |\gamma_{2,3}|/\gamma_1$, $\phi = (2\phi_1 - \phi_2 - \phi_3) \text{sign } m$, $\Delta = (\delta/\gamma_1) \text{sign } m$. If $\Gamma_2 = \Gamma_3 = \Gamma$, then, as in the preceding case, the difference between the intensities of the second and third wave tends to zero in the course of time. Meanwhile, we obtain

$$\dot{a}_1 = a_1 + a_1 a_2^2 \sin\phi, \quad \dot{a}_2 = -\Gamma a_2 - (a_1^2 a_2/2) \sin\phi,$$

$$\dot{\phi} = \Delta + (2a_2^2 - a_1^2) \cos\phi - a_1^2 + a_2^2 . \quad (4.30)$$

Putting $X = a_1 \cos(\phi/2)$, $Y = a_1 \sin(\phi/2)$, $Z = a_2^2/2$, we transform (4.30) to

$$\dot{X} = X - \frac{\Delta}{2} Y + Y(Z + X^2), \quad \dot{Y} = Y + \frac{\Delta}{2} X + X(3Z - X^2), \quad \dot{Z} = -2\Gamma Z - 2XYZ. \quad (4.31)$$

As above, only $\Gamma > 1$ have a physical meaning; for them, phase volume contracts. The numerical study of (4.31) for such Γ showed that, in a certain parameter range, the solution is chaotic, its correlation function decreases, and the point mapping of the plane Z=const into itself is strongly extended along the Y-axis; therefore, the mapping can approximately be reduced to a one-dimensional mapping. The point mapping, time series and the attractor for $\Delta = 2.3$, $\Gamma = 1.26$ are given in Fig. 9.51.

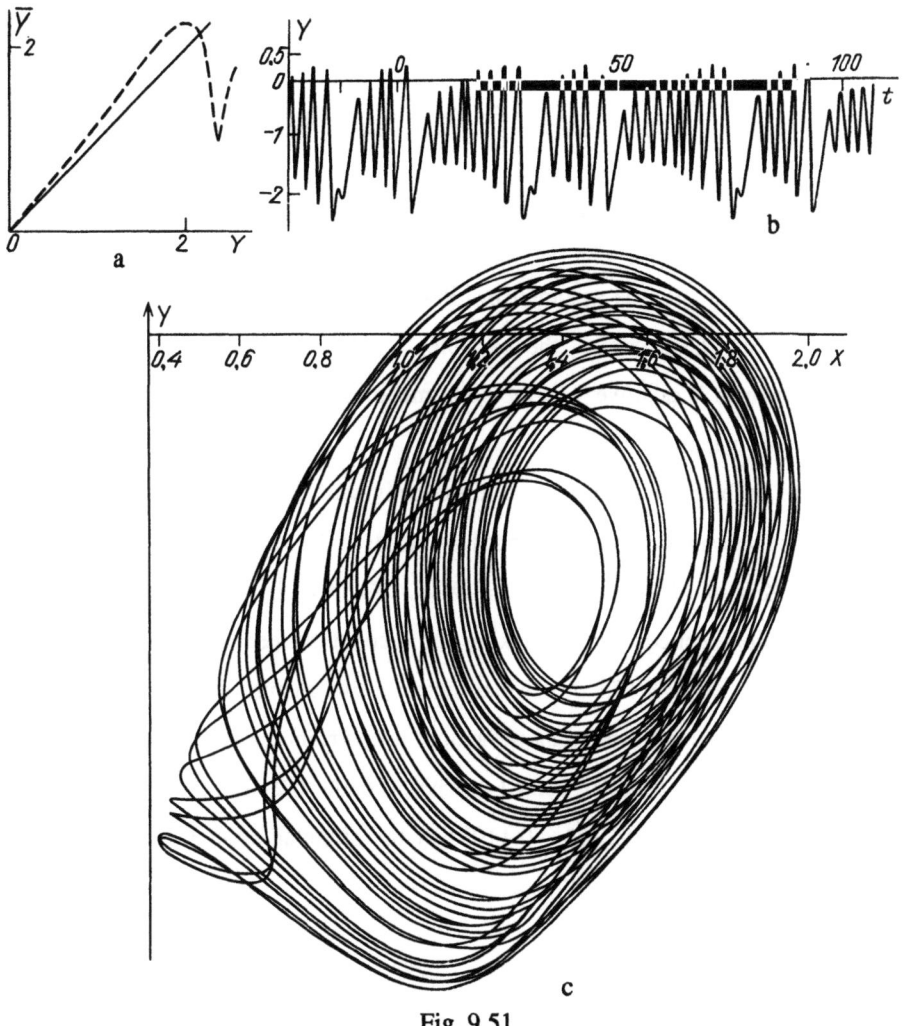

Fig. 9.51

A somewhat different model for 3-wave interaction, describing the Raman effect[17] in a dielectric, is investigated in [584]. Equations for complex wave amplitudes if $2\omega_1 = \omega_2 + \omega_3$, ω_1 is the frequency of the wave excited by pumping, and ω_2, ω_3 are the frequencies of the anti-Stokes and Stokes components, are

$$\dot{A}_1 = \gamma_1 A_1 + \beta_1 \left(|A_2|^2 - |A_3|^2 \right) A_1,$$
$$\dot{A}_{2,3} = \gamma_{2,3} A_{2,3} + \beta_{2,3} \left(|A_1|^2 A_{2,3} + A_1^2 A_{3,2}^* \right), \qquad (4.32)$$

---

[17] The term is actually incorrect, since the effect was discovered by L. I. Mandelshtam.

[388]. Introducing real amplitudes and phases, i.e., putting $A_k = a_k \exp(i\phi_k)$, we obtain

$$\dot{a}_1 = \gamma_1 a_1 + \beta_1 (a_2^2 - a_3^2) a_1,$$

$$\dot{a}_{2,3} = \gamma_{2,3} a_{2,3} + \beta_{2,3} (a_{2,3} + a_{3,2} \cos\phi) a_1^2,$$

$$\dot{\phi} = -\left(\beta_2 \frac{a_3}{a_2} + \beta_3 \frac{a_2}{a_3}\right) a_1^2 \sin\phi, \qquad (4.33)$$

where $\phi = 2\phi_1 - \phi_2 - \phi_3$. It was assumed in [584] that $\phi(t) \equiv 0$ and $\gamma_2 = 1$, $\gamma_{1,3} = -1$, $\beta_1 = 5$, $\beta_2 = -9$, $\beta_3 = 1$. For these values, the variables a_k change chaotically, the maximal Lyapunov exponent is positive, and the point mapping is approximately reducible to a one-dimensional mapping.

The problem of motion of a rigid body with linear feedback control for stabilizing the motion is considered in [555], and leads to the relations

$$\dot{X} = -0.4X + Y + 10YZ, \quad \dot{Y} = -X - 0.4Y + 5XZ, \quad \dot{Z} = \alpha Z - 5XY \qquad (4.34)$$

obtained from the Euler equations by taking into account the feedback circuit. Numerical solution of (4.34) for $\alpha = 0.175$ shows that there are two chaotic attractors in the phase space (Fig. 9.52): one corresponding to random jumps from the singular point (0.2390, 0.0308, 0.2103) to the singular point (–0.2390, 0.0308, 0.2103); the other, from the singular point (0.0315, –0.1224, –0.1103) to the singular point (–0.0315, 0.1224, –0.1103).

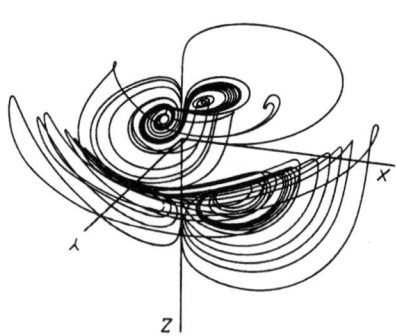

Fig. 9.52

5. Action of a harmonic external force on periodic and chaotic oscillation generators

A periodic external force can have an effect on the generator of one of two kinds: On the one hand, it can make periodic oscillations chaotic [141, 147, 173, 216, 428, 472, 669]; on the other, it can lead to synchronization of chaotic oscillations [28, 36, 37, 54, 76, 78, 151, 214, 215, 223, 428].

In the sequel, we will consider both effects. The phenomenon of chaotization of oscillations under the action of an external force applied to a generator with inertial non-linearity was apparently discovered by L. N. Kaptsov [173] in the numerical simulation of truncated equations for oscillation amplitude and phase, and voltage of the detector. The equations are

$$\dot{a} = \frac{1}{2}(1 - bV)a - a_0 \Delta_c \sin \varphi,$$

$$\dot{\varphi} = \Delta - \frac{a_0}{a} \Delta_c \cos \varphi, \quad \dot{V} = -\gamma V + \gamma K a \tag{5.1}$$

[216], where $a_0 = 1/bK$ is the oscillation amplitude of the autonomous generator. Chaotization of oscillations in a generator with inertial non-linearity can also occur under parametric action of an external force with frequency of the order of that of the relaxation process [216]. The phenomenon was experimentally and numerically studied in [74, 75].

In experimental investigation, the circuit diagram of a generator with detector was used (Fig. 9.45). The parametric action $U = U_0 \cos 2\pi \nu t$ was fed to the amplifier I, and added to detector voltage V. The parametric action leads to amplitude modulation of the generator. For small U_0, the amplitude modulation is maximal if the frequency of the force is close to that of the relaxation process [216], determined by the excess η over the oscillation threshold and by the detector's relaxation time $R_d C_d = \gamma^{-1}$. By non-linearity, the dependence of the modulation amplitude on frequency is hysteretic, which may lead to jumps.

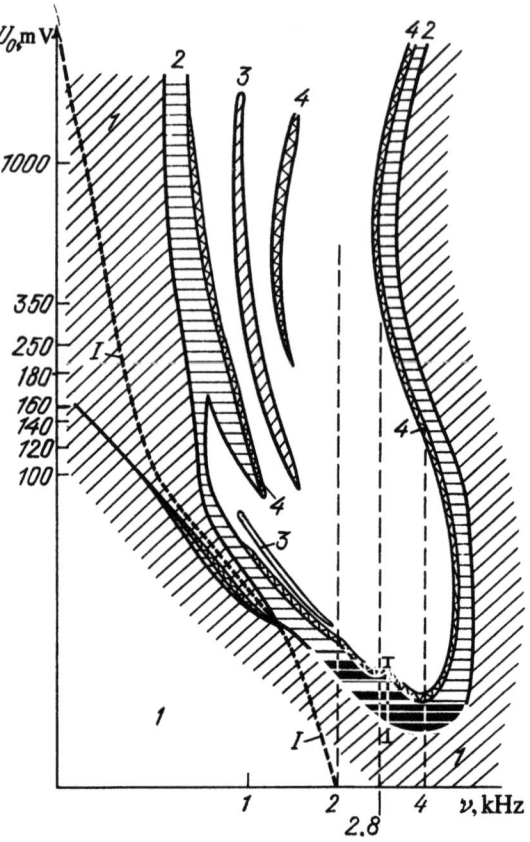

Fig. 9.53

In Fig. 9.53, the regions of various regime of amplitude modulation are represented. The line marked by I is associated with a jump of the modulation amplitude from smaller (on the left) to larger values. The shaded regions are associated with regular modulation. In the regions marked by 1, 2, 3, 4, the modulation frequency is ν, $\nu/2$, $\nu/3$, $\nu/4$, respectively. (Regions of synchronisms of order higher than 1/4 are not shown.) In the unshaded regions, modulation is irregular. For the parameters studied, exit from the regions of regular modulation occurred via an infinite sequence of period-doubling bifurcations; e.g., for frequency $\nu =3$ kHz close to the frequency of the relaxation process and with amplitude U_0 increasing in the strip marked in Fig. 9.53 by the vertical bar, the evolution of the time series of the signal of the generator is given in Fig. 9.54.

Examples of mechanical, physical, chemical, and biological systems 353

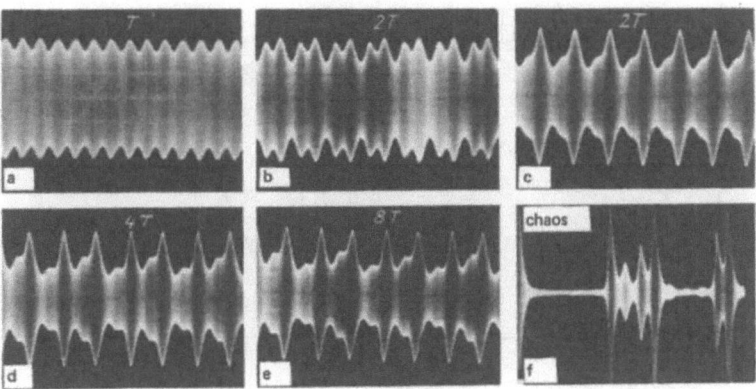

Fig. 9.54

To determine the boundaries of transition to chaos and to estimate the degree of chaos, the synchronization threshold was used. The generator with inertial non-linearity under investigation, in the regime of chaotic amplitude modulation, was fed a signal from an additional generator, at frequency close to that of self-excited oscillations of the generator under investigation (130 kHz). Meanwhile, it turned out that, starting from a certain critical amplitude value B of the additional generator, the amplitude modulation of the generator in question became regular. Each time the frequency ν was varied so that the critical value might be minimal. Consequently, the graph of the dependence on U_0 of the minimum critical value B (by the definition given in Sect.3, Chapt.8, we call this value the *synchronization threshold* B_{thr}) was constructed for fixed ν. The dependences for three values of ν, 2.0 kHz, 2.8 kHz, 4.0 kHz, are given in Fig. 9.55 a-c. That of B_{thr} on U_0 near the boundary of transition to chaos for ν = 3 kHz is shown in more detail in Fig. 9.55 d. This dependence is of power character, which is corroborated by its graph constructed in Fig. 9.56 in a logarithmic scale, with similar dependences for the two other values of ν. The exponent r in $B_{thr} = (U_0 - U_{0\infty})^r$ turns out to be different for different ν (if ν = 2.5 kHz, r = 0.5; if ν = 3 kHz, r = 0.4; if ν = 3.9 kHz, r = 0.3). Whatever its origin, the reason for this difference, which attests to no universality, remains unclear.

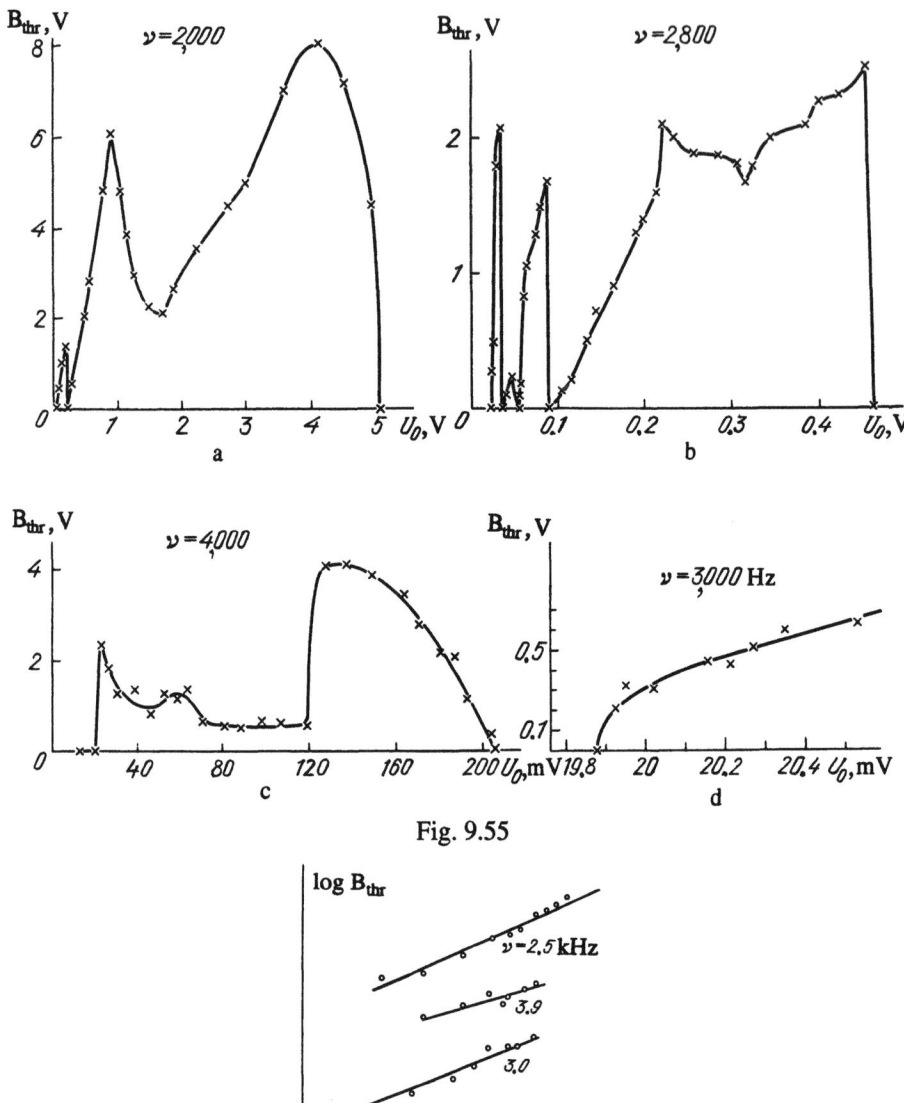

Fig. 9.55

Fig. 9.56

Numerical simulation of the processes in the generator under investigation was carried out within the framework of the truncated equations for amplitude and voltage of the detector, viz.,

$$\dot{a} = \frac{1}{2}[\eta - \alpha a^2 - b(V + U_0 \cos 2\pi\nu t)]a, \quad \dot{V} = -\gamma V + \gamma K a, \qquad (5.2)$$

where α is the coefficient specifying the non-inertiality of the non-linearity of the amplifier. They were solved for

$\eta = 2.5 \cdot 10^6$, $\alpha = 100$, $b = 10^6$, $\gamma = 180$, $K = 1.3$.

The nature of the solution is qualitatively associated with the transition to chaos, observed experimentally. The positive Lyapunov exponent λ was calculated by the Benettin algorithm for (5.2) in the region of chaos. Its dependence on U_0 for $\nu = 2.8 \cdot 10^3$ is given in Fig. 9.57. Qualitatively, the correlation is modelled on the dependence of the synchronization threshold B_{thr} on U_0 for the corresponding frequency ν. Thus, physical and numerical results corroborate the conjecture that the synchronization threshold may serve a convenient information parameter for chaotic motions.

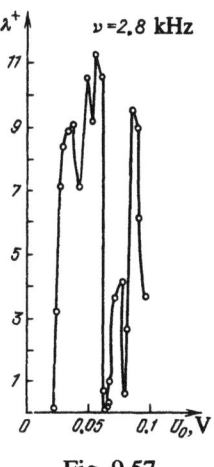

Fig. 9.57

A numerical and experimental investigation of the self-oscillatory system described by the equation

$$\ddot{x} - \mu(1-x^2)\dot{x} + x^3 = B \cos \omega t \qquad (5.3)$$

was performed in [669]. In the autonomous mode (B=0), the oscillations are periodic with frequency $\omega_0 = 1.617$ and amplitude $A = 1.824$. Depending on ω, synchronization of different orders, and quasi-periodic oscillations associated with a two-dimensional torus in the phase space, are possible for $B \neq 0$. For sufficiently large B, oscillations can be chaotic; e.g. for $\omega=4$, B=1, quasi-periodic oscillations occur, associated with a closed invariant curve (Fig. 9.58 a) on the cutting surface t=0 [mod $2\pi / \omega$]. Their power spectrum contains four components at frequencies ω_1, $4\omega_1 - \omega$, ω, $3\omega_1$, where $\omega_1 = 1.614 \approx \omega_0$ (Fig. 9.58 b). As B increases, the oscillations become chaotic. The point mapping on the cutting plane and the corresponding spectrum for B=17 are given in Fig. 9.58 c, d. Other sets of B, ω with chaotic oscillations are indicated in [669]. It is stressed that the cubic term for (5.3), making the equation different from van der Pol equation with external action, is essential to make the oscillations chaotic if μ is small.

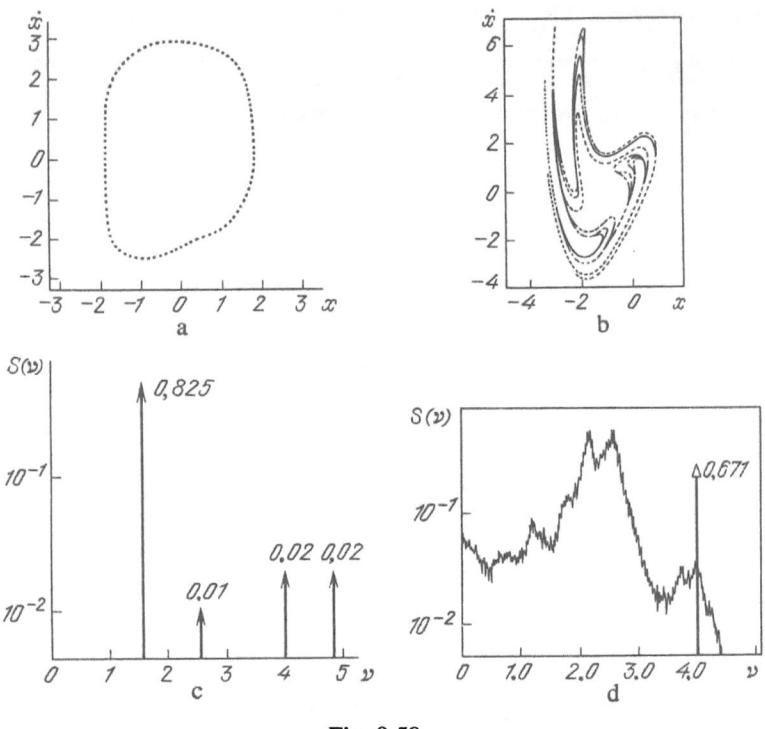

Fig. 9.58

For μ large and associated with the functioning of the autonomous generator in the relaxation regime, chaotic behaviour is possible also in the usual van der Pol generator with external action, which is described by the equation

$$\ddot{x} - \mu(1-x^2)\dot{x} + x = B\cos\omega t \qquad (5.4)$$

[141, 147, 472]. Thus, the chaotic solution of (5.4) was discovered in [147] for μ=3, B=2.5, ω=2.7.

A more general case was considered in [149], where the equation

$$\ddot{x} - \mu(1-x^2)\dot{x} + x + \xi x^3 = B\cos\omega t$$

was investigated via analog and numerical simulation. It was shown that, in a wide range of parameter values μ, ξ, B, ω, the equation has chaotic solutions with a broad spectrum, chaos arising because of formation and destruction of two-dimensional tori.

Transition to the chaos via appearance and destruction of quasi-periodic regimes was also observed under a harmonic external action on a Lorenz system [428] and on the generator with inertial non-linearity [28, 34–36].

The external action was given parametrically, i.e., the system

$$\dot{X} = \sigma(Y-X), \quad \dot{Y} = -Y + (r + a\cos\omega t)X - XZ, \quad \dot{Z} = -bZ + XY$$

was investigated, with $\sigma=10$, $b=8/3$. The parameter r was made to vary near the value $r_{tr} = 166.06$ associated with the intermittency threshold of the autonomous system, while ω varied near twice the oscillation frequency in the laminar phase ($\omega \approx 11$). Above the intermittency threshold (for $r = 166.07$), the external action with amplitude $a = 1$ and in a narrow frequency interval near 11.02 synchronized the chaotic oscillations, and led to intermittent transition from the torus to chaos outside it (Fig. 9.59a, b, c, d). For $r \lesssim r_{tr}$ (r=166), $\omega=10$ and small amplitudes a, quasi-periodic oscillations were observed (Fig. 9.59e), being disrupted and becoming chaotic when a exceeds the critical value $a_{cr} \approx 0.05775$. The transition from the torus to chaos was intermittent (Fig. 9.59f).

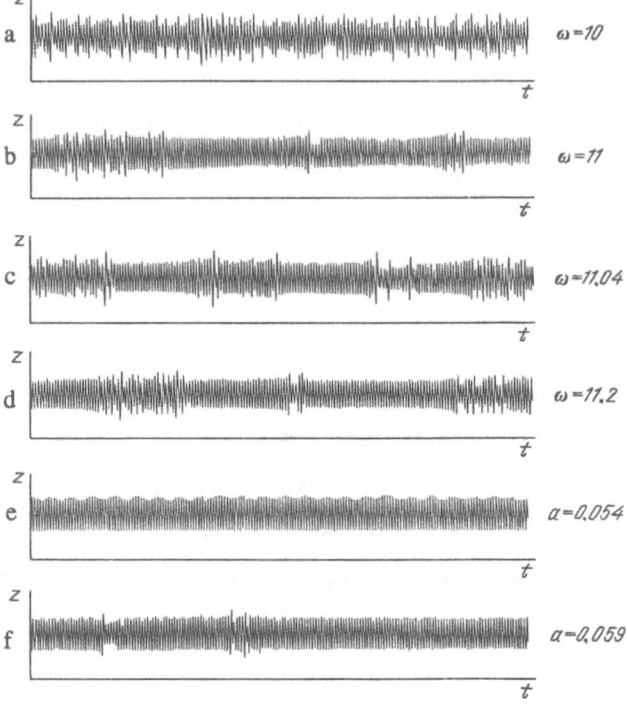

Fig. 9.59

The action of a harmonic external force on a generator with inertial non-linearity was numerically and experimentally investigated in [28, 34, 35] within the framework of the model

$$\dot{X} = \mu X + Y - XZ + B \sin \omega t, \quad \dot{Y} = -X, \quad \dot{Z} = \gamma[\vartheta(X)X^2 - Z] \quad (5.5)$$

(cf. (4.20)). For a non-resonant effect ($\omega \neq 1$) and small μ such that the basic periodic regime is stable for the autonomous generator, beats arise associated in the phase space, with a two-dimensional torus (Fig. 9.60, physical experiment; Fig. 9.61a, numerical experiment). If the frequencies are incommensurate, then the torus' coil is quasi-periodic (Fig. 9.60a); otherwise, periodic (Fig. 9.60 b, c). The characteristic realization of the process X(t) for $\omega < 1$ and $\omega > 1$ is illustrated in Fig. 9.62 (physical experiment). As μ increases, quasi-period-doubling bifurcations of the torus emerge (Fig. 9.63 a, b, physical experiment; Fig. 9.61b, numerical experiment), and a chaotic attractor arises (Fig. 9.63c,

Examples of mechanical, physical, chemical, and biological systems 359

Fig. 9.61c). Similar results also hold if the force amplitude B increases. It is shown in [35] that if B is fixed, the number of torus doublings until it is destroyed is finite, and determined by B: The smaller B, the larger the number of doublings. The spectrum of Lyapunov exponents for the attractor associated with Fig. 9.63c is $\lambda = (0.03196, 0, -0.00523, -0.228)$. The Lyapunov dimension was determined by the Kaplan-Yorke formula; $d_L = 3.117$ (on the basis of these data).

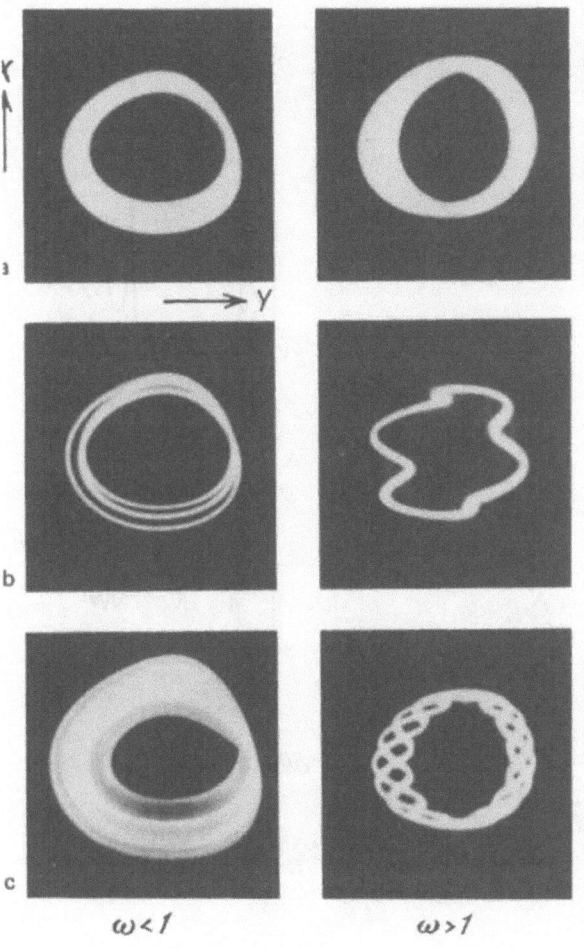

Fig. 9.60

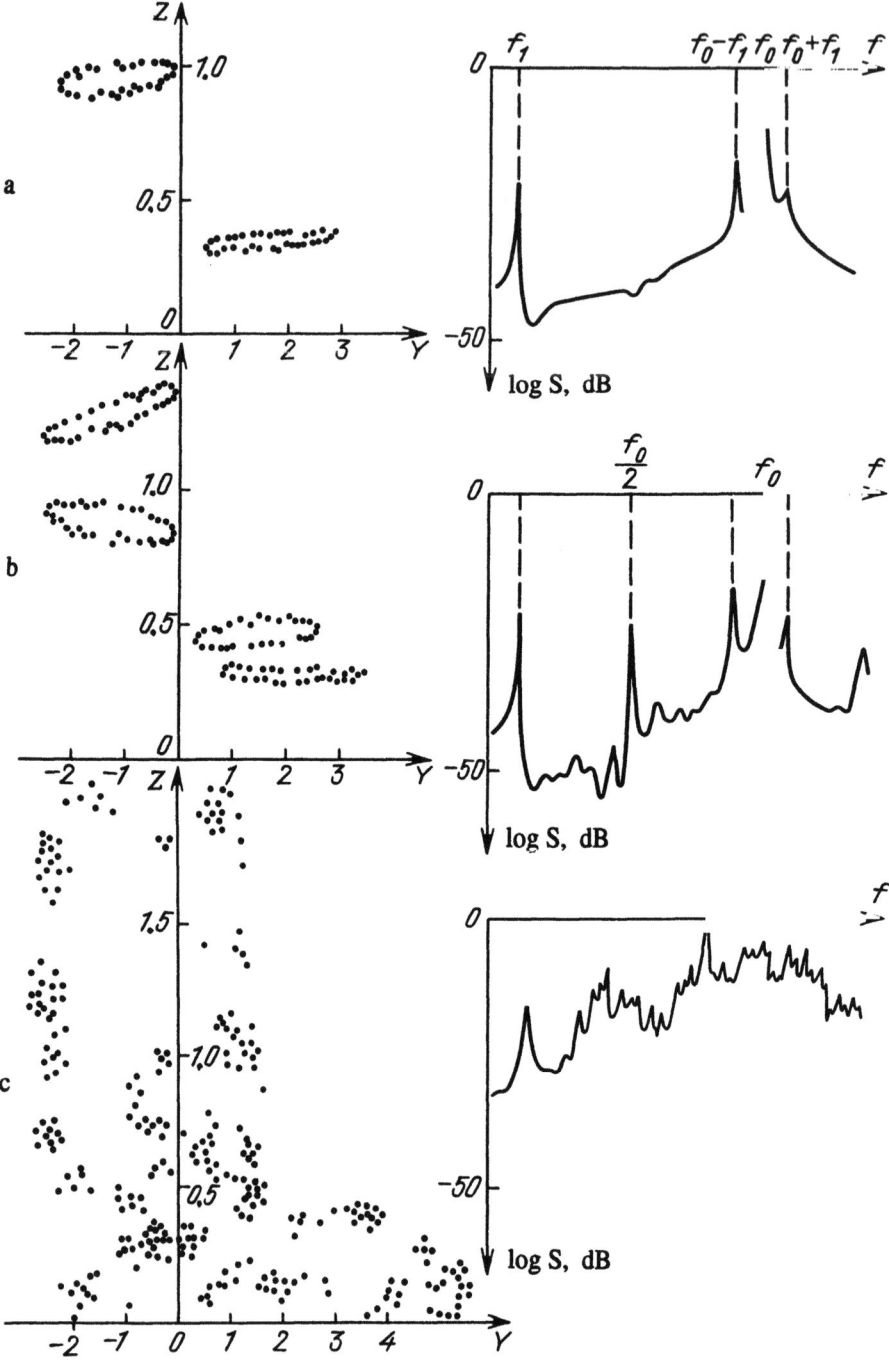

Fig. 9.61

Examples of mechanical, physical, chemical, and biological systems

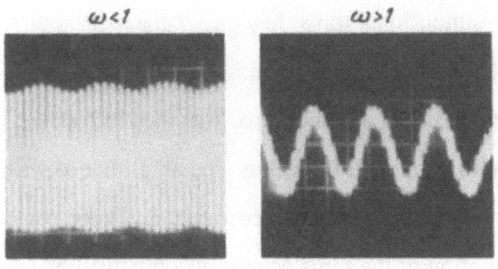

Fig. 9.62

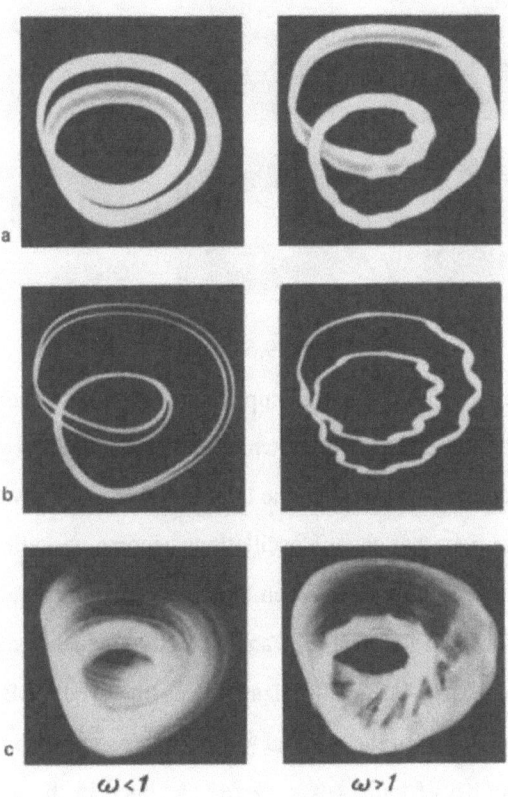

Fig. 9.63

A different picture was observed as μ decreases, starting with the values associated with a two-dimensional torus. In a physical experiment, the regime is attained as follows: With no external action, μ increases until there is no stable two-revolution cycle (e.g., μ = 1.4, γ = 0.3). An external force was then introduced to provide for stable synchronization at the fundamental frequency. The amplitude B was smoothly decreased until a stable two-dimensional torus was generated, and kept constant in further study. Meanwhile, hard disruption of the torus occurs, accompanied by emergence of a chaotic attractor (Fig. 9.64). The phenomenon was especially manifest in the parameter γ range from 0.1 to 0.5.

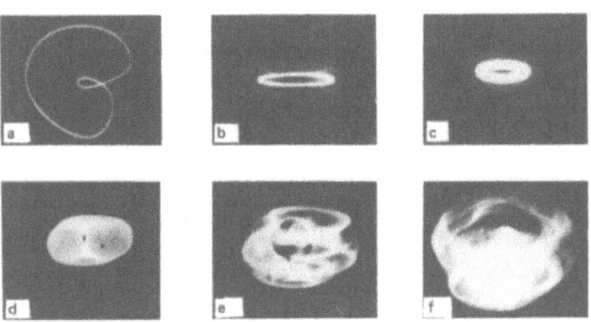

Fig. 9.64

A harmonic external force was applied in [36] to a system of two coupled generators with inertial non-linearity. Meanwhile, in a certain parameter range, transition to chaos via appearance and destruction of the three-dimensional torus was observed.

The converse, i.e., that chaotic oscillations become periodic under a harmonic external action, was observed in [28]. When the autonomous generator operated in the chaotic regime (e.g., for μ = 1.45, γ = 0.21), and when the frequency of the external force was close to natural (e.g., ω = 1), synchronization was observed for $B > B_0 \approx 1.14$, with oscillation period equal to that of the external force. Regions of synchronization for (5.5), with γ = 0.3 and a number of values of μ, were calculated in the graduation thesis of G. G. Shatalova[18]. An example of one such a region is given in Fig. 9.65a.

---

[18]Shatalova G. G. Synchrotization phenomenon in chaotic systems under the action of a

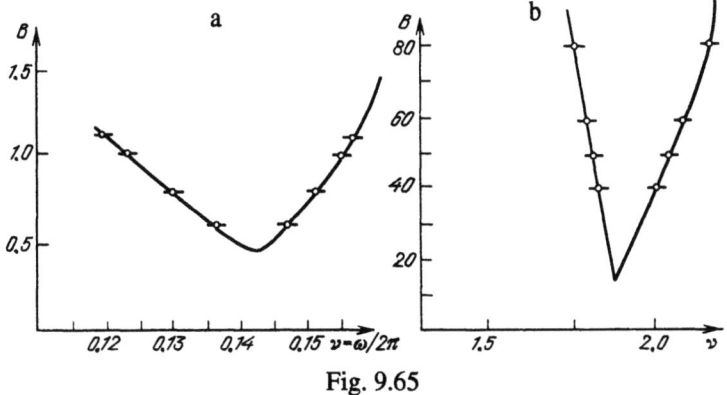

Fig. 9.65

Similar phenomena were also observed under an external action on a Lorenz system. Oscillation synchronization in a Lorenz system was apparently first investigated in [151] on an analog computer, using simulated equations of the form

$$\dot{X} = -\sigma(X-Y), \quad \dot{Y} = -Y + rX - X\tilde{Z},$$
$$\dot{\tilde{Z}} = -b\tilde{Z} + XY, \quad \tilde{Z} = Z + B\cos(2\pi v),$$

where $\sigma=10$, $b=8/3$, $r = 55$. The regions of synchronization discovered in the parameter (B, v)-plane are represented in Fig. 9.66. In the region marked by I, the period T of synchronous oscillations coincides with that of the external force, $T_0 = 1/v$; in the region marked by II, T is twice as large as T_0. The synchronization threshold with respect to the force amplitude is essential.

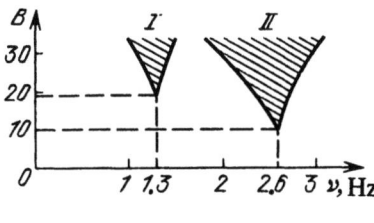

Fig. 9.66

Oscillation synchronization in a Lorenz system under the action of an external force was investigated by G. G. Shatalova, by introducing a harmonic external force in the first equation. Thus, the system of equations

---

periodic external force. Diploma work - Moscow, Physics Depatrment of Moscow University, 1985 (*Russian*).

$$\ddot{x}+2\delta\dot{x}+\omega_0^2 x = -\omega_0 y - \left(\frac{3}{2}\omega_0 x + \frac{x^2}{2} + y\right)x + B\cos(2\pi\nu t),$$

$$\dot{y}+by = ax + \frac{a}{2\omega_0}x^2 \tag{5.6}$$

was (numerically) simulated, where $2\delta = \sigma + 1 = 11$, $\omega_0 = \sqrt{b(r-1)}$, $b = 8/3$, $a = \omega_0(2\delta - b) = (52/3)\omega_0$. One of the synchronization regions for $r = 40$ is given in Fig. 9.65b.

A more detailed study of synchronization of chaotic oscillations was made in [223], where a system (4.2) with inertial self-excitation, non-linear functions of the form (4.18b), given by the equations

$$\ddot{x}+2\delta\dot{x}+\omega_0^2 x = -ky - nx^3 + B\cos(2\pi\nu\tau),$$

$$\dot{y}+\gamma y = ax - lx^2 - hx^3, \tag{5.7}$$

was considered.

The system was simulated on a computer for the parameters $2\delta = 0.75$, $\omega_0 = \gamma = a = l = h = 1$, $k = 20$, $n = 17.5$. The synchronization regions found are given in Fig. 9.67. The chaotic attractor in Fig. 9.38a for an autonomous system is replaced in the synchronization regions, depending on B, ν, by two-, three-, four- or six-revolution cycles. The two- and three-revolution cycles are given in Fig. 9.68a,b; the regions of the three- and six-revolution ones, in Fig 9.67, are shaded. Note that transition inside the synchronization regions from two- to four-revolution and from three- to six-revolution cycles is associated with period-doubling bifurcations. The two- and three-revolution cycles in the region marked by I are associated with synchronization at the fundamental frequency of the external force (the period of oscillation being equal to that of the external force); in the region marked by II, with synchronization at the second subharmonic (the period of oscillation being twice as large as that of the external force). To have an idea of the transitions from cycles of one type to another as the frequency ν changes, bifurcation diagrams were constructed, depicting the dependence on ν of the coordinate x of the cycles as they intersect the plane $\dot{x} = -0.2$ from above for two fixed amplitude values B (Fig. 9.69). It can be seen that, for $B = 2$, a cascade of period-doubling bifurcations occurs near the boundaries of the synchronization regions, leading to chaos. Doubling bifurcations in the synchronization regions do not start any such series, therefore, do not lead to chaos.

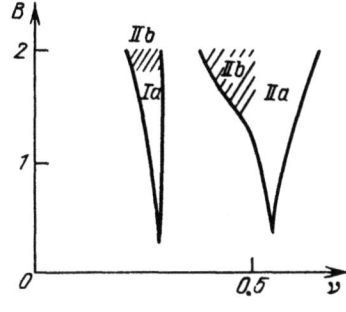

Fig. 9.67

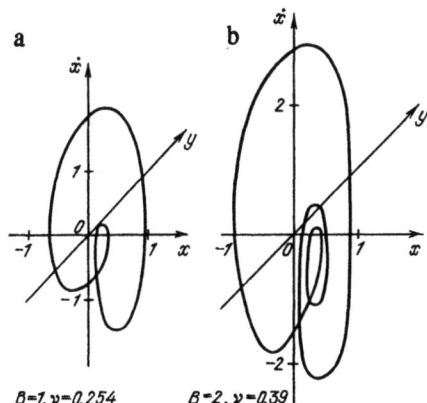

Fig. 9.68

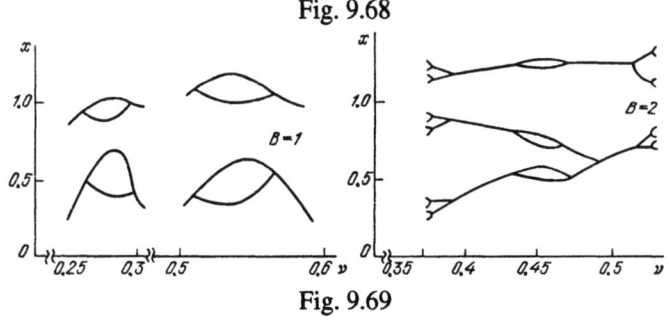

Fig. 9.69

Numerical computations show that transitions from synchronization regions into the region of chaos occurs differently for various amplitudes of the external force. For small B, the transitions occur via beats, or via quasi-periodic oscillations. Let us track one transition for B= 1 when ν varies from 0.249 to 0.255. As ν decreases up to the value $\nu_1 = 0.2535$ associated with leaving the synchronization region, beats appear, leading to peaks in the low-frequency part of the spectrum, at $f_m = m\nu_b$ (m = 1, 2, 3, ...), and at $f_n = v+nv_b$ ($n = \pm 1, \pm 2, \pm 3, \ldots$) near the frequency of the external force (Fig. 9.70a). As v decreases, the number of distinguishable frequencies increases; v_b increases slightly. For $v = v_b = 0.2504$, a quasi-period $T_t = 1/v_b$ doubling bifurcation of the torus occurs; subsequently, frequencies $mv_b/2$ and $v+nv_b/2$ appear in the spectrum. For $v = v_3 = 0.24903$, the spectrum becomes practically continuous, though with peaks at v, $2v$ (Fig. 9.70b). As v decreases further, the peaks vanish.

For $B > B_\alpha \approx 1.3$, the transition from the synchronization region into the region of chaos occurs via a sequence of cycle's period-doubling bifurcations. Thus, for $B = 2$, the first three bifurcations could be tracked (Fig. 9.69). The cycle's phase portrait after the first doubling bifurcation ($B=2$, $v = 0.38$) is given in Fig. 9.71.

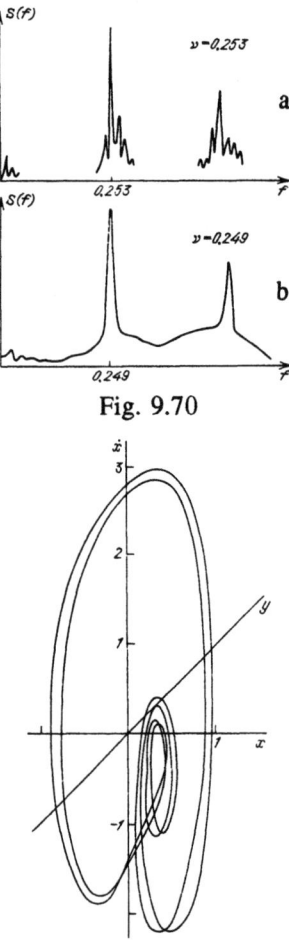

Fig. 9.70

Fig. 9.71

Examples of mechanical, physical, chemical, and biological systems

As discussed in Sect.3, Chapt.8, papers [214, 215] conjectured the ralation of the synchronization threshold to Kolmogorov's metric entropy coinciding with the positive Lyapunov exponent if the system is three-dimensional. The hypothesis was checked in the systems (5.5), (5.6), (5.7), and also in (4.2) with non-linear functions of the form (4.18a). Taking external force into account, (4.2) is of the form

$$\ddot{x} + 2\delta\dot{x} + \omega_0^2 x = -ky - gxy + B\cos\omega t ,$$
$$\dot{y} + \gamma y = ax + lx^2 .$$
(5.8)

Numerical simulation of (5.8) was carried out for the parameters $\gamma = 0.27$, $\omega_0 = a = l = 1$, $k = 1.73$, $2\delta = 0.5, 1.1$.

Examples of synchronization regions for $2\delta = 0.5$, $g = 12.2$, and $2\delta = 1.1$, $g = 30$ are given in Fig. 9.72.

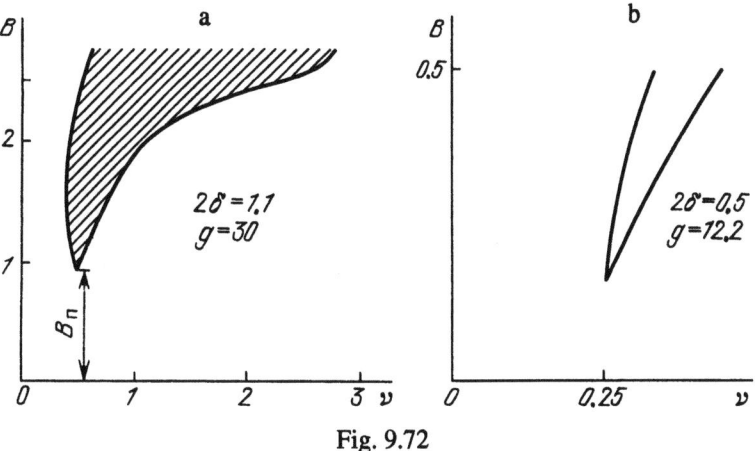

Fig. 9.72

The ralation of the synchronization threshold to the entropy K equal to the positive Lyapunov exponent is shown in Fig. 8.11 for all systems investigated (Δ for (5.5), $\times$ for (5.6), o for (5.7), ● for (5.8)). In Fig. 9.73, the dependences of λ^+ on the corresponding parameters are shown. The points, i.e., values of λ^+, are plotted. They are calculated by the formula (3.1), Chapt.8, on the basis of the values B_{thr} measured. Up to the errors of the numerical experiment, which are mostly due to the error in determining the synchronization threshold, with rare exception, the points lay on the curves associated with the dependences.

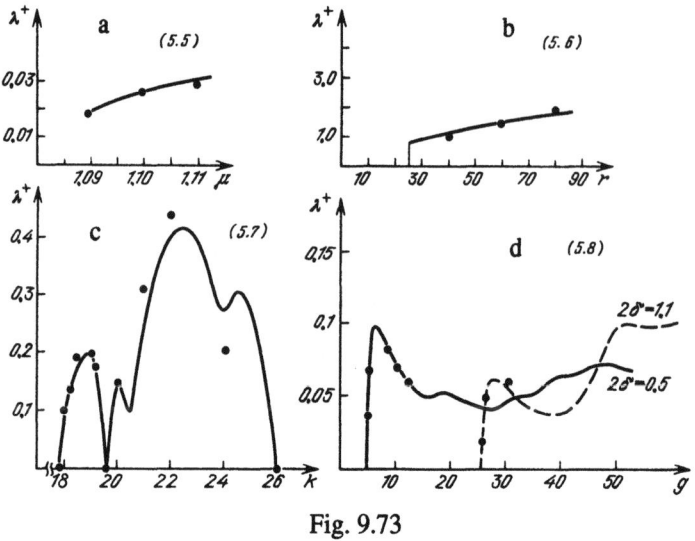

Fig. 9.73

6. Interaction of oscillation generators of various kinds

When generators of periodic and chaotic oscillations interact, they can also become either periodic or chaotic. We first of all prove this by taking the example of two coupled generators: one described by equations (4.2), (4.18b), and the other by the van der Pol equation [223]. The equations describing the interacting generators are

$$\ddot{x} + 2\delta\dot{x} + \omega_0^2 x = -ky - nx^3 + m_1 u ,$$
$$\dot{y} + \gamma y = ax - lx^2 - hx^3 ,$$
$$\ddot{u} - \mu(1 - u^2)\dot{u} + \omega^2 u = m_2 x .$$
(6.1)

In the numerical simulation of the equations (6.1), the parameter values $2\delta = 0.75$, $\kappa = 20$, $n = 17.5$, $\omega_0 = \gamma = a = l = h = 1$, $\mu = 0.1$ were fixed; the others varied. The case $m_2 = 0$ is equivalent to the above (see the preceding section), while that where $m_1 = 1$ is associated with B= 2. We fix $m_1 = 1$, $\omega = \pi$, and follow the behaviour of the system for varying m_2.

For $m_2 < m_{tr} \approx 1.2$, the solution of (6.1) is of period T = 4, twice as large as that of oscillations of the autonomous van der Pol oscillator.

This means that the chaotic oscillations of the first generator are synchronous with periodic oscillations of the second. For $m_2 > m_{tr}$, desynchronization occurs and

Examples of mechanical, physical, chemical, and biological systems

beats, i.e., quasi-periodic oscillations, develop. Their domain of existence takes a small part of the parameter m_2 range. As m_2 increases further, the first generator is in the chaotic oscillation regime, in turn, imposing its frequencies on the van der Pol generator, and thereby chaotizing the latter. The power spectra of both generators and the mutual correlation function for $m_2 = 3$ are given in Fig. 9.74. Note that the function has maximum 0.5 for $\tau = 0$. As m_2 increases still further, the spectra of both generators approach each other in width and peak location, whereas the maximum of the mutual correlation function increases.

When the frequency ω of the autonomous van der Pol generator is outside the synchronization regions for $m_2 = 0$ of the first generator, the former chaotizes, and its spectrum broadens substantially even for small m_2. As m_2 increases further, the spectrum width increases until it reaches that of the first generator in the autonomous regime.

Interesting phenomena are observed when two generators of chaotic oscillations interact. As an example, consider the two coupled generators described by (4.2) with non-linear functions (4.18a), (4.18b) [223]. The system of equations describing the interaction is

$$\ddot{x}_1 + 2\delta_1 \dot{x}_1 + \omega_1^2 x_1 = -k_1 y_1 - n x_1^3 + m_1 x_2,$$
$$\dot{y}_1 + \gamma_1 y_1 = a_1 x_1 - l_1 x_1^2 - h x_1^3,$$
$$\ddot{x}_2 + 2\delta_2 \dot{x}_2 + \omega_2^2 x_2 = -k_2 y_2 - g x_2 y_2 + m_2 x_1, \quad (6.2)$$
$$\dot{y}_2 + \gamma_2 y_2 = a_2 x_2 + l_2 x_2^2.$$

It is modelled for the parameters

$$2\delta_1 = 0.75, n = 17.5, k_1 = 20,$$
$$\gamma_1 = a_1 = l_1 = h = a_2 = l_2 = \omega_1 = \omega_2 = 1,$$
$$2\delta_2 = 0.5, \gamma_2 = 0.1, k_2 = 10,$$
$$g = 20, m_1 = m_2 = m.$$

When there is no coupling (for m=0), the oscillations of both generators are chaotic and uncorrelated. The power spectra are different in width and location of peaks (Fig. 9.75a). As m increases, their widths become the same, and the peaks close in frequency. In this sense, we can speak of mutual synchronization of generators of chaotic oscillations. We illustrate by the power spectra of the processes $x_1(t)$, $x_2(t)$ for m=1 (see Fig. 9.75b). Comparing Figs. 9.75 a and b, we see that, for m=1, the spectrum of the second generator is substantially shifted into the low-frequency part, and the spectrum width is much narrower than for m=0. The mutual correlation function for $x_1(t)$, $x_2(t)$ is given in Fig. 9.76. Its first maximum is placed for $\tau \neq 0$, which indicates a certain "phase difference" between the oscillations of both generators. As another example, consider the interaction of two identical chaotic oscillation generators with the equations

$$\ddot{x}_{1,2} + 2\delta \dot{x}_{1,2} + \omega^2 x_{1,2} = -k y_{1,2} - n x_{1,2}^3 + m x_{2,1},$$
$$\dot{y}_{1,2} + \gamma y_{1,2} = a x_{1,2} - l x_{1,2}^2 - h x_{1,2}^3,$$

(6.3)

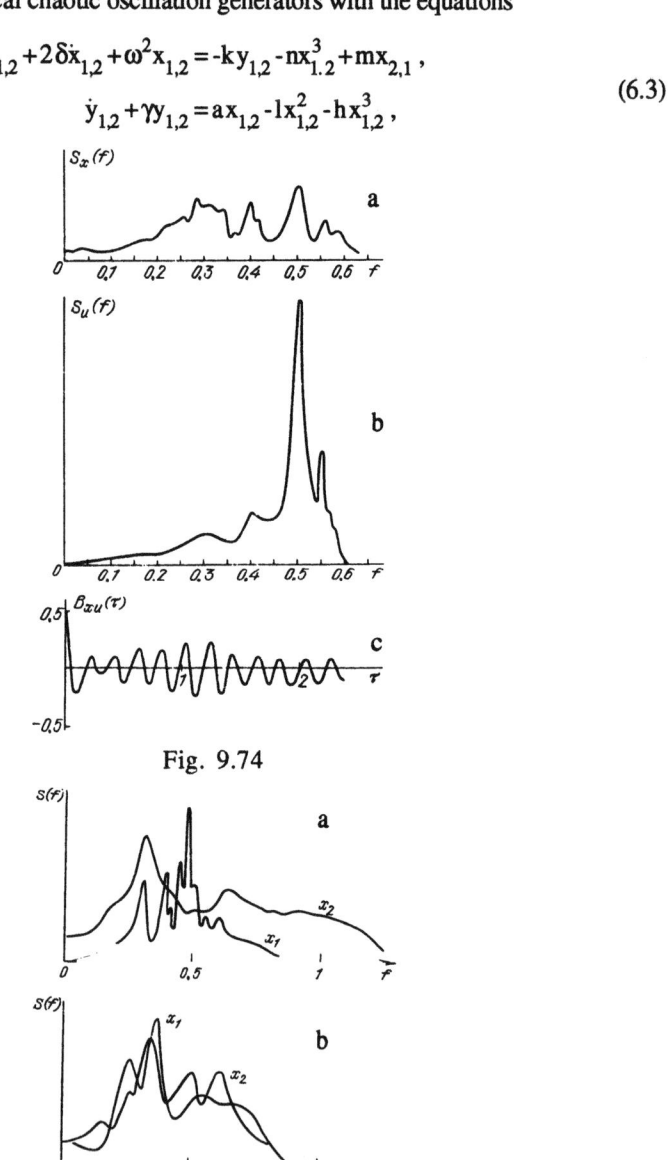

Fig. 9.74

Fig. 9.75

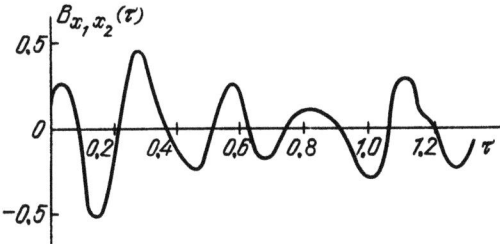

Fig. 9.76

where $2\delta = 0.75$, $k = 20$, $n = 17.5$, $\omega = \gamma = a = l = h = 1$. If there is no coupling (for m=0) and the initial conditions are different, then the oscillations of both generators are not correlated. For m≠0, the generators synchronize in the above sense. Meanwhile, the peaks of the spectrum are shifted into the low-frequency part (Fig. 9.77a). The mutual correlation function of the oscillations for m=3 is given in Fig. 9.77b.

Similar effects are also observed when the generators of chaotic oscillations described by equations (4.2), (4.18a) interact.

The paper [34] supplies a numerical study of interaction of two inductively coupled generators with inertial non-linearity. If the parameters of both generators are identical, then the model's equations are

$$(1-\Gamma^2)\dot{X}_{1,2} = \mu X_{1,2} + Y_{1,2} - X_{1,2}Z_{1,2} + \Gamma(\mu X_{2,1} + Y_{2,1} - X_{2,1}Z_{2,1}),$$
$$\dot{Y}_{1,2} = -X_{1,2}, \quad \dot{Z}_{1,2} = \gamma[\nu(X_{12})X_{1,2}^2 - Z_{1,2}], \quad (6.4)$$

where $0 \leq \Gamma \leq 1$ is the inductive coupling coefficient.

Despite the total identity, quasiperiodic oscillations with fundamental frequencies f_0, f_1 develop in the system for increasing parameter μ and fixed $\gamma = 0.3$, $\Gamma = 0.3$ (the oscillations developing for $\mu = \mu^* = 1.58$). The corresponding power spectrum is given in Fig. 9.78a. Despite $f_1 \approx (2/5)f_0$, the process is quasi-periodic, i.e., the corresponding torus coil is not closed. This is manifested by the fact that the set of Lyapunov exponents contains two zeros. As μ increases further, the quasi-period becomes twice as large (Fig. 9.78b), and the oscillations become chaotic (Fig. 9.78c). There are finitely many doublings of the torus, and they are the fewer the larger the coupling coefficient Γ; this is consistent with the results in [538]. The chaotic attractor arising after breakdown of the torus is asymmetric, i.e., at least one of the inequalities $X_1 \neq X_2$, $Y_1 \neq Y_2$, $Z_1 \neq Z_2$ holds. It is interesting that the power spectrum for chaotic oscillations arising in this way contains no explicit peaks, and is quite wide.

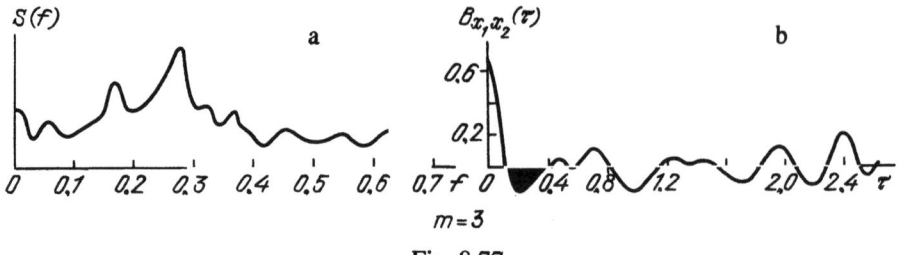

Fig. 9.77

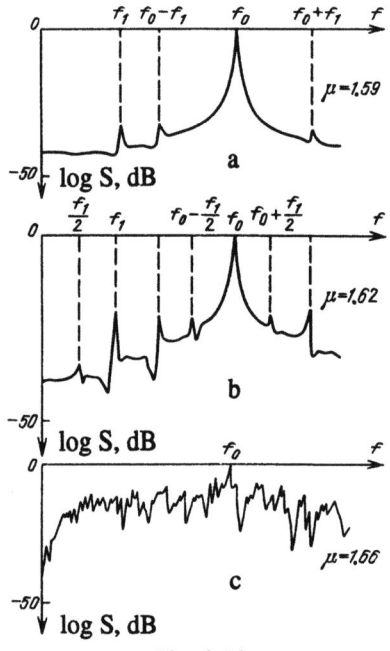

Fig. 9.78

The dependence of the Lyapunov dimension d_L and metric entropy K on μ, determined from the set of Lyapunov exponents, is given in Fig. 9.79. It can be seen that the dimension of the chaotic attractor exceeds three, similarly to the case of an external harmonic force acting on one generator (see Sect.5), whereas entropy increases monotonically with μ.

Oscillation chaotization may also occur when the interacting generators with no coupling make periodic oscillations. Examples of the kind can be found in [169, 322, 323]. In the latter two papers, a system simulating the interaction of two neuron populations is investigated, each consisting of neurons of two types: *excitatory* (x_j) and *inhibitory* (y_j) [322]. The system is of the form

$$\dot{x}_1 = -x_1 + \tanh(u_1 + h_1 x_1 - b y_1) + \gamma(x_2 - x_1),$$
$$\dot{y}_1 = -\varepsilon y_1 + \varepsilon \tanh(v + p x_1),$$
$$\dot{x}_2 = -x_2 + \tanh(u_2 + h_2 x_2 - b y_2) + \gamma(x_1 - x_2), \quad (6.5)$$
$$\dot{y}_2 = -\varepsilon y_2 + \tanh(v + p x_2),$$

where $u_j = u_0 \cdot h_j + b$, $v = v_0 \cdot p$ (j=1, 2).

Equations (6.5) were numerically studied for fixed $u_0=0.8$, $v_0=4$, $b=0.5$, $p=5$, $\varepsilon=0.005$, $\gamma=0.01$, $h_2=1.141$, and varying parameter h_1 in the strip $1.11<h_1<1.16$. In the absence of coupling (for $\gamma=0$), each population is for these parameters in the self-oscillatory activity regime, i.e., the corresponding equations describe a periodic oscillation generator. For $h_1=h_2$, the generators are identical, and the coupled system also makes periodic oscillations. The latter become chaotic for $h_1 \neq h_2$. It turns out that, for $h_1<1.1225$, the system has a stable three-revolution cycle. For $h_1 \approx 1.1225$, the cycle merges with an unstable three-revolution one, and vanishes. Intermittent stochasticity arises as a result. Near the transition, the point mapping on the Poincaré cutting surface can be approximately reduced to a one-dimensional discontinuous mapping.

A system of capacity-coupled generators with inertial non-linearity, making periodic close-to-relaxation oscillations in the autonomous regime, has been considered [169]. It was shown numerically and experimentally that the oscillations became chaotic on account of this coupling.

7. Certain discrete models of turbulence

As noted above, a Lorenz system is the simplest (3-mode) model of convective turbulence. In the classical problem of a plane liquid layer heated from below, the system is distinguished from a more complete system by restriction to the first spatial harmonics of the velocity components, and the zero, first and second spatial harmonics in the temperature spectrum [217]. It is obvious that, because of these restrictions, a Lorenz system is adequate only near the threshold for appearance of a convective roler, i.e., for r close to one. For larger r, we have to take into account higher spatial harmonics, and the

Lorenz equations become inadequate. This was done in [574], where the nature of the solution depends substantially on the number of modes taken into account.

Besides the Lorenz equations, a number of discrete models of the Navier-Stokes equation are considered in the literature. A five-dimensional model was described in [409, 462]; a seven-dimensional one, in [461]; a twelve-dimensional one, in [463]. The results obtained are substantially different. Thus, in the case of dimension five, as the Reynolds number increases, transition to chaos occurs only via period-doubling bifurcations. Meanwhile, the point mapping on the cutting surface can be reduced to a one-dimensional one. However, in the cases of dimensions seven and twelve, transition to chaos occurs via the appearance and subsequent destruction of quasi-periodic motions associated in the phase space with two-dimensional tori.

That the results are substantially different if the dimension of the model increases shows that the models do not give a satisfactory description of the solution of the original Navier-Stokes equation. However, to study the models is interesting and useful from the viewpoint of the general theory of dynamical systems, of the study of their complicated behaviour, and of possible transitions to chaotic regimes.

We now show how the equations for the models are obtained. Consider the two-dimensional flow of an incompressible fluid under the action of a spatially periodic external force $\mathbf{f}$, which can be represented by finitely many terms in the Fourier series, viz.,

$$\mathbf{f} = \sum_{\mathbf{k} \in L} (\mathbf{k}/|\mathbf{k}|) f_{\mathbf{k}} \exp(i\mathbf{k}\mathbf{x}), \qquad (7.1)$$

where $\mathbf{x}=\{x_1, x_2\}$, $\mathbf{k}=\{h_1, h_2\}$ is a wave vector with integer components, $\mathbf{k}_\perp=\{h_2,-h_1\}$, and L the set of 2N vectors such that if $\mathbf{k} \in L$, then also $-\mathbf{k} \in L$. The coefficients $f_{\mathbf{k}}$ satisfy the condition

$$f_{\mathbf{k}} = -f_{-\mathbf{k}}^*. \qquad (7.2)$$

Then the flow is described by the continuity equation

$$\operatorname{div} \mathbf{U} = 0 \qquad (7.3)$$

and the Navier-Stokes equation

$$\frac{\partial \mathbf{U}}{\partial t} + (\mathbf{U}\nabla)\mathbf{U} = \nabla p + \nu \Delta \mathbf{U} + \mathbf{f}, \qquad (7.4)$$

where $\mathbf{U}=\{u, v\}$ is the velocity vector, ν kinematic viscosity, and p pressure.

Examples of mechanical, physical, chemical, and biological systems 375

The solution of (7.3), (7.4) can be represented as

$$U(x) = \sum_{k \in L} (k_\perp / |k|) \gamma_k \exp(ikx) \tag{7.5}$$

in accordance with (7.1), where $\gamma_k = -\gamma_{-k}^*$. The solution of (7.5) satisfies (7.3) automatically. Substituting (7.5), (7.1) in (7.4) and eliminating p, we obtain

$$\dot{\gamma}_k = -i \sum_{\substack{k_1 + k_2 = k, \\ k_1, k_2 \in L}} \frac{(k_{1\perp} k_2)(|k_2|^2 - |k_1|^2)}{2|k_1||k_2||k|} \gamma_{k_1} \gamma_{k_2} - \nu |k|^2 \gamma_k + f_k . \tag{7.6}$$

Let L be the set of vectors $k_1 = (1, 1)$, $k_2 = (3, 0)$, $k_3 = (2, -1)$, $k_4 = (1, 2)$, $k_5 = (0, 1)$, $k_6 = (2, 2)$, $k_7 = (1, -2)$, and suppose force **f** is independent of time, containing only one spatial harmonic with wave vector k_3, i.e., $f = (k_{3\perp}/|k_3|) f_{k_3} \exp(ik_3 x)$. In this case, equations (7.6) have the particular solution

$$\gamma_{k_1} = aX_1, \quad \gamma_{k_2} = -iaX_2, \quad \gamma_{k_3} = aX_3, \quad \gamma_{k_4} = iaX_4,$$
$$\gamma_{k_5} = aX_5, \quad \gamma_{k_6} = iaX_6, \quad \gamma_{k_7} = iaX_7,$$

where $a = 10\nu\sqrt{2}$, and $X_1, X_2, ..., X_7$, are real quantities satisfying the equations

$$\dot{X}_1 = -2X_1 + 4\sqrt{5}X_2X_3 + 4\sqrt{5}X_4X_5,$$
$$\dot{X}_2 = -9X_2 + 3\sqrt{5}X_1X_3 + 3\sqrt{5}X_6X_7,$$
$$\dot{X}_3 = -5X_3 + 9X_1X_7 - 7\sqrt{5}X_1X_2 + R, \tag{7.7}$$
$$\dot{X}_4 = -5X_4 - \sqrt{5}X_1X_5, \quad \dot{X}_5 = -X_5 - 3\sqrt{5}X_1X_4,$$
$$\dot{X}_6 = -8X_6 - 4\sqrt{5}X_2X_7, \quad \dot{X}_7 = -5X_7 + \sqrt{5}X_2X_6 - 9X_1X_3,$$

where the dot denotes differentiation with respect to time $\tau = \nu t$, and $R = f_{k_3}/a$ is an analog of the Reynolds number. The equations for the five-dimensional model are obtained from (7.6) if we confine ourselves to the five wave vectors k_1, k_2, k_3, k_4, k_5 in L [409, 462]. However, to obtain the twelve-dimensional model, we have to take twelve vectors.

We restrict ourselves to the results obtained in [461] for the seven-dimensional model (7.7). For $R = R_0 \approx 227.1$, the four symmetrically located limit cycles become unstable, and turn into four two-dimensional tori with frequencies f_1 (cycle frequency) and $f_2 = 1/T_t$, where T_t is the quasi-period of the torus. The projection onto the (X_6, X_3)-plane of the invariant curve on the section hyperplane $x_1 = 0$, associated with one of such tori, as well as the spectral densities of the Poincaré mapping for x_3 and of the solution of (7.7) for $x_1(\tau)$, where R= 269, are given in Fig. 9.80a. For $R = R_1$, $275<R_1<276$, the first quasi-period doubling bifurcation of the torus arises (Fig. 9.80b); for $R = R_2$, $294<R_2<294.5$, the second (Fig. 9.80c). The third doubling bifurcation of the torus was not manifest. As R increases, starting with the value R_2, the invariant curve becomes still more irregular (Fig. 9.81). Chaos sets in for $R = R_3 \approx 299.25$. The evolution of the Poincaré mapping and of the mapping's spectrum as R changes from 300 to 310 is represented in Fig. 9.82. It can be seen that, for R = 300, the frequency $f_2/4$ is fundamental in the spectrum; for R= 304, $f_2/2$; for R = 310, f_2. The evolution is similar to the one observed as chaos is generated via a sequence of cycle period-doubling bifurcations (e.g., see [221, 344, 564]). As R increases up to 325, the chaotic attractor continually increases in size, preserving its structure. Thus, when R = 326.25, the Poincaré mapping is of the form of a closed curve describing three loops (Fig. 9.83a). Since there is continuous background in the spectrum (Fig. 9.83b), in addition to sharp peaks, the attractor is still chaotic. The peaks correspond to frequencies nf', where $f' \approx f_2/3$. For $R > R_4 \approx 326.25$, synchronization regions develop. They are associated with different rational Poincaré rotation numbers $\rho = f'/f_1$. The dependence of ρ on R in the strip $325<R<347$ is given in Fig. 9.84. The regions of synchronization of various orders can be seen clearly. Finally, for $R>R_5$, where $347<R_5<348$, the structure of the attractor is similar to the above for $R<325$ (Fig. 9.85).

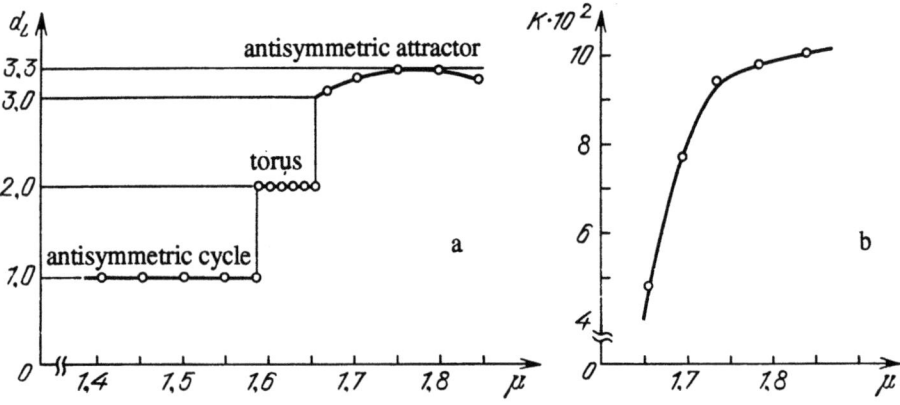

Fig. 9.79

Examples of mechanical, physical, chemical, and biological systems

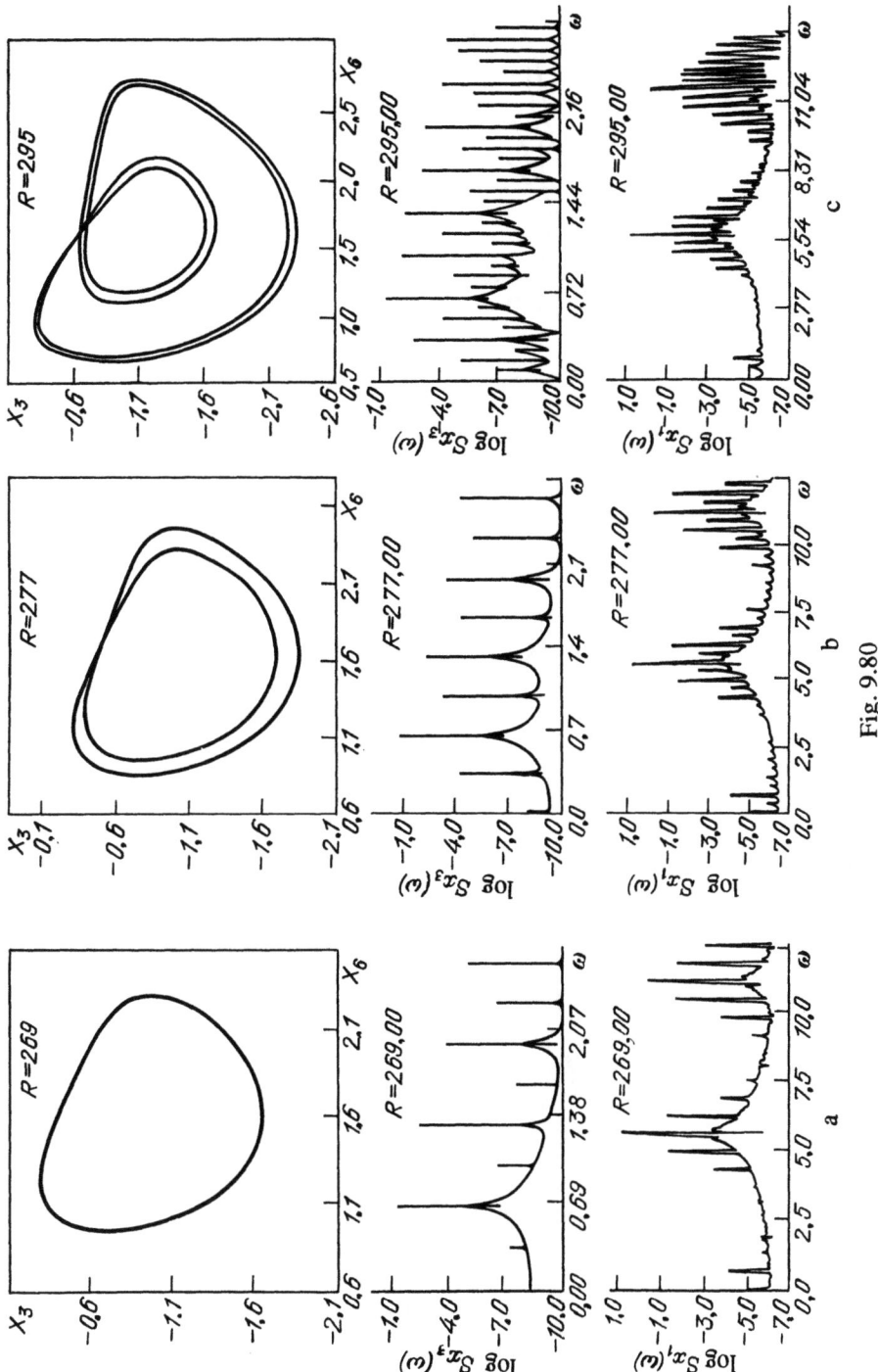

Fig. 9.80

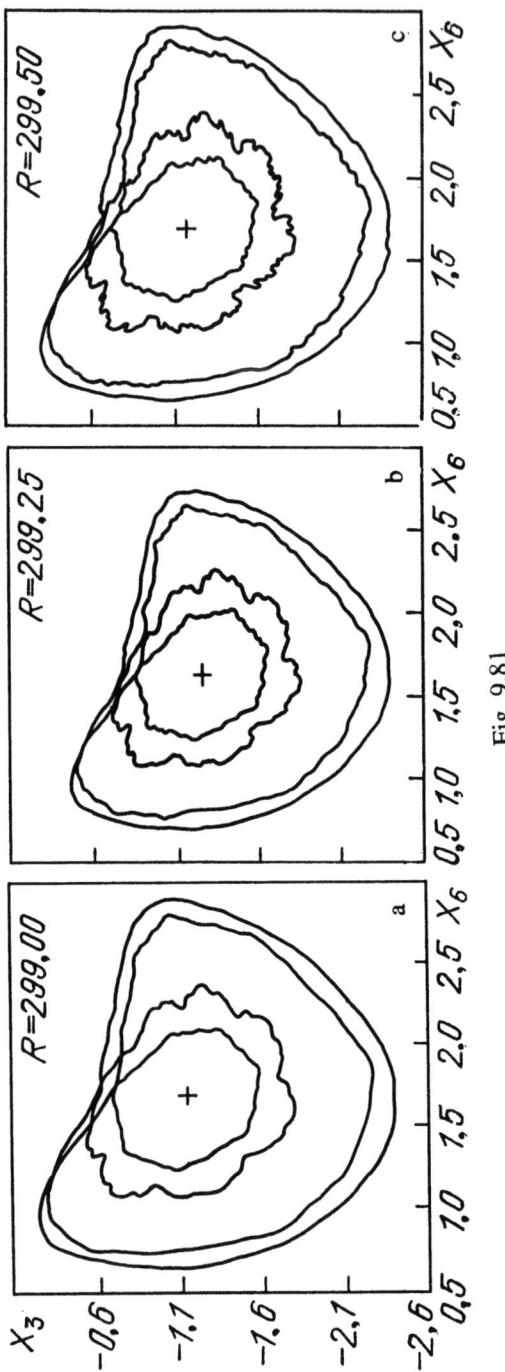

Fig. 9.81

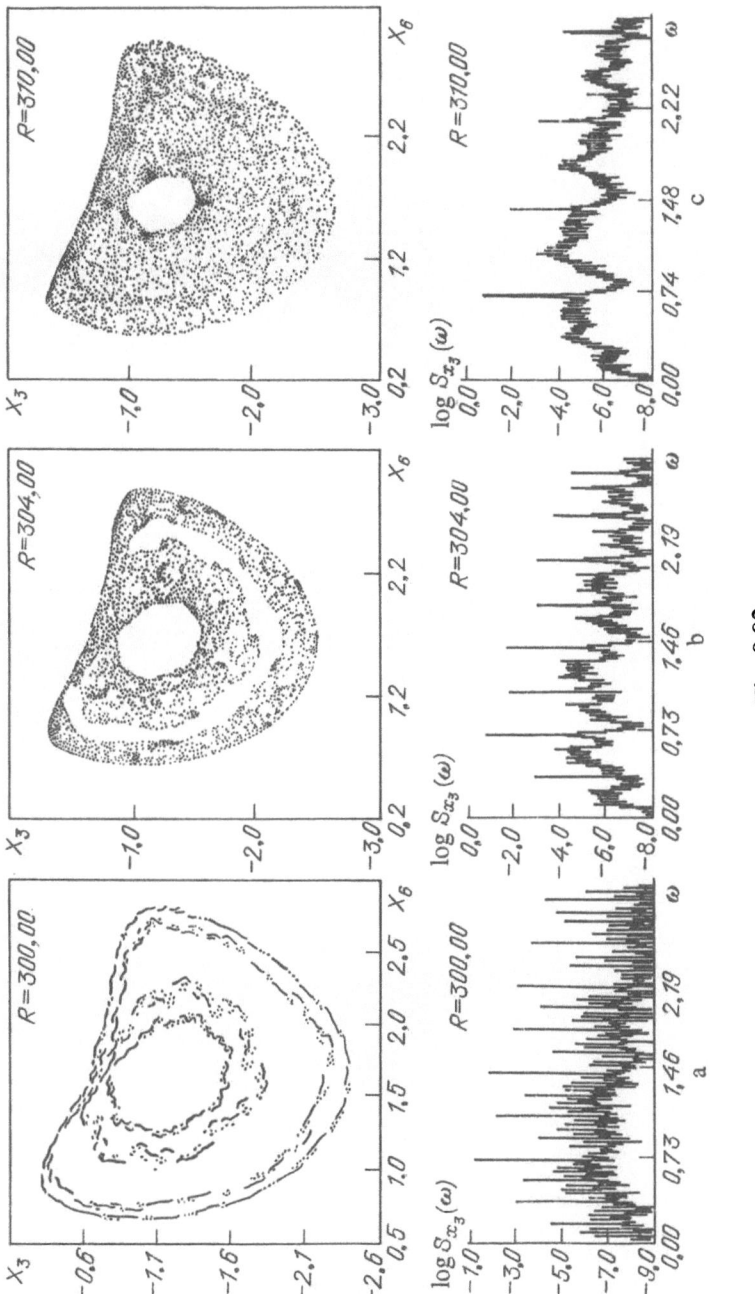

Fig. 9.82

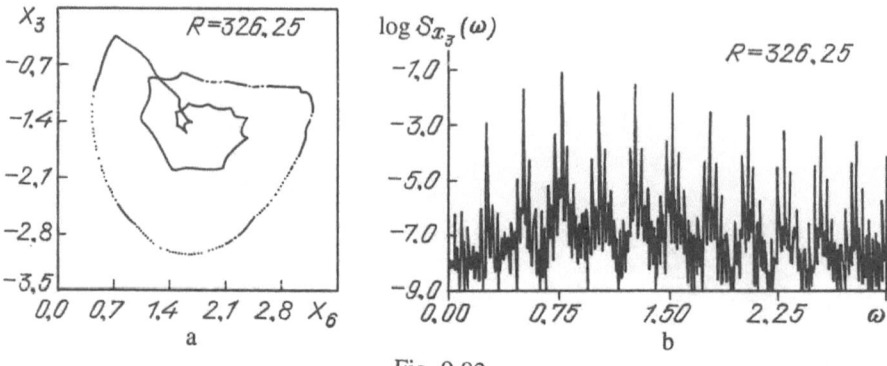

Fig. 9.83

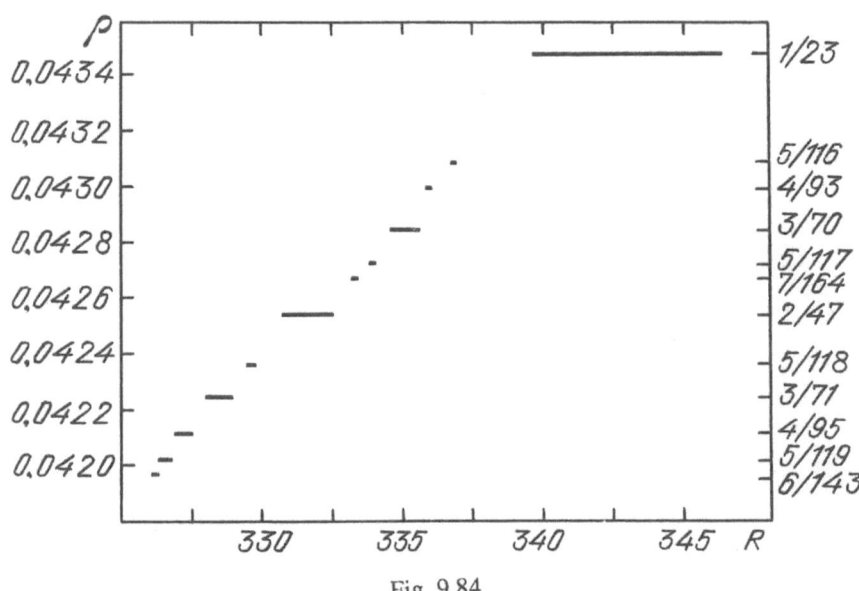

Fig. 9.84

Examples of mechanical, physical, chemical, and biological systems 381

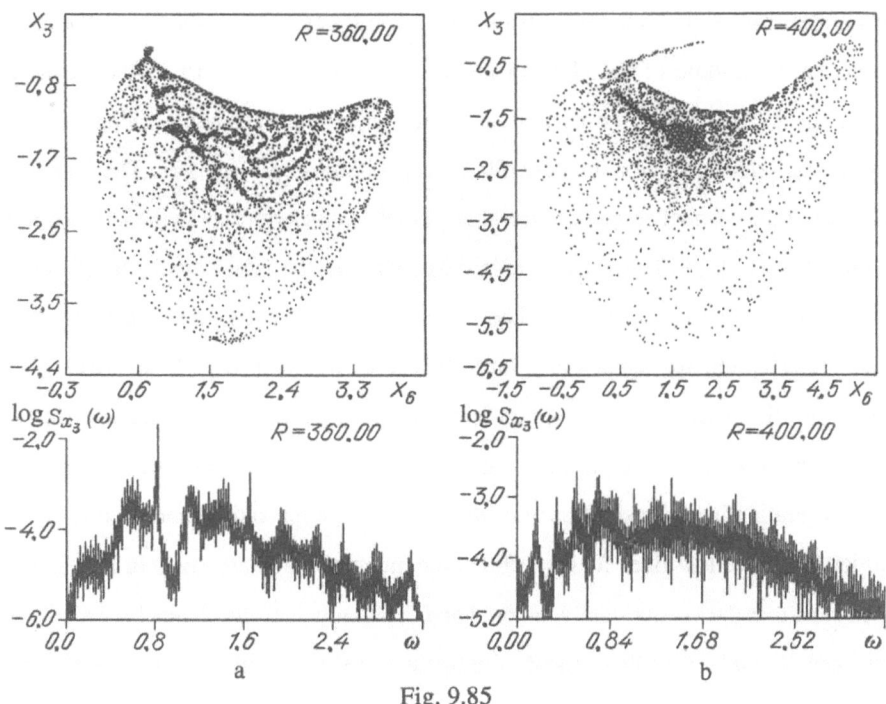

Fig. 9.85

The Lyapunov dimension d_L of the attractor, calculated by the Kaplan-Yorke formula, turns out to be slightly larger than two for $R_3<R<300$. As R increases, d_L also increases, becoming slightly larger than three when $R>300$. For $R_4<R<R_5$, the dimension again decreases until values slightly exceeding two. For $R>R_5$, the Lyapunov dimension increases again, exceeding three, and increasing continuously as R increases. However, the increase rate is low. For $R = 500$, still $d_L = 3.28$.

In contrast to the papers [409, 461-463] in which low-dimensional models for the Navier-Stokes equation are considered, [300], [701] investigate the multi-dimensional model obtained when solving the equation by the finite-difference method with boundary conditions associated with a flow between closely spaced parallel planes (*Poiseulle flow*) [217]. If p is eliminated from (7.3), (7.4), f=0, and the stream function ψ_n introduced, so that u=∂ψ/∂y, v=-∂ψ/∂x, then we obtain the equation

$$\frac{\partial \Delta \psi}{\partial t} = -\frac{\partial \psi}{\partial y}\frac{\partial \Delta \psi}{\partial x} + \frac{\partial \psi}{\partial x}\frac{\partial \Delta \psi}{\partial y} + \frac{1}{R}\Delta\Delta\psi \qquad (7.8)$$

for ψ [217], where x, y are dimensionless coordinates along and across the channel in terms of the channel width h, t dimensionless time in terms of h/u_0, where u_0 is the liquid velocity in the middle of the channel in laminar flow, and $R=u_0h/\nu$ is the Reynolds number [217]. The flow was simulated on channel's part of dimensionless length l=6.28. A mesh was selected to contain 129x129 elements, time step being 10^{-3}. The boundary conditions at the channel entrance corresponded to the velocity profile in the laminar flow, i.e., $u(0,y,t)=4y(1-y)$, $v(0,y,t)=0$. As the initial conditions, the same velocity profile in the whole channel, however, with a small perturbation, eas chosen (not more than 1%). Due to the perturbation, for sufficiently large Reynolds numbers ($R_{cr}=5,780$ [217]), the velocity profile broke down, and a non-laminar flow was established, characterized by spatially irregular, but time-periodic, large-scale vortices. For R=6,500, the period of vortex generation in dimensionless time was T=3.65. The relaxation time for this flow is of order 200. The flow lines for R=6,500 at the moment t=300 are given in Fig. 9.86. It can be seen that the vortices are generated and exist only near the channel's walls. That vortex generation is periodic is manifested by the cycle of complicated shape but closed, in the phase space with coordinates $u_1(t)$, $v_1(t)$, $u_1(t+\tau)$, $v_1(t+\tau)$, $u_1(t+2\tau)$, $v_1(t+2\tau)$, where $u_1(t)=u(5,0.48,t)$, $v_1(t)=v(5,0.48,t)$, $\tau=0.07$, constructed by the Packard-Takens procedure. The projection of this cycle onto the ($u_1(t)$, $u_1(t+2\tau)$)-plane is given in Fig. 9.87a.

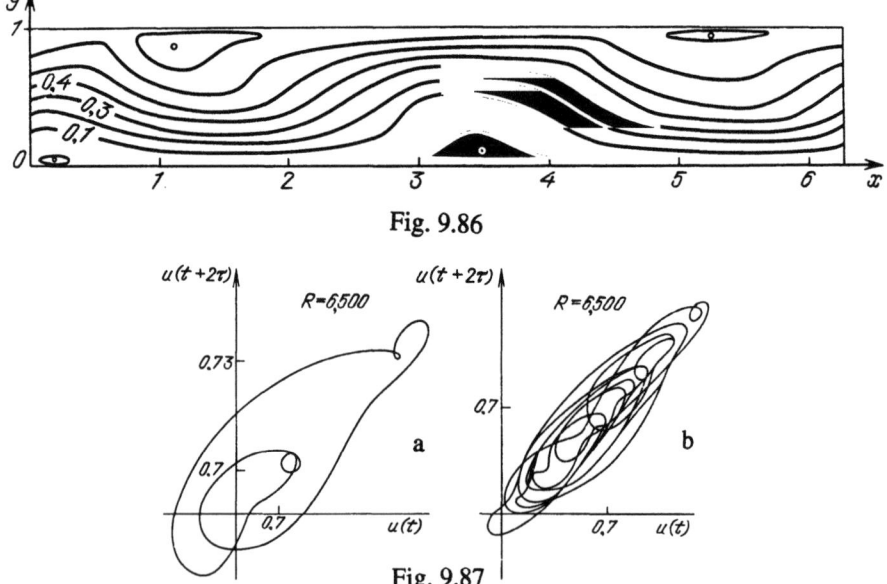

Fig. 9.86

Fig. 9.87

It is noted in [300] that the velocity profile after the relaxation process is close to the one observed experimentally (deviation not exceeding 8%). Turbulent resistance (with difference of about 12%) also turns out to be close to experimental values.

Despite the time-periodicity of the flow, the liquid particles diverge exponentially, i.e., mixing occurs. The divergence can be specified by the quantity

$$\lambda = \left\langle \lim_{d(0)\to 0} \frac{1}{t} \ln \frac{d(t)}{d(0)} \right\rangle$$

similar to a Lyapunov exponent, where $d(t)$ is the distance between particles at a moment t and the angle brackets denote the averaging over all channel points. For $R = 6,500$, $\lambda = 0.36$ was obtained.

Periodicity of vortex generation is disrupted even under a weak external action on the flow. Thus, for a small vibration of the channel walls $(v(x,0,t) = v(x,1,t) = 0.01 \sin 2\pi t)$, the phase portrait is as in Fig. 9.87b, i.e., a chaotic attractor arises. The latter's correlation dimension in the above six-dimensional phase space turns out to be 3.7. The divergence coefficient for particles λ with walls vibrating increases, and attains 0.6 (spread of values at different channel points being of order 0.15). Similar phenomena are also observed if the velocity modulation is periodic at the channel entrance.

8. Examples of models for chemical kinetics

In 1920, A. Lotka [565] hypothesized a chemical reaction with possible reagent concentration oscillations. The equations are

$$\dot{X} = k_1 X - a_1 XY, \quad \dot{Y} = -k_2 Y + a_2 XY.$$

A similar model was subsequently used by V. Volterra to account for oscillations in the number of competing species of animals and plants [114], called a *predator-prey pair*.

In 1951, an oscillatory chemical reaction in a homogeneous system was discovered experimentally by B. P. Belousov. However, his results were published only in 1959 [80]. Such reactions were then investigated by A. M. Zhabotinsky in [152]. The possibility of oscillatory chemical reactions (in particular, in homogeneous systems) was not put in doubt since then. At present, quite a few periodic reactions are known [1, 152, 199, 319, 356]. To understand them better, certain simple mathematical models were worked out. Two, termed the *Brusselator* and *Oregonator*, became most widespread [356]. The Brusselator describes the chemical reaction between six substances A, X, Y, B, D E, viz.,

$$A \to X, \quad B + X \to Y + D, \quad 2X + Y \to 3X, \quad X \to E, \qquad (8.1)$$

and is termed so due to its investigation in Brussels by I. Prigogine and co-workers [129, 294, 595]. Denoting the reagent concentrations in (8.1) by the same letters and regarding A, B as fixed, we obtain the Brusellator equation as a special case of the system suggested by A. Turing in 1952, viz.,

$$\dot{X} = A - (B+1)X + X^2 Y, \quad \dot{Y} = BX - X^2 Y \qquad (8.2)$$

[666]. For $B > 1 + A^2$, equations (8.2) have a stable periodic solution [217]. If A is harmonically time-modulated, i.e., $A = A_0 + a \cos \omega t$, then the oscillations in X and Y can become chaotic [491, 535], in which case equations (8.2) describe a non-linear oscillator with external harmonic action. The property established in [519], and showing how the total intensity of a continuous spectrum depends on the parameter after transition to chaos due to an infinite sequence of period-doubling bifurcations (see Sect.4, Chapt.8), was corroborated numerically [535] in considering (8.2) with A modulated. In numerical simulation, $A_0 = 0.4$, $B = 1.2$, $a = 0.05$ were selected. Meanwhile, the frequency of self-excited oscillations of the autonomous system was $\omega_0 = 0.375$. Period-doubling bifurcations and transition to chaos were observed as the modulation frequency ω increased. The bifurcation sequence started when $\omega = 0.712$, and terminated when $\omega = \omega_\infty = 0.7877$. As ω increased, the intensity of the continuous spectrum increased as in Fig. 9.88. The linear part of the dependence (solid line) is well consistent with the formula

$$N(\omega) = N_0 (\omega - \omega_\infty)^\zeta, \text{ where } \zeta = 1.5247..., \text{ which was established in [519].}$$

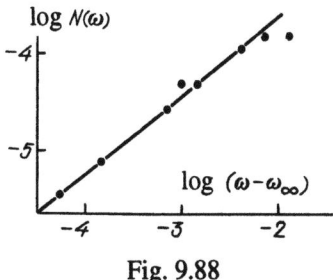

Fig. 9.88

The role of modulation of A, leading to chaos in the reaction (8.1), can be played by an additional substance Z, with concentration changing along with changes in X, Y, and, in turn, affecting their variation. E.g., consider the scheme of the reaction

$$A \to X, \quad B+X \to Y+D, \quad 2X+Y \to 3X,$$
$$X \to Z, \quad Z+Y \to X \tag{8.3}$$

[217]. The corresponding equations for concentration changes are

$$\dot{X} = A - (B+1)X + X^2Y + YZ,$$
$$\dot{Y} = BX - X^2Y - YZ, \tag{8.4}$$
$$\dot{Z} = r(X - YZ).$$

Note that the system thus described belongs to the class of self-oscillatory systems with inertial self-excitation [52, 53, 391]. In fact, introducing the new variables $x = X+Y-X_0-Y_0$, $Y = Z-Z_0 + r(1+Z_0)x$, $z = X-X_0$ where $X_0 = A$, $Y_0 = (B-1)/A$, $Z_0 = A^2/(B-1)$ are stationary values of X, Y, Z, we transform (8.4) to the form

$$\dot{x} = -z,$$
$$\dot{y} + \gamma y = ax + \varphi(x, y, z), \tag{8.5}$$
$$\dot{z} = -2\delta z + \omega_0^2 x + ky - f(x, y, z),$$

with

$$\gamma = r(B-1)/A, \quad a = r[r(B-1+A^2)/A - A^2/(B-1)],$$
$$2\delta = A^2B/(B-1) - B + 3, \quad \omega_0^2 = A^2B/(B-1) - r(B-1+A^2)/A,$$
$$k = (B-1)/A, \quad \varphi(x,y,z) = -r(x-z)[y - r(B-1+A^2)x/(B-1)],$$
$$f(x,y,z) = -(x-z)[y + z^2 + 2Az - r(B-1+A^2)x/(B-1)] - (B-1)z^2/A.$$

Eliminating z, we obtain equations of the form (4.2), which describe oscillations in systems with inertial self-excitation.

Though system (8.4) has not yet been studied numerically, one can expect that the changes in X, Y, Z are chaotic in a certain parameter range.

In [631, 637, 638], a system of two coupled Brusselators of the form (8.2) was considered, viz.,

$$\dot{X}_{1,2} = A_{1,2} - (B_{1,2}+1)X_{1,2} + X_{1,2}^2 Y_{1,2} + C_x(X_{2,1} - X_{1,2}),$$
$$\dot{Y}_{1,2} = B_{1,2}X_{1,2} - X_{1,2}^2 Y_{1,2} + C_y(Y_{2,1} - Y_{1,2}),$$
(8.6)

where C_x, C_y are coupling constants. For $A_1 = 1.6$, $B_i = A_i^2 + 2$ ($i = 1, 2$), $C_x = C_y = C$, and the varying parameters A_2, C, in the numerical simulation of (8.6) in [631], small C, generally speaking, oscillations in variables $X_{1,2}$, $Y_{1,2}$ are quasi-periodic with frequency ratio ρ determined by A_2. Such oscillations with $\rho = 0.61791$ and spectrum of Lyapunov exponents (0, 0, -1.081, -1.997) were observed; e.g., for C=0.01, A_2=0.84896. As C increased, both resonances of different orders and transitions from quasi-periodic to chaotic regimes were observed. They were controlled by the spectrum of Lyapunov exponent; e.g., for C=0.125, A_2=0.5550, the spectrum (0.037, -0.002, -0.616, -1.591) was obtained. A positive Lyapunov exponent testifies to the chaotic regime.

In [637, 638], the Brusselators were assumed to be identical, while the coefficients C_x, C_y different. Meanwhile, $A_1=A_2=A=2$, $C_x/C_y=q=0.1$. The parameters C_x, $B=B_1=B_2$ were made to vary. It turned out that, for small C_x transition to chaos occurs via the formation and destruction of the torus [637]; for large C_x, via a sequence of cycle period-doubling bifurcations [638]. To illustrate the transitions, see Tables 9.7, 9.8. They specify the spectra of Lyapunov exponents and the Lyapunov dimension of the attractor in relation to the parameter C_x for fixed B. For $C_x \in [0.05246, 0.05244]$ (B=5.5), with dimension of the attractor equal to one, resonance in the torus with frequency ratio 3:1 was observed. For $C_x \in [0.9, 1.1933]$ (B=5.9), several stable and unstable periodic motions were fixed; meanwhile, as C_x increased, the period of the stable motions doubled. In Table 9.8, values of C_x in the region of chaos were chosen.

c_x /	λ_1	λ_2	λ_3	λ_4	d_L	
0,05247	0,000	0,000	−0,011	−2,684	2	
0,05245	0,000	−0,010	−0,010	−2,681	1	
0,05242	0,000	0,000	−0,034	−2,677	2	
0,05239	0,000	0,000	−0,049	−2,673	2	$B = 5,5$
0,05238	0,009	0,000	−0,063	−2,671	2,143	
0,0523	0,043	0,000	−0,172	−2,658	2,250	
0,0522	0,068	0,000	−0,42	−2,57	2,162	
0,0521	0,075	0,000	−0,58	−2,47	2,129	

Table 9.7

c_x	λ_1	λ_2	λ_3	λ_4	d_L	
1,20	0,078	0,00	−2,68	−30,87	2,029	
1,21	0,107	0,00	−2,69	−30,67	2,040	$B = 5,9$
1,26	0,200	0,00	−2,82	−32,17	2,071	

Table 9.8

Another familiar model of an oscillatory chemical reaction is the so-called *Oregonator* [105, 141, 246, 356, 452] first suggested by R. Field and R. Noyes [452] at the University of Oregon (hence the name). The model is a much simplified version of the one suggested in 1972 [451] to describe the main properties of the *Belousov–Zhabotinsky reaction*. The scheme of the reaction called the Oregonator is

$$A + Y \xrightarrow{k_1} X + P, \quad X + Y \xrightarrow{k_2} 2P, \quad A + X \xrightarrow{k_3} 2X + 2Z,$$
$$2X \xrightarrow{k_4} A + P, \quad Z \xrightarrow{k_5} hY,$$

(8.7)

where A=[BrO$_3^-$], X=[HBrO$_2$], Y=[Br$^-$], Z=[Ce^{4+}], P=[HOBr].

Units of concentration X, Y, Z and time scale can be selected so that (8.7) becomes

$$\dot{X} = \alpha(Y - XY + X - gX^2),$$
$$\dot{Y} = \alpha^{-1}(-Y - XY + fZ),$$
$$\dot{Z} = \mu(X - Z)$$

(8.8)

[452], where α is of order 10^2, μ of order 10^{-1}, t_0 of order 10^{-3}, g of order 10^{-5} to 10^{-3}, f of order 1. It is known [141, 246, 452] that the Oregonator reflects the main features of the self-oscillatory Belousov-Zhabotinsky reaction. The form of oscillations of X, Y, Z, and

the projections of the corresponding limit cycle, obtained in [452] for α=77.27, μ=0.1610, g=8.375·10⁻⁶, f=1, are given in Fig. 9.89 (labels pointing to time). D. Ruelle suggested as early as 1973 [624] that oscillations in concentrations in the Belousov-Zhabotinsky reaction could be chaotic, and not only periodic. Furthermore, the hypothesis was corroborated by a qualitative study [661, 668], by analog [660] and numerical [667] simulation, as well as by experiment [437, 520, 616, 617, 623, 634, 652, 656, 673].

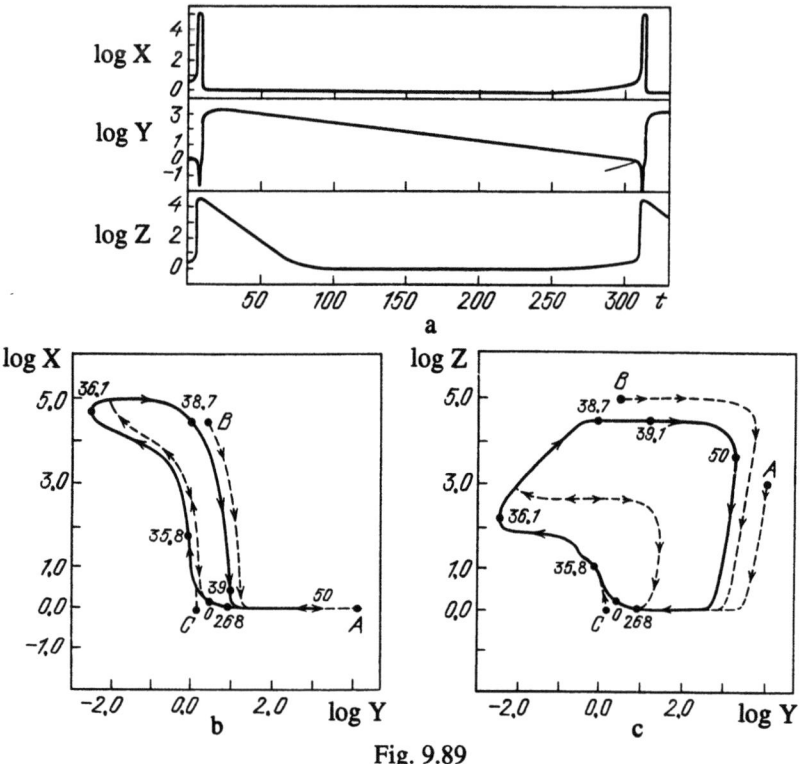

Fig. 9.89

J. Tyson [668] suggested the scheme and equations of the "modified" Oregonator:

$$A+Y \to X+P, \quad X+Y \to 2P, \quad A+X \to 2X+2Z,$$
$$2X \to A+P, \quad X+Z \to hY+Q,$$
$$\dot{X} = \alpha(Y-XY+X-\varepsilon XZ-gX^2),$$
$$\dot{Y} = \alpha^{-1}(-Y-XY+fZ),$$
$$\dot{Z} = \mu(X-\varepsilon XZ-Z),$$

(8.9)

where $\varepsilon \approx 2 \cdot 10^{-5}$. Despite being different from (8.8) by the small terms εXZ, equations

(8.9) have qualitatively different solutions.

Another modification considered in [667] consists in admitting reactions in both directions in (8.7). The reaction rates in the reverse direction are denoted by $k_{-1}, k_{-2}, ..., k_{-5}$, respectively. Taking into account the fact that the reaction proceeds in a continuous-flow reactor, the equations for this model can be conveniently written as

$$\dot{X} = w_1 - w_2 + w_3 - 2w_4 - X/\tau,$$
$$\dot{Y} = -w_1 - w_2 + w_5 - Y/\tau,$$
$$\dot{Z} = w_3 - w_5 - Z/\tau,$$
$$\dot{P} = w_1 + 2w_2 + w_4 - w_5 - P/\tau,$$
(8.10)

where

$$w_1 = k_1 AY - k_{-1} XP, \quad w_2 = k_2 XY - k_{-2} P^2, \quad w_3 = k_3 AX - k_{-3} X^2 Z,$$
$$w_4 = k_4 X^2 - k_{-4} AP, \quad w_5 = k_5 Z - k_{-5} Y,$$

and τ is the reaction time equal to the ratio of the reactor volume to the flow rate. Equations (8.10) were studied numerically in [667] for different values of τ and

$$k_1 = 1.34, \ k_{-1} = 10^4, \ k_2 = 1.6 \cdot 10^9, \ k_{-2} = 5 \cdot 10^{-5},$$
$$k_3 = 8 \cdot 10^3, \ k_{-3} = 4.8 \cdot 10^{11}, \ k_4 = 4 \cdot 10^7, \ k_{-4} = 1.6 \cdot 10^{-10},$$
$$k_5 = 90, \ k_{-5} = 10^{-5}, \ A = 0.06.$$

It was suggested in [333, 654] that the Bautin system [15, 71] should be used. Remarkably, its exact solution is known, and it is a convenient model of the Belousov-Zhabotinsky reaction. The system is given by the equations

$$\dot{X} = aY + E_1 X - X^3 - XY^2, \quad \dot{Y} = -aX + E_2 Y - Y^3 - YX^2. \quad (8.11)$$

Some versions are given in [333, 654]; in particular, the non-autonomous system obtained from (8.11) by replacing the constant parameters E_1, E_2 and a by

$$E_1(t) = e_1 - \varepsilon z_1 \sin \omega t, \quad E_2(t) = e_2 + \varepsilon z_2 \sin \omega t,$$
$$a(t) = a_0 + \varepsilon \sin \omega t.$$

Apparently, complicated irregular self-excited oscillation regimes are possible.

First experiments with chaotic oscillations in reactant concentrations during the Belousov-Zhabotinsky reaction in a continuous-flow reactor are discussed in [520, 623, 634]. To account for the Hudson experiments [520], the model

$$\dot{x} = hx + y + 0.1z, \quad \dot{y} = -x,$$
$$\varepsilon \dot{z} = -4(x+z+x^3) + \tanh[100(1+4z-16x)]$$
(8.12)

was suggested [605], where ε is a small parameter. The numerical solution of equations (8.12) for $\varepsilon=0.1$ and different h is qualitatively associated with the Hudson experimental data for different flow rates g, measured in ml/min. The correspondence is illustrated in Fig. 9.90 [605] with the experimental data given on the left, while the solutions z(t) of (8.12) stand on the right.

J. Roux and H. Swinney with co-workers investigated the reaction in more detail [105, 616, 617, 652, 656, 667]. For certain flow rates, they observed transition from periodic oscillations in concentrations (Fig. 9.91a) to chaotic ones (Fig. 9.91b); oscillations in the potential of bromide ions are shown on the left, the corresponding power spectra stand in the middle, and the two-dimensional attractors in coordinates $B(t_k)$, $B(t_k+\tau)$, constructed by the Packard-Takens algorithm, stand on the right. Note that the two-dimensional attractor substantially depends on the time delay τ. As τ varies, its deformation for the cases in Fig. 9.91a, b is represented in Figs. 9.92, 9.93, respectively. The point mapping on the cutting plane perpendicular to the diagram in Fig. 9.94a and passing through the dashed line (Fig. 9.94b) is constructed in [652] on the basis of the three-dimensional attractor with respect to coordinates $B(t_i)$, $B(t_i+\tau)$, $B(t_i+2\tau)$, whose projection is given in Fig. 9.94a. Since all points of the mapping lie on one straight line, the mapping is one-dimensional. The dependence of a subsequent value of the coordinate X along the line on the one immediately preceding it, which is well approximated by the formula $X_{n+1} = aX_n \exp(-bX_n)$, is given in Fig. 9.94c.

Examples of mechanical, physical, chemical, and biological systems 391

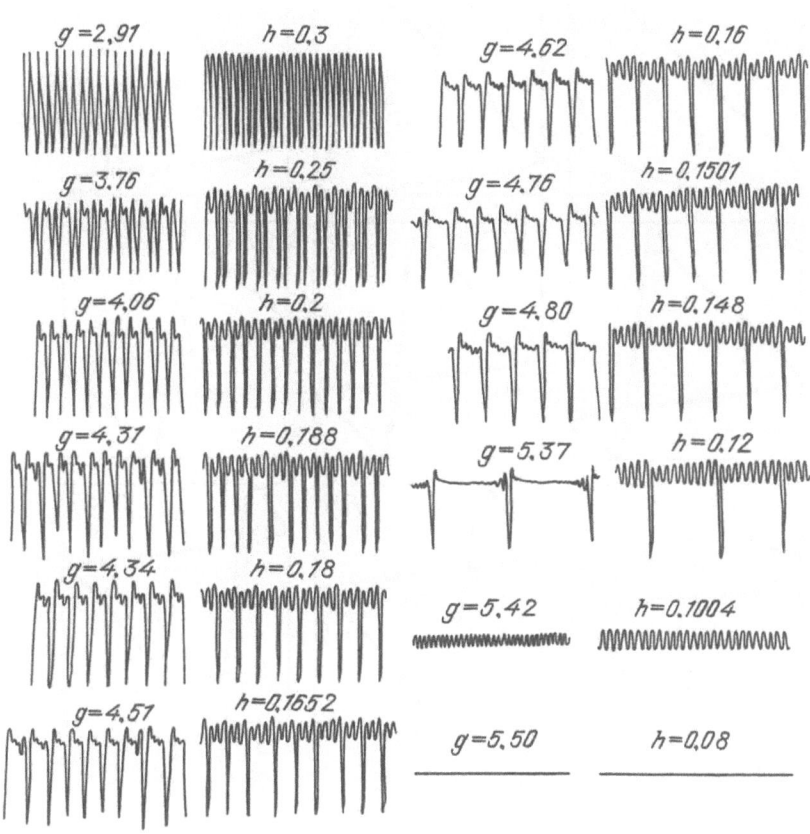

Fig. 9.90

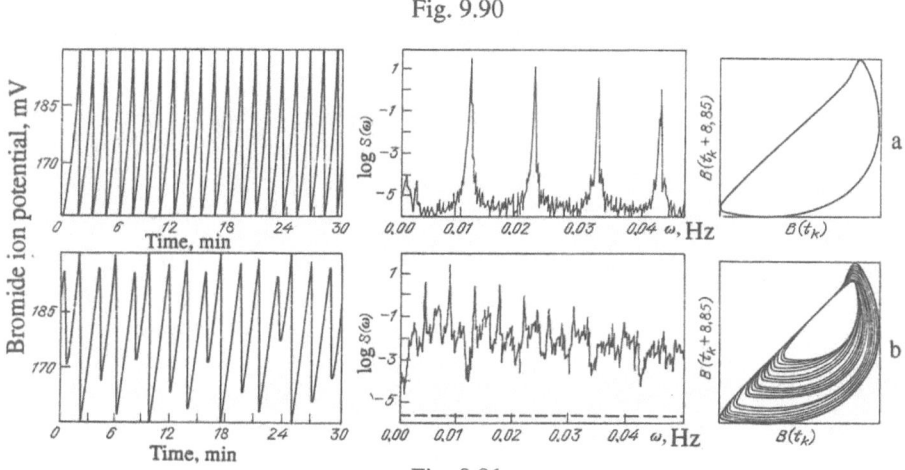

Fig. 9.91

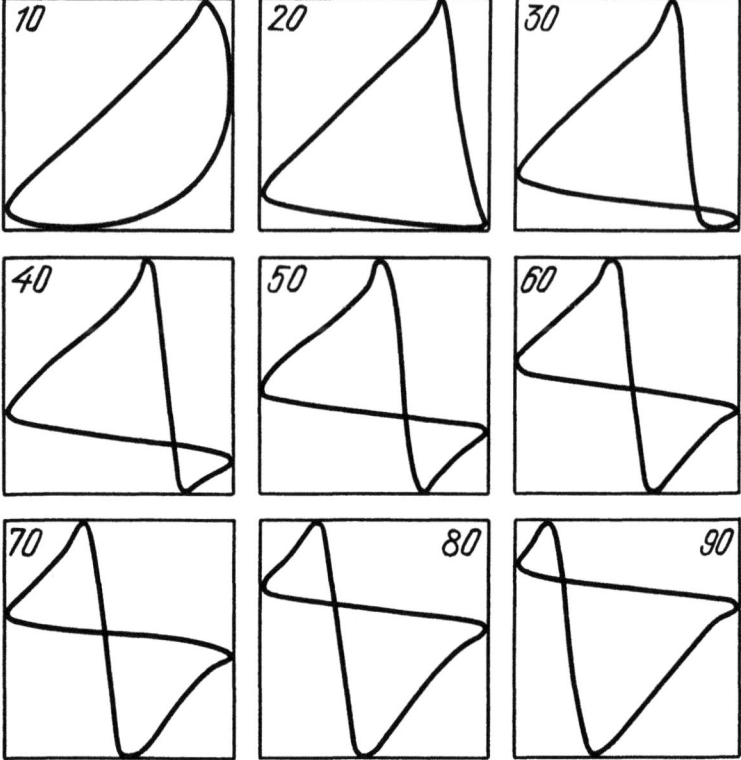

Fig. 9.92

Examples of mechanical, physical, chemical, and biological systems

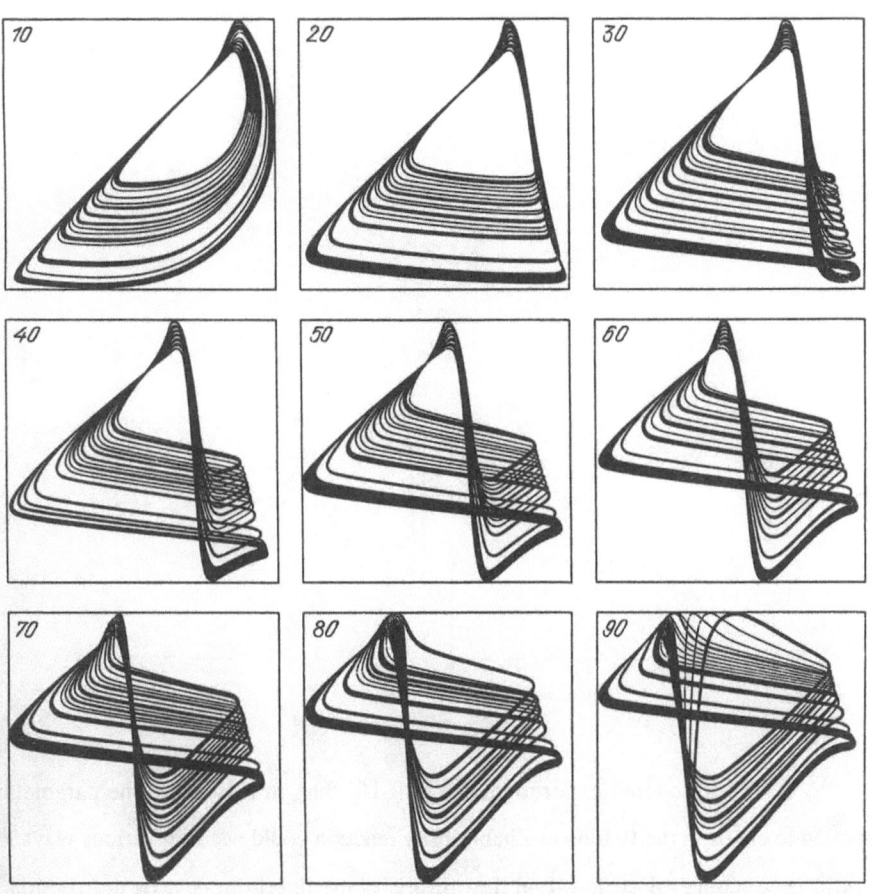

Fig. 9.93

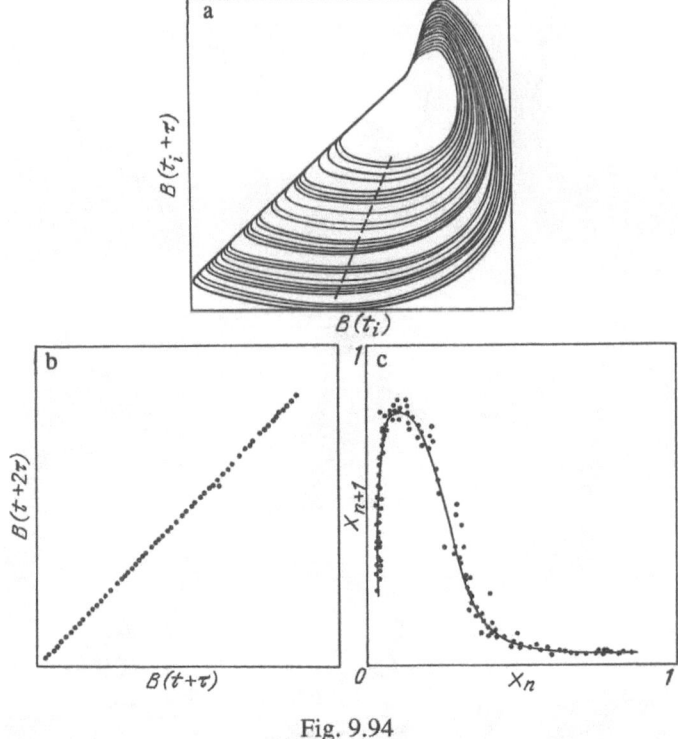

Fig. 9.94

It was discovered experimentally in [616] that, in relation to the parameters, transition to chaos in the Belousov-Zhabotinsky reaction could occur in various ways: via an infinite sequence of period-doubling bifurcations, intermittency, or destruction of quasi-periodic motion. The universality of transitions to chaos in systems of different nature and complexity is confirmed by this once again.

Besides the Belousov-Zhabotinsky reaction, many other oscillatory chemical reactions are now known, with possible chaotic changes in reactant concentrations.

Chaotic changes in oxygen concentration in peroxidase catalyzed oxidation of NADH were described in [596]. Equations of a metal-catalyzed chemical reaction between hydrogen and oxygen were studied numerically [367]. It was shown that they could have a solution in the form of chaotic oscillations. Complicated oscillatory regimes in open biochemical systems were discovered in [415, 640], with three first-order equations as a model. The chaotic behaviour of a system of coupled chemical oscillators was considered in [393].

O. Rössler [619-622] suggested a number of models for hypothetical chemical reactions with possibly chaotic changes in reactant concentrations. The models are:

(1)

$$\begin{array}{c} Y \xleftarrow{k_5} X \xrightarrow{k_7} Z \end{array} \quad \text{(with } k_6, k_3, k_4, k_8', e_0, k_1, k_2' \text{)} \tag{8.13}$$

The corresponding equations for reactant concentrations are

$$\dot{X} = k_1 + k_2 X - \frac{k_3 Y + k_4 Z}{X + K} X$$

$$\dot{Y} = k_5 X - k_6 Y, \quad \dot{Z} = k_7 X - \frac{k_8}{Z + K'} Z \tag{8.14}$$

[619-622], where $k_2 = k_2' - k_5 - k_7$, $k_8 = k_8' e_0$. The stereoplot of (a) the attractor, (b) its projections on the XY-, XZ-, ZY-planes, and (c) the time series of the processes X(t), Y(t), Z(t) in the region of chaos are given in Fig. 9.95. O. Rössler called such chaos *spiral*, or *amplitudinal*.

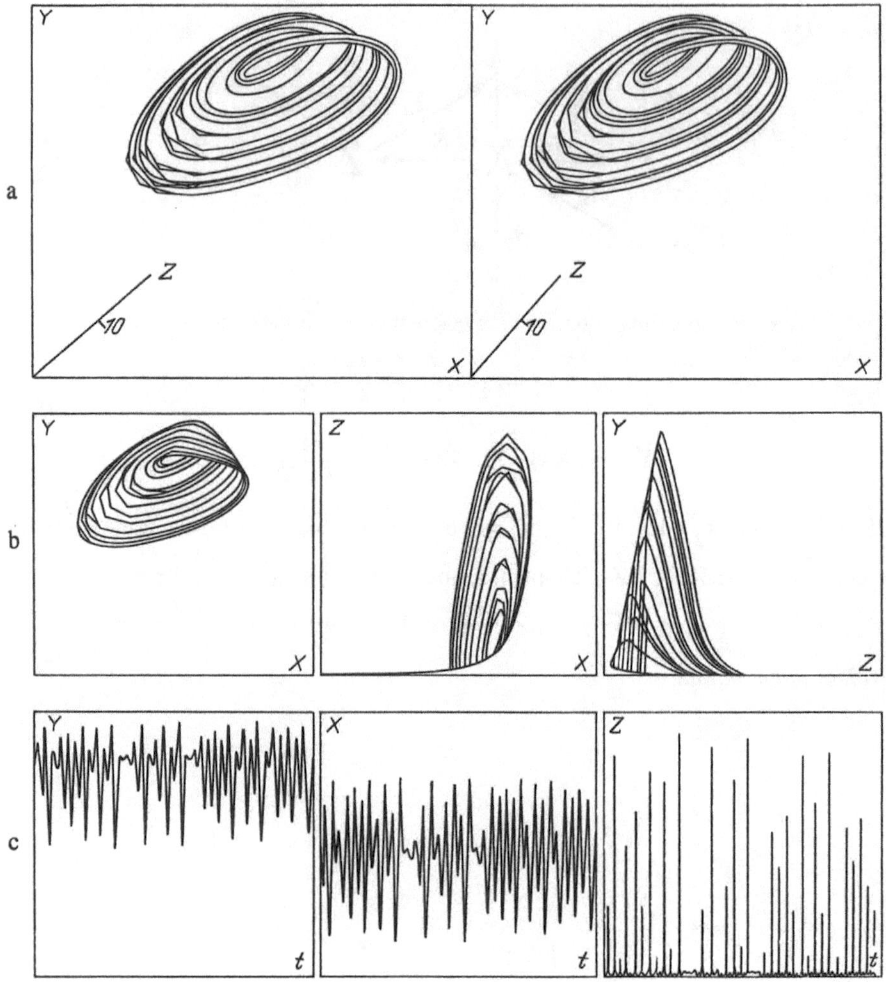

Fig. 9.95

(2)

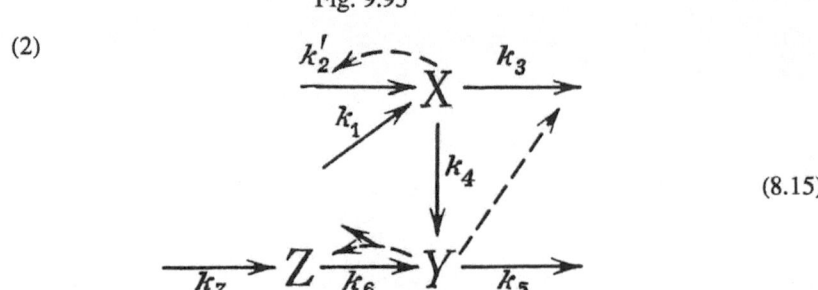

(8.15)

Reaction (8.15) is described by the equations

$$\dot{X} = k_1 + k_2 X - k_3 \frac{XY}{X+K},$$
$$\dot{Y} = k_4 X - k_5 Y + k_6 Y^2 Z, \quad \dot{Z} = k_7 - k_6 Y^2 Z,$$
(8.16)

where $k_2 = k_2' - k_4$ [621]. As well as equations (8.14), (8.16) have chaotic solutions in a certain parameter range. An example of one solution is given in Fig. 9.96. It was called *screw-type chaos* by O. Rössler [620].

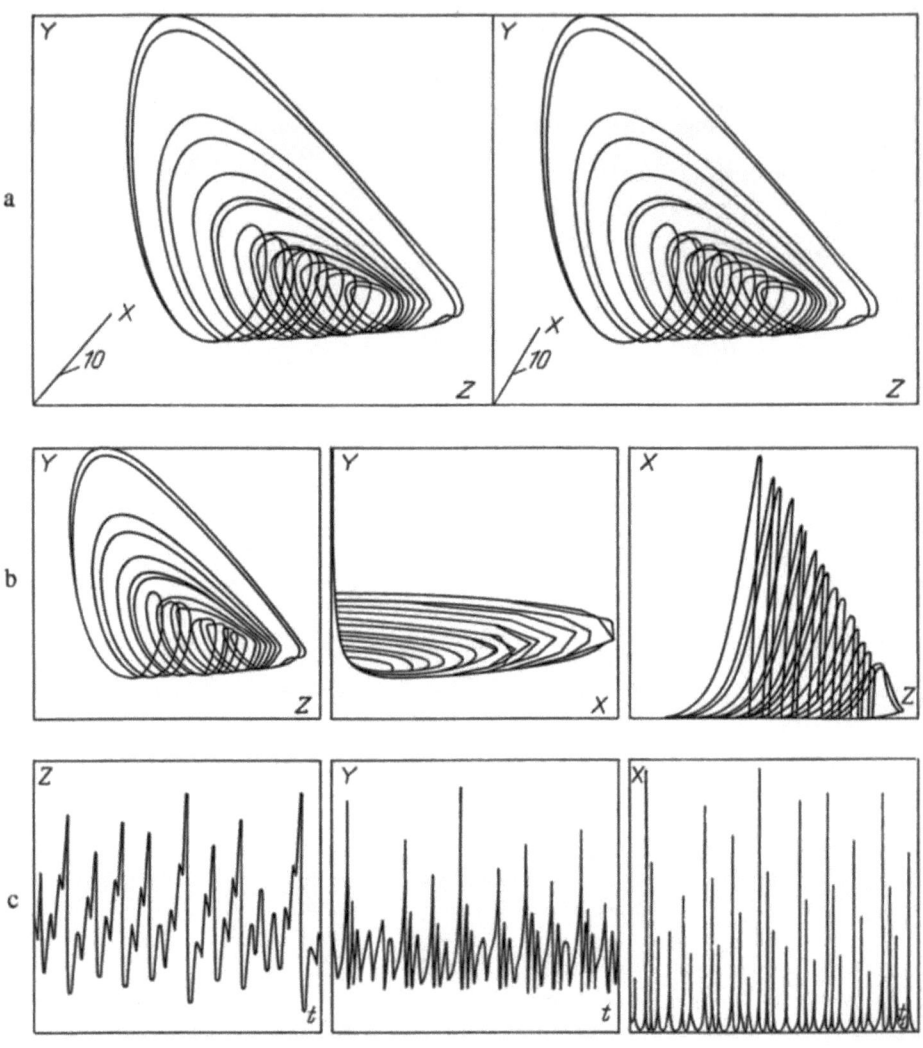

Fig. 9.96

(3) Besides (8.13), (8.14), O. Rössler suggested a model for a hypothetical reaction described by the system of fourth-order differential equations

$$\dot{X} = -Y - Z, \quad \dot{Y} = X + 0.25Y + W,$$
$$\dot{Z} = 3 + XZ, \quad \dot{W} = -0.5Z + 0.05W,$$
(8.17)

[622]. The shape of the chaotic attractor in the stereoplot for the model and the time series

of the process Z(t) are plotted in Fig. 9.97. System (8.17) is interesting, since it has two positive Lyapunov exponents. O. Rössler called this type of chaos *hyperchaos*.

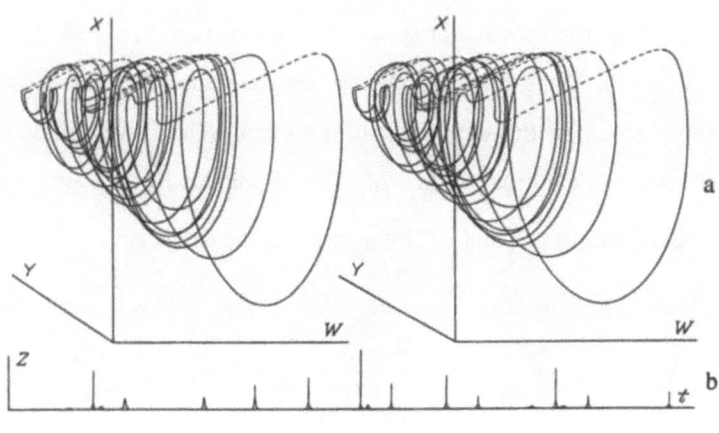

Fig. 9.97

If chemical reactions proceed in large volumes, then diffusion is substantial [319, 666]; therefore, the corresponding equations involve partial derivatives [217, 319, 356, 453, 666]. Diffusion can also help chaos to appear [306]. Thus, it was established numerically in [550] that chaos was a consequence of diffusion only, since only periodic oscillations are possible in the homogeneous system described by two first-order differential equations

$$\frac{\partial x}{\partial t} - D_x \Delta x = F_1(x,y), \quad \frac{\partial y}{\partial t} - D_y \Delta y = F_2(x,y),$$

where Δ is the Laplacian. This can be written as one equation with complex coefficients

$$\frac{\partial w}{\partial t} = [1 + ic_0 - (1 + ic_2)|w|^2]w + (1 + ic_1)\Delta w$$

[549, 684] near the threshold for self-excitation. The experiments carried out in [658] also testify to this.

The problems in chemical kinetics are also close to ecological ones on the interaction of different species of animals and plants. Classical for such interaction is the Lotka-Volterra model

$$\dot{n}_i = n_i \left(\gamma_i + \sum_{j=1}^{m} v_{ij} n_j \right) \tag{8.18}$$

[114, 566], where n_i is *i*th population size (i=1,2,...,m). In was noted in [664] that a sequence of period-doubling bifurcations leading to chaos could arise in the model for m=3 and one varying parameter.

We illustrate by a model of an ecological system consisting of two *predator–prey* pairs. The scheme for species interaction is in Fig. 9.98 [5, 383]. The plants M_1, M_1' (prey), living (e.g., by photosynthesis) consume some substance M_0 (biogen) whose stock in the environment is limited. The consumption of M_0 is compensated owing to the decay of M_1, M_1' and of animals M_2, M_2' after their death. M_2, M_2' (predators) subsist on plants. The equations for the model are

$$\dot{x}_1 = -\varepsilon_1 x_1 - \gamma_1 \frac{x_1 x_2}{1+ax_1} + \beta_1 \frac{x_1 x_0}{1+bx_0}, \quad \dot{x}_2 = -\varepsilon_2 x_2 + \gamma_2 \frac{x_1 x_2}{1+ax_1},$$

$$\dot{x}_3 = -\varepsilon_3 x_3 - \gamma_3 \frac{x_3 x_4}{1+ax_3} + \beta_3 \frac{x_3 x_0}{1+bx_0}, \quad \dot{x}_4 = -\varepsilon_4 x_4 + \gamma_4 \frac{x_3 x_4}{1+ax_3}, \quad (8.19)$$

where ε_i are the indices of mortality, γ_i the consumption coefficients, β_i the photosynthesis coefficients, a, b the saturation factors, x_0 the amount of the biogenic element M_0, x_1, x_3, the biogen contents in M_1, M_1', and x_2, x_4 that in M_2, M_2'. The total amount of elements is

$$\sum_{i=0}^{4} x_i = M = \text{const.} \quad (8.20)$$

With certain restrictions on the parameters, the system of equations (8.19), (8.20) admits a steady–state solution associated with non–zero biogen contents in all the interacting species. For $\varepsilon_1 = \varepsilon_2 = \beta_1 = \beta_3 = \gamma_2 = \gamma_4 = 1$, $\varepsilon_3 = \varepsilon_4 = \gamma_1 = \gamma_3 = 2$, $b = a$, selected in numerical simulation of (8.19), (8.20), the domain of existence of the solution on the (a,M)-plane lies above the curve ABC (Fig. 9.99) [383]. The regions where the solution is stable are shaded obliquely. In crossing the bifurcation curves DB and FH, periodic oscillations are excited softly. The limit cycle increases monotonically until it reaches the phase space boundaries, as the bifurcation curves are made more distant. Note that the phase space of (8.19), (8.20) is bounded, being the four-dimensional pyramid formed by the coordinate hyperplanes $x_i=0$ (i=1, 2, 3, 4) and the hyperplane

$$\sum_{i=1}^{4} x_i = 0 \quad (x_0 = 0)$$

As the boundaries are attained, the cycle bends and undergoes bifurcations. Cycle-doubling bifurcations are discovered numerically, leading to chaos. Its regions are shaded horizontally in Fig. 9.99. Examples of a complicated limit cycle (a), and of a chaotic attractor (b) are given in Fig. 9.100, which also shows the time series (c) and the power spectrum (d) of the process $x_3(t)$ in the chaotic regime. It is interesting to note that, in spite of the fourth order of the system in question, quasi-periodic motions were observed for no a, M.

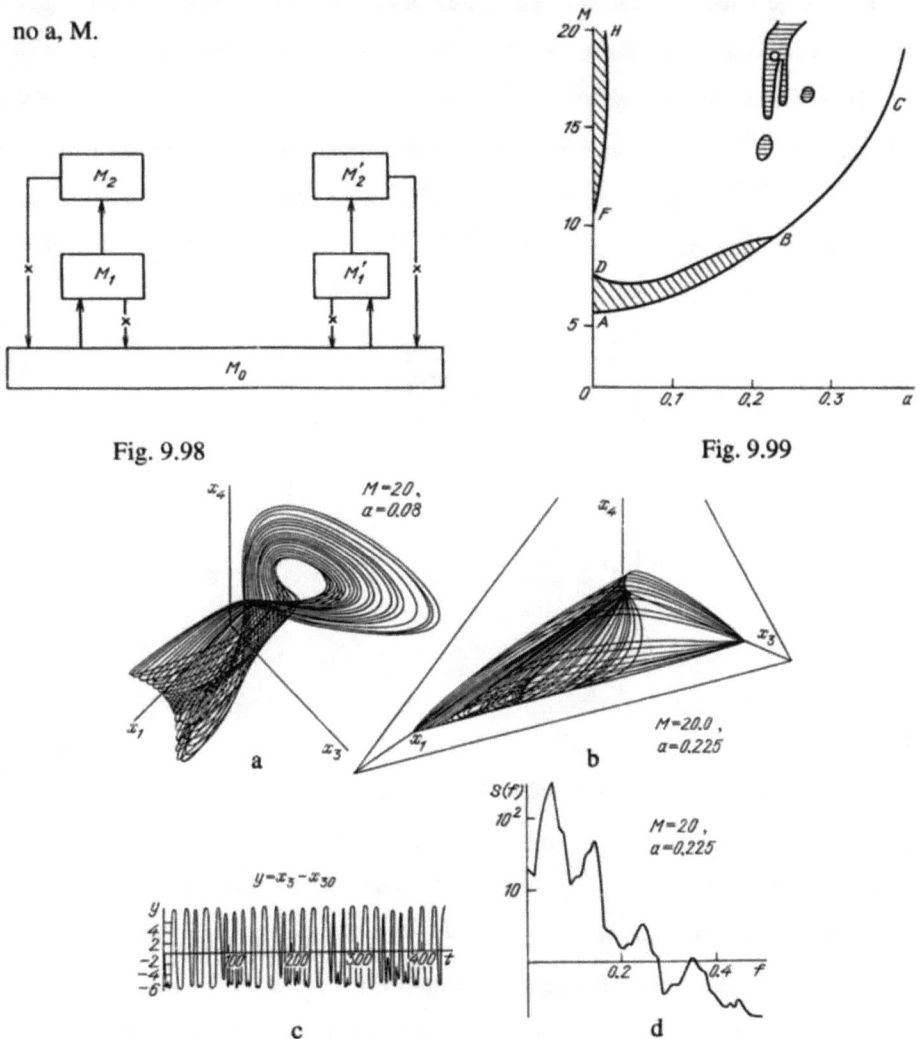

Fig. 9.98   Fig. 9.99

9. Systems with delay and other continuous systems

Chaotic oscillations in systems with delay, a particular case of continuous systems, have been studied sufficiently extensively. First of all, a large number of works devoted to numerically and experimentally studying generators should be pointed out, mostly in the microwave band with delayed feedback [21, 26, 37, 76-78, 128, 144-146, 176-183, 186-189, 209, 212, 226]. It was shown that strong chaotic oscillations in a wide parameter range could be excited. V. Ya. Kislov and co-workers were the first to design a generator of this kind, the *shumotron*[19], on the basis of the travelling-wave tube with delay line in the feedback circuit [188]. The block diagram of the generator is given in Fig. 9.101 [21, 176, 177, 183]; it contains a non-linear active element, a travelling-wave tube, a resonant filter, and a delay line. The filter with Q-factor $f_0/\Delta f = 200$ bounds the active element amplification band Δf, and enables us to return to the central frequency smoothly. The number of generator's proper modes in the amplification band is of order 10, where τ is the delay time. The papers [176-183] investigated the generator in particular detail.

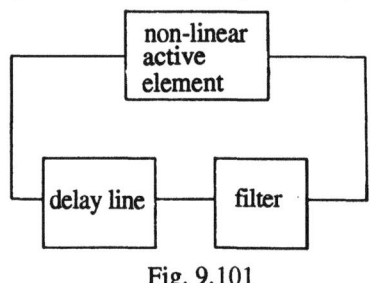

Fig. 9.101

As the feedback depth $\lambda = 10 \log(P_{inp}/P_{outp})$ increases, where P_{inp}, P_{outp} are the wave-power at the input and output of the active element, respectively, periodic self-oscillations start, close to harmonic ones in form, and at frequency f associated with the mode of maximum linear increment. From a value $\lambda = \lambda_1$ onwards, the regime is replaced by self-modulation with modulation period 2τ. The existence of a large number of harmonics of fundamental frequency $f_1 = \sqrt{2}\,\tau$ in the spectrum of the envelope is characteristic of the regime. In a certain region of detunings $\sigma = \tau\,(f_0 - f)$, a further increase in λ is accompanied by consecutive appearance in the envelope spectrum of another two frequencies f_2, f_3 and of their linear combinations $m_1f_1+m_2f_2+m_3f_3$ (Fig. 9.102, where the values m_1, m_2, m_3 are enclosed in parentheses). For $\lambda = \lambda_{cr}$ the components of the discrete spectrum are widened abruptly; simultaneously, a "noise

---

[19] Noise generator (*Russian*).

Examples of mechanical, physical, chemical, and biological systems 403

pedestal" appears, increasing together with peaks widening as λ increases. Transition to chaos via the three-frequency modulation regime is illustrated in Fig. 9.103, with the spectra of the modulation signal on the left, the modulation signal at the input of the active element (upper curves) in the middle, and the projection of the phase diagram on the right.

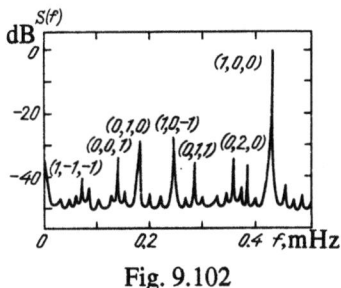

Fig. 9.102

The three-frequency quasi-periodic modulation regime is only observed in a narrow region of detunings σ and frequencies f. For most fixed values of f, generation of frequency f_3 is not observed as σ varies. Periodic self-modulation (Fig. 9.104a) is then replaced by quasi-periodic with two incommensurate frequencies f_1, f_2 as λ increases (Fig. 9.104b). Transition from two-frequency quasi-periodic modulation to chaos occurs via the synchronization regime (via resonances in the torus). Depending on σ, the transition may look differently. In one case, a resonance with frequency ratio $f_1:f_2=7:3$ (Fig. 9.104c) appears, accompanied by hysteresis relative to λ, which attests to the "hard" appearance of the resonance. As λ increases until λ_{cr}, modulation abruptly becomes chaotic (Fig. 9.104d). In another case, after quasi-periodic modulation, two adjacent peaks at f_2, f_1-f_2 approach each other as λ increases, "collapsing" into one narrow peak at frequency $f_1/2$ for certain $\lambda = \lambda_S$ related to resonance in the torus with

frequency ratio $f_1:f_2 = 2:1$ (Fig. 9.105b). Further, a sequence of cycle period-doubling bifurcations (Fig. 9.105 c, d) arises, terminating when a chaotic attractor forms. As λ increases further, a sequence of inverse bifurcations in the chaotic oscillations comes up (Fig. 9.105 e-g). Thus, the route to chaos is as in the Feigenbaum scenario.

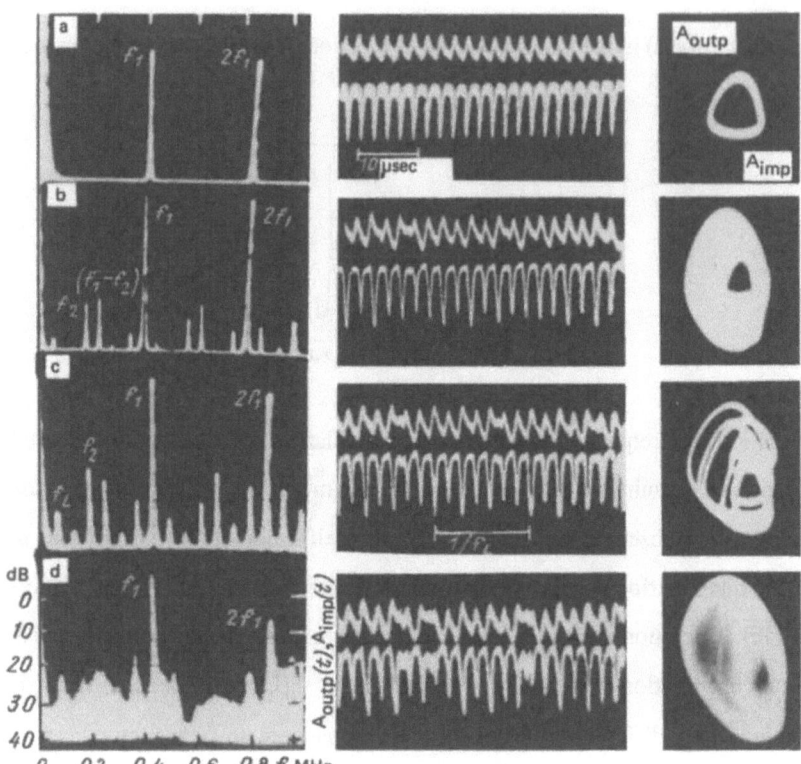

Fig. 9.103

In certain region of σ, f, transition to chaotic self-modulation occurs via intermittency. Depending on the parameters, transitions to chaos through intermittency from the mode of steady generation (Fig. 9.106a), of periodic (Fig. 9.106b) and quasi-periodic (Fig. 9.106c) self-modulation were observed in the experiment.

The region of chaos is not homogeneous as λ increases, including bifurcations of three types:

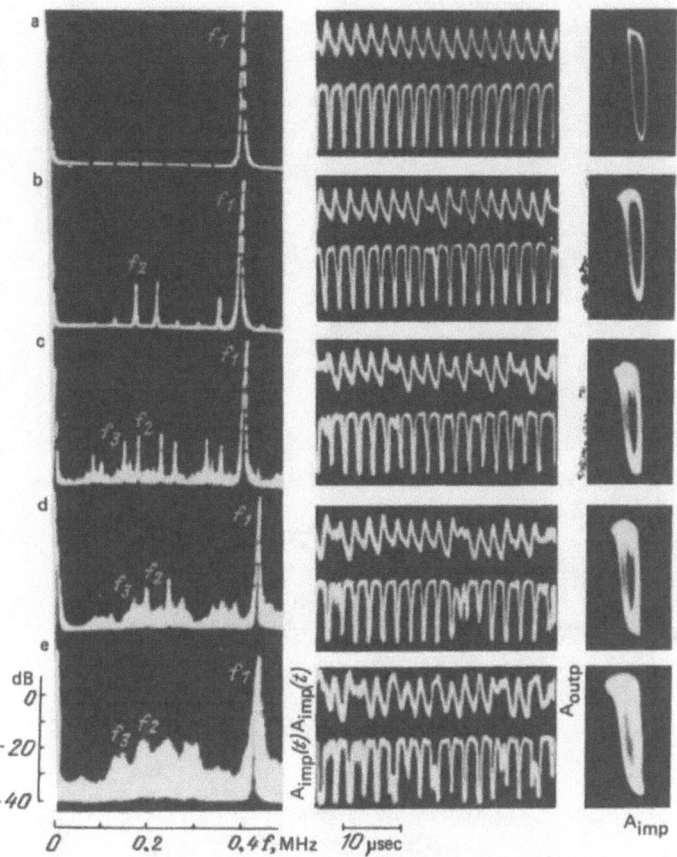

Fig. 9.104

(1) Generation of a stable and an unstable limit cycle from condensation of trajectories in the phase space, and subsequent appearance of a "stability window".

(2) Hard disruption of chaotic modulation, and establishment of steady generation at the frequency of the adjacent mode.

(3) Chaos-chaos intermittency from one regime of chaotic modulation into another. The transition is illustrated in Fig. 9.107 with the phase portrait, spectrum and probability distribution for one of the chaotic attractors (a), and for merger of two chaotic attractors (b); see also the shape of the modulation signal (c).

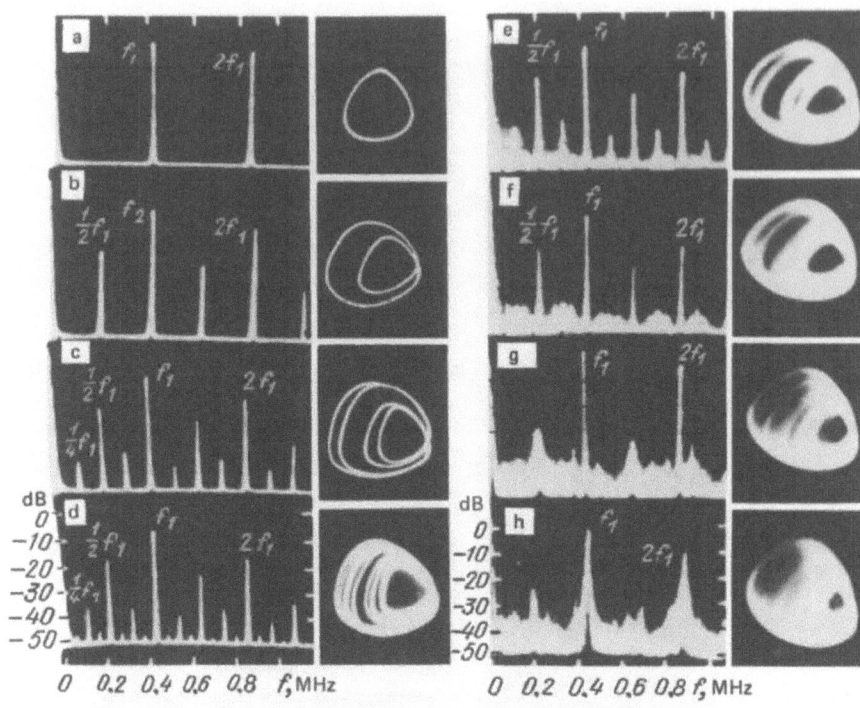

Fig. 9.105

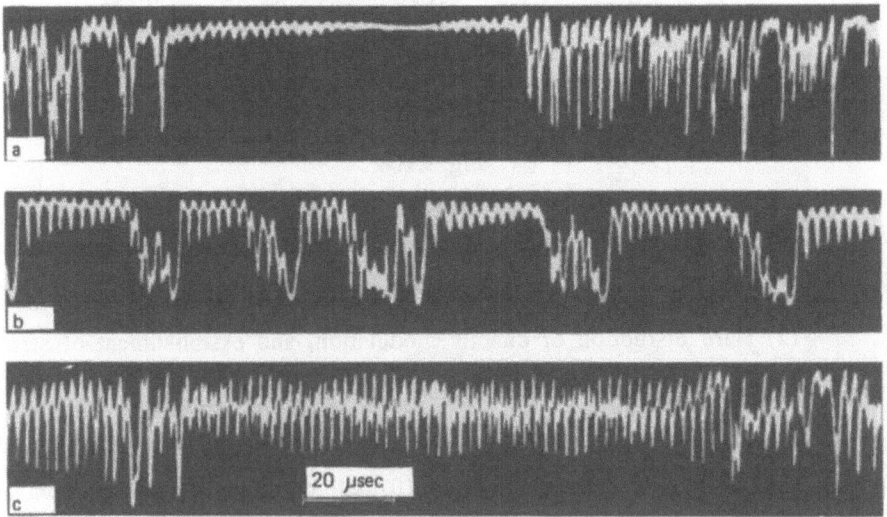

Fig. 9.106

Examples of mechanical, physical, chemical, and biological systems 407

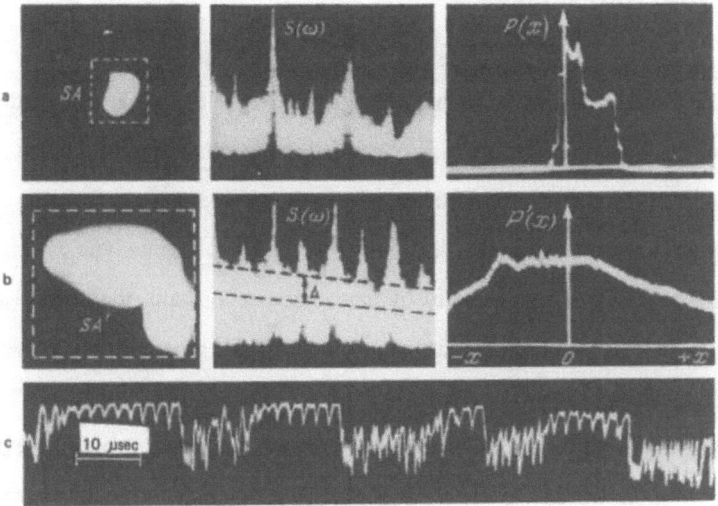

Fig. 9.107

Non-linear delayed feedback may lead to chaotic oscillations even in such a simple system as a linear oscillatory circuit (Fig. 9.108) [146]. The oscillation equation is

$$\ddot{x} + 2\delta\dot{x} + \omega_0^2 x = B f[x(t-\tau)], \qquad (9.1)$$

where $2\delta = R/L$, $\omega_0^2 = 1/LC$, $\beta = K/LC$. In [146], (9.1) is replaced by the corresponding difference equation with computerized solution. The two cases are considered: (1) $f(x)=\sin x$ for $x \in [0,\pi]$, $f(x)=0$ for $x \notin [0,\pi]$, i.e., where the function $f(x)$ does not change sign, and (2) $f(x)=\sin x$, i.e., where $f(x)$ is alternating. It was discovered in both cases that, as the amplification factor K increases, periodic oscillations with discrete spectrum turn into chaotic with continuous spectrum. However, the latter's width and intensity turn out to be substantially larger than those of the former.

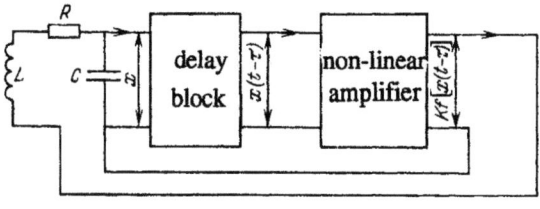

Fig. 9.108

The excited oscillations can be chaotized also if delayed feedback is introduced additionally into the self-oscillatory system; e.g., a van der Pol oscillator with non-linear natural frequency [226]. It's equation is

$$\ddot{x} - \mu(1-x^2)\dot{x} + \omega_0^2[1+f(x)]x = \omega_0^2 kx(t-\tau(t)), \qquad (9.2)$$

taking delayed feedback into account, where $\mu > 0$, ω_0 is the linear part of the natural frequency of the generator, f(x) a non-linear function not containing the linear term, k the delayed feedback factor, $\tau(t)$ time delay, and in the general case, a function of time t. Equation (9.2) can serve as the simplest model of the Doppler autodyne used in a short-range radar set [364]. If the velocity of the object is constant, then $\tau=\tau_0+at$, where $\tau_0=2L_0/c$, $a=2v/c$, L_0 is the distance to the object at the moment t=0, v the object's velocity, and c velocity of light.

Introducing dimensionless time t'= $\omega_0 t$, we obtain the expressions $\mu'=\mu/\omega_0$, $\tau'_0=\omega_0\tau_0 = 4\pi L_0/\lambda$, where λ is the wavelength of the radar signal. It can be seen that, in reality, τ_0 is very large, whereas a small. Accordingly, the study of the behaviour of the solution for (9.2) for a=0 is of great practical and theoretical interest. In the numerical simulation of the equation, $\omega_0=1$, $\mu=0.1$, $f(x)=Gx^2$, $\tau = \tau_0 + at$, while the parameters G, k, τ_0, a are made to vary. First of all, consider the results of solving (9.2) when a=0. If the values G=0.1, $\tau_0=30$ are fixed and the additional feedback factor k varies, then the evolution of the solution is as follows: For small k (0< k< $k_0 \approx$ 0.09), the solution is periodic with frequency $f_0=0.17 \approx 1/2\pi$, approximately that of the oscillation generator without additional feedback. For k=k_0, the solution becomes unstable, and a stable quasi-periodic solution with fundamental frequencies f_0, $f_1 = 0.029 \approx 1/\tau_0$ arises. The power spectrum for k=0.4 is in Fig. 9.109a. As k increases further, terms quasi-period doubling bifurcations arise. One is observed for k=$k_1\approx 0.42$. The power spectrum after the first bifurcation is shown in Fig. 9.109b. The other bifurcation occurs for k=$k_2\approx 0.75$. If k=$k_3\approx 1.3$, a picture is observed similar to the third bifurcation. However, we cannot state this with certainty, since amplitudes of arising subharmonics are comparable to the background noise. As k increases further (k >$k_4\approx 1.7$), the noise increases sharply, which indicates chaos. The shape of the spectrum for k=4 is given in Fig. 9.109c.

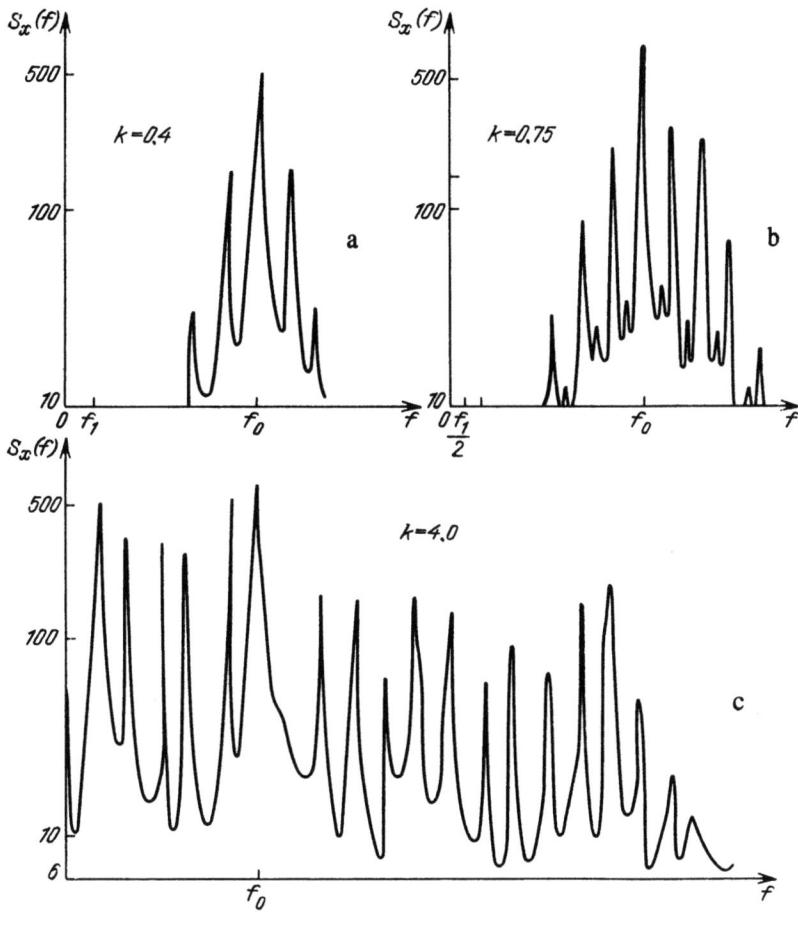

Fig. 9.109

Note that, as the parameter k varies, the frequencies f_0, f_1 vary slightly. Frequency f_1 is mostly determined by the time delay τ_0, which is manifested by the graph in Fig. 9.110. Behaviour of the system similar to the above is also observed for other values of τ_0 and the non-linearity parameter G. However, the concrete values of the coefficient k for which torus generation and its quasi-period doubling bifurcations occur, substantially depend on the parameters. As the time delay increases, the corresponding bifurcation values of k decrease, which is corroborated by Fig. 9.111, where the dependence of the values k_0, k_1 and k_2 (for G=0.1) on τ_0 is given. As G increases, the bifurcation values of k also decrease, which is manifested by the data in Table 9.9 for τ_0=40.

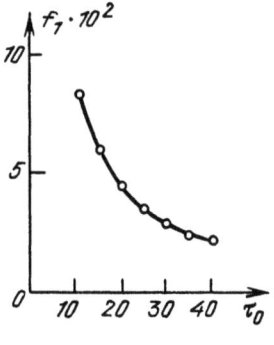

Fig. 9.110

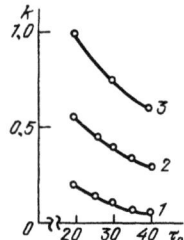

Fig. 9.111

G	k_0	k_1	k_2
0,1	0,03	0,3	0,6
1,0	0,003	0,025	0,045

Table 9.9

The nature of the transition to chaos is preserved qualitatively if we take τ_0 or G as bifurcation parameter. The difference is only in the substantial variation of the frequency f_1 as τ_0 varies (Fig. 9.110); as G varies, f_0 mostly changes.

To determine the effect of the parameter a, equation (9.2) was solved for the sufficiently large, necessarily unreal, value of $a = 10^{-2}$, in which case transition to chaos occurred differently from the case a=0. Consider the transition for G=0.1, τ_0=20 and increasing parameter k. Already for small k, quasi-periodic oscillations with two fundamental frequencies f_0=0.17, f_D=0.002≈af_0 are generated, where f_D is the Doppler frequency shift. In the phase space, the oscillations are associated with a two-dimensional torus. For k=k_1≈0.3, the torus loses stability, and a three-dimensional one arises, related to quasi-periodic oscillations with three fundamental frequencies f_0, f_D, f_1=0.05≈1/τ_0. As

Examples of mechanical, physical, chemical, and biological systems

the coefficient k increases further, the oscillations abruptly become chaotic, which is manifested by a sharp increase in the background noise of the spectrum.

The observed transitions to chaos can be characterized, along with the spectrum, the dimension by change in the attractor, for which dynamical systems of different dimensions s are constructed by the Packard-Takens method on the basis of the time series of the process x(t). The correlation dimension v_s is determined, rapidly tending to a constant value v as s increases. The value of v is taken to be the attractor's dimension.

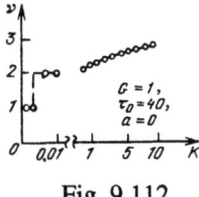

Fig. 9.112

We illustrate the dependence of v on k in Fig. 9.112. As expected, $v=1$ in the region of periodic oscillations; $v=2$ in the region of quasi-periodic oscillations associated with the two-dimensional torus; in the region of chaos, starting with two, v increases smoothly as k increases.

Many papers are devoted to numerical and experimental investigations of systems described by the equation

$$\frac{dx(t)}{dt} = -bx(t) + f(x(t-\tau)); \tag{9.3}$$

e.g., a ring cavity with non-linear absorbent and external light field [465, 521-523] (Fig. 9.113). We derive (9.3) for this system, assuming that the field in the absorbent is described by the equations of the semi-classical theory for two-level atoms with homogeneous line broadening [185], of the form

$$\frac{\partial^2 e}{\partial t^2} - c^2 \frac{\partial^2 e}{\partial z^2} + 4\pi \frac{\partial^2 p}{\partial t^2} = 0,$$

$$\frac{\partial^2 p}{\partial t^2} + 2\gamma_1 \frac{\partial p}{\partial t} + \Omega^2 p + \frac{2\Omega d^2}{\hbar} Ne = 0, \tag{9.4}$$

$$\frac{\partial N}{\partial t} + \gamma(N - N_0) = \frac{2}{\hbar \Omega} e \frac{\partial p}{\partial t}$$

[110], where e is the strength of the light field in the medium, p polarization, N population difference, Ω atomic transition frequency, d the dipole moment, γ_1 the quantity inverse to trnaversal relaxation time, γ the quantity reciprocal to the longitudinal

relaxation time, z the coordinate along the resonator, and N_0 the population difference with no field present. We rewrite (9.4) with the variables $\xi = t - z/c$ and z:

$$2c\frac{\partial^2 e}{\partial \xi \partial z} - c^2 \frac{\partial^2 e}{\partial z^2} + 4\pi \frac{\partial^2 p}{\partial \xi^2} = 0,$$

$$\frac{\partial^2 p}{\partial \xi^2} + 2\gamma_1 \frac{\partial p}{\partial \xi} + \Omega^2 p + \frac{2\Omega d^2}{\hbar} Ne = 0, \qquad (9.5)$$

$$\frac{\partial N}{\partial \xi} + \gamma(N-N_0) = \frac{2}{\hbar\Omega} e \frac{\partial p}{\partial \xi},$$

and seek for a solution in the form

$$e(\xi, z) = \frac{1}{2}\hat{E}(\xi+\frac{z}{c}, z)e^{-i\omega\xi} + \text{complex conjugate},$$

$$p(\xi, z) = \frac{1}{2}\hat{P}(\xi+\frac{z}{c}, z)e^{-i\omega\xi} + \text{complex conjugate}, \qquad (9.6)$$

where $\hat{E}(\xi+\frac{z}{c}, z)$, $\hat{P}(\xi+\frac{z}{c}, z)$ are two slowly varying complex-valued functions of ξ and z. We assume that the population difference $N(\xi+\frac{z}{c}, z)$ is a slowly varying function, too. Substituting (9.6) in (9.5), and only retaining infinitesimals of the first order, we obtain

$$\frac{\partial \hat{E}}{\partial z} = 2\pi ik\hat{P}, \quad \frac{\partial \hat{P}}{\partial \xi} = (i\delta - \gamma_1)\hat{P} - \frac{id^2}{\hbar}N\hat{E},$$

$$\frac{\partial N}{\partial \xi} = -\gamma(N-N_0) + \frac{i}{2\hbar}(\hat{P}^*\hat{E} - \hat{P}\hat{E}^*), \qquad (9.7)$$

where $k = \omega/c$ is the wave number, and $\delta = \omega - \Omega$ the detuning between the incident light and atomic transition frequencies.

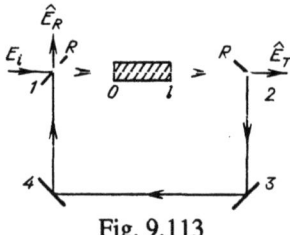

Fig. 9.113

For $\gamma_1 \gg \gamma c/L$, which is usually true for solid-state absorbers, the polarization

$$\hat{P} = \frac{id^2 N\hat{E}}{\hbar(i\delta-\gamma_1)} = -\frac{id^2 N\hat{E}}{\hbar(\delta^2+\gamma_1^2)}(\gamma_1+i\delta) \qquad (9.8)$$

can be eliminated from (9.7). Substituting (9.8) in the first equation of (9.7), and integrating, we obtain

$$\hat{E}(\xi+z/c, z) = \hat{E}(\xi, 0) \exp[2\theta\, W(\xi, z)(i\delta+\gamma_1)/(\delta^2+\gamma_1^2)], \qquad (9.9)$$

where $\theta = \pi d^2 k/\hbar$, $W(\xi, z) = \int_0^z N(\xi+\frac{z'}{c}, z')\,dz'$. Now, substituting (9.8) and (9.9) in the third equation of (9.7), and integrating from 0 to z with respect to z, we have the equation

$$\frac{\partial W}{\partial \xi} = -\gamma(W - N_0 z) - \frac{d^2}{4\hbar^2\theta} E^2(\xi, 0) \left\{ \exp\left(\frac{4\theta\gamma_1 W}{\delta^2+\gamma_1^2}\right) - 1 \right\} \qquad (9.10)$$

for $W(\xi, z)$, where $E(\xi, z) = |\hat{E}(\xi, z)|$.

We now turn to the old variables t, z in (9.10), put z="l", and let $\phi(t) = W(t-\tau, l)/N_0 l$, where $\tau = L/c$ is the time of transverse of the resonator. The equation for $\phi(t)$ then takes the form

$$\frac{\partial \phi}{\partial t} = -\gamma(\phi-1) - \frac{d^2}{4\hbar\theta l} E^2(t-\tau, 0)\left[\exp\left(\frac{4\theta\gamma_1 N_0 l}{\delta^2+\gamma_1^2}\phi\right) - 1\right]. \qquad (9.11)$$

To obtain the equation for $\hat{E}(t, 0)$, we write out the boundary condition

$$\hat{E}(t, 0) = \sqrt{1-R}\, E_i(t) + R\, e^{ikL} \hat{E}(t-\frac{L-l}{c}, l), \qquad (9.12)$$

which is a direct consequence of Fig. 9.113. Substituting (9.12) in (9.9), we get

$$\hat{E}(t, 0) = \sqrt{1-R}\, E_i(t) + R\, e^{ikL} \hat{E}(t-\tau, 0)\exp\left(\frac{2\theta N_0 l(i\delta+\gamma_1)}{\delta^2+\gamma_1^2}\phi\right). \qquad (9.13)$$

The system of equations (9.11), (9.13) determines the field amplitude at the point z=0, and the function $\phi(t)$.

Introducing dimensionless variables

$$t' = \gamma t, \quad \tau' = \gamma\tau = \gamma L/c,$$

$$x(t') = 2\theta N_0 l \delta \frac{\phi(t')-1}{\delta^2+\gamma_1^2},$$

$$y(t') = \frac{d}{\hbar}\left[\frac{\delta}{2\gamma(\delta^2+\gamma_1^2)}\right]^{1/2} \hat{E}(t', 0),$$

and omitting the primes, we obtain the equations

$$\frac{dx(t)}{dt} = -x(t) + [1 - B^2(x(t))/R^2]|y(t-\tau)|^2, \qquad (9.14)$$

$$y(t) = A + B(x(t))\, y(t-\tau)\, \exp\left[i(x(t)+x_0)\right] \qquad (9.15)$$

for the variables $x(t)$ and $y(t)$, where

$$A = \frac{d}{\hbar}\sqrt{\frac{\delta(1-R)}{2\gamma(\delta^2+\gamma_1^2)}}\, E_i\, ,$$

$$B(x) = R\exp\left[\frac{\gamma_1}{\delta}x + \frac{2\theta\gamma_1 N_0 l}{\delta^2+\gamma_1^2}\right],$$

$$x_0 = \left(\frac{2\theta\delta N_0 l}{\delta^2+\gamma_1^2}+kL\right)[\bmod 2\pi]\, .$$

If the external field is weak enough and the absorbent sufficiently long when $(1-R)E_i^2 \ll 4\pi\hbar\gamma kl\ |N_0|$, the quantity ϕ is close to one and the function $B(x)$ changes only slightly. If, in addition, $kl|N_0| \gg \hbar(\delta^2+\gamma_1^2)/2\pi d^2\gamma_1$, then $B\ll 1$ (since $N_0 < 0$), and (9.15) can be solved by the method of successive approximations. To a first approximation with respect to B, we have

$$y(t-\tau) = A + AB(x(t-\tau))\,\exp\left[i(x(t-\tau)+x_0)\right], \qquad (9.16)$$

substituting which in (9.14) and neglecting terms of order B^2, we obtain the equation in closed form for x

$$\frac{dx(t)}{dt} = -x(t) + A^2\{1+2B(x(t-\tau))\cos[x(t-\tau)+x_0]\}, \qquad (9.17)$$

containing terms with retarded argument. The relation is of the same form as (9.3), where $b = 1$, $f(x) = A^2[1 + 2B(x)\cos(x+x_0)]$.

The steady-state solution of (9.17) is determined by the transcendental equation

$$x_S = A^2[1 + 2B(x_S)\cos(x_S+x_0)], \qquad (9.18)$$

which can easily be solved graphically.

Note that, even for a weak dependence of B on x, generally speaking, B cannot be regarded as constant, since $\dfrac{dB}{dx} = \dfrac{\gamma_1}{\delta}B$. Only under a non-resonant force acting on the medium when $\delta \gg \gamma_1$, we can put $B = \text{const}$ as in [522, 523]. Below, we consider precisely this case.

To study the stability of the solution of (9.18), put $x = x_s + \zeta$ in (9.17), and linearize with respect to ζ, viz.,

$$\frac{d\zeta(t)}{dt} = -\zeta(t) - 2A^2 B\sin(x_s + x_0)\zeta(t-\tau) ; \qquad (9.19)$$

hence the characteristic equation

$$p + 1 + 2C\exp(-p\tau) = 0, \qquad (9.20)$$

where $C = A^2 B\sin(x_s + x_0)$. Using the method of D-decomposition [280], we construct the regions where the steady-state solution of (9.18) is stable on the (C, τ)-parameter plane (Fig. 9.114). It can be seen from the diagram that, for $|C| \le 0.5$, the solution is always stable; for $C < -0.5$, aperiodically unstable for any time delay τ; finally, for $C > 0.5$, stable for small $\tau \le \tau^*(C)$, and oscillatorily unstable for $\tau > \tau^*(C)$.

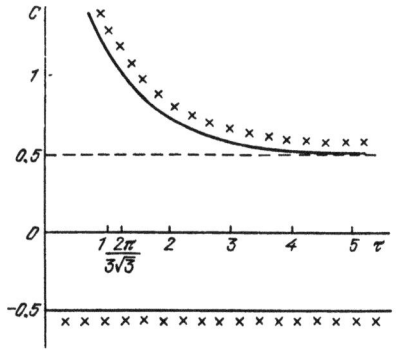

Fig. 9.114

It was shown in [522] that, in the transition through the oscillatory instability limit, periodic, and then chaotic, oscillations of the functions $x(t)$, $y(t)$ emerge (Fig. 9.115a,c). The corresponding power spectra of the process $|y(t)|$ are given in Fig. 9.115b,d. It is interesting to note that, in the chaotic oscillation regime, energy is pumped into stable modes for which $\text{Re } p < 0$.

Equation (9.3) was given a numerical study in [523] for $b = 1$, $\tau = 40$, $f(x) = \pi\mu(1 - \sin x)$, where μ is the bifurcation parameter. The graph from which the stationary values x_S for this case are read off is given in Fig. 9.116. It can be seen that $x_S < \pi/2$, i.e., $C = 0.5\pi\mu \cos x_S > 0$, which means that the steady-state solution can only

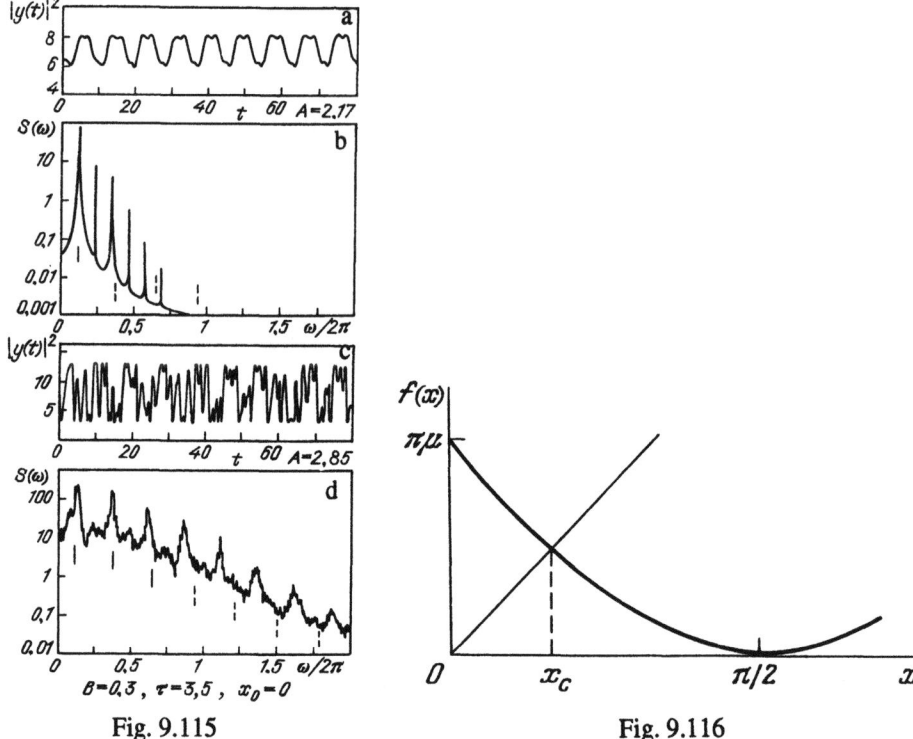

Fig. 9.115 Fig. 9.116

oscillatorily lose stability in this system. It follows from (9.20) that, at the stability limit, where $p = i\omega$, the equations $\omega = -\tan \omega\tau$, $2C = -1/\cos \omega\tau$ should hold. Since $C > 0$ and $\tau \gg 1$, the values $\omega_1 \approx \pi/\tau$, $\omega_3 \approx 3\pi/\tau$, $\omega_5 \approx 5\pi/\tau$, etc., satisfy them. The corresponding values of C are $C_1 \approx 1/2 + \pi^2/4\tau^2$, $C_3 \approx 1/2 + 9\pi^2/4\tau^2$, $C_5 \approx 1/2 + 25\pi^2/4\tau^2$, etc. Hence, taking into account (9.18), we infer that the first mode with frequency ω_1 loses stability when $\mu = \mu_1 = 0.376$, and $x_S = 0.557$ at the stability limit. The following (i.e., third) mode with frequency ω_3 loses stability for $\mu_3 = 0.387$ ($x_S = 0.565$); the fifth, for $\mu_5 = 0.407$ ($x_S = 0.579$); the seventh, for $\mu_7 = 0.437$ ($x_S = 0.599$), etc. Numerical study has shown that periodic oscillations with period $T_0 \approx 2\tau$, developing for $\mu = \mu_A = \mu_1$, lose stability for certain μ, and a cascade of period-doubling bifurcations occur, terminating when $\mu = \mu_F \approx 0.696$. The solution becomes chaotic; however, until $\mu = \mu_B \approx 0.777$, the average oscillation frequency is still $\omega_0 \approx \pi/\tau$ (Fig. 9.117a). For $\mu > \mu_B$, a sequence of transformations of different type arises, viz., the average oscillation frequency changes in jumps ($\omega_0 \to 3\omega_0 \to 5\omega_0 \to 7\omega_0$) as the parameter μ increases (Fig. 9.117b,c,d). As μ decreases, the process is reverse; however, hysteresis occurs

(Fig. 9.118a). The hysteresis testifies to two or several attractors at once, set apart by a certain separatrix surface. Similar results are obtained for fixed μ and increasing parameter τ (Fig. 9.118b).

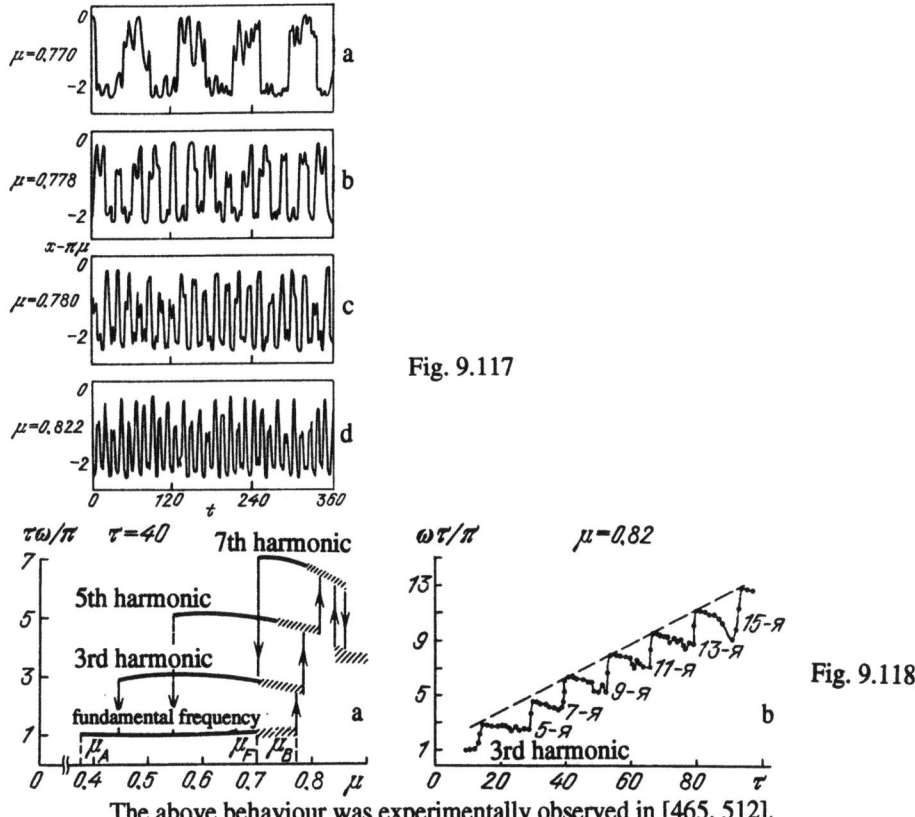

Fig. 9.117

Fig. 9.118

The above behaviour was experimentally observed in [465, 512].

Equation (9.3) also describes the change in amplitude of the signal generated in the annular circuit consisting of a resonant filter, a delay line and a non-linear amplifier with amplitude response $A_{outp} = f(A_{inp})$, assuming that the natural frequency of one of the closed circuit modes coincides with the resonant frequency of the filter and that there is no phase non-linearity [182, 186, 335]. For the numerical simulation of (9.3), $f(x) = \mu - x^2$, $b\tau = 15$ were taken in [182]. The results are given in Fig. 9.119. It can be seen that, as μ increases from 1.65 to 1.80, the average oscillation frequency becomes three times as large. The transition is similar to that in [523]; however, it has not been studied in more detail.

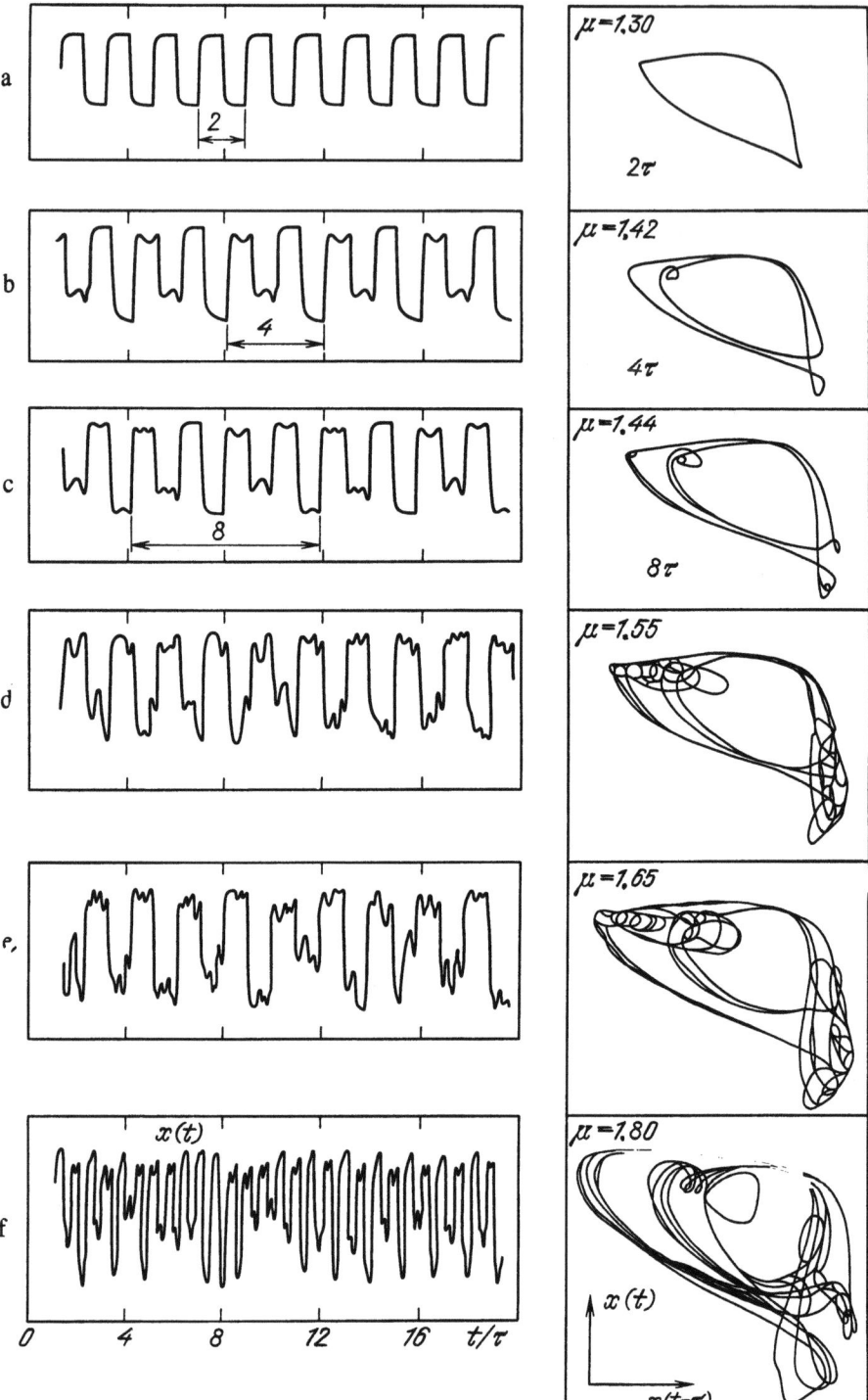

Fig. 9.119

Equation (9.3) with

$$f(x) = \frac{ax}{1+x^n} \qquad (9.21)$$

was used in [568] as a mathematical model for the process of regeneration of white blood cells of patients with chronic leukemia. It is known from medical records that the number of white blood cells varies with time chaotically (Fig. 9.120a) [568]. Numerical study of (9.3), (9.21) shows that, for certain choice of parameters, viz., when b = 0.1/day, a = 0.2/day, n = 10 and τ = 10 days, the solution is chaotic, and the dependence of the blood concentration x on the time t (Fig. 9.120b) is similar to that in Fig. 9.120a. For smaller τ, x(t) varies periodically. The transition from the periodic solution to the chaotic one as τ increases occurs via a sequence of period-doubling bifurcations.

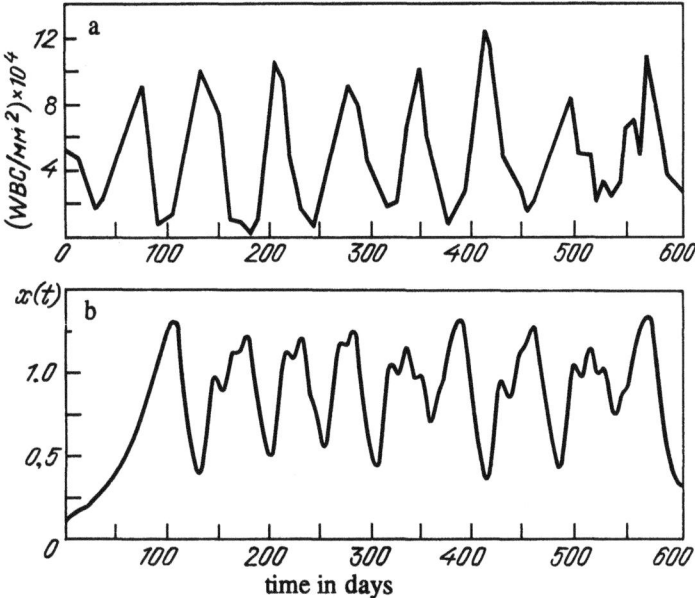

Fig. 9.120

Equations (9.3), (9.21) for b = 0.1, a = 0.2, n = 10, and various τ were studied in more detail numerically [224, 439, 477, 610]. For τ < 4.53, the steady-state solution $x = x_0 = 1$ is stable. In the strip 4.53 < τ < 13.3, x(t) varies periodically with period T = 3τ. For $\tau = \tau_{cr}$ = 13.3, a cascade of doubling bifurcations starts, and terminates when τ = 16.8 The evolution of the time series of the process x(t) and of its power spectrum as τ increases, beginning at τ = 14, is given in Fig. 9.121a [224, 439]. The corresponding evolution of the attractor under projection onto the (x(t), x(t - τ))-plane is illustrated by Fig. 9.121b. As the time delay τ increases, the attractor's fractal dimension d and the number n of positive Lyapunov exponents on the average increases (Fig. 9.122). Meanwhile, there are regions of τ in which the solution is periodic and dimension decreases to one (Fig. 9.122b). In spite of increasing dimension, the Kolmogorov metric entropy K remains approximately the same (Fig. 9.122c), due to a decrease in the maximal Lyapunov exponent with increasing τ (Fig. 9.122c). The spectrum of Lyapunov exponents, the dimension of the attractor, and the entropy were found for a discrete dynamical system, using the Packard-Takens recipe, which was repeatedly stressed above. Taking the system (9.3), (9.21) as an example, different dimensions of the attractor were compared in [439, 477, 610]. The results are in Table 9.10.

τ	ν	d_M	d	d_L
17	1,95	2	2,13	2,10
23	2,44	2	2,76	2,82
30	3,00	3	2,84	3,58
100	7,50	6		10

ν is correlation dimension; d_M is dimension, calculated by Mori's formula; d is fractal dimension, calculated by the definition; d_M is Liapunov dimension, calculated by the Kaplan-Yorke formula.

Table 9.10

Examples of mechanical, physical, chemical, and biological systems

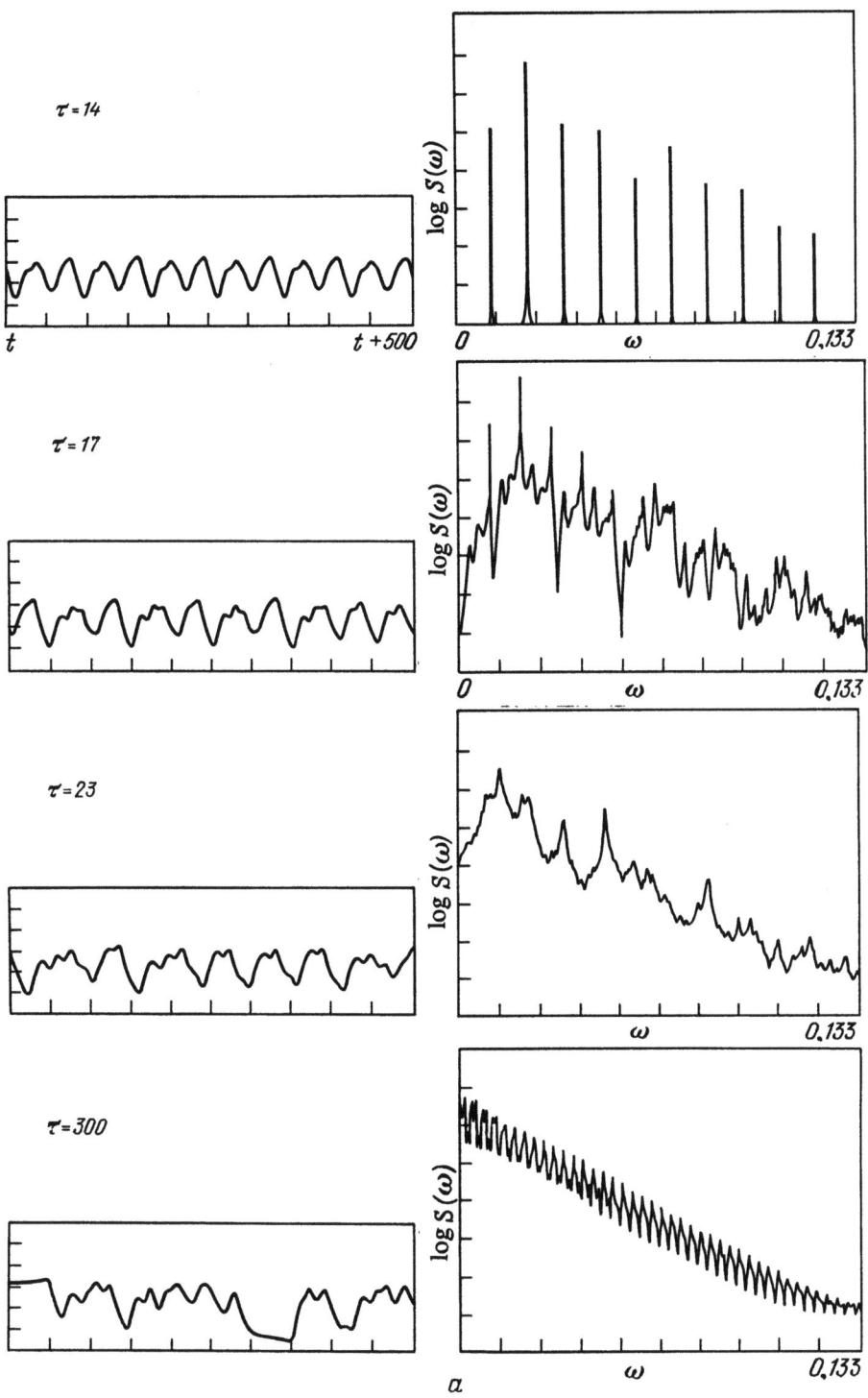

Fig. 9.121a

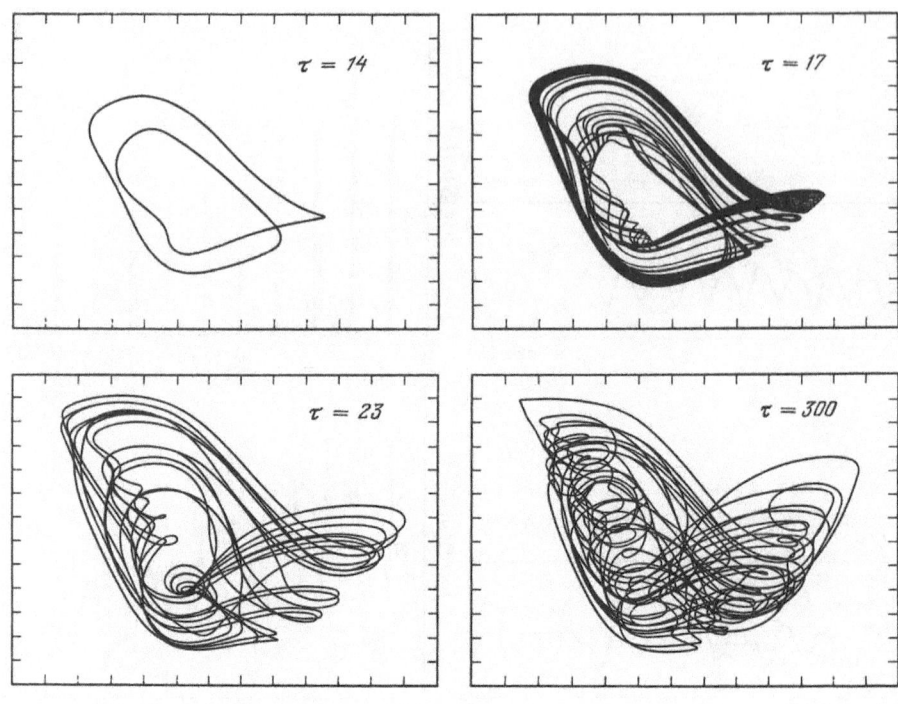

Fig. 9.121b

Examples of mechanical, physical, chemical, and biological systems

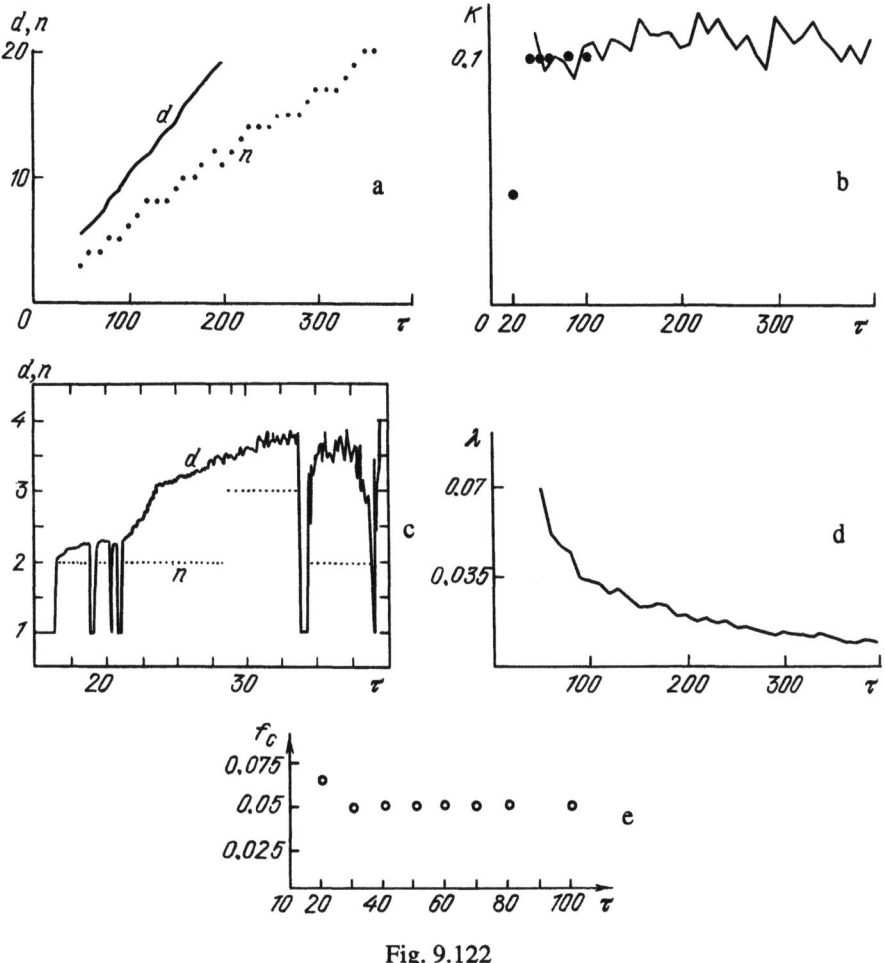

Fig. 9.122

With the purpose of studying synchronization, the system (9.3), (9.21) was harmonically and externally acted upon, i.e., the equation

$$\frac{dx(t)}{dt} = -0.1\, x(t) + \frac{0.2\, x(t-\tau)}{1+[x(t-\tau)]^{10}} + B\cos 2\pi\nu t$$

was solved numerically [224]. The regions of synchronization in the (B, ν)-parameter plane were found for different τ, and the synchronization threshold B_{thr} was determined. It turns out that $B_{thr} = 0.06 \pm 0.008$ for $\tau = 20$; B_{thr} is practically unchanged, equal to $B = 0.14 \pm 0.01$ for $\tau = 30, 40, 50, 60, 80, 100$ (the dots in Fig. 9.122c). It is interesting to note that the frequency associated with the synchronization threshold is practically unchanged as τ varies, starting with $\tau = 30$ (Fig. 9.122e), and equals $f_s = 2/(3\tau_{cr})$, i.e., the system "remembers", as it were, the critical value of τ when its complicated behaviour started.

Systems of two Lotka-Volterra equations with time delay

$$\dot{N}_i(t) = [\varepsilon_i - a_{ii} N_i(t - \tau_{ii}) - a_{ij} N_j(t - \tau_{ij})] N_i(t) \qquad (i, j = 1, 2, i \neq j), \qquad (9.22)$$

were studied in [200, 201, 646]. These describe the dynamics of populations of two competing species, saturation being taken into account. In [646], $\varepsilon_i > 0$, $a_{ii} > 0$, $a_{ij} > 0$, $\tau_{ii} = \tau_i$, $\tau_{ij} = 0$, and the following two cases are considered: (1) $\varepsilon_1 = 1.5$, $\varepsilon_2 = 2$, $a_{11} = a_{22} = 1$, $a_{12} = 1$, $a_{21} = 2$; (2) $\varepsilon_1 = \varepsilon_2 = 2$, $a_{11} = a_{22} = 2$, $a_{12} = a_{21} = 1$. In the former case, the equilibrium state associated with the stationary existence ($N_1 = 0.5$, $N_2 = 1$) of both species is unstable for $\tau_i = 0$; in the latter ($N_1 = N_2 = 2/3$), it is stable for $\tau_i = 0$. The second case is considered in more detail. The results are as follows: When τ_1, τ_2 increase, the steady state loses stability, and periodic oscillations in the populations occur. As τ_1, τ_2 increase further, regions develop in which the regimes are chaotic, and transition to chaos occurs via a sequence of periodic-doubling bifurcations (Fig. 9.123); e.g., for $\tau_1 = 16$ and increasing τ_2, the cascade of period-doubling bifurcations begins with $\tau_2 \approx 0.77$. After the cascade has terminated, a chaotic attractor in the stereoplot of Fig. 124a appears. Simultaneously, there is a stable limit cycle in the phase space. Depending on the initial conditions, oscillations in the population are either periodic or chaotic. As τ_2 increases, the cycle bifurcates and generates another chaotic attractor (Fig. 9.124b). In [201], (9.22) is used for describing the dynamics of the development of insects with two active states of development: larvae and imagos (which are adult insects reproducing and populating the area). In the equations, N_1 is the imago population, N_2 the larva population, $\varepsilon_1 > 0$, $\varepsilon_2 = 0$, $a_{ii} > 0$, $a_{ij} < 0$, $\tau_{11} = \tau_1$ is the mean lifetime of the imago population, $\tau_{22} = 0$, $\tau_{12} = \tau_2$ the time interval between larva and imago states, $\tau_{21} = 1 - \tau_2$ the interval between imago and larva states. If $|a_{21}| N_1 \gg \varepsilon_1$, then the derivative $\dot{N}_2$ can be neglected in the second equation of (9.22), and its solution takes the form

$$N_2(t) = N_1(t - (1 - \tau_2)) \equiv N(t - (1 - \tau_2)) \qquad (9.23)$$

for $a_{22} = |a_{21}|$. Substituting (9.23) in (9.22), we obtain

$$\dot{N}(t) = [\varepsilon - a_1 N(t - \tau) + a_2 N(t - 1)] N(t), \qquad (9.24)$$

where $\varepsilon = \varepsilon_1$, $\tau = \tau_1$, $a_1 = a_{11}$, $a_2 = |a_{12}|$.

Examples of mechanical, physical, chemical, and biological systems

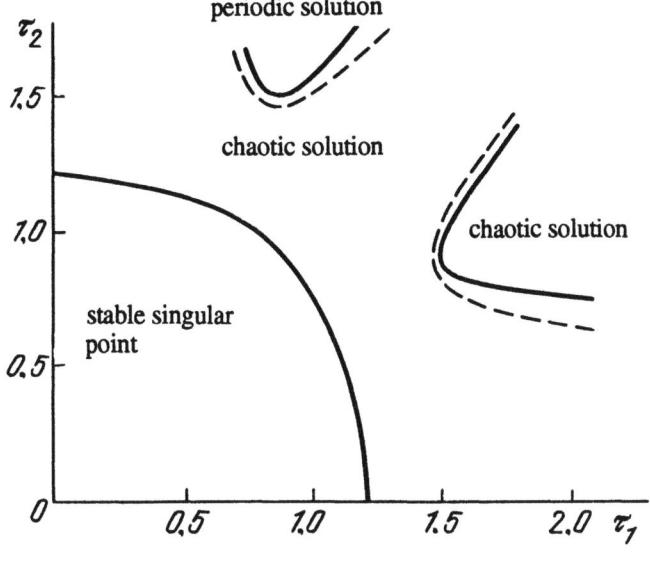

Fig. 9.123

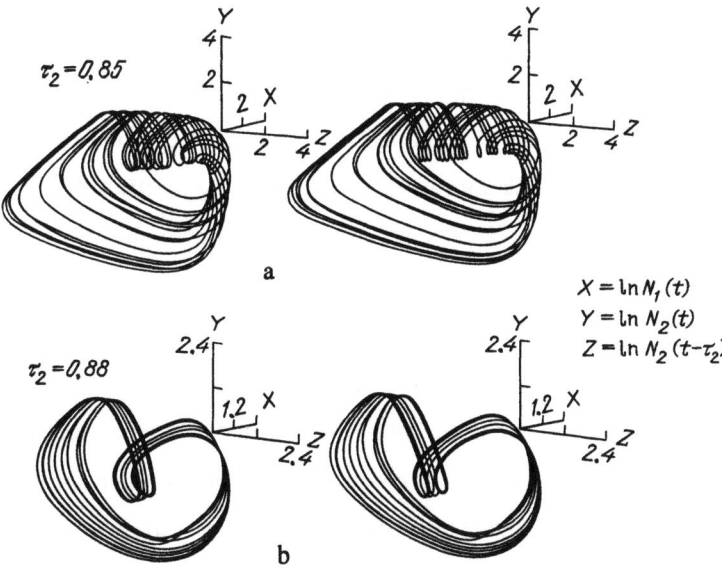

Fig. 9.124

Equation (9.24) was solved on a computer for $\varepsilon = 3$, $\tau = 0.125$, $a_1 = 3/(1-a)$, and $a = a_2/a_1$ was made to vary. The steady-state solution $N = 1$ loses stability for $a \approx 0.506$, and a stable limit cycle is generated. The latter's amplitude increases sharply as a increases, while the period is approximately constant, and equals 1.1. When a = 0.665, a cascade of period-doubling bifurcations starts, terminating when a = 0.730. When a ≈ 0.587, another stable periodic regime appears, with period approximately twice as less as the preceding one (T ≈ 0.55). When a = 0.64, the stable regime turns into the beat regime, and then into chaos. In a sense, the latter is the above two periodic regimes combined: There are intervals on the time scale with distances of orders 0.55 and 1.1 between the peaks. As a increases further, windows with periodic regimes of large period develop; e.g. the period is approximately 38 for a = 0.71.

In contrast to systems with delay, studies of continuous systems described by partial differential equations are extremely limited. Here, we can only point to [51, 165], where the dimension of an attractor is estimated analytically for certain partial differential equations; to [25, 38, 41, 123, 213, 306], in which circuits that simulate one-dimensional dissipative media are studied; and also to few papers which discover chaotic regimes in solving partial differential equations numerically. One of these [300] was already mentioned in Sect.7. The paper [687], in which equations similar to the Korteweg-de Vries equation are solved, and [396, 406, 509, 524], in which the a.c. driven damped sine-Gordon equation is modelled, also belong to the above studies. On the other hand, there are comparatively many experimental works devoted to the observation and investigation of chaotic oscillations in hydrodynamics (e.g., [395, 411, 469, 470, 561, 569]), lasers [376-378, 488, 492, 505, 525, 592, 674, 675], non-linear optics [431, 454, 525, 591, 594], and some other systems [2]. However, most results should be sharpened.

Papers [395, 411, 569] are interesting in that, apparently for the first time, they show that the dimension of an attractor in certain hydrodynamical systems can be quite small. The authors resorted to the Packard-Takens method for processing experimental data. The liquid flow between two rotating concentric cylinders with radius ratio 0.875 is considered in [411]. Figs. 9.125a, b illustrate types of attractors with respect to coordinates V(t), V(t+τ), where V(t) is the radial component of the liquid velocity, and also the corresponding Poincaré sections for a number of values of the ratio of the Reynolds number R, which is proportional to the angular rotation velocity of the internal cylinder, to the critical Reynolds number R_c for which Taylor vortices develop. The Poincaré sections are obtained by cutting phase trajectories in the three-dimensional (V(t),

$V(t+\tau), V(t+2\tau))$-space by the plane parallel to the $V(t+2\tau)$-axis, and passing through the dashed line in Fig. 9.125a. Both the attractor capacity d and the correlation exponent ν are determined. It can be seen in Fig. 9.126 that these quantities turn out to be close.

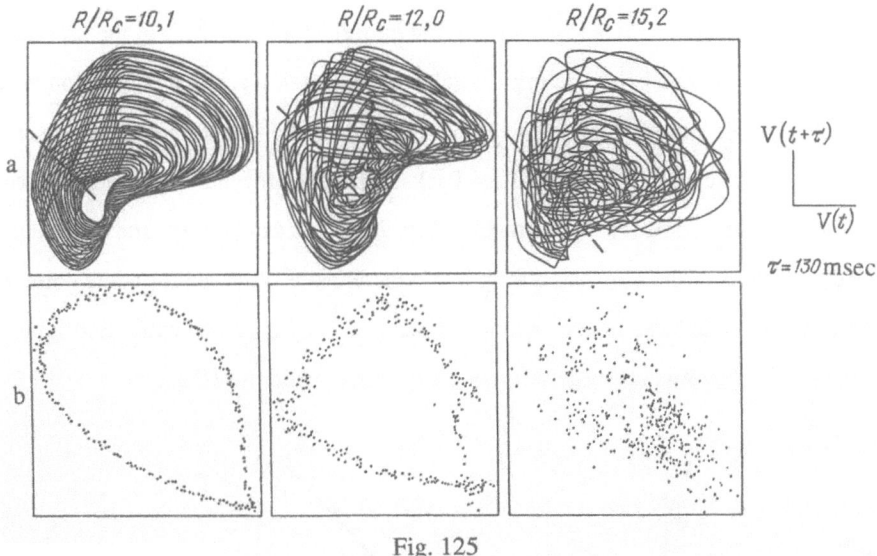

Fig. 125

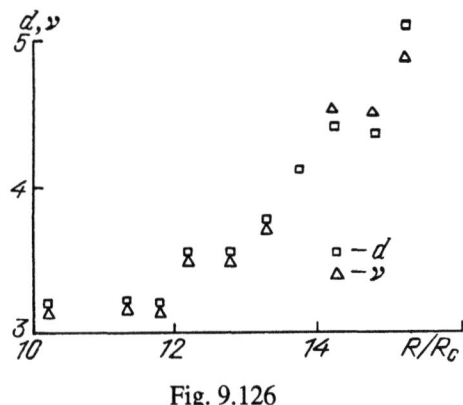

Fig. 9.126

The Couette flow between two spheres with radii in the relation $(r_2 - r_1)/r_1 = 1.006$ and outer radius $r_2 = 150.16 \pm 0.03$ mm were studied in [395]. Only the inner sphere was rotating. In a broad band of Reynolds numbers $R = \Omega r_1^2 / v_0$, where Ω is the angular rotation velocity of the inner sphere, and v_0 kinematic viscosity, the velocity spectra were measured, and the dimension of the attractor of the dynamical system (constructed according to Packard-Takens) calculated. It was found that, as R increases, the dimension increases on the average, too; however, it remains comparatively small.

Convective flow in heating and strong unipolar injection of ions was studied in [569]. It turned out that dimension of the attractor, estimated on the basis of the correlation exponent ν, is 2.8±0.1 in the first case for $Ra/Ra_c = 235$; in the second, for $U/U_c = 5.4$ ($U_c = 50V$ being the critical value of potential, for which convection arose), 5.1±0.3. The dependence of ν on the dimensions of the phase spaces of dynamical systems, constructed on the basis of the Packard-Takens procedure, for the above cases (2, 3 and for white noise) are illustrated in Fig. 9.127. As can be seen from the diagram, small deviations of the dimension from n as n increases is observed for white noise, too, due to the boundedness of the number of samples; however, the saturation is not so explicit as for experimental time series of the processes under investigation.

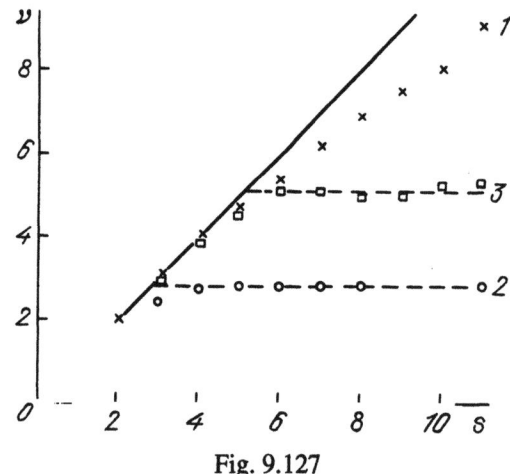

Fig. 9.127

Irregular moving strata were studied in [2] in low-temperature plasma in the positive column of a gas discharge [217]. On the self-excitation boundary, the strata are sinusoidal in shape; as the boundary becomes more distant, their spectrum is enriched, and becomes continuous. An example of dependence on time of side surface glow's intensity, illustrating the stratum shape in the tube part, is given in Fig. 9.128a; the power spectrum, in Fig. 9.128b. Processing the signals by the Packard-Takens method, and using the procedure for finding the correlation exponent ν, enabled us to estimate the

Examples of mechanical, physical, chemical, and biological systems 429

corresponding dimension of the attractor. In the periodic regime, ν turns out to be close to one (Fig. 9.129a, c); in the regime associated with Fig. 9.128, ν ≈ 5.5 (Fig. 9.129b, d). The variation of ν, depending on the dimension of the system constructed in the periodic (c) and chaotic (d) regimes is illustrated in Fig. 9.129c, d.

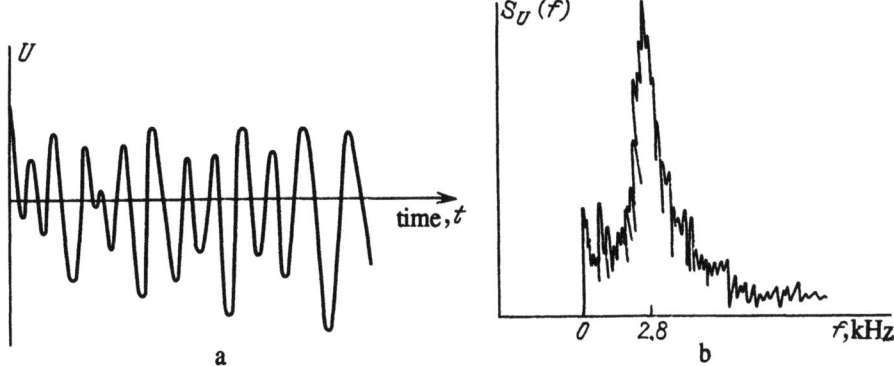

Fig. 9.128

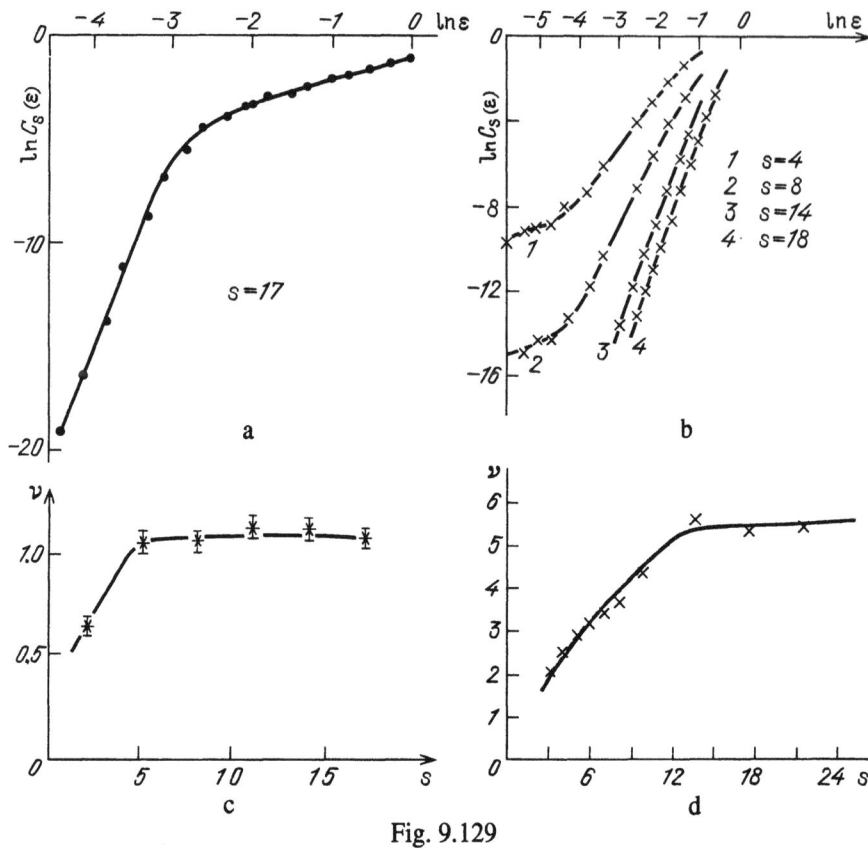

Fig. 9.129

10. Stochasticity in quantum systems[20]

In contrast to classical systems, whose motion can be either fully deterministic, or stochastic, or chaotic, determinism in known not to occur in quantum systems, by the definition itself of quantum random variables described by non-commuting operators. If the state of a quantum system is such that some set of commuting variables is deterministic in nature, then the other variables (not commuting with the latter) are substantially uncertain, statistically speaking. The simplest limits of minimum uncertainty in a system of non-commuting variables are determined by the uncertainty principle. In quantum systems, classical stochastic or chaotic motions are always accompanied by quantum indeterminism leading to the indeterminacy of the concept of the classical

---

[20]Written by B.A. Grishanin.

trajectory. Systems stochastic in the classical limit h → 0 [156, 691] are usually said to be *quantum stochastic*. In the quantum case, one confines himself to Hamiltonian systems only, since the quantization of non-Hamiltonian classical systems is not unique. It is nothing else but associating canonical momenta and coordinates of the operators $\hat{p}$, $\hat{q}$, with the standard commutation relations $\hat{q}_i \hat{p}_j - \hat{p}_j \hat{q}_i = \hbar \delta_{ij}$.

The first difficulty arising in an attempt to generalize the classical results of the theory of chaotic and stochastic systems to the quantum case is related to different traditional mathematical forms of classical and of quantum mechanics. If a system with n degrees of freedom is described in classical mechanics by equations for numerical variables $q_1, ..., q_n, p_1, ..., p_n$, then a similar system is given by the Schrödinger equation for the wave function $\psi(q_1, ..., q_n, p_1, ..., p_n, t)$ in the Schrödinger representation of quantum mechanics. To establish a correspondence between quantum and classical equations on this basis is difficult. Using the traditional method, this is done by considering quasi-classical approximation relating the wave-function to quasi-classical trajectories. However, this relation is simple only for totally integrable systems for which a separate quantization of the action integrals can be carried out, corresponding to separation of variables according to the rules associated with Bohr-Sommerfeld quantization of stationary orbits [247]. However, for systems with stochastic motions arising in the classical limit, there exists no simple correspondence between stationary wave functions and classical trajectories.

Despite all this, attempts were made to relate the stochastic behaviour of quantum systems in the classical limit to the properties of the stationary power spectrum in a quasi-classical approximation. It was shown in [601, 602] that a stochastic motion of a classical system could be associated with the irregular power spectrum of a quantum system. By the analogy with classical billiards, the law of repulsion of random energy levels is derived in the form of a probability distribution for the distance between neighbouring levels ΔE, viz.,

$$p(\Delta E) \sim |\Delta E|^{\text{const}/\ln k}, \qquad (10.1)$$

where k is a dimensionless parameter such that, for k > 1, the corresponding classical system is stochastic [155]. It follows from (10.1) that, for k > 1 and as $\Delta E \to 0$, the probability $p(\Delta E) \to 0$. The above tests of stochasticity are corroborated by a number of numerical experiments. It is noted meanwhile that the statistical properties of motion are determined by those of eigenfunctions, and not only by the structure of the power spectrum [160].

A clearer qualitative idea of the correspondence between classical and quantum theories is formed if they are treated identically. This possibility is related to the use, instead of the Schrödinger equations, of the equivalent Heisenberg formulas for the dynamical variables. These coincide in form with the Hamilton equations, and are different from the latter in their operator nature and non-commutativity of the canonical momenta and coordinates. Still greater similarity between the formalisms of the two kinds is attained in describing quantum dynamical variables by numerical functions of classical phase variables $\mathbf{X} = (q,p)$. This is possible after introducing a linear basis $\hat{e}(\mathbf{X})$ into the space of quantum dynamical variables, by resorting to the representation

$$\hat{A} = \int f(\mathbf{X})\hat{e}(\mathbf{X})\, d\mathbf{X} \qquad (10.2)]$$

which associates each $\hat{A}$ with its function $f(\mathbf{X})$. In the classical limit as $\hbar \to 0$, the use of bases associated with δ-functions so that $\hat{e}(\mathbf{X}) \to \delta(\hat{\mathbf{X}} - \mathbf{X})$, $\hat{\mathbf{X}} = (\hat{q}, \hat{p})$, is the most explicit. The Wigner representation is one such representation used widely [678]. It is important to note that, taking into account the finiteness of the dimension of the Hilbert space in relation to the finite phase volume Ω ($N = \Omega/(2\pi\hbar)^r$, where r is the number of degrees of freedom), the operators representable by functions $f_\Omega(\mathbf{X})$ vanishing outside Ω are of the form of square $N \times N$ matrices, and require finitely many basis elements for their description. This means that the set of the basis elements $\hat{e}(\mathbf{X})$, where $\mathbf{X} \in \Omega$, is filled over: it has continuum power. The limit transition $\Omega \to \infty$ makes it evident that the basis $\hat{e}(\mathbf{X})$ is filled over also in infinite phase volume. This circumstance means that the representation $\hat{A} \to f(\mathbf{X})$ may be not unique, which can be eliminated on selecting the corresponding simplest rule, i.e., putting $f(\mathbf{X}) = (2\pi\hbar)^r \text{Tr } \hat{A}\hat{e}(\mathbf{X})$ in the Wigner representation [136].

Examples of mechanical, physical, chemical, and biological systems 433

With the aid of the Wigner representation, a change in both the quantum and the classical variables (also characterized by the functions f(X)) is described by the equation

$$\frac{\partial f}{\partial t} = \mathcal{L}\, f, \tag{10.3}$$

where $\mathcal{L}$ is a linear operator called a *Liouville operator* (or a *Liouvillian*). In the quantum case, it is specified by the Wigner representation of the quantum Poisson brackets, i.e.,

$$\mathcal{L} = \frac{i}{\hbar} N\left[\mathcal{H}\left(X + \frac{i\hbar}{2}\sigma\frac{\partial}{\partial X^T}\right) - \mathcal{H}\left(X - \frac{i\hbar}{2}\sigma\frac{\partial}{\partial X^T}\right)\right], \tag{10.4}$$

where $i\hbar\,\sigma$ is the commutation matrix of the canonical variables, and N the ordering symbol placing the operators $\partial/\partial X^T$ to the right of multiplication by the variables X. In the classical limit, or, as $\hbar \to 0$, $\mathcal{L}$ equals $\mathcal{L}_0$ where $\mathcal{L}_0$ denotes the classical Poisson brackets

$$\mathcal{L}_0 = -\frac{\partial \mathcal{H}}{\partial X}\sigma\frac{\partial}{\partial X^T} = \frac{\partial \mathcal{H}}{\partial p}\frac{\partial}{\partial q} - \frac{\partial \mathcal{H}}{\partial q}\frac{\partial}{\partial p}. \tag{10.5}$$

Under $\mathcal{L}_0$, the transformation F(X) is reduced to the motion of the initial values of X along the classical trajectories, since the differential operator is of order one. In the quantum case, $\mathcal{L} - \mathcal{L}_0 = \mathcal{L}_D \neq 0$ for non-linear systems, and the motion is accompanied by additional quantum diffusion described by an operator $\mathcal{L}_D$ whose expansion in terms of $\partial/\partial X$ only involves odd powers, starting with the third. The diffusion is reversible, viz., $\mathcal{L}^+ = -\mathcal{L}$, i.e., $\mathcal{L}$ is an anti-Hermitian operator. The eigenvalues are purely imaginary, i.e., $\lambda = i\omega$, whereas the eigenfunctions satisfy the condition $f_{-\omega}(X) = f_\omega^*(X)$. If the Hamiltonian $\mathcal{H}$ does not depend on time, then the evolution by a moment t is described by the operator $S(t) = \exp(\mathcal{L}t)$, reversibility meaning that return to the initial state can also be attained by the dynamical evolution with Liouvillian $\mathcal{L}^+ = -\mathcal{L}$.

The time evolution operator $S(t) = T \exp\left(\int_0^t \mathcal{L}\, dt\right)$ can be represented as $S(t) = S_D(t)S_0(t)$, where S_0 describes motion along the classical trajectories, while

$$S_D(t) = T \exp\left[\int_0^t S_0(\tau)\, \mathcal{L}_D\, S_0^{-1}(\tau)\, d\tau\right] \tag{10.6}$$

is the additional quantum diffusion due to anharmonicity of the system. This representation makes explicit the basic laws in the dynamics of quantum systems which are stochastic as $\hbar \to 0$. Quantum diffusion leads to blurring a stochastic motion as t increases until the complete suppression of stochasticity. The mechanism due to quantum diffraction from stochastic trajectories was studied in [490, 599], determining a lower time bound for the stochastic motion to stop.

As noted above, the continuity of $\mathfrak{L}$ and f(X), which represent the dynamical variables, is fictitious in the quantum case, because the basis ê(X) is filled over. A consequence is discreteness and boundedness of the eigenvalue spectrum of $\mathfrak{L}$ in moving inside a bounded region of the phase space of the system with Hamiltonian independent of time explicitly. The motion described by the operator S(t), made up of finitely many harmonic components, is quasi-periodic, and returns to the original state as close as we please. Thus, the stochasticity in quantum systems is "transient" in principle [164, 402-405, 412, 645], which was checked in numerical calculations; in particular, for a model of the classical kicked rotator. In contrast to classical systems, for which the numerical calculation of a stochastic motion after a time t in reverse direction does not make the system return to the original state due to round-off errors, this come-back meanwhile occurs in the quantum case, because the number of quantum states is finite if the errors are sufficiently small [164].

Temporal change in the state of the system both in the classical and the quantum case can be described equivalently in two ways: either by varying the variables which specify the physical quantities or by changing the probability distribution determining the state. In the quantum case, the two methods to describe temporal evolution are called the *Heisenberg* an the *Schrödinger representation*, respectively. The Heisenberg representation was described above. Transition to the Schrödinger representation is carried out via the replacement $\mathfrak{L} \to \mathfrak{L}^+$, where $\mathfrak{L}^+$ is simply reduced to the replacement $\partial/\partial X \to -\partial/\partial X$ in the Wigner representation, since the $\partial/\partial X$ are anti-Hermitian operators.

The quantum probability distribution in the Wigner representation is specified by expanding the density matrix $\hat{\rho}$ with respect to the basis $(2\pi\hbar)^r$ ê(X), viz.,

$$\hat{\rho} = \int w(X)[(2\pi\hbar)^r \hat{e}(X)]dX.$$

The function w(X) so introduced is of the same dimension as the probability density in the phase space; it is also real and normalized to unity. It is called the *Wigner* (probability) *density function*.

With the aid of w(**X**), mean values of physical quantities represented by functions f(**X**) according to (10.2) can be calculated by the classical formula.

$$\langle \hat{A} \rangle = \int f(\mathbf{X}) w(\mathbf{X}) d\mathbf{X}.$$

In contrast to a classical probability density, a Wigner density function is not necessarily positive. However, in reality, this occurs practically in all cases where the use of the Wigner representation is technically convenient. The following quantum property of w(**X**) is more essential: In contrast to classical distributions, the function cannot be localized in an arbitrarily small region of the phase space, i.e., be of the form $\delta(\mathbf{X} - \mathbf{X}_0)$ related to the deterministic state with phase variables exactly equal to $\mathbf{X}_0$. For any density matrix $\hat{\rho}$, the Wigner density function w(**X**) is such that the canonical variable correlation matrix

$$\mathcal{K} = \int \mathbf{X} \mathbf{X}^T w(\mathbf{X}) d\mathbf{X}$$

satisfies the Heisenberg matrix inequality [136]. That deterministic states are impossible is related to the definition of canonical variables as non-commuting operators $\hat{\mathbf{X}}$.

Consider the simplest example of a quantum stochastic system studied in many papers [237], i.e., the classical kicked rotator gaining a generalized momentum V(θ) which depends on the angle of rotation θ. Its Hamiltonian is

$$\hat{\mathcal{H}} = \hat{p}^2/2I + V(\hat{\theta}) \Delta_{t_0}(t),$$

where $\hat{p}, \hat{\theta}$ are the generalized momentum and position operators, I is the moment of inertia and $\Delta_{t_0}(t) = \sum_n \delta(t - nt_0)$ a periodic sequence of δ-functions describing the instantaneous kicks.

According to (10.4), the Liouvillian describing the change in the Wigner density function is represented by the expression

$$\mathcal{L}^+ = -\frac{p}{I}\frac{\partial}{\partial \theta} - \frac{i}{\hbar} \Delta_{t_0}(t) \left[V\left(\theta + \frac{i\hbar}{2}\frac{\partial}{\partial p}\right) - V\left(\theta - \frac{i\hbar}{2}\frac{\partial}{\partial p}\right) \right] \quad (10.7)$$

in the Wigner representation. Meanwhile, since the system is kinematically symmetric, viz., the function V(θ) periodic with respect to the angle θ, and the values of θ's differing only by an integer multiple of 2π are indistinguishable, the Hilbert state space is specified by the set of basis functions $\psi_m(\theta) = \exp(im\theta)/\sqrt{2\pi}$. The most general form of the

Wigner density, associated with this structure, is determined by the expression

$$w(p,\theta) = \sum_n \sum_m \frac{1}{2\pi} \rho_{mn} e^{i(m-n)\theta} \delta(p - \hbar\frac{m+n}{2}), \quad (10.8)$$

where ρ_{mn} are the entries of the positive semi-definite, normalized ($\Sigma \rho_{mn} = 1$) Hermitian matrix describing the density matrix $\hat{\rho}$ in the basis of functions ψ_m. The states associated with exactly determined momenta are described by the Wigner density function

$$w_m(p,\theta) = \delta(p - \hbar m)/2\pi,$$

and are completely indeterminate as far as the angle is concerned.

An example of a similar Wigner distribution (for a plane rotator) is given in Fig. 9.130. The volume where the envelope of the Wigner density function is localized is not less than $2\pi\hbar$ (the shaded strip). The momentum distribution is localized at the values $\hbar(m+n)/2$, i.e., quantized by $\hbar/2$. Taking the periodicity of $V(\theta)$ in (10.7) into account, this structure is invariant under the corresponding transformation $S^+(t)$. In fact, expanding $V(\theta)$ in terms of the functions $\exp(im\theta)$, we see that the second term in (10.7), determining the momentum-related change, is associated with combinations of shifts through $\hbar m/2$, since $\exp[\pm\frac{m\hbar}{2}\frac{\partial}{\partial p}]$ is the operator of shift through $\pm m\hbar/2$ with respect to p.

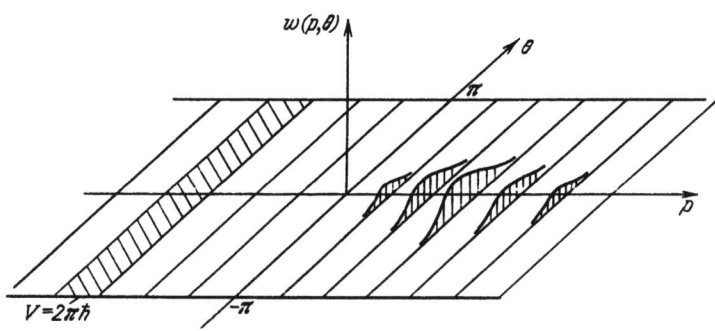

Fig. 9.130

Turning to dimensionless variables $p' = pt_0/I$, $t' = t/t_0$, $v(\theta) = V(\theta)/V_0$, where V_0 is the maximal generalized momentum gained, we have

$$\mathcal{L} \to \mathcal{L}' t_0, \quad \Delta_{t_0}(t) \to \Delta(t') = \sum_n \delta(t' - n), \quad \hbar \to \xi = \frac{\hbar t_0}{I},$$

Examples of mechanical, physical, chemical, and biological systems

$$w(p', \theta) = \sum_n \sum_m \frac{1}{2\pi} \rho_{mn} e^{i(m-n)\theta} \delta(p' - \xi \frac{m+n}{2}), \quad (10.9)$$

Thus, the role of the Planck constant is now played by the dimensionless exponent ξ of quantization degree, determining the phase space quantization volume.

Omitting the primes, we obtain

$$\mathcal{L}^+ = -p\frac{\partial}{\partial \theta} - iK\Delta(t)\frac{1}{\xi}\left[v\left(\theta + i\frac{\xi}{2}\frac{\partial}{\partial p}\right) - v\left(\theta - i\frac{\xi}{2}\frac{\partial}{\partial p}\right)\right], \quad (10.10)$$

in dimensionless variables from (10.7), where $K = V_0 t_0 / I$ is the maximal dimensionless momentum gained.

For the case $v(\theta) = \cos\theta$, we derive from the above that

$$\mathcal{L}^+ = -p\frac{\partial}{\partial \theta} - K\Delta(t)\sin\theta \frac{1}{\xi}\left[\exp\left(\frac{\xi}{2}\frac{\partial}{\partial p}\right) - \exp\left(-\frac{\xi}{2}\frac{\partial}{\partial p}\right)\right]. \quad (10.10^*)$$

Since the functions (10.9) in question are singular with respect to the momenta p, it is important that they are unbounded in (10.10), (10.10*), due to momentum quantization, which does not enable us to turn immediately to a quasi-classical approximation. In fact, the operator $\frac{\xi}{2}\frac{\partial}{\partial p}$ cannot be regarded as small for any ξ; nor can the shift operator $\exp\left[\frac{\xi}{2}\frac{\partial}{\partial p}\right]$ be expanded in terms of it. However, it is possible to do as follows: Describe the replacement of the true singular function $w(p, \theta)$ by a certain continuous function $\tilde{w}(p, \theta)$ for which the original one $w(p, \theta)$ of form (10.9) can be reconstructed simply and uniquely. The most natural and widely used is the following procedure for the replacement $w \to \tilde{w}$: The function (10.9) is of the structure

$$w(p, \theta) = \sum_l w_l(\theta) \delta(p - l\frac{\xi}{2}),$$

where

$$w_l(\theta) = \sum_{m+n=l} \frac{1}{2\pi} \rho_{mn} \exp[i(m-n)\theta].$$

Spread each $\delta(p - l\frac{\xi}{2})$ to an interval of width ξ, i.e., replace the function by a continuous one $f_l(p)$ such that

$$\int f_l(p) \, dp = \int \delta(p - l\frac{\xi}{2}) \, dp = 1 \; ;$$

e.g., $f_l(p) = \sin[\pi(2p - l\xi)/\xi] / \pi(p - l\xi/2)$.

Then

$$\widetilde{w}(p,\theta) = \sum_l w_l(\theta) \sin[\pi(2p-l\xi)/\xi] / \pi(p-l\xi/2).$$

At the points $p=l\xi/2$, the function $\widetilde{w}(p,\theta)$ equals $2w_l(\theta)/\xi$. It is obvious that $\widetilde{w}(p,\theta)$ satisfies the normalization condition, as required for the Wigner density function.

Under $\mathcal{L}^+$, $\widetilde{w}(p,\theta)$ is transformed so that the corresponding values $w_l(\theta)$ can be expressed in terms of themselves, and, after reconstructing $w(p,\theta)$, the latter is always of the form (10.9). If $w_l(\theta)$ changes slightly in transition from l to $l+1$, then $\widetilde{w}(p,\theta)$ changes smoothly on scales of order ξ. Therefore, we can distinguish functions associated with such quasi-classical states in the class of functions $\widetilde{w}(p,\theta)$ for which $(\xi/2)\, \partial/\partial p$ is a small operator.

Separating $\mathcal{L}^+$ determined by expression (10.10*) into the classical part of the form (10.5) and a quantum contribution, we obtain

$$\mathcal{L}^+_{\bar{\sigma}} - p\partial/\partial\theta - K\Delta(t)\sin\theta\, \partial/\partial p ,\qquad(10.11)$$

$$\mathcal{L}^+_D = -K\Delta(t)\sin\theta\, \frac{1}{\xi}\left[2\sinh\left(\frac{\xi}{2}\frac{\partial}{\partial p}\right) - \xi\frac{\partial}{\partial p}\right] .\qquad(10.12)$$

the component (10.11) describes the classical motion in one time interval from $t=n+0$ to $t=n+1+0$, including the impulse jump through $K\sin\theta$ at the moment $t=n+1$. The component is associated with the transition operator

$$S_0^+ = T\exp\left\{\int_{n+0}^{n+1+0}\left[-K\sin\theta\,\delta(\tau-n)\frac{\partial}{\partial p} - p\frac{\partial}{\partial\theta}\right]d\tau\right\} .$$

Taking into account that $\partial/\partial p$, $\partial/\partial\theta$ are infinitesimal shift operators with respect to the corresponding variables (i.e., shifting by the infinitesimal $d\tau$), after factorizing the exponential function by the standard formula of transition to the interaction representation, we obtain

$$S_0^+ = T_{\widetilde{p}}(-K\sin\widetilde{\theta})\, T_{\widetilde{\theta}}(-p) .\qquad(10.13)$$

In (10.13), $T_\alpha(\beta)=\exp(\beta\partial/\partial\alpha)$ is the operator of shift with respect to the subscript variable by the quantity in the argument; θ, p are the values of the variables θ, p at the moment $t=n+0$; and $\widetilde{\theta}$, $\widetilde{p}$ are the corresponding values at the moment $t=n+1+0$. The operator $T_{\widetilde{\theta}}(-p)$ describes the free motion of the rotator until a kick. Then $\widetilde{\theta}=\theta+p$; the momentum is not altered. On the contrary, the operator $T_{\tilde{p}}(-K\sin\tilde{\theta})$ describes a momentum jump during the kick when the angle $\tilde{\theta}$ does not change, and $\tilde{p}=p+K\sin\tilde{\theta}$.

The operator S_0^+ is associated with superposition of the above transformations, called the <u>standard mapping</u>, and well studied in classical theory [157, 237].

The complete quantum transition operator associated with the sum of the infinitesimal operators (10.11), (10.12) can be represented as $S^+ = S_K^+ T_{\tilde{\theta}}(-p)$, where

$$S_K^+ = \exp\left\{K\sin\tilde{\theta}\frac{1}{\xi}\left[T_{\tilde{p}}(-\frac{\xi}{2}) - T_{\tilde{p}}(\frac{\xi}{2})\right]\right\} \qquad (10.14)$$

is the state transformation operator $\tilde{w}(p,\theta)$ during the kick. As $\xi \to 0$, on the set of quasi-classical states, S_K^+ degenerates into the shift operator $T_{\tilde{p}}(-K\sin\tilde{\theta})$.

Distinguishing the classical and quantum factors in (10.14) in accordance with (10.12), we have

$$S_D^+ = S_K^+ T_{\tilde{p}}(K\sin\tilde{\theta}) \qquad (10.15)$$

for the pure quantum contribution of the form (10.6).

In contrast to $S_K^+ = S_D^+ T_{\tilde{p}}(-K\sin\tilde{\theta})$, the operators S_D^+, $T_{\tilde{p}}(-K\sin\tilde{\theta})$ are not individually reducible to superposition of the shifts $T_{\tilde{p}}(\pm\xi/2)$; therefore, they violate the quantum structure of the density $w(p,\theta)$ corresponding to $\tilde{w}(p,\theta)$, and represented in Fig. 9.130. The correct quantum transformation which is valid for any density $\tilde{w}(p,\theta)$, and not only for a quasi-classical one, is merely the superposition of these operators.

To clarify the limits within which quasi-classical analysis can be applied, it suffices to retain the terms in the expansion of the shift operators $T_{\tilde{p}}(\pm\xi/2) = \exp[\pm(\xi/2)\partial/\partial p]$ in terms of ξ up to order three inclusive in (10.14). Substituting the expression for S_K^+ in (10.15), we get

$$S_D^+ = \exp[(K\xi^2/24)\sin\tilde{\theta}\partial^3/\partial\tilde{p}^3 + O(\xi^4)], \qquad (10.16)$$

where $\tilde{\theta}$, $\tilde{p}$ vary from one moment t=n until another in accordance with the classical standard mapping.

It is obvious that the degree of quantization of motion substantially depends, thanks to the operator $\partial/\partial\tilde{p}$ in (10.16), on the rate of change of the function $\tilde{w}(p,\theta)$ with respect to the variable p. The function is determined by the scales of motion details described, viz., $\tilde{w}(p,\theta)$ can be non-local only on these scales, the non-locality following inevitably from the uncertainty principle. The less the described scale of detail, the greater $\partial/\partial \tilde{p}$ in (10.16), and the more severe the restrictions placed on the parameter $\xi \ll 1$ in order that the quasi-classical description be applicable; e.g., let $\Delta \gg \xi$ be the size of the stability zone on the phase plane, which is a characteristic relative to p. Then the localization scale for $\tilde{w}(p,\theta)$ necessary to describe the motion qualitatively correctly must not be less than Δ. Therefore, the quantity $\partial/\partial \tilde{p}$ is at least of order $1/\Delta$ if there is no finer structure of motion dependence on the coordinates inside the region Δ. The quantum contribution operator is determined by superposition of the operators (10.16) in time $t \gg 1$. Assuming that the indices of the exponential function in each operator can be regarded as independent random variables of equal order in absolute value, and, expanding all the functions in series, we get

$$\chi = \frac{K\xi^2}{24} \frac{\sqrt{t}}{\Delta^3} \overline{(\sin^2\tilde{\theta})}^{1/2}$$

for the resulting deformation deviation from the identity operator. Assuming χ to be of order 1, we hence for the time scale derive the estimate

$$t^* \sim \left(\frac{24\Delta^3}{K\xi^2}\right) \Big/ \overline{\sin^2\tilde{\theta}}$$

for stability loss due to quantum transition into the region exhibiting stochastic behaviour. The inverse transition can occur similarly.

In the above example of a non-autonomous system, its Hamiltonian does not guarantee motion in a bounded region in the phase space; nor does it guarantee that hence the motion should be quasi-periodic. In the classical case, for $K > K_c \approx 0.97164$, the transition to infinite momentum diffusion should occur [237]. However, the quantum nature of the motion becoming essential for sufficiently large time intervals leads to suppressing the diffusion according to numerical experiments. If the system were autonomous with bounded phase volume, then suppression would be an obvious consequence of phase space quantization. Note that analysis of the evolution of the Wigner density function in the example under consideration is also contained in [404], and describes in a qualitatively correct manner the nature of density variation for large time values. In conclusion, the authors would like to express their gratitude to A.P. Krylova for help in preparing the bibliography list, and V.V. Alexeyev, V.S.

Anishchenko and D.I. Trubetskov for photography of the experimental results.

BIBLIOGRAPHY

1. Auto-Wave Processes in Diffusion Systems // Applied Physics Institute, Gorky, 1981 (*Russian*).
2. *Aleksandrov B.P., Landa P.S. and Shvilkin B.N.* A method for determining the number of degrees of freedom from experimental data, considering the example of moving strata // Fizika plazmy. - 1986. - V.12, No 1.-PP. 120-123 (*Russian*).
3. *Alekseev A.S.* An electronic model of a two-position regulator of temperature with a lead zone // *DAN SSSR*. - 1952. - V.87, No 3.-PP. 393-396 (*Russian*).
4. *Alekseev A.S.* A two-position regulator of temperature with a lead zone // In Memory of A.A.Andronov.- *AN SSSR*, Moscow, 1955. - PP. 45-76; *Trudy Gork. fiz.-tekhn. in-ta i radiofaka GGU, uchenye zapiski.* - 1957.- V.35- PP.105 - 201 (*Russian*).
5. *Alekseev V.V. and Kornilovsky A.I.* Auto-stochastic processes in biophysical systems // *Biofizika*.- 1982.-V.27, No 5.-PP.890-894 (*Russian*).
6. *Alekseev V.M.* Quasi-random dynamical systems, I, II, III // *Mat.sb.*-1968.-V76, No 1.-pp.72-134; 1968.-V.77, No 4.-PP.545-601; 1969.-V.78, No 1.-PP.3-50 (*Russian*).
7. *Alekseev V.M.* Quasi-random oscillations and qualitative problems in celestial mechanics. *Ninth Mathematical Summer School* (Kaciveli, 1971). *Izdanie Inst. Mat. Akad. Nauk Ukrain. SSR*, Kiev, 1972.-PP.212-341 (*Russian*).
8. *Alekseev V.M.* Symbolic dynamics. *Eleventh Mathematical School (Summer School Kolomyya, 1973) (Russian).-PP.5-210.-Izdanie Inst. Mat. Akad. Nauk Ukrain. SSR*, Kiev, 1976.
9. *Alekseev V.M. and Yakobson M.V.* Symbolic dynamics and hyperbolic systems (Appendix to R. Bowen's *Methods for Symbolic Dynamics)*, Moscow, Mir, 1979.-PP.196-240 (*Russian*).

10. *Andreychikov I.P., Petrovskaya N.V. and Yudovich V.I.* Bifurcations and stochastic motions in certain hydrodynamic systems // VINITI.-Moscow, 1981, No 3485-80 *(Russian)*.

11. *Andreychikov I.P. and Yudovich V.I.* Numerical analysis of bifurcations of limit cycles of a Lorenz system // VINITI.-Moscow, 1981.-No 3630-80 *(Russian)*.

12. *Andronov A.A.* Mathematical problems in self-oscillation theory // Collected Works.-Moscow, AN SSSR, 1956.-PP.84-124 *(Russian)*.

13. *Andronov A.A.* The theory of point transformations of Poincaré-Brouwer-Birkhoff and nonlinear oscillation theory // *Vestnik AN SSSR*.-1944.-No 6 *(Russian)*.

14. *Andronov A.A. and Vitt A.A.* On the synchronization theory of van der Pol // Andronov A.A. Collected Works.-Moscow, AN SSSR, 1956.-PP.51-64 *(Russian)*.

15. *Andronov A.A., Vitt A.A. and Khaykin S.E.* Theory of Oscillations. Translated from the Russian by F. Immirzi. Pergamon Press. Oxford-N.Y.-Toronto, Ont., 1966.

16. *Andronov A.A. and Leontovich E.A.* Certain cases of dependence of limit cycles on parameters // *Uchenyye zapiski GGU*.-1939.-Vyp.6, No 3.-PP.3-33 *(Russian)*.

17. *Andronov A.A. and Leontovich E.A.* Generation of limit cycles from a structurally unstable focus or center and a limit cycle // DAN SSSR-V 99.-PP.885-888.-1954.- *(Russian)*.

18. *Andronov A.A. and Leontovich E.A.* Dynamical systems of first degree of structural instability on the plane // *Mat. Sb.*-1965.-V.68, No 3.-PP.328-372 *(Russian)*.

19. *Andronov A.A., Leontovich E.A., Gordon I.I. and Mayer A.G.* Theory of Bifurcations of Dynamic Systems on Plane.-Moscow, Nauka, 1967 *(Russian)*.

20. *Andronov A.A. and Pontryagin L.S.* Systèmes grossiers // DAN SSSR.-1937.-V.14, No 5.-PP.247-251 *(Russian)*.

21. *Anisimova Yu.V., Dmitriev A.S., Zalogin N.N., Kalinin V.I., Kislov V.Ya. and Panas A.I.* On a mechanism of transition to chaos in the system "electron beam-electromagnetic wave" // *Pis'ma ZHETF*.-1983.-V.37, No 8.-PP.387-389 *(Russian)*.

22. *Anishchenko V.S.* Interaction of strange attractors (SA), intermittency of "chaos-

chaos" type // *Pis'ma ZHETF.*-1984.-V.10, vyp.10.-PP.629-632 (*Russian*).

23. *Anishchenko V.S.* Stochastic Oscillations in Radio-Physical Systems. Parts 1, 2.- Saratov: *SGU* Press, 1985; 1986 (*Russian*).

24. *Anishchenko V.S.* Disruptor of quasi-periodic oscillations and chaos in dissipative systems // *ZHTF.*-1986.-V.56, No 2.-PP.225-237 (*Russian*).

25. *Anishchenko V.S., Aranson I.S., Postnov D.E. and Rabinovich M.I.* Spatial synchronization and bifurcation of chaos development in a chain of coupled generators // *DAN SSSR.*-1986.-V.286, No 5.-PP.1120-1124 (*Russian*).

26. *Anishchenko V.S. and Astakhov V.V.* Experimental study of self-oscillation stochastization in external feedback amplifiers // Lectures on microwave electronics and radiophysics, Bk.5.-Saratov: *SGU* Press, 1980. PP. 118-133 (*Russian*).

27. *Anishchenko V.S. and Astakhov V.V.* Experimental study of the mechanism for the appearance and structure of a strange attractor in a generator with inertial non-linearity // *Radio-tekhnika i elektronika.*-1983.V.28, No 6.-PP.1109-1115 (*Russian*).

28. *Anishchenko V.S. and Astakhov V.V.* Bifurcation phenomena in an auto-stochastic generator under external regular action // *ZHTF.*-1983.-V53,*vyp*.11.-PP.2165-2170 (*Russian*).

29. *Anishchenko V.S., Astakhov V.V. and Letchford T.E.* Multifrequency and stochastic self-oscillations in a generator with inertial non-linearity // *Radiotekhnika i elektronika.*-1982.-V.27, No 10.-PP.1972-1978 (*Russian*).

30. *Anishchenko V.S., Astakhov V.V. and Letchford T.E.* Stochastic attractor in a model of a generator with inertial non-linearity // *Voprosy elektroniki SVCh.*-Saratov: *SGU* Presss, 1983.-PP.31-44 (*Russian*).

31. *Anishchenko V.S., Astakhov V.V. , Letchford T.E. and Safonova M.A.* On bifurcations in a three-dimensional two-parameter autonomous oscillatory system with a strange attractor // *Izv. VUZov. Radiofizika.*-1983.-V.26, No 2.-PP.169-176 (*Russian*).

32. *Anishchenko V.S., Astakhov V.V. , Letchford T.E. and Safonova M.A.* On the problem of quasi-hyperbolic stochasticity in an inertial auto-generator // *Izv.*

VUZov. Radiofizika.-1983.-V.26, No 7.-PP.832-842 (*Russian*).

33. *Anishchenko V.S., Letchford T.E. and Safonova M.A.* Effect of dissipative nonlinearity on bifurcations in auto-stochastic systems // Radiotekhnika i elektronika.-1984.-V.29, No 7.-PP.1355-1361 (*Russian*).

34. *Anishchenko V.S., Letchford T.E. and Safonova M.A.* Stochasticity and the disruptor of quasiperiodic motion due to doubling in a system of coupled generators // Radiophys. and Quantum Electronics.-1984.-V.27, No 5.-PP.381-390.

35. *Anishchenko V.S., Letchford T.E. and Safonova M.A.* Synchronization effects and bifurcations of synchronous and quasiperiodic oscillations in a nonautonomous generator // Radiophys. and Quantum Electronics.-1985.-V.28, No 9.-PP.766-776.

36. *Anishchenko V.S., Letchford T.E. and Safonova M.A.* Critical phenomena in harmonic modulation of double frequency oscillations. Translations to chaos through a 3-torus / Pis'ma ZhTf.-1985.-V.11, vyp.9.-PP.536-541 (*Russian*).

37. *Anishchenko V.S. and Postnov D.E.* Transition to stochasticity in an inertial delay generator. Problem of finite-dimensional representation. // ZhTf-1985.-V.55, No 1.-PP.162-167 (*Russian*).

38. *Anishchenko V.S., Postnov D.E. and Safonova M.A.* Dimension and physical properties of chaotic attractors in a coupled generator chain // Pis'ma ZhTf.-1985.-V.11,vyp.24.-PP.1505-1509 (*Russian*).

39. *Anosov D.V.* Roughness of geodesic flows on compact Riemannian manifolds of negative curvature // Soviet Math. Dokl.-1962.-V.3.-PP.1068-1070; Ergodic properties of geodesic flows on closed Riemannian manifolds of negative curvature // Soviet Math. Dokl.-1963.-V.4.-PP.1153-1156.

40. *Anosov D.V.* Geodesic flows on closed Riemannian manifolds with negative curvature. Trans. from the Russian by S.Feder. Providence, American Mathematical Society, 1969.

41. *Aranson I.S., Rabinovich M.I. and Starobinets I.M.* Transitions in chaos. Nonlinear and turbulent processes in physics, V.3 (Kiev, 1983), PP.1139-1160, Harwood Academic Publ., 1984.

42. *Arnold V.I.* Loss of stability of self-induced oscillations near resonance and versal

deformations of equivariant vector fields // *Funktsional. Anal. i Prilozhen.*-1977.-V.11, No 2.-PP.1-10 *(Russian)*.

43. *Arnold V.I.* Additional Chapters in Theory of Ordinary Differential Equations.-Moscow: Nauka, 1978 *(Russian)*.

44. *Astakhov V.V.* Study of dynamic processes in self-oscillatory systems with inertial non-linearity // Questions of Microwave Electronics.-Saratov: *SGU* Press, 1983. PP. 45-50 *(Russian)*.

45. *Afraimovich V.S.* The annulus principle and quasiattractors // Ninth international conference on nonlinear oscillations // Kiev: *Naukova dumka*, 1985.-PP.21-24 Strange attractors and quasiattractors. Nonlinear and turbulent processes in physics, V. 3 (Kiev, 1983) *(Russian)*.

46. *Afraimovich V.S, Bykov V.V. and Shilnikov L.P.* The origin and structure of the Lorenz attractor // *Dokl. Akad. Nauk SSSR.*-1977.-V.234, No 2.-PP.336-339 *(Russian)*.

47. *Afraimovich V.S , Rabinovich M.I. and Ugodnikov A.D.* Critical points and "phase transitions" in the stochastic behaviour of a nonautonomous anharmonic oscillator // *Pis'ma ZhETF.*-1983.-V.38,*vyp.*2.-PP.64-67 *(Russian)*.

48. *Afraimovich V.S and Shilnikov L.P.* Small periodic perturbations of autonoous systems // *Dokl. Akad. Nauk SSSR.*-1974.-V.214, No 4.-PP.739-742 *(Russian)*.

49. *Afraimovich V.S and Shilnikov L.P.* Certain global bifurcations connected with the disappearance of a fixed point of saddle-node type // *Dokl. Akad. Nauk SSSR.*-1974.-V.219, No 6.-PP.1281-1284 *(Russian)*.

50. *Afraimovich V.S and Shilnikov L.P.* Invariant two-dimensional tori, their breakdown and stochasticity // Methods of the qualitative theory of differential equations.-Gorky: *GGU* Press, 1983.-PP.3-26 *(Russian)*.

51. *Babin A.V. and Vishik M.I.* Attractors of evolution partial differential equations and estimates of their dimension // *UMN* .-1983.-V.38, *vyp.* 4.-PP.113-187 *(Russian)*.

52. *Babitsky V.I. and Landa P.S.* Self-excited vibrations in systems with inertial excitation// Sov. Phys. *Dokl.*-1982.-V.27, No 10.-PP. 826-827.

53. *Babitsky V.I. and Landa P.S.* Self-oscillatory systems with inertial excitation //

Dynamics of Systems.-Gorky: *GGU* Press, 1983.-PP.147-181 (*Russian*).

54. *Babitsky V.I. , Landa P.S., Olkhovoy A.F. and Perminov S.M.* Stochastic behaviour of self-oscillatory systems with inertial self-excitation // Dynamics of Systems. Gorky: *GGU* Press, 1985.-PP.14-49 (*Russian*).

55. *Babitsky V.I., Landa P.S. and Perminov S.M.* Oscillations in machines, taking into account the inertial behaviour of rotation of the velocity regulator // *Mashinovedenie.*-1985.-No 2.-PP.31-35 (*Russian*).

56. *Balescu R.* Equilibrium and Nonequilibrium Statistical Mechanics. New York, Wiley, 1975.

57. *Barenblatt G.I.* Similarity, Self-Similarity and Intermediate Asymptotics. Translated from the Russian by Norman Stein. Consultants Bureau, New York-London, 1909.

58. *Barsuk L.O., Belosludtsev N.M., Neimark Yu.I. and Salganskaya N.M.* Stability of the fixed point of a transformation in the critical case and certain singular bifurcations // *Izv. VUZov. Radiofizika.*-1968.-V.11, No 11.-PP.1632-1641 (*Russian*).

59. *Batalova Z.S.* Study of vibratory pile driving with soil drug // *Inzh. zhurnal, mekhanika tverdogo tela.*-1966.-No 2.-PP.165-173 (*Russian*).

60. *Batalova Z.S.* On the motions of a rotor under the action of external harmonic force // *Inzh. zhurnal, mekhanika tverdogo tela.*-1967.-No 2.-PP.66-73 (*Russian*).

61. *Batalova Z.S.* Resonance levels of certain Hamiltonian systems // Dynamics of Systems, *vyp.*19.-Gorky: *GGU* Press,1980.-PP.60-79 (*Russian*).

62. *Batalova Z.S., Bezdenezhnykh A.N., Neimark Yu.I. and Romashova I.B.* Bifurcations leading to chaotization of the motion of a rotor // Dynamics of Systems, *vyp.*19.-Gorky: *GGU* Press,1980.-PP.110-125 (*Russian*).

63. *Batalova Z.S.. and Belyakova G.V.* On oscillatory pendular motion with a vibrating suspension point // Dynamics of Systems (Optimization and Adaptation).-Gorky: *GGU* Press, 1982.-PP.145-170 (*Russian*).

64. *Batalova Z.S. and Belyakova G.V.* On the structure of the phase space of the

equations of pendular motion with a vibrating suspension point // *VINITI.*- Moscow, 1984.-No 3539-84 (*Russian*).

65. *Batalova Z.S. and Bukhalova N.V.* Hierarchy of the structure of the phase space of the equation of motions of a pendulum with an oscillating axis of rotation .// Dynamics of Systems.-Gorky: *GGU* Press,1983.-PP.85-111 (*Russian*).

66. *Batalova Z.S. and Bukhalova N.V.* Dynamics of a pendulum with a vibrating vertical axis of rotation.-Preprint / *NIRFI.*-Gorky, 1984.-No 187 (*Russian*).

67. *Batalova Z.S., Bukhalova N.V. and Neimark Yu.I.* Bifurcations of the phase portrait of an imbalanced rotor on a harmonically vibrating base, leading to a chaotization of its motions // Dynamics of Systems (Numerical Methods for Study of Dynamic Systems).-Gorky: *GGU* Press,1982.-PP.112-126 (*Russian*).

68. *Batalova Z.S., Dubrovina I.A., Neimark Yu. I. and Orlova E.E.* Numerical study of the solutions of the differential equations of Lorenz convective turbulence // Dynamics of Systems, *vyp.*16.-Gorky: *GGU* Press, 1979.-PP.59-81 (*Russian*).

69. *Batalova Z.S., Dubrovina I.A., Neimark Yu. I. and Orlova E.E. and Orlova E.E.* On the structure of the phase space and bifurcations in the discrete Lorenz model of convective turbulence. // *J. Appl. Math. Mech.* -1981.-V.45, No.4.- PP.468-473.

70. *Batalova Z.S. and Neimark Yu. I.* A certain dynamical system with homoclinic structure // Oscillation Theory, Applied Mathematics and Cybernetics, *vyp.*1.- Gorky: *GGU* Press, 1973.-PP.131-147 (*Russian*).

71. *Bautin N.N.* On the theory of synchronization // *ZhTF.*-1939.-V.9, No 6.-PP. 510-513 (*Russian*).

72. *Bautin N.N.* On the behaviour of dynamic systems with small violations of the Routh-Hurwitz stability conditions // *PMM.*-1948.-V.12, No 5.-PP. 613-632 (*Russian*).

73. *Bautin N.N.* The behaviour of dynamical systems near the boundaries of the domain of stability.-2nd edition.-*Nauka*, 1984 (*Russian*).

74. *Bezaeva L.G., Kaptsov L.N. and Landa P.S.* Investigation of chaotic oscillation modulation in a generator with inertial nonlinearity under external parametric action // *Radiotekhnika i elektronika.*-1987.-V.32, No 3.-PP.647-650 (*Russian*).

75. *Bezaeva L.G., Kaptsov L.N. and Landa P.S.* Threshold for synchronization as a stochasticity test in a generator with inertial non-linearity // *ZhTF*.-1986.-V.56, No 9.-PP.1849-1853 *(Russian)*.

76. *Bezruchko B.P., Bulgakova L.V., Kuznetsov S.P. and Trubetskov D.I.* Experimental study and theoretical analysis of stochastic self-oscillations in a backward wave tube // Lectures on microwave electronics and radiophysics, Bk.5.-Saratov: *SGU* Press, 1980.-PP.25-77 *(Russian)*.

77. *Bezruchko B.P., Bulgakova L.V., Kuznetsov S.P. and Trubetskov D.I.* Stochastic self-oscillations and instability in a backward wave tube // *Radiotekhnika i elektronika*.-1983.-V.28, No 6.-PP.1136-1139 *(Russian)*.

78. *Bezruchko B.P., Kuznetsov S.P. and Trubetskov D.I.* Experimental study of stochastic self-oscillations in the dynamic system "electronic beam-backward electromagnetic wave" // *Pis'ma ZhETF*.-1979.-V.29, *vyp*.3.-PP.180-184 *(Russian)*.

79. *Beletsky V.V. and Khentov A.A.* Rotatory Motions of Magnetic Satellite.-Moscow: Nauka, 1985 *(Russian)*.

80. *Belousov B.P.* Periodical Reaction and Its Mechanism *in* Abstracts from Radiation Medicine.-Moscow: Medgiz, 1959.-PP.145-148; *Khimiya i zhizn'*.-1982.-No 7.-PP.65-68 *(Russian)*.

81. *Belykh V.N.* On bifurcations of separatrices of a saddle point of the Lorenz system // *Differentsialnye uravneniya*.-1984.-V.20, No 10.-PP.1666-1674 *(Russian)*.

82. *Belyustina L.N. and Belykh V.N.* A nonautonomous phase control system with a small parameter that contains invariant tori and structurally stable homoclinic curves // *Izv. VUZov. Radiofizika*.-1972.-V.15, No 7.-PP.1039-1048 *(Russian)*.

83. *Belyustina L.N. and Belykh V.N.* The global structure of the decomposition of the cylindrical phase space of a certain nonautonomous system //*Differentsialnye uravneniya*.-1973.-V.9, No 4.-PP.595-608 *(Russian)*.

84. *Belyustina L.N. and Belykh V.N.* Homoclinic structures generated by the simplest model of phase self-tuning. Phase synchronization.-Moscow: *Svyaz'*, 1975.-PP.97-106 *(Russian)*.

85. *Belyustina L.N., Kiveleva K.G. and Fraiman L.A.* Qualitatively numerical method

in the study of three-dimensional nonlinear phase synchronization system // Phase synchronization system.-Moscow: *Radio i svyaz'*, 1982.-PP.21-45 (*Russian*).

86. *Belyustina L.N. and Otrokova I.N.* Study of homoclinic structure appearance by a qualitatively numerical method // Dynamics of Systems (mathematical methods for oscillation theory).-Gorky: *GGU* Press, 1979.-PP.82-94 (*Russian*).

87. *Belykh V.N. and Maksakov V.P.* Qualitative investigation of a discontinuous cylinder mapping in phase synchronization theory. // Methods for qualitative differential equation theory.-Gorky: *GGU* Press, 1982.-PP.135-149 (*Russian*).

88. *Birkhoff G.* Dynamical Systems. Am.Math.Soc., 1927.

89. *Blekhman I.I.* Synchronization of Dynamical Systems.-Moscow: *Nauka*, 1971 (*Russian*).

90. *Blekhman I.I.* Synchronization in Nature and Engineering.-Moscow:*Nauka*, 1980 (*Russian*).

91. *Bogolyubov N.N.* Perturbation theory in nonlinear mechanics // *Sb.Instituta stroit. mekhaniki AN USSR.*-1950.-V.14.-PP.9-34 (*Russian*).

92. *Bogolyubov N.N. and Mitropolsky Yu.* Asymptotic Methods in the theory of Nonlinear Oscillations.-Moscow:*Nauka*, 1974 (*Russian*).

93. *Bolotin V.V., Vorobyev V.I., Semenov V.A. and Chernov V.K.* On parametric stabilization of unstable forms of equilibrium of mechanical systems // *MTT.*-1979.-No 1.-PP.36-44 (*Russian*).

94. *Bonch-Bruevich V.L.* Stochastic self-oscillations as one of possible noise 1/f sources // *Dokl.Akad.Nauk SSSR.*-1984.-V.278, No 2.-PP.335-339 (*Russian*).

95. *Bryuno A.D.* Analytic form of differential equations // *Tr.Mosk.mat.ob-va.*-1971.-V.25.-PP.119-262;1972.-V.26.-PP.199-239. -Sets of analyticity of a normalizing transformation.-Preprint/*IPM AN SSSR.*-Moscow, 1974.-No 93 (*Russian*).

96. *Bulgakov B.V.* Oscillations.-Moscow: Gostekhizdat, 1954 (*Russian*).

97. *Bunimovich L.A.* On rate of decrease of correlations in dynamic systems // Problems of nonlinear and turbulent processes in physics: *Tr. II Mezhdunar.rabochey gruppy.* P.2./Eds. A.S. Davydov, V.M. Chernousenko.-

Kiev:*Naukova dumka*, 1985.-PP.28-30 (*Russian*).

98. *Bunimovich L.A. and Sinai Ya.G.* Attractor stochasticity in the Lorentz model // Nonlinear waves.-Moscow:*Nauka*, 1979.-PP.212-226 (*Russian*).

99. *Bunimovich L.A. and Sinai Ya.G.* Rate of decrease of correlations in one-dimensional ecological models // Thermodynamics and kinetics of biological processes.-Moscow: *Nauka*, 1980.-PP.226-235 (*Russian*).

100. *Butenin N.V., Neimark Yu.I. and Fufaev N.A.* Introduction to the Theory of Nonlinear Oscillations.-Moscow: *Nauka*, 1976 (*Russian*).

101. *Bylov B.F., Vinograd R.E., Grobman D.M. and Nemytsky V.V.* Theory of Lyapunov exponents and its application to problems of stability.-Moscow: *Nauka*, 1966 (*Russian*).

102. *Vainberg D.V and Pisarenko G.S* Mechanical Oscillations and their role in engineering.-Moscow: *Nauka*, 1965 (*Russian*).

103. *Pol B.van der.* Selected scientific papers.-Amsterdam, North-Holland Publ.Co., 1960.

104. *Ventsel A.D. and Freidlin M.I.* Small random perturbations of dynamical systems // *UMN*.-1970.-V.25, vyp.1.-PP.2-35 (*Russian*).

105. *Vidal C.* Dynamical instabilities in the Belousov-Zhabotinskii *in* Synergetics: Far from Equilibrium. Springer, Berlin, 1975.

106. *Wilson K.G.* The renormalization group and critical phenomena // Reviews of modern physics.-1983.-V.55.-No 3.-PP.583-600.

107. *Vinogradova M.B., Rudenko O.V. and Sukhorukov A.P.* Wave Theory.-Moscow: *Nauka*, 1979 (*Russian*).

108. *Wittenburg J.* Dynamics of systems of rigid bodies.-B.G.Teubner, Stuttgart, 1977.

109. *Vlasova O.F. and Zaslavsky G.M.* Effect of dissipation on the appearance of chaos for two overlapping resonances.-Preprint/*SO AN SSSR*.-Krasnoyarsk, 1983.-No 245F (*Russian*).

110. Wave and Fluctuation Processes in Lasers / Zeiger S.G., Klimontovich Yu.L., Landa P.S., Lariontsev E.G. and Fradkin E.E.-Moscow: *Nauka*, 1974 (*Russian*).

111. *Volosov V.M. and Morgunov B.I.* Averaging Method in the Theory of Nonlinear Oscillatory Systems.-Moscow: *MGU Press*, 1971.-PP.249-277 (*Russian*).

112. *Volkenshtein M.V.* Biophysique. Translated from Russian by E. Oumanski.-Mir,-Moscow, 1985.

113. *Volter B.V. and Salnikov I.E.* Stability of work regimes of chemical reactors.-Moscow: *Khimiya*, 1981 *(Russian)*.

114. *Volterra V.* Leçons sur le théorie mathématique de la lutte pour la vie.-Paris, Gauthier-Villars et Cie, 1931.

115. *Vul E.B., Sinai Ya.G. and Khanin K.M.* Feigenbaum universality and thermodynamic formalism // *UMN*.-1984.-V.39, vyp.3.-PP.3-37 *(Russian)*.

116. *Vulfson I.I. and Kolovsky M.Z.* Nonlinear problems in the dynamics of machines.-Moscow: Mashinostroeniye, 1968 *(Russian)*.

117. *Gavrilov N.K.* Three-dimensional dynamical systems that have a structurally unstable homoclinic contour // *Mat.zametki*.-1973.-V.14, No 5.-PP.687-696 *(Russian)*.

118. *Gavrilov N.K.* Bifurcations of the periodic motion near an inner resonance 1:3 // Stability studies and oscillation theory.-Yaroslavl', 1977.-PP.192-199 *(Russian)*.

119. *Gavrilov N.K. and Shilnikov L.P.* Three-dimensional dynamical systems that are close to systems with a structurally unstable homoclinic curve, I, II // *Mat.sbornik*.-1972.-No 8.-PP.575-492; 1973.-No 1.-PP.139-156 *(Russian)*.

120. *Gaponov-Grekhov A.V. and Rabinovich M.I.* Mandelshtam and the modern theory of nonlinear oscillations and waves. // Soviet Phys.Uspekhi.-1979.-V.22, No 8.-PP.590-614.

121. *Gaponov-Grekhov A.V. and Rabinovich M.I.* Chaotic dynamics of simple systems // *Priroda*.-1981.-No 2.-PP.54-65 *(Russian)*.

122. *Gaponov-Grekhov A.V. and Rabinovich M.I.* Nonlinear Physics. Stochasticity and Structures.-Preprint / *IPF AN SSSR*.-Gorky, 1983.-No 87 // The Physics of the 20th Century: Development and Prospects.-Moscow: *Nauka*, 1984.-PP.219-280 *(Russian)*.

123. *Gaponov-Grekhov A.V., Rabinovich M.I. and Starobinets I.M.* Dynamic model for spatial development of turbulence // *Pis'ma ZhETF*.-1984.-V.39, vyp.12.-PP.561-564 *(Russian)*.

124. *Gaushus E.V.* Study of dynamical system by the method of point transformations.

-Moscow: *Nauka*, 1976 *(Russian)*.

125. *Gelfer I.S. and Neimark Yu.I.* Probability densities corresponding to piecewise-linear mappings with Markov partition. // Dynamics of systems-Gorky: GGV Press, 1983.-PP.52-85 *(Russian)*.

126. Hydrodynamic instabilities and the transition to turbulence. Eds. H. Swinney, J. Gollub.

127. *Gilmore R.* Catastrophe theory for scientists and engineers.-New York, Wiley, 1981.

128. *Ginzburg N.S. and Kuznetsov S.P.* Periodic and stochastic self-modulation modes in electronic generators with distributed interaction // Relativistic radio-frequency electronics.-Gorky: *IPF AN SSSR*, 1981.-PP.101-144 *(Russian)*.

129. *Glansdorff P. and Prigogine I.* Thermodynamic theory of structure, stability and fluctuations.-New York, Wiley, 1972.

130. *Golberg A.I., Sinai Ya.G. and Khanin K.M.* Universal properties of sequences of period-tripling bifurcations // *UMN*.-1983.-V.38, No 1.-PP.159-160 *(Russian)*.

131. *Goldshtik M.A. and Shtern V.N.* Structural turbulence in dissipative systems.-Preprint / *ITF SO AN SSSR*.-Novosibirsk, 1981.-No 74-81 *(Russian)*.

132. *Gorelik G.S.* Oscillations and Waves.-Moscow: *Fizmatgiz*, 1959 *(Russian)*.

133. *Gradshtein I.S. and Ryzhik I.M.* Tables of integrals, sums, series and products.-Moscow: *Nauka*, 1971.-PP.259-262 *(Russian)*.

134. *Grasyuk A.Z. and Oraevsky A.N.* Transient processes in a molecular generator // *Radiotekhnika i elektronika*.-1964.-V.9, No 3.-PP.524-532 *(Russian)*.

135. *Grinberg A.I.* Dynamic models for national economy.-Moscow: *Ekonomika*, 1985 *(Russian)*.

136. *Grishanin B.A.* Quantum electrodynamic theory for radiophysics students.-Moscow: *MGU* Press, 1981 *(Russian)*.

137. *Gruzdev V.G. and Neimark Yu.I.* A symbolic description of motion in the neighbourhood of a not structurally stable homoclinic structure and of its chance in transition to close systems // Dynamics of systems, vyp.8.-Gorky: *GGU* Press, 1975.-PP.13-33.- A symbolic description of motion in the neighbourhood of a

structurally unstable homoclinic curve // *Ukr.mat.zhurnal.*-1976.-V.28, No 6.-PP.723-734 (*Russian*).

138. *Gorbikov S.P. and Neimark Yu.I.* Basic regimes of motion in vibratory conveying with tossing // *Izv. AN SSSR. Mekhanika tverdogo tela.*-1981.-No 4.-PP.39-50 (*Russian*).

139. *Gubankov V.N., Ziglin S.L., Konstaninyan K.I., Koshelets V.P. and Ovsyannikov G.A.* Stochastic oscillations in tunnel Josephson junctions // *ZhETF.*-1984.-V.86, vyp.1.-PP.343-351 (*Russian*).

140. *Gubankov V.N., Konstantinyan K.I., Koshelets V.P. and Ovsyannikov G.A.* Chaotic behaviour of instability in tunnel Josephson junctions // *Pis'ma ZhTF.*-1982.-V.8, No 21.-PP.1332-1335 (*Russian*).

141. *Guckenheimer J.* Instabilities and chaos in non-hydrodynamic systems *in* Hydrodynamic instabilities and the transition to turbulence. Topics in Applied Physics, V.45. Eds:H.Swinney, J.Gollub. Springer, Berlin, 1981.

142. *Gurtovnik A.S. and Neimark Yu.I.* On synchronization of dynamic systems // J.Appl.Math.Mech.-1975.-V.38, PP.749-758.

143. Dynamic Systems.- 1, 2. Modern Problems in Mathematics Ser., Fundamental Trends.-Moscow: *VINITI*, 1985 (*Russian*).

144. *Dikhtyar V.B.* Stochastic self-oscillations in a system of coupled self-generators with delay // *Radiotekhnika i elektronika.*-1982.-V.27, No 2.-PP.310-320 (*Russian*).

145. *Dikhtyar V.B. and Kislov V.Ya.* Stochastic oscillations in resonant self-generators with delay // Nonlinear waves. Stochasticity and turbulence.-Gorky, 1980.-PP.37-45 (*Russian*).

146. *Dikhtyar V.B. and Starkov S.O.* Auto-stochastization of oscillations in generators described by a nonlinear difference equation reducible to a second-order delay differential equation // *Radiotekhnika i elektronika.*-1982.-V.27, No 12.-PP.2457-2463 (*Russian*).

147. *Dmitriev A.S. and Kislov V.Ya.* Strange attractor in a van der Pol non-autonomous equation // *Radiotekhnika i elektronika.*-1982.-V.27, No 12.-PP.2454-2456 (*Russian*).

148. *Dmitriev A. S. and Kislov V. Ya.* Stochastic oscillations in a self-generator with internal first-order delay // *Radiotekhnika i elektronika.*-1984.-V.29, No 12.-PP.2389-2398 *(Russian)*.

149. *Dmitriev A. S. , Kislov V. Ya. and Spiro A. G.* Chaotic oscillation in a non-autonomous oscillator with a reactive nonlinearity // *Radiotekhnika i elektronika.*-1983.-V.28, No 12.-PP.2430-2439 *(Russian)*.

150. *Druzhilovskya T. Ya. and Neimark Yu. I.* Stochastic self-oscillations of a non-linear oscillator with impact energy absorber // J.Appl.Math.Mech.-1982.-V.46, No 6.-PP.740-745.

151. *Dudnik E .N., Kuznetsov Yu. I., Minakova I. I. and Romanovsky Yu. M.* Synchronization in system with strange attractor // *Vestnik MGU. Ser.3.*-1983.-V.24, No 4.-PP.84-87 *(Russian)*.

152. *Zhabotinsky A. M.* Concentrational self-oscillations.-Moscow: *Nauka*, 1974 *(Russian)*.

153. *Zheludev N. I., Makarov V. A., Matveeva A. V. and Svirko Yu. P.* Structure of chaos in excitation of a nonlinear oscillator by a harmonic external force // *Vestnik MGU. Ser.3.*-1984.-V.25, No 5.-PP.106-109 *(Russian)*.

154. *Zaslavsky G. M.* Stochastic wave processes.-Preprint / *NIRFI.*-Gorky, 1973.-No 41 *(Russian)*.

155. *Zaslavsky G. M.* Statistics of energy levels in destroying integral of motion // *ZhETF.*-1977.-V.73, *vyp* 6(12).-PP.2089-2097 *(Russian)*.

156. *Zaslavsky G. M.* Stochasticity of quantum systems // Non-linear waves. Self-Organization.-Moscow: *Nauka*, 1983.-PP.96-106; Statistical Irreversibility in Non-Linear Systems.- Moscow: *Nauka*, 1970. (Modern Problems in Physics.) *(Russian)*.

157. *Zaslavsky G. M.* Stochasticity of dynamical systems.- Moscow: *Nauka*, 1984 *(Russian)*.

158. *Zaslavsky G. M. and Chirikov B. V.* Stochastic instability of non-linear oscillations // Soviet Physics. Uspekhi.-1972.-V.14.-PP.549-568.

159. *Zeldovich Ya. B. and Sokolov D. D.* Fractals, similarity, intermediate asymptotics

// Soviet Physics. Uspekhi.-1985.-V.28, No 7.-PP.608-616.

160. *Izrailev F. M.* Distribution of distances between quasi-energy levels for quantum systems in the classical limit.- Preprint / *IYaF SO AN SSSR*.- Novosibirsk, 1984.- No 84-63 *(Russian)*.

161. *Izrailev F. M., Rabinovich M. I. and Ugodnikov A. D.* Approximate description of three-dimensional dissipative systems with stochastic behaviour.- Preprint / *IPF AN SSSR*.- Gorky, 1981.-No 17 *(Russian)*.

162. *Izrailev F. M. and Chirikov B. V.* Statistical properties of a non-linear string // *DAN SSSR*.-1966.-V.166, No 1.-PP.57-59 *(Russian)*.

163. *Izrailev F. M. and Chirikov B. V.* Stochasticity of the simplest dynamical model with separated phase space.- Preprint / *IYaF SO AN SSSR*.-Novosibirsk, 1968.- No 191 *(Russian)*.

164. *Izrailev F. M., Chirikov B. V. and Shepelyansky D. L.* Transient stochasticity in quantum mechanics.- Preprint / *IYaF SO AN SSSR*.-Novosibirsk, 1980.-No 80-120 *(Russian)*.

165. *Ilyashenko Yu. S.* The dimension of arrtactors of K-contracting systems in an infinite-dimensional space // *Vestnik MGU. Ser.1*.-1983.-V.24, No 3.-PP.52-59 *(Russian)*.

166. *Yorke J. and Yorke E.* Chaotic Behaviour and Fluid Dynamics *in* Hydrodynamics // Hydrodynamic instabilities and the transition to turbulence. Topics in Applied Physics, V. 45. Eds: H. Swinney, J. Gollub. Springer, Berlin, 1981.-PP.77-95.

167. *Joss G. and Joseph D.* Elementary stability and bifurcation theory. New York, 1980.

168. *Kadomstev B. B. and Ryazanov A. I.* What is Synergetics? // *Priroda*.- 1983.- No 8.-PP.2-11 *(Russian)*.

169. *Kalyanov E. V. and Lebedev M. N.* Stochastic oscillations in a system of coupled generators with inertiality // *Radiotechnika i elektronika*.- 1985.-V.30, No 8.- PP.1570-1576 *(Russian)*.

170. *Kantorovich L. V. and Krylov V. I.* Approximate methods for higher analysis.- Moscow-Leningrad: *Gostekhizdat*, 1949 *(Russian)*.

171. *Kapitsa P. L.* Dynamical stability of a pendulum with a vibrating suspension point // *ZhETF.-* 1951.-V.21, *vyp.* 5.-PP.588-597 (*Russian*).

172. *Kapranov M. V., Kuleshov V. N. and Utkin G. M.* Oscillation theory in radio engineering.- Moscow: *Nauka*, 1984 (*Russian*).

173. *Kaptsov L. N.* Spike mode operation in a non-autonomous generator with inertial non-linearity // *Radiotekhnika i elektronika.-* 1975.-V.20, *vyp.* 12.-PP.2496-2499 (*Russian*).

174. *Karpman V. I.* Non-linear waves in dispersive media. Translated from the Russian by F. F. Cap. Pergamon Press, Oxford-New York-Toronto, Ont., 1975.

175. *Katok A. V.* Dynamical systems with hyperbolic structure. Ninth Mathematical Summer School (Kaciveli, 1971) Kiev: 1972.-PP. 126-211 (*Russian*).

176. *Kats V. A.* Structure stochastization and transitions in chaos in a self-generator with delay // Lectures on Microwave Electronocs and Radiophysics, Bk 2.- Saratov: *SGU* Press, 1983.-PP.49-64 (*Russian*).

177. *Kats V. A.* Transitions in chaos, initiated by external harmonic effect, in a distributed self-generator with delay (experiment) // *Ibid.*-PP.65-68 (*Russian*).

178. *Kats V. A.* Experimental demonstration of universal properties of the Feigenbaum doubling bifurcation sequence in the transition to chaos in a distributed generator with delay // *Pis'ma ZhTF.-* 1984.-V.10, No 11.-PP.684-689 (*Russian*).

179. *Kats V. A.* Transition to "developed" chaos in the model of a distributed generator with delay // Certain problems in modern physics, P.1.- Saratov: *SGU* Press, 1984.-PP.31-37 (*Russian*).

180. *Kats V. A.* Mechanism generating chaos in a backward wave distributed oscillator // *Ibid.*-P.2.-PP.28-33 (*Russian*).

181. *Kats V. A. and Kuznetsov S. P.* Transition to chaos through period-doubling bifurcations in a model generator incorporating delayed feedback. Numerical experiment // *Ibid.*-PP.45-48 (*Russian*).

182. *Kats V. A.* Appearance of chaos and evolution of chaos in a distributed generator with delay // *Izv. VUZov. Radiofizika.-* 1985.-V.28, No 2.-PP.161-176 (*Russian*).

183. *Kats V. A. and Trubetskov D. I.* Appearance of chaos in destroying quasi-periodic

regimes, and in transition through intermittency in a distributed generator with delay // *Pis'ma ZhETF*.- 1984.-V.39, *vyp.* 3.-PP.116-119 *(Russian)*.

184. *Kashchenevsky L. Ya.* Stochastic self-oscillations in solid friction // *Inzhenerno-fizichesky zhurnal.*- 1984.-V.47, No 1.-PP.143-147 *(Russian)*.

185. Quantum Radiphisics, V.I, V.II.- Moscow: *Sov. Radio*, 1972 *(Russian)*.

186. *Kislov V. Ya.* Theoretical analysis of noise-like oscillations in electron-wave systems and self-generators with delay // Lectures on Microwave Electronics and Radiophysics, Bk. 5.- Saratov: *SGU* press, 1980.-PP.25-77 *(Russian)*.

187. *Kislov V. Ya., Zalogin N. N. and Myasin E. A.* Study of stochastic self-oscillatory processes in self-generators with delay // *Radiotekhnika i elektronika.*- 1979.-V.24, No 6.-PP.1118-1130 *(Russian)*.

188. *Kislov V. Ya., Myasin E. A. and Bogdanov E. V.* UHF broad-band vibration generator.- Application No 964513/19-09 dated July 31, 1968 *(Russian)*.

189. *Kislov V. Ya., Myasin E. A. and Zalogin N. N.* On non-linear stochastization of self-oscillations in an electron-wave generator with delayed feedback // *Radiotekhnika i elektronika.*- 1980.-V.25, No 10.-PP.2160-2168 *(Russian)*.

190. *Kiyashko S. V., Pikovsky A. S. and Rabinivich M. I.* Auto-generator of a radio-frequency range with stochastic behaviour // *Radiotekhnika i elektronika.*- 1980.-V.25, No 2.-PP.336-343 *(Russian)*.

191. *Klimontovich Yu. L.* Brownian motion in self-oscillatory systems and in phase transitions // Thermodynamics and kinetics of biological processes.- Moscow: *Nauka*, 1980.-PP.100-118 *(Russian)*.

192. *Klimontovich Yu. L.* Statistical Physics.- Moscow: *Nauka*, 1983.-PP.330-333, 521-553 *(Russian)*.

193. *Klyatskin V. I.* Statistical Description of Dynamical Systems with Fluctuating Parameters.- Moscow: *Nauka*, 1975 *(Russian)*.

194. *Klyatskin V. I.* Stochastic Equations and Waves in Random-Inhomogeneous Media.- Moscow: *Nauka*, 1980 *(Russian)*.

195. *Kogan V. P.* Loss of stability of a fixed point in the neigbourhood of 1:3 frequency ratio resonance.-*VINITI*, No 8068-84 *(Russian)*.

196. *Kogan V. P.* Bifurcaitons of fixed points of two-dimensional mappings in the neighbourhood of 1:3 frequency ratio resonance.-*VINITI*, 1985, No 3882-85 *(Russian)*.

197. *Kogan V. P. and Neimark Yu. I.* Bifurcations of the phase pattern in the neighbourhood of a fixed point under 1:3 frequency ratio // Dynamics of systems (optimization and adaptation).- Gorky: *CGU* Press, 1982.-PP.115-144 *(Russian)*.

198. *Kogan V. P. and Neimark Yu. I.* Bifurcations series and nested structures // Dynamics of systems (stability, self-oscillations and stochasticity).- Gorky: *CGU* Press, 1981.-PP.78-89 *(Russian)*.

199. Oscillatory processes in biological and chemical systems,V.1.- Moscow: *Nauka*, 1967; V. 2.- Pushchino-na-Oke: *ONTI NTsBI AN SSSR*, 1971 *(Russian)*.

200. *Kolesov Yu. S.* Mathematical models of ecology // Studies in stability and the theory of oscillations.- Yaroslavl: *YaGU* Press, 1979.-PP.3-40 *(Russian)*.

201. *Kolesov Yu. S. and Kubyshkin E. P.* Some properties of the solution of differential-difference equations that model the dynamics of populations of insects. Studies in stability and the theory of oscillations. // Radiophysics and Quantum Electronics.- 1985.-V.28, No 4.-PP.285-287.

202. *Kolmogorov A. N.* New metric invariant of transitive dynamical systems and automorphisms on Lebesgue spaces // *DAN SSSR*.- 1958.-V.119.-PP.861-864 *(Russian)*.

203. *Kolmogorov A. N.* On entropy per unit time as a metric invariant of automorphisms // *DAN SSSR*.- 1959.-V.124.-PP.754-755 *(Russian)*.

204. *Kolmogorov A. N., Petrovsky G. I. and Piskunov N. S.* Study of the diffusion equation with substance source, and its application to biological problems // *Byull. MGU. Matematika i mekhanika*.- 1937.-V.1, *vyp.* 6.-PP.1-26; *Voprosy kibernetiki, vyp.* 12.- Moscow: *Izd-vo AN SSSR*, 1975.-P.3; *Kolmogorov A. N.* Selected Works. *Matematika i mekhanika*.- Moscow: *Nauka*, 1985.-PP.221-246 *(Russian)*.

205. *Kornfeld I. P., Sinai Ya. G. and Fomin S. V.* Ergodic Theory.- Moscow: *Nauka*, 1980 *(Russian)*.

206. *Krasnoselsky M. A.* Topological methods for the theory of non-linear integral equations.- Moscow: *Gostekhizdat*, 1959 *(Russian)*.

207. *Krylov N. S.* Foundations of Statictical Physics.- Moscow-Leningrad: *Izd-vo AN SSSR*, 1950 *(Russian)*.

208. *Kryukov B. I. and Seredovich G. I.* On the "strange" behaviour of solutions of the Duffing equation // *DAN SSSR*.- 1981.-V.258, No 2.-PP.311-314 *(Russian)*.

209. *Kuznetsov S. P.* Complicated dynamics of generators with delayed feeback (review) // *Izv. VUZov. Radiofizika*.- 1982.-V.25, No 12.-PP.1410-1428 *(Russian)*.

210. *Kuznetsov S. P.* On the effect of external periodic perturbation on a system illustrating the order-disorder transition via period-doubling bifurcations // *Pis'ma ZhETF*.- 1984.-V.39, *vyp.* 3.-PP.113-116 *(Russian)*.

211. *Kuznetsov S. P. and Kolosova, G. M.* On a certain type of universal behaviour of coupled systems. // Soviet Phys. JETP.- 1984.-V.59.-No 3.-PP.582-585.

212. *Kuznetsov S. P., Perelman A. N. and Trubetskov D. I.* Self-modulation and stochastic models in a travelling-wave klystron with external feedback // *ZhTF*.- 1983.-V.53,*vyp.* 1.-PP.163-166 *(Russian)*.

213. *Kuznetsov S. P. and Pikovsky A. S.* Universality of period doubling bifurcations in a one-dimensional dissipative medium // Radiophys. and Quantum Electronics.- 1985.-V.28, No 3.-PP.205-214.

214. *Kuznetsov Yu. I., Landa P. S., Olkhovoy A. F. and Perminov S. M.* Threshold for synchronization as a characteristic of the chaos-order phase transition.- Preprint / Faculty of Physics of Moscow University.- Moscow, 1984.-No 9/1984 *(Russian)*.

215. *Kuznetsov Yu. I., Landa P. S., Olkhovoy A. F. and Perminov S. M.* Relationship between the amplitude threshold of synchronization and entropy in stochastic self-excited systems // Sov. Phys. Dokl.- 1985.-V.30, No 3.-PP.221-223.

216. *Landa P. S.* Self-oscillations in Lumped Parameter Systems.- Moscow: *Nauka*, 1980 *(Russian)*.

217. *Landa P. S.* Self-oscillations in Continuous Systems.- Moscow: *Nauka*, 1983 *(Russian)*.

218. *Landa P. S.* The effect of noise on transitions to chaos through intermittency // *Vestnik MGU. Ser. 3.-* 1987.-V.28, No 5.-PP.25-27 *(Russian)*.

219. *Landa P. S. and Neimark Yu. I.* Chaotic Oscillations in Non-Equilibrium Physical and Chemical Systems: *Tr. Instituta neftekhimicheskogo sinteza AN SSSR.-* Moscow: *Nauka*, 1987 *(Russian)*.

220. *Landa P. S., Olkhovoy A. F. and Perminov S. M.* Study of stochastic self-oscillations in physical systems with inertial self-excitation // *Izv. VUZov. Radiofizika.-* 1983.-V.26, No 5.-PP.566-572 *(Russian)*.

221. *Landa P. S., Olkhovoy A. F. and Perminov S. M.* Evolution of self-excited vibration spectrum in region exhibiting stochastic behaviour // *Vestnik MGU. Ser. 3.-* 1984.-V.25, No 3.-PP.74-77 *(Russian)*.

222. *Landa P. S. and Perminov S. M.* Routes to stochastic self-oscillations in systems with inertial self-excitation // *Vestnik MGU Ser. 3.-* 1985.-V.26, No 2.-PP.27-30 *(Russian)*.

223. *Landa P. S. and Perminov S. M.* Interaction of periodic and stochastic auto-oscillations // Electronics.- 1985.-V.28, No 4.-PP.285-287.

224. *Landa P. S. and Perminov S. M.* Synchronization of chaotic oscillations in the Mackey-Glass system // *Izv. VUZov. Radiofizika.-* 1987.-V.30, No 3.-PP.437-439 *(Russian)*.

225. *Landa P. S. and Perminov S. M.* Chaotic oscillations of a non-linear oscillator with parametric action // *VINITI.-* Moscow, 1986.-No 7098-86 *(Russian)*.

226. *Landa P. S., Perminov S. M., Shatalova G. G. and Damgov V. N.* Stochastic auto-oscillations in a generator with additional delayed feedback // *Raditekhnika i elektronika.-* 1986.-V.31, No 4.-PP.730-733 *(Russian)*.

227. *Landa P. S. and Stratonovich R. L.* Steady-state distribution of the probabilities for one of the simplest strange attractors // *DAN SSSR.-* 1982.-V.267, No 4.-PP.832-836 *(Russian)*.

228. *Landa P. S. and Stratonovich R. L.* Probability characteristics of stochastic oscillations of an adjustable pendulum // Mech. Solids.- 1984.- No 4.-PP.22-27.

229. *Landa P. S. and Stratonovich R. L.* On the theory of intermittency // *Izv. VUZov.*

Radoifizika.- 1987.-V.30, No 1.-PP.65-69 *(Russian)*.

230. *Landau L. D.* On the problem of turbulence // *DAN SSSR.*- 1944.-V.44, No 8.-PP.339-342 *(Russian)*.

231. *Landau L. D.* The mechanics of continuous media.- Moscow: *Gostekhizdat*, 1953; *Hydrodynamics.*- Moscow: *Nauka*, 1986 *(Russian)*.

232. *Landau L. D. and Lifshits E. M.* Statistical Physics. Translated from the Russian by E. Peierls and R. F. Peierls. Pergamon Press Ltd, London-Paris; Addison-Wesley Publ. Comp., Inc., Reading, Mass., 1958.

233. *Leonov G. A.* On the global stability of the Lorenz system // *PMM.*- 1983.-V.47, *vyp.* 5.-PP.861-863 *(Russian)*.

234. *Leonov N. N.* On the point mapping of a straight line into a straight line // *Izv. VUZov. Radiofizika.*- 1959.-V.2, No 6.-PP.942-956 *(Russian)*.

235. *Leonov N. N.* On the theory of a discontinuous mapping of a straight line onto a straight line // *Izv. VUZov. Radiofizika.*- 1960.-V.3, No 5.-PP.872-886 *(Russian)*.

236. *Leontovich-Andronova E. A. and Neimark Yu. I.* On one singular bifurcation of a dynamical system in the three-dimensional space, related to the vanishing of a periodic solution, in *Sb. dokl. politekhn. in-ta.*- Tashkent, 1964.-PP.132-139 *(Russian)*.

237. *Lichtenberg A. and Lieberman M.* Regular and stochastic motion.- Springer, 1983.

238. *Lukashchuk S. N., Predtechensky A. A., Falkovich G. E. and Chernykh A. I.* On the calculation of dimensions of attractors from experimental data.- Preprint / *SO AN SSSR.*- Novosibirsk, 1985.-No 280 *(Russian)*.

239. *Lyusternik L. A. and Sobolev V. I.* Elements of functional analysis. Translated from the Russian by Anthony E. Labarre, Jr., Herber Izbicki, H. Ward Crawley, Frederick Ungar Publishing Co., New York, 1961.

240. *Lyapunov A. M.* Collected Works., Vols. 1, 2.- Moscow: *Izd-vo AN SSSR*, 1954-1956 *(Russian)*.

241. *Ma Sh.* Modern theory of critical phenomena.- Benjamin, 1976.

242. *Mayer A. G.* Structurally stable transformation of a circle // *Uch. zap. GGU.*- 1939.- *Vyp.* 12.-PP.215-230 *(Russian)*.

243. *Malkin I. G.* The Methods of Lyapunov and Poincaré in the theory of non-linear oscillations.- Moscow: *Gostekhizdat*, 1949 *(Russian)*.

244. *Malkin I. G.* Theory of dynamical stability.- Moscow: *Nauka*, 1966 *(Russian)*.

245. *Mandelshtam L. I.* Lectures on vibrations (1930-1932).- Moscow: *Izd-vo AN SSSR*, 1955 *(Russian)*.

246. *Marsden J. and McCraken M.* The Hopf Bifurcation and its applications.- New York, 1976.

247. *Maslov V. P. and Fedoryuk M. V.* Quasi-classical approximation for the equations of quantum mechanics.- Moscow: *Nauka*, 1976 *(Russian)*.

248. *Matorin I. I., Pikovsky A. S. and Khanin Ya. I.* Multistability and auto-stochasticity in a laser with inertial active medium in periodic modulation of loss // *Kvantovaya elektronika*.- 1984.-V.11, No 10.-PP.2096-2103 *(Russian)*.

249. *Matorin I.I and Khanin Ya. I.* Auto-stochastic phenomena in lasers // Lectures on microwave electronics and radiophysics, Bk. 2.- Saratov: *Izd-vo SGU*, 1983.-PP.23-31 *(Russian)*.

250. *Melnikov V. K.* On the stability of a center for time-periodic perturbations // *Tr. Moskovskogo mat. ob-va*.- 1963.-V.12.-PP.3-52 *(Russian)*.

251. *Migulin V. V., Medvedev V. I., Mustel E. R. and Parygin V. N.* Foundations of Oscillation Theory.- Moscow: *Nauka*, 1978 *(Russian)*.

252. *Mitropolsky Yu. A.* The method of averaging in non-linear mechanics.- Kiev: *Naukova dumka*, 1971 *(Russian)*.

253. *Mikhlin S. G.* Variational methods in mathematical physics. Traslated by T. Boddington. The Macmillian Co., New York, 1964.

254. *Mishenkov F. M., Motorov B. M. and Motorova E. A.* Stability of Natural Thermal Mass Transport.- Moscow: *Atomizdat*, 1976 *(Russian)*.

255. *Moser J.* Lectures on Hamiltonian Systems.- Memoirs of the American Mathematical Society, 1968.

256. *Molchanov A. M.* On the resonant structure of the solar system // Modern problems in celestial mechanics and astrodynamics.- Moscow: *Nauka*, 1973.-PP.32-41 *(Russian)*.

257. *Monin A. S.* On the natute of turbulence // *UFN*.- 1978.-V.125, *vyp.* 1.-PP.97-

122; Hydrodynamic instability // *UFN.*- 1986.-V.150, *vyp.* 1.-PP.61-105 (*Russian*).

258. *Motorova E. A. and Neimark Yu. I.* On the stability of a non-linear distributed model for natural circulation // *Avtomatika i telemekhanika.*- 1974, No 3.-PP.28-36 (*Russian*).

259. *Neimark Yu. I.* Method of point mappings in the theory of non-linear oscillations. I, II, III // *Izv. VUZov. Radiofizika.*- 1958.-No 1.-PP.5-6,41-66; No 2.-PP.95-117; Nos. 5-6.-PP.146-165 (*Russian*).

260. *Neimark Yu. I.* Some cases of the dependece of periodic motions on parameters // *DAN SSSR.*- 1959.-V.129, No 4.-PP.736-739 (*Russian*).

261. *Neimark Yu. I.* The method of point transformations in the theory of non-linear oscillations, *in* Qualitative methods in the theory of non-linear vibrations (Proc. Internat. Sympos. Non-Linear Vibrations, Vol. II, 1961).-PP.268-307 (*Russian*).

262. *Neimark Yu. I.* Some methods for investigating dynamical systems, in *Tr. 2-go Vses. syezda po teor. i prikl. mekh., Vyp.* 2.- Moscow: *Nauka*, 1965.-PP.97-111 (*Russian*).

263. *Neimark Yu. I.* Motions close to doubly asymptotic motion // *DAN SSSR.*- 1967.-V.172, No 5.-PP.1021-1024 (*Russian*).

264. .*Neimark, Yu. I.* Existence and structural stability of invariant manifolds of pointwise mappings // *Izv. VUZov. Radiofizika.*- 1967.-V.10, No 3.-PP.311-320 (*Russian*).

265. *Neimark Yu. I.* Integral manifolds of differential equations.- *Izv. VUZov. Radiofizika.*- 1967.-V.10, No 3.-PP.321-334 (*Russian*).

266. *Neimark Yu. I.* Structure of motions of a dynamical system in the neighbourhood of a homoclinic curve. Fifth Math. Summer School (Uzhgorod, 1967): The Theory of Ordinary Differential Equations and Non-linear Oscillations.- Kiev, 1968.- PP.400-433 (*Russian*).

267. *Neimark Yu. I.* Method of point mappings in the theory of non-linear oscillations // Mechanics in the USSR for the Past 50 Years.- Moscow: *Nauka*, 1968.-PP.137-156 (*Russian*).

268. *Neimark Yu. I.* On the class of dynamical systems: *Tr. Mezhdunar. simp. po nelin. koleb.*, V. 2.- Kiev: *Izd-vo AN SSSR*, 1970.-PP.363-376 *(Russian)*.
269. *Neimark Yu. I.* The method of point transformations in the theory of non-linear oscillations.- Moscow: *Nauka*, 1972 *(Russian)*.
270. *Neimark Yu. I.* Stochasticity of dynamical systems // Theory of oscillations, applied mathematics and cybernetics.- Gorky: *Izd-vo GGU*, 1973.-PP.3-11 *(Russian)*.
271. *Neimark Yu. I.* Stochastic motions of dynamical systems // Dynamics of systems, *vyp.* 7.- Gorky: *Izd-vo GGU*, 1974.-PP.3-50 *(Russian)*.
272. *Neimark Yu. I.* On the appearance of stochasticity in dynamical systems // *Izv. VUZov. Radiofizika.*- 1974.-V.17, No 4.-PP.602-607 *(Russian)*.
273. *Neimark Yu. I.* Synchronization and stochasticity // Phase Synchronization.- Moscow: *Svyaz'*, 1975.-PP.64-82 *(Russian)*.
274. *Neimark Yu. I.* Homoclinic Structures ans Resonances // Dynamics of Systems, *vyp.* 9.- Gorky: *Izd-vo GGU*, 1976.-PP.53-59 *(Russian)*.
275. *NeimarkYu. I.* Symbolic dynamics that is generated by homoclinic structures // Differential equations.- 1976.-No 2.-PP.256-262 *(Russian)*.
276. *Neimark Yu. I.* Singular integral manifolds and appearance of stochasticity in dynamical systems: *Tr. konf. pamyati N. G. Chetaeva* (Irkutsk, 1977).- Novosibirsk: *Nauka*, 1979.-PP.128-134 *(Russian)*.
277. *Neimark Yu. I.* Invariant manifolds and stochastic motions of dynamical systems // Problems in asymptotic non-linear oscillation theory.- Kiev: *Naukova dumka*, 1977.-PP.160-168 *(Russian)*.
278. *Neimark Yu. I.* Theory of non-linear oscillations and stochastic notions of dynamical systems // Dynamics of systems, *vyp.* 12.- Gorky: *Izd-vo GGU*, 1977.-PP.74-95 *(Russian)*.
279. *Neimark Yu. I.* Synchronization and stochasticity: *VII Intern. Konf. über nichtlineare Schwingungen, b. 1, 2.*- Berlin: *Akademie*, 1977.-PP.133-146.
280. *Neimark Yu. I.* Dynamical Systems and Controlled Processes.- Moscow: *Nauka*, 1978 *(Russian)*.
281. *Neimark Yu. I.* On the appearance of stochasticity in dynamical systems: *Tr.*

Vses. konf. Asimptoticheskiye metody v teorii nelineiny kolebaniy.- Kiev: *Naukova dumka,* 1979.-PP.133-141 *(Russian).*

282. *Neimark Yu. I.* Chaotic and stochastic self-oscillations. // Qualitative methods for investigating non-linear differential equations and non-linear oscillations (Uzhgorod, 1980).- Kiev, Institute for Mathematics of AN USSR, 1981.-PP.99-115 *(Russian).*

283. *Neimark Yu. I., Batalova Z. S., Belyakova G. V., Ezhevskaya N. E., Kogan L. V. and Khodova A. F.* Algorithms and Programmes for Numerical Study of Dynamical Systems: *Izd-vo GGU,* 1983 *(Russian).*

284. *Neimark Yu. I., Batalova Z. S., Vasin Yu. G. and Breido M. D.* Pattern Recognition and Medical Diagnostics.- Moscow: *Nauka,* 1972 *(Russian).*

285. *Neimark Yu. I., Kogan V. P. and Gurtovnik A. S.* On smoothness with respect to variables and parameter of invariant surfaces of point mappings in a Banach space // Dynamics of Systems, Optimization and Adaption, *vyp.* 14.- Gorky: *Izd-vo GGU,* 1978.-PP.115-142 *(Russian).*

286. *Neimark Yu. I., Kogan N. Ya. and Savelyev V. P.* Dynamical Models for Control Theory.- Moscow: *Nauka,* 1985 *(Russian).*

287. *Neimark Yu. I. and Fufaev N. A.* The method of point mappings and its application to the study of dynamical systems // *Uspekhi mekhaniki.-* 1980.-V.3, *vyp.* 4.-PP.23-53 *(Russian).*

288. *Neimark Yu. I. and Shilnikov L. P.* A condition for the generation of periodic motions // *DAN SSSR.-* 1965.-V.160, No 6.-PP.1261-1264 *(Russian).*

289. *Neimark Yu. I. and Shilnikov L. P.* A case of generation of periodic motions // *Izv. VUZov.-Radiofizika.-* 1965.-V.8, No 2.-PP.330-340 *(Russian).*

290. *Neishtad L. P.* Bifurcations of the phase portrait of one system arising in the problem of self-oscillation stability loss near 1:4 frequency ratio resonance // *PMM.-* 1978.-V.48, *vyp.* 5.-PP.830-840 *(Russian).*

291. Non-linear Waves // *Izv. VUZov. Radiofizika.-* 1974.-V.17, No 4. *(Russian).*

292. Non-linear Waves // *Izv. VUZov. Radiofizika.-* 1976.-V.19, Nos 5-6 *(Russian).*

293. Non-linear Waves. Stochasticity and turbulence.- Gorky: *IPF AN SSSR,* 1980

(Russian).

294. *Nicolic G. and Prigogine I.* Self-organisation in non-equilibrium systems.- MIT Press, 1971.

295. *Nitecki Z.* Differentiable Dynamics. Cambridge (Mass.). London, the MIT Press, 1971.

296. *Oraevsky A. N.* Masers, lasers and strange attractors // *Kvantovaya elektronika.-* 1981.-V.8, No 1.-PP.130-142 *(Russian).*

297. *Oseledets V. I.* Multiplicative ergodic system. Characteristic Lyapunov exponents of dynamical systems // *Tr. Mosk. mat. ob-va.-* 1968.-V.19.-PP.179-210 *(Russian).*

298. *Patashinsky A. Z. and Pokrovsky V. L.* Fluctuation theory of phase transitions.- Moscow: *Nauka,* 1982 *(Russian).*

299. *Pervozvansky A. A.* Random processes in non-linear control systems. Translated by Scripta Technica. Academic Press, New York-London, 1965.

300. *Perminov S. M., Dobrovolsky V. A., Anikeev G. I. and Genkin M. D.* Mixing process in liquid in a plane channel // *ZhTF.-* 1987.-V.57, No 1.-PP.171-173 *(Russian).*

301. *Pesin Ya. B.* Characteristic Lyapunov exponents, and smooth ergodic theory // *UMN.-* 1977.-V.32, No 4.-PP.55-112 *(Russian).*

302. *Petrovskaya N. V. and Yudovich V. I.* Homoclinic loops of the Salzmann-Lorenz system / *VINITI.-* Moscow, 1980.-No 2380-79 *(Russian).*

303. *Pikovsky A. S.* On the statistical properties of the simplest models of stochastic self-oscillations // *Izv. VUZov. Radiofizika.-* 1980.-V.23, No 7.-PP.883-884 *(Russian).*

304. *Pikovsky A. S.* On the influence of fluctuations on the transition to intermittent stochasticity.- Preprint.- Gorky, 1981.-No 39 *(Russian).*

305. *Pikovsky A. S.* On the behaviour of the strange attractor spectrum at a critical point // *Izv. VUZov. Radiofizika.-* 1982.-V.25, No 7.-PP.846-848 *(Russian).*

306. *Pikovsky A. S.* Chaotic autowaves // *Pis'ma ZhTF.-* 1985.-V.11, vyp. 11.- PP.672-675 *(Russian).*

307. *Pliss V. A.* Non-local problems in the theory of oscillations.- Moscow: *Nauka,*

1964 (*Russian*).

308. *Pokrovsky L. A.* Solution of a system of Lorenz equations in the asymptotic limit of a large Rayleigh number // *Teor. i mat. fizika.*- 1985.-V.62, No 2.-PP.272-290 (*Russian*).

309. *Polak L. S. and Mikhailov A. S.* Self-organization in non-equilibrium physicochemical systems.- Moscow: *Nauka*, 1983 (*Russian*).

310. *Pontryagin L. S., Andronov A. A. and Vitt A. A.* On statistical consideration of dynamical systems // *ZhETF.*- 1933.-V.3, *vyp.* 3.-PP.165-180; Collected Works of A. A. Andronov.- Moscow: *Izd-vo AN SSSR*, 1956.-PP.142-160 (*Russian*).

311. *Poincaré H.* Sur les courbes définies par les équations différentielles.- Paris, 1886.

312. *Poincaré H.* Les méthodes nouvelles - Gauthier-Villars, Paris, 1899.

313. *Rabinovich M. I.* Stochastic self-oscillations in radiophysics and hydrodynamics. Experiments and models // Non-linear waves. Stochasticity and turbulence.- Gorky: *IPF AN SSSR*, 1980.-PP.5-23 (*Russian*).

314. *Rabinovich M. I. and Trubetskov D. I.* Introduction to Theory of Oscillations and Waves.- Moscow *Nauka*, 1984 (*Russian*).

315. *Rabinovich M. I. and Fabrikant A. L.* Stochastic Self-Modulation of Waves in Non-Equilibrium Media // *ZhETF.*- 1979.-V.77, *vyp.* 2(8).-PP.617-629 (*Russian*).

316. *Riemann B.* Gesammelte mathematische Werke und Wissenschaftlicher Nachlass.- Dover Publications, Inc., New York, 1953.

317. *Richtmyer R.* Principles of advanced mathematical physics. New York, Springer, 1978.

318. *Rozenvasser E. N.* Oscillations of Non-linear Systems.- Moscow: *Nauka*, 1969 (*Russian*).

319. *Romanovsky Yu. M., Stepanova N. V. and Chernavsky D. S.* Mathematical biophysics.- Moscow: *Nauka*, 1984 (*Russian*).

320. *Rytov S. M.* Introduction to Statistical Radiophysics.- Moscow: *Nauka*, 1966 (*Russian*).

321. *Rand D.* Topological classification of Lorenz attractors // Mathematical Proceedings of the Cambridge Philosophical Society.-V.83-1978.-PP.451-460.

322. *Sbitnev V. I.* Stochasticity in a systems of two coupled vibrators // Non-linear Waves. Stochasticity and Turbulence.- Gorky: *IPF AN SSSR*, 1980.-PP.46-56 *(Russian)*.

323. *Sbitnev V. I.* Stochasticity in a system of coupled oscillators: *Tr. IX Mezhd. konf. po nelin. koleb.*, V. 3.- Kiev: *Naukova dumka*, 1984.-PP.477-479 *(Russian)*.

324. *Sinay Ya. G.* On the concept of entropy for a dynamical system // *DAN SSSR.-* 1959.-V.124, No 4.-PP.768-771 *(Russian)*.

325. *Sinay Ya. G.* Markov partitions and Y-diffeomorphisms // *Funkts. analiz i yego prilozheniya.-* 1968.-V.2, No 1.-PP.64-89 *(Russian)*.

326. *Sinay Ya. G.* Dynamical systems with elastic reflections // *UMN.-* 1970.-V.25, No 2.-PP.141-192 *(Russian)*.

327. *Sinay Ya. G.* Theory of phase transitions.- Moscow: *Nauka*, 1980 *(Russian)*.

328. *Sinay Ya. G.* Randomness of the non-random // *Priroda.-* 1981.-No 3.-PP.72-80 *(Russian)*.

329. Synergetics / Ed. B. B. Kadomtsev.- Moscow: *Mir*, 1984 *(Russian)*.

330. *Smale S.* A structurally stable differentiable homeomorphism with an infinite number of periodic points: Qualitative methods in the theory of non-linear vibrations, Vol II, 1961.-PP.365-366.

331. *Smale S.* Structurally stable systems are not dense // Amer. J. Math.- 1966.-V.88.-PP.491-496.

332. *Smale S.* Differentiable dynamical systems // *UMN.-* 1970.-V.25, No 1.-PP.113-185 *(Russian)*.

333. *Smoes M.-L.* Chemical waves in the oscillatory Zhabotinskii systems. Transition from temporal to spatio-temporal organization / Synergetics: Far from Equilibrium.- Berlin, Springer, 1979.-PP.80-96.

334. *Sobolev S. A.* On mixed problems for partial differential equations with two independent variables // *DAN SSSR.-* 1958.-V.122, No 4.-PP.555-559 *(Russian)*.

335. *Solntsev V. A. and Andreyevskaya T. M.* Conditions for amplitude self-modulation in a self-generator with delay // *Radiotekhnika i elektronika.-* 1983.-V.28, No 3.-

PP.561-568 (*Russian*).

336. *Sonechkin D. M.* Stochasticity in models for general atmospheric circulation.- Leningrad: *Gidrometeoizdat*, 1984 (*Russian*).

337. *Starzhinsky V. M.* Applied methods for non-linear oscillations.- Moscow: *Nauka*, 1977 (*Russian*).

338. *Stanley H.* Introduction to phase transitions and critical phenomena.- Oxford: Clarendon Press, 1971.

339. *Stanley H., Coniglio A., Klein W., Nakamishi H., Render S., Reynolds P. and Shilifer C.* Critical Phenomena-Past, Present and Future.- Dynamics of Synergetic Systems.- Springer, Berlin, 1980.

340. *Stoker J.* Non-linear vibrations in mechanical and electrical systems.- New York: Interscience, 1950.

341. Strange Attractors. Eds. Ya. G. Sinay and L. P. Shilnikov.- Moscow: *Mir, 1981* (*Russian*).

342. *Stratonovich R. L.* Selected problems in the theory of fluctuations in radioengineering.- Moscow: *Sov. Radio*, 1961 (*Russian*).

343. *Stratonovich R. L.* Correlators of processes in elementary systems with strange attractors // *DAN SSSR.*- 1982.-V.267, No 2.-PP.355-359 (*Russian*).

344. *Stratonovich R. L. and Nikolayevsky E. S.* On the bifurcations sequence of strange attractor motion in splitting them in the case of simple systems // *DAN SSSR.*- 1984.-V.276, No 2.-PP.363-366 (*Russian*).

345. *Strelkov S. P.* Introduction to the theory of vibrations.- Moscow: *Nauka*, 1964 (*Russian*).

346. *Rayleigh J.* The Theory of Sound.- New York: Dover Publ., 1945.

347. *Teodorchik K. F.* Self-oscillatory Systems.- Moscow: *Gostekhizdat*, 1952 (*Russian*).

348. *Timoshenko S. P.* Vibration problems in engineering.- New York: Wiley, 1974.

349. *Tondl A.* Non-linear oscillations in mechanical systems.- Moscow: *Mir*, 1973 (*Russian*).

350. *Tondl A.* Self-excited Vibrations.- Bechovice, 1970.

351. *Truzhenikova V. N.* Conditions for generating a periodic motion from an

equilibrium state // *Izv. VUZov. Radiofizika.*- 1962.-V.5, No 2 *(Russian)*.

352. *Whitham G. B.* Linear and non-linear waves.- New York: Wiley, 1977.

353. *Feynman R.* The Feynman Lectures on Physics.- Reading, Mass., 1964.

354. *Fufaev N. A.* Magnetic interrupter theory: In memory of A. A. Andronov.- Moscow: *Izd-vo AN SSSR*, 1955.-PP.334-382 *(Russian)*.

355. *Haken H.* Transition phenomena in stochastic non-linear systems in physics, chemistry and biology. Proceedings of the Workshop, Bielefeld, October 5-11, 1980. Springer Verlag 1981.-PP.12-19.

356. *Haken H.* Synergetics.- Berlin: Springer, 1978; Advanced Synergetics.- Berlin: Springer, 1983.

357. *Khandokin P. A. and Khanin Ya. I.* Auto-stochastic regime of generating a solid-state ring laser with low-frequency modulation of loss // *Kvantovaya elektronika.*- 1984.-V.11, No 7.-PP.1483-1487 *(Russian)*.

358. *Kharkevich A. A.* Self-Oscillations.- Moscow: *Gostekhizdat*, 1953 *(Russian)*.

359. *Hayashi Ch.* Non-linear Oscillations in Phisical Systems.- New York: McGraw-Hill, 1964.

360. *Hale J.* Oscillations in non-linear systems.- New York: McGraw-Hill, 1963.

361. *Khentov A. A.* Satellite synchronization // Dynamics of systems.- Gorky: *Izd-vo GGU*, 1974.-PP.51-102 *(Russian)*.

362. *Khinchine A. Ya.* Correlation theory of stationary stochastic processes // *UMN.*- 1938.- *Vyp.* 5.-PP.42-51 *(Russian)*.

363. *Hopf E.* Statistik der geodätischen Linkien in Mannigfaltigkeiten nerativer Krümmung // *Ber. Verl. Sächs. Akad. Wiss.*- 1939.-V.91.-PP.261-304.

364. *Khotuntsev Yu. L. and Tamarchak D. Ya.* Synchronizad generators and semiconductor autodyne receivers.- Moscow: *Radio i svyaz'*, 1982 *(Russian)*.

365. *Chetaev A. N.* Neuron Networks and Markov Chains.- Moscow: *Nauka*, 1985 *(Russian)*.

366. *Chirikov B. V.* Stochastic Wave Processes.- Preprint / *NIRFI.*- Gorky, 1973.- No 42 *(Russian)*.

367. *Chumakov G. A. and Slin'ko M. G.* Kinetic turbulence (chaos) of the rate of metal-catalyzed reaction of hydrogen with oxygen // *DAN SSSR.*- 1982.-V.266, No 5.-

PP.1194-1198 *(Russian)*.

368. *Sharkovsky A. N.* Co-existence of cycles of a continuous mapping of the line into itself // *Ukr. mat. zhurnal.*- 1964.-V.26, No 1.-PP.61-71 *(Russian)*.

369. *Shilnikov L. P.* Some cases of generation of periodic motions from singular trajectories // *Math. sb.*, 1963.-V.61(104), No 4.-PP.433-446 *(Russian)*.

370. *Shilnikov L. P.* A case of the existence of a denumerable set of periodic motions // *DAN SSSR.*- 1965.-V.160, No 3.-PP.558-561 *(Russian)*.

371. *Shilnikov L. P.* Generation of a periodic motion from the trajectory going from the state of equilibrium of the saddle-saddle type into the same state // *DAN SSSR.*- 1966.-V.170, No 1.-PP.49-52 *(Russian)*.

372. *Shilnikov L. P.* Existence of a countable set of periodic motions in a neighbourhood of a homoclinic curve // *DAN SSSR.*- 1967.-V.172, No 2.-PP.298-301 *(Russian)*.

373. *Shilnikov L. P.* On a problem of Poincaré-Birkhoff // *Mat. sb.*- 1967.-V.74(116), No 3.-PP.378-397 *(Russian)*.

374. *Shilnikov L. P.* On the problem of the structure of the neighbourhood of a homoclinic tube of an invariant torus // *DAN SSSR.*- 1968.-V.180, No 2.-PP.286-289 *(Russian)*.

375. *Yudovich V. I.* Asymptotic behaviour of the limit cycles of a Lorenz system for large Rayleigh numbers / *VINITI.*- Moscow, 1979.- *VINITI*, No 2611-78 *(Russian)*.

376. *Abraham N.B.* A New Focus on Laser Instabilities and Chaos // Laser Focus.- 1983, No 5.-PP.73-81.

377. *Abraham N.B., Dangoisse D., Glorieux P. and Mandel P.* Observation of Undamped Pulsations in a Low-Pressure, Far-Infrared Laser and Comparison with a Simple Theoretical Model // J. Opt. Soc. Amer. B.- 1985.-V.2, No.1.-PP.23-34.

378. *Abraham N.B., Lugiato L.A., Mandel P., Narducci L.M. and Bandy D.K.* Steady-State and Unstable Behavior of a Single-Mode Inhomogeneously Broadened Laser.- *Ibid.* PP.35-46.

379. *Aceves A., Adachihara H., Jones C., Lerman J.C., McLaughlin D.W., Moloney*

J.W. and Newell A.C. Chaos and Coherent Structures in Partial Differential Equations // Physica D.- 1986.-V.18, No 1-3.-PP.85-112.

380. *Adler R.L., Konheim A.G. and McAndrew M.H.* Topological Entropy // Trans. Amer. Math. Soc.- 1965.-V.114, No 2.-PP.309-319.

381. *Afraimovich V.S. and Sil'nikov L.P.* On Strange Attractors and Quasi-Attractors // Nonlinear Dynamics and Turbulence.- Boston - London - Melbourne, 1983.-PP.1-34.

382. *Alekseev V.M. and Yakobson M.V.* Symbolic Dynamics and Hyperbolic Dynamic System // Phys. Rep.- 1981.-V.75, No 5.-PP.290-325.

383. *Alekseev V.M. and Kornilovsky A.N.* Ecosystems Stochasticity Model // Ecological Modelling.- 1985.-V.28.-PP.217-229.

384. *Andrade R. and Rauh A.* Nonlinear Stability Analysis of the Lorenz Model Lyapunov's Direct Method // Z.Phys.- 1983.-B.50, No 2.-S.151-159.

385. *Arecchi F.T., Badii R. and Politi A.* Low-frequency Phenomena in Dynamical Systems with Many Attractors // Phys. Rev. A.- 1984.-V.29, No 2.-PP.1006-1009.

386. *Arecchi F.T. and Lisi F.* Hopping Mechanism Generating 1/f Noise in Nonlinear Systems // Phys. Rev. Lett.- 1982.-V.49, No 2.-PP.94-98.

387. *Arecchi F.T., Meucci R., Puccioni G. and Tredicce J.* Experimental Evidence of Subharmonic Bifurcations, Multistability and Turbulence in a Q-switched Gas Laser // Phys. Rev. Lett.- 1982.-V.49, No 17.-PP.1217-1220.

388. *Armstrong J.A., Bloembergen N., Ducuing J. and Pershan P.S.* Interactions between Light Waves in a Nonlinear Dielectric // Phys. Rev.- 1962.-V.127, No 6.-PP.1918-1939.

389. *Arneodo A., Coullet P.H. and Spiegel E.A.* Cascade of Period Doublings of Tori // Phys. Lett.- 1983.-V.94A, No 1.-PP.1-4.

390. *Aronson D.G., Chory M.A., Hall G.F. and McGehee R.P.* Biffurcations from an Invariant Circle for Two-Parameter Families of Maps of the Plane: a Computer-Assisted Study // Comm. Math. Phys.- 1982.-V.83, No 3.-PP.303-354.

391. *Babitzky V.I. and Landa P.S.* Auto-oscillation Systems with Inertial Self-Excitation // ZAMM.- 1984.-B.64, No 8.-S.329-339.

392. *Babitzky V.I., Landa P.S., Olhovoy A.F. and Perminov S.M.* Stochastical

Behavior of Auto-Oscillation Systems with Inertial Self-Excitation // ZAMM.- 1986.-B.66, No 2.-S.73-81.

393. *Bar-Eli K.* On the Stability of Coupled Chemical Oscillators // Physica D.- 1985.- V.14, No 2.-PP.242-252.

394. *Beiersdorfer P. and Wersinger J.M.* Topology of the Invariant Manifolds of a Period-doubling Attractor for Some Forced Nonlinear Oscillators // Phys. Lett.- 1983.-V.96A, No 6.-PP.269-272.

395. *Belyaev Yu.N., Monakhov A.A., Scherbakov S.A. and Yavorskaya I.M.* Some Routes to Turbulence in Spherical Couette Flow // Laminar - Turbulent Transition. IUTAM Symp. Novosibirsk 1984.- Berlin - Heidelberg: Springer, 1985.-PP.669-676.

396. *Benett D., Bishop A.R. and Trullinger S.E.* Coherence and Chaos in the Driven, Damped Sine-Gordon Chain // Z.Phys. B.- 1982.-B.47, No.3 .-S.265-277.

397. *Benettin G., Froeschle C. and Scheidecker J.P.* Kolmogorov Entropy of a Dynamical System with an Increasing Number of Degrees of Freedom // Phys. Rev. A.- 1979.-V.19, No 6.-PP.2454-2460.

398. *Benettin G. and Galgani L.* Lyapunov Characteristic Exponents and Stochasticity // Intrinsic Stochasticity in Plasma /Ed. by G. Laval, D. Gressilon.- Orsay: Les éditions de physique courtaboeuf, 1979.-PP.93-114.

399. *Benettin G., Galgani L., Giorgilli A. and Strelcyn J.M.* Lyapunov Characteristic Exponents for Smooth Dynamical Systems and for Hamiltonian Systems; a Method for Computing All of Them. P. 1, 2 // Meccanica.- 1980.-V.15, No 1.-PP.9-20, 21-30.

400. *Benettin G., Galgani L. and Strelcyn J.M.* Kolmogorov Entropy and Numerical Experiments // Phys. Rev. A.- 1976.-V.14, No 6.-PP.2338-2342.

401. *Ben-Jacob E., Goldhirsch I., Imry Y. and Fishman S.* Intermittent Chaos in Josephson Junctions // Phys. Rev. Lett.- 1982.-V.49, No 22.-PP.1599-1602.

402. *Berman G.P., Iomin A.M. and Zaslavsky G.M.* Method of Quasiclassical Approximation for C-number Projection in Coherent States Basis // Physica D.- 1981.-V.4D, No 1, 2.-PP.113-121.

403. *Berman G.P. and Kolovsky A.R.* Correlation Function Behavior in Quantum

Systems which are Classically Chaotic // Physica D.- 1983.-V.8D, No 1, 2.- PP.117-141.

404. *Berman G.P. and Kolovsky A.R.* Dynamics of Classically Chaotic Quantum Systems in Wigner Representation // Physica D.- 1985.-V.17D, No 2.-PP.183-197.

405. *Berman G.P. and Zaslavsky G.M.* Quantum Mapping and the Problem of Stochasticity in Quantum Systems // Physica A.- 1982.-V.111A, No 1.-PP.17-44.

406. *Bishop A.R., Fesser K., Lomdahl P.S. and Trullinger S.E.* Influence of Solitons in the Initial State on Chaos in the Driven Damped Sine-Gordon System // Physica D.- 1983.-V.7D, No 1-3.-PP.259-279.

407. *Bishop A.R., and Lomdahl P.S.* Nonlinear Dynamics in Driven, Damped Sine-Gordon Systems // Physica D.- 1986.-V.18D, No 1-3.-PP.54-66.

408. *Bohr T., Bak, P. and Jensen M.H.* Transition to Chaos by Interaction of Resonances in Dissipative Systems. II Josephson Junctions, Charge-Density Waves, and Standard Maps // Phys. Rev. A.- 1984.-V.30, No 4.-PP.1970-1981.

409. *Boldrighini C. and Franceschini V.A.* Five-Dimensional Truncation of the Plane Incompressible Navier-Stokes Equations // Comm. Math. Phys.- 1979.-V.64, No 2.-PP.159-170.

410. *Born M.* Vorhersagbarkeit in der klassischen Mechanik // Z.Physik.- 1958.-B.153, No 3.-S.372-388.

411. *Brandstäter A., Swift J., Swinney H.L. and Wolf A.* Low-Dimensional Chaos in a Hydrodynamic System // Phys. Rev. Lett.- 1983.-V.51, No 16.-PP.1442-1445.

412. *Casati G., Chirikov B.V., Israilev F.M. and Ford J.* Stochastic Behavior of a Quantum Pendulum under a Periodic Perturbation // Lect. Notes in Phys.- Berlin: Springer.- 1979.-V.93.-PP.334-352.

413. Chaos and Statistical Methods /Ed. Y. Kuramoto.- Berlin: Springer, 1984.

414. *Chandrasekhar S.* Hydrodynamic and Hydromagnetic Stability.- Oxford: Clarendon Press, 1961.

415. *Chay T.R.* Chaos in a Three-Variable Model of an Excitable Cell // Physica D.- 1985.-V.16D, No 2.-PP.233-242.

416. *Chenciner A. and Iooss G.* Bifurcations de Tores Invariants // Arch. Rat. Mech.

Anal.- 1979.-V.69, No 2.-PP.109-198.

417. *Chenciner A. and Iooss G.* Persistance et Bifurcation de Tores Invariants // Arch. Rat. Mech. Anal.- 1979.-V.71, No 4.-PP.301-306.

418. *Chirkov B.V.* A Universal Instability of Many-Dimensional Oscillator Systems // Phys. Rep.- 1979.-V.52, No 5.-PP.263-379.

419. *Chui S. and Ma K.B.* Nature of Some Chaotic States for Duffing's Equation // Phys. Rev. A.- 1982.-V.26, No 4.-PP.2262-2265.

420. *Collet P., Eckmann J.P. and Koch H.* Period Doubling Bifurcations for Families of Maps on R^n // J. Stat. Phys.- 1981.-V.25, No 1.-PP.1-14.

421. *Collet P., Eckmann J.P. and Lanford O.E.* Universal Properties of Maps on an Interval // Comm. Math. Phys.- 1980.-V.76, No 3.-PP.211-254.

422. *Crawford J.D. and Cary J.R.* Decay of Correlations in a Chaotic Measure-Preserving Transformation // Physica D.- 1983.-V.6D, No 2.-PP.223-232.

423. *Croquette V. and Poiton C.* Cascade of Period Doubling Bifurcations and Large Stochasticity in the Motions of a Compass // J. Physique.- 1981.-V.42, No 24.-PP.537-539.

424. *Crutchfield J.P., Farmer J.D. and Huberman B.A.* Fluctuations and Simple Chaotic Dynamics // Phys. Rep.- 1982.-V.92, No 2.-PP.45-82.

425. *Crutchfield J.P., Farmer J.D., Packard N., Shaw R., Jones G. and Donnely R.J.* Power Spectral Analysis of a Dynamical System // Phys. Lett.- 1980.-V.76A, No 1.-PP.1-4.

426. *Crutchfield J.P. and Packard N.H.* Symbolic Dynamics of One-Dimensional Maps: Entropies, Finite Precision and Noise // Int. J. Theor. Phys.- 1982.-V.21, No 6-7.-PP.433-446.

427. *Crutchfield J.P. and Packard N.H.* Symbolic Dynamics of Noisy Chaos // Physica D.- 1983.-V.7D, No 1-3.-PP.201-223.

428. *Dadio H.* Resonance and Intermittent Transition from Torus to Chaos in Periodically Forced System near Intermittency Threshold // Progr. Theor. Phys. Japan.- 1983.-V.70, No 3.-PP.879-882.

429. *Davidson R.C.* Methods in Nonlinear Plasma Theory, Ch. 6.- N. Y.: Academic, 1972.

430. *Derrida B., Gervois A. and Pomeau Y.* Universal Metric Properties of Bifurcations of Endomorphisms // J. Phys. A.- 1979.-V.12, No 3.-PP.269-296.

431. *Destine M.W., Gibbs H.M., Hopf F.A. and Kaplan D.L.* Bifurcation Gap in a Hybrid Optically Bistable System // Phys. Rev. A.-1982.-V.26, No 6.-PP.3720-3722.

432. *D'Humieres D., Beasley M.R., Huberman B.A. and Libchaber A.* Chaotic States and Routes to Chaos in the Forced Pendulum // Phys. Rev. A.-1982.-V.26, No 6.-PP.3483-3496.

433. *Eckmann J.P.* Roads to Turbulence in Dissipative Dynamical Systems // Rev. Mod. Phys.- 1981.-V.53, No 4, P. 1.-PP.643-654.

434. *Eckmann J.P. and Ruelle D.* Ergodic Theory of Chaos and Strange Attractors // Rev. Mod. Phys.- 1985.-V.57, No 3, P. 1.-PP.617-656.

435. *Eckmann J.P., Thomas L. and Wittwer P.* Intermittency in the Presence of Noise // J. Phys. A.- 1981.-V.14A, No 12.-PP.3153-3168.

436. *Elgin J.N. and Forster D.* Mechanism for Chaos in the Duffing Equation // Phys. Lett.- 1983.-V.94A, No 5.-PP.195-197.

437. *Epstein I.R.* Oscillations and Chaos in Chemical Systems // Physica D.- 1983.-V.7D.-PP.47-56.

438. *Farmer J.D.* Spectral Broadening of Period-Doubling Bifurcation Sequences // Phys. Rev. Lett.- 1981.-V.47, No 3.-PP.179-182.

439. *Farmer J.D.* Chaotic Attractor of an Infinite-Dimensional Dynamical System //Physica D.- 1982.-V.4D.-PP.366-393.

440. *Farmer J.D.* Information Dimension and Probabilistic Structure of Chaos // Z. Naturforsch - 1982.-B.37a, H.11.-S.1304-1325.

441. *Farmer J.D., Crutchfield J., Frochling H., Packard N. and Shaw R.* Power Spectra and Mixing Properties of Strange Attractors // Ann.- N. Y. Acad. Sci.- 1980.- No 357.-PP.453-472.

442. *Farmer J.D. , Ott E. and Yorke J.A.* The Dimension of Chaotic Attractors // Physica D.- 1983.-V.7, No 1-3.-PP.153-180.

443. *Fauve S. and Libchaber A.* Rayleigh - Bernard Experiment in a Low Prandtl

Number Fluid, Mercury // Proc. Intern. Symp. on Synergetics at Schloss Elmau, Bavaria, 1981. /Ed. H. Haken.- Berlin - Heidelberg - N.Y.: Springer.- 1981.-PP.25-35.

444. *Feigenbaum M.J.* Quantitative Universality for a Class of Nonlinear Transformations // J. Stat. Phys.- 1978.-V.19, No 1.-PP.25-52.

445. *Feigenbaum M.J.* The Universal Metric Properties of Nonlinear Transformations // J. Stat. Phys.- 1979.-V.21, No 6.-PP.669-706.

446. *Feigenbaum M.J.* The Onset Spectrum of Turbulence // Phys. Lett.- 1979.-V.74A, No 6.-PP.375-378.

447. *Feigenbaum M.J.* The Transition to Aperiodic Behavior in Turbulent Systems // Comm. Math. Phys.- 1980.-V.77, No 1.-PP.65-86.

448. *Feigenbaum M.J.* Universal Behavior in Nonlinear Systems // Los Alamos Sci.- 1980.-V.1, No 1.-PP.4-27; Physica D.- 1983.-V.7D, No 1-3.-PP.16-39.

449. *Feigenbaum M.J., Kadanoff L.P. and Shenker S.J.* Quasiperiodicity in Dissipative Systems: a Renormalization Group Analysis // Physica D.- 1982.-V.5D, No 2, 3.-PP.370-386.

450. *Fermi E., Pasta I. and Ulam S.* Studies of Non-Linear Problems.- Los Alamos Scient. Lab. Report LA - 1940, 1955.

451. *Field R.J., Körös E. and Noyes R.M.* Oscillations in Chemical Systems. II. Thorough Analysis of Temporal Oscillation in the Bromate - Cerium - Malonic Acid System // J. Amer. Soc.-1972.-V.94, No 25.-PP.8649-8664.

452. *Field R.J., Noyes R.M.* Oscillations in Chemical Reaction. [V. Limit Cycle Behavior in a Model of a Real Chemical Reaction] // J. Chem. Phys.- 1974.-V. 60, No 5.-PP.1877-1884.

453. *Fife P.C.* Mathematical Aspects of Reacting and Diffusing Systems // Lect. Notes in Biomath. No 29.- Berlin: Springer, 1979.

454. *Firth W.J. and Wright E.M.* Oscillation and Chaos in a Fabry - Perot Bistable Cavity with Gaussian Input Beam // Phys. Lett.- 1982.-V.92, No 5.-PP.211-216.

455. *Fischer R.* The advance of advantageous genes // Ann. of Eugenics.- 1937.-V.7.-PP.355-369.

456. *Ford J. and Lanford G.H.* Stochastic Behavior of Resonant Nearly Linear

Oscillator Systems in the Limit of Zero Nonlinear Coupling // Phys. Rev.- 1970.- V.1A, No 1.-PP.59-70.

457. *Fowler A.C., Gibbon J.D. and McGuinness M.J.* The Complex Lorentz Equations // Physica D.- 1982.-V.4D, No 1.-PP.139-163.

458. *Fowler A.C. and McGuinness M.J.* Hysteresis in the Lorentz Equations // Phys. Lett.- 1982.-V.92A, No 3.-PP.103-106.

459. *Franaszek M.* Effect of Random Noise on the Deterministic Chaos in a Dissipative System // Phys. Lett.- 1984.-V.105A, No 8.-PP.383-386.

460. *Franceschini V.* A Feigenbaum Sequence of Bifurcations in the Lorenz Model // J. Stat. Phys.- 1980.-V.22, No 3.-PP.397-406.

461. *Franceschini V.* Bifurcations of Tori and Phase Locking in a Dissipative System of Differential Equations // Physica D.- 1983.-V.6D, No 3.-PP.285-304.

462. *Franceschini V. and Tebaldi C.* Sequences of Infinite Bifurcations and Turbulence in a Five-Mode Truncation // J. Stat. Phys.- 1979.-V.21, No 6.-PP.707-726.

463. *Franceschini V. and Tebaldi C.* Breaking and Disappearance of Tori //Comm. Math. Phys.- 1984.-V.94, No 2.-PP.317-329.

464. *Frosch H. and Büttner H.* Two Coupled Impact Oscillators // Z. Phys. B: Condenced Matter.- 1985.-B.58, No 4.-S.323-328.

465. *Gibbs H.M., Hopf F.A., Kaplan D.L. and Shoemaker R.L.* Observation of Chaos in Optical Bistability // Phys. Rev. Lett.- 1981.-V.46, No 7.-PP.474-477.

466. *Glass L., Guevara M.R. and Shrier A.* Bifurcations and Chaos in a Periodically Stimulated Cardiac Oscillator // Physica D.- 1983.-V.7D, No 1-3.-PP.89-101.

467. *Goldberger A.L., Bhargava V., West B.J. and Mandell A.J.* Nonlinear Dynamics of Heartbeat // Physica D.- 1985.-V.17D, No 2.-PP.207-214.

468. *Goldschmidt V.M.* Turbulent Transport: Some General Comments // Lect. Notes in Phys. V.76.- Berlin: Springer, 1978.-PP.1-21.

469. *Gollub J.P., Benson S.V. and Steinman J.A.* Subharmonic Route to Turbulent Convection // Ann. N. Y. Acad. Sci- 1980.-No 357.-PP.22-27.

470. *Gollub J.P. and Meyer C.W.* Symmetry-Breaking Instabilities on a Fluid Surface // Physica D.- 1983.-V.6D, No 3.-PP.337-346.

471. *Gorman M. and Widmann P.J.* Nonlinear Dynamics of a Convection Loop: a

Quantitative Comparison of Experiment with Theory // Physica D.- 1986.-V.19D, No 2.-PP.255-267.

472. *Grasman J., Nijmeijer H. and Veling E.J.M.* Singular Perturbations and a Mapping on an Interval for the Forced Van-der-Pol Relaxation Oscillator // Physica D.- 1984.-V.13D, No 1, 2.-PP.195-210.

473. *Grassberger P.* On the Hausdorff Dimension of Fractal Attractors // J. Stat. Phys.- 1981.-V.26, No 1.-PP.173-179.

474. *Grassberger P.* Generalized Dimension of Strange Attractors // Phys. Lett.- 1983.-V.97A, No 6.-PP.227-231.

475. *Grassberger P., Ben-Mizrachi A. and Procaccia I.* Characterization of Experimental (Noisy) Strange Attractors // Phys. Rev. A.-1984.-V.29, No 2.-PP.975-977.

476. *Grassberger P. and Procaccia I.* Characterization of Strange Attractors // Phys. Rev. Lett.- 1983.-V.50, No 5.-PP.346-349.

477. *Grassberger P. and Procaccia I.* Measuring the Strangeness of Strange Attractors // Physica D.- 1983.-V.9D, No 1.-PP.189-208.

478. *Grassberger P. and Procaccia I.* Estimation of the Komogorov Entropy from a Chaotic Signal // Phys. Rev. A.- 1983.-V.28, No 4.-PP.2591-2593.

479. *Grassberger P. and Procaccia I.* Dimensions and Entropies of Strange Attractors from a Fluctuating Dynamics Approach // Physica D.- 1984.-V.13, No 1, 2.-PP.34-54.

480. *Greenside H.S., Wolf A., Swift J. and Pignotaro T.* Impracticality of a Box-Counting Algorithm for Calculating the Dimensionality of Strange Attractors // Phys. Rev. A.- 1982.-V.25, No 6.-PP.3453-3456.

481. *Grossman S. and Thomae S.* Invariant Distributions and Stationary Correlation Functions of One-Dimensional Discrete Processes // Z. Naturfosch.- 1977.-B.32a, H.12.-S.1353-1363.

482. *Gubankov V.N., Konstantinyan K.I., Kosheletz V.P. and Ovsyannikov G.A.* Chaos in Josephson Tunnel Junctions // IEEE Trans. Mag.- 1983.-V.19, No 3.-PP.637-639.

483. *Guckenheimer J.M. and Holmes Ph.* Nonlinear Oscillations, Dynamical Systems,

and Bifurcation of Vector Fields.- N.Y. - Berlin - Heidelberg - Tokyo: Springer, 1983.

484. *Gumowski I. and Mira C.* Recurrences and Discrete Dynamic Systems // Lect. Notes in Math. No 809.- Berlin: Springer, 1980.

485. *Hadamard J.* // Bull. Soc. Math. France.- 1901.-V.29.-PP.224-228.

486. *Haken H.* Cooperative Phenomena in Systems Far from Thermal Equilibrium and in Nonphysical Systems // Rev. Mod. Phys.- 1975.-V.47, No 1.-PP.67-121.

487. *Haken H.* Synergetics - a Field Beyond Irreversible Thermodynamics // Lect. Notes in Phys. V.84.- Berlin: Springer, 1978.-PP.140-168.

488. *Halas H.J., Liu S.N. and Abraham N.B.* Route to Mode Locking in a Three-Mode He-Ne 3.39 μm Laser Including Chaos in the Secondary Beat Frequency // Phys. Rev. A.- 1983.-V.28A, No 5.-PP.2915-2920.

489. *Hale J.* Infinite Dimensional Dynamical Systems // Lect. Notes in Math.-V.1007.- Berlin: Springer, 1983.-PP.379-400.

490. *Hanson J.D., Ott E. and Antonsen T.M.* Influence of Finite Wavelength on the Quantum Kicked Rotator in the Semiclassical Regime // Phys. Rev. A.- 1984.- V.29A, No 2.-PP.819-825.

491. *Hao Bai-lin and Zhang Shu-yu.* Subharmonic Stroboscopy as a Method to Study Period-Doubling Bifurcations // Phys. Lett.- 1982.-V.87A, No 6.-PP.267-270.

492. *Hauck R., Hollinger F. and Weber H.* Chaotic and Periodic Emission of High Power Solid State Lasers // Opt. Comm.- 1983.-V.47, No 2.-PP.141-145.

493. *Hausdorff F.* Dimension und Äußeres Maß // Math. Ann.- 1918.-B.79, H.2.- S.157-179.

494. *Hauser P.R., Tsallis C. and Curado M.F.* Criticality of the Routes to Chaos of the $1 - \alpha |x|^z$ Map // Phys. Rev. A.- 1984.-V.30A, No 4.-PP.2074-2079.

495. *Hayashi C.* The Method of Mapping with Reference to the Doubly Asymptotic Structure of Invariant Curves // Int. J. Non.-Linear Mech.- 1980.-V.15, No 415.- PP.341-348.

496. *Hayashi C., Abe M., Oshima K. and Kawakami H.* The Method of Mapping as Applied to the Solution for Certain Types of Nonlinear Differential Equations // *Trudy IX Mezhdun. konferentsii po nelineynym kolebaniyam* (Kiev, 1981), V.1.-

Kiev, Naukova dumka, 1984.-PP.40-44.

497. *Hayashi C. and Kawakami H.* Bifurcations and the Generation of Chaotic States in the Solutions of Nonlinear Differential Equations // Teor. i prikl. mekh. 4 Nats. kongr. Varna, 1981. Dokl., V.I.- Sofia, 1981.-PP.537-542.

498. *Hedlund G.A.* The Dynamics of Geodesic Flows // Bull. Amer. Math. Soc.- 1939.-V.45, No 4.-PP.241-260.

499. *Helleman R.H.G.* Self-Generated Chaotic Behavior in Nonlinear Mechanics // Fundamental Problems in Statistical Mechanics. V.5. /Ed. E.G.D. Cohen.- Amsterdam. - N.Y.: North-Holland, 1980.-PP.165-233.

500. *Henon M.* A Two-Dimensional Mapping with a Strange Attractor // Comm. Math. Phys.- 1976.-V.50, No 1.-PP.69-77.

501. *Henon M. and Heiles C.* The Applicability of the Third Integral of Motion; Some Numerical Experiments // Astron. J.- 1964.-V.69, No 1.-PP.73-79.

502. *Hentschel H.G.E. and Proccacia I.* The Infinite Number of Generalized Dimensions of Fractals and Strange Attractors //Physica D.- 1983.-V.8D, No 3.- PP.435-444.

503. *Hirsch J.E., Huberman B.A. and Scalapino D.J.* Theory of Intermittency // Phys. Rev. A.- 1982.-V.25A, No 1.-PP.519-532.

504. *Hirsch J.E., Nauenberg M. and Scalapino D.J.* Intermittency in the Presence of Noise: a Renormalization Group Formulation // Phys. Lett.- 1982.-V.87A, No 8.- PP.391-393.

505. *Hoffer L.M., Chyba T.H. and Abraham N.B.* Spontaneous Pulsing, Period Doubling, and Quasi-Periodicity in a Unidirectional, Single-Mode, Inhomogeneously Broadened Ring Laser // J. Opt. Soc. Amer. B.- 1985.-V.2, No 1.-PP.102-107.

506. *Holmes P.* Nonlinear Oscillator with a Strange Attractor // Phil. Trans. Roy. Soc. London.- 1979.-V. 292, No 1394.-PP.419-448.

507. *Holmes P. J.* Averaging and Chaotic Motions in Forced Oscillations // SIAM J. Appl. Math.- 1980.-V. 38, No 1.-PP.65-80.

508. *Holmes P. and Whitley D.* On the Attracting Set for Duffing's Equation // Physica D.- 1983.-V. 7D, No 1-3.-PP.111-123.

509. *Holyst J. A., Zagorski A. and Sukiennicki A.* On the Possibility of Deterministic Chaos in a Driven Damped Heisenberg Chain // Phys. Status Solidi.- 1985.-B. 127, No 1.- S. 29-32.

510. *Hopf E.* Abzweigung einer periodischen Lösung von einer stationaren Lösung eines differential Systems: Ber. Math.- Phys. Sachsische Akademie der Wissenschaften. B. 94.- Leipzig, 1942.- S. 1-22.

511. *Hopf E.* A Mathematical Example Displaying the Features of Turbulence // Comm. Pure Appl. Math.- 1948.-V. 1.-PP.303-322.

512. *Hopf F. A., Kaplan D. L., Gibbs H. M. and Shoemaker R. L.* Bifurcations to Chaos in Optical Bistability // Phys. Rev.- 1982.-V. 25A, No 4.-PP.2172-2182.

513. *Hu B.* Introduction to Real-Space Renormalization-Group Methods in Critical and Chaotic Phenomena // Phys. Rep.- 1982.-V. 91, No 5.-PP.233-295.

514. *Hu B. and Mao J. M.* Period Doubling: Universality and Critical Point Order // Phys. Rev. A.- 1982.-V. 25A, No 6.-PP.3259-3261.

515. *Hu B. and Rudnick J.* Exact Solutions to the Feigenbaum Renormalization-Group Equations for Intermittency //Phys. Rev. Lett.- 1982.-V. 48, No 24.-PP.1645-1648.

516. *Hu B. and Satija I. I.* A Spectrum Of Universality Classes in Period Doubling and Period Tripling // Phys. Lett.- 1983.-V. 98A, No 4.-PP.143-146.

517. *Huberman B. A. and Crutchfield J. P.* Chaotic States of Anharmonic Systems in Periodic Fields // Phys. Rev. Lett.- 1979.-V. 43, No 23.-PP.1743-1747.

518. *Huberman B. A. and Rudnick J.* Scaling Behaviour of Chaotic Flows // Phys. Rev. Lett.- 1980.-V. 45, No 3.-PP.154-157.

519. *Huberman B. A. and Zisook A. B.* Power Spectra of Strange Attractors // Phys. Rev. Lett.- 1981.-V. 46, No 10.-PP.626-632.

520. *Hudson J. L., Hart M. and Marinko D.* An Experimental Study of Multiple Peak Periodic and Nonperiodic Oscillations in the Belousov-Zhabotinsky Reaction // J. Chem. Phys.- 1979.-V. 71, No 4.-PP.1601-1606.

521. *Ikeda K.* Multiple-Valued Stationary State and Its Instability of the Transmitted Light by a Ring Cavity System // Opt. Comm.- 1979.-V. 30, No 2.-PP.257-261.

522. *Ikeda K., Daido H. and Akimoto O.* Optical Turbulence: Chaotic Behaviour of

Transmitted Light from a Ring Cavity // Phys. Rev. Lett.- 1980.-V. 45, No 9.-PP.709-712.

523. *Ikeda K., Kondo K. and Akimoto O.* Successive Higher-Harmonic Bifurcations in Systems with Delayed Feedback // Phys. Rev. Lett.- 1982.-V. 49, No 20.-PP.1467-1470.

524. *Imada M.* Chaos Caused by the Soliton-Soliton Interaction // J. Phys. Soc. Japan.- 1983.-V. 52, No 6.-PP.1946-1956.

525. Instabilities in Active Optical Medium // J. Opt. Soc. Amer.- 1985.-V. B2, No 1.

526. *Ito H.* Successive Subharmonic Bifurcations and Chaos in a Nonlinear Mathieu Equation // Progr. Theor. Phys. Japan.- 1979.-V. 61, No 3.-PP.815-824.

527. *Ivanov D. V., Khanin Ya. I., Matorin I. I. and Pikovsky A. S.* Chaos in a Solid-State Laser with Periodically Modulated Losses // Phys. Lett.- 1982.-V. 89A, No 5.-PP.229-230.

528. *Izrailev F. M., Rabinovich M. I. and Ugodnikov A. D.* Approximate Description of Three-Dimensional Dissipative Systems // Phys. Lett.- 1981.-V. 86A, No 6, 7.-PP.321-325.

529. *Leffries C. and Perez J.* Direct Observation of Crises of the Chaotic Attractor in a Nonlinear Oscillator // Phys. Rev. A.- 1983.-V. 27, No 1.-PP.601-603.

530. *Jensen J. H., Christiansen P. L. and Scott A. C.* Chaos in the Beeler-Reuter System for the Action Potential of Ventricular Myocardial Fibres // Physica D.- 1984.-V. 13, No 1, 2.-PP.269-277.

531. *Jensen M. H., Bak P. and Bohr T.* Complete Devil's Staircase, Fractal Dimension, and Universality of Mode-Locking Structure in the Circle Map // Phys. Rev. Lett.- 1983.-V. 50, No 21.-PP.1637-1639.

532. *Jensen R. V. and Oberman C. R.* Calculation of the Statistical Properties of Strange Attractors // Phys. Rev. Lett.- 1981.-V. 46, No 24.-PP.1547-1550.

533. *Jones C. A., Weiss N. O. and Cattaneo F.* Nonlinear Dynamos a Complex Generalization of the Lorenz Equations // Physica D.- 1985.-V.14D, No 2.-PP.161-176.

534. *Kadanoff L. P.* Roads to Chaos // Physics Today.- 1983.-V. 36, No 12.-PP.46-53.

535. *Kai T.* Universality of Power Spectra of a Dynamical System with an Infinite

Sequence of Period-Doubling Bifurcations // Phys. Lett.- 1981.-V. 86A, No 5.- PP.263-266.

536. *Kai T. and Tomita K.* Statistical Mechanics of Deterministic Chaos // Progr. Theor. Phys. Japan.- 1980.-V. 64, No 5.-PP.1532-1550.

537. *Kaneko K.* Doubling of Torus // Progr. Theor. Phys. Japan.- 1983.-V. 69, No 6.- PP.1806-1810.

538. *Kaneko K.* Oscillation and doubling of Torus // Progr. Theor. Phys. Japan.- *1984.*-V. 72, No 2.-PP.202-215.

539. *Kaplan J. L and Yorke J. A.* Preturbulence: a Regime Observed in a Fluid Flow Model of Lorenz.- Research Supported by NSF Grant NCS76-24432, 1977.-PP.1-49.

540. *Kaplan J. L and Yorke J. A.* Chaotic Behaviour of Multi-Dimensional Difference Equations // Lect. Notes in Math.- 1979.- No 730.-PP.204-227.

541. *Kawai H. and Tye S.-H. H.* Approach to Chaos: Universal Quantitative Properties of One-Dimensional Maps // Phys. Rev. A.- 1984.-V. 30, No 4.-PP.2005-2023.

542. *Kawakami H.* The Bifurcation Pattern of Periodic Solutions Observed in Duffing's Equation // *Trudy IX Mezdunar. konf. po nelineynym kolebaniyam* (Kiev, 1981).- Kiev: Naukova dumka, 1984.-PP.162-165.

543. *Kerr W. C., Williams M. B, Bishop A. R., Fesser K., Lomdahl P. S. and Trullinger S. E.* Symmetry and Chaos in the Motion of the Damped Driven Pendulum // Z. Phys. B: Condensed Matter.- 1985.- B. 59, No 1.-S. 103-110.

544. *Khandokhin P. A. and Khanin Ya. I.* Instabilities in a Solid-State Ring Laser // J. Opt. Soc. Amer. B.- 1985.-V. 2, No 1.-PP.226-231.

545. *Klische W. and Weiss C. O.* Instabilities and Routes to Chaos in a Homogeneously Broadened One- and Two-Mode Ring Laser // Phys. Rev. A.- 1985.-V. 31, No 6.- PP.4049-4051.

546. *Koch B. P and Leven R. W.* Subharmonic and Homoclinic Bifurcations in a Parametrically Forced Pendulum // Physica D.- 1985.-V. 16D, No 1.-PP.1-13.

547. *Koch B. P., Leven R. W., Pompe B. and Wilke C.* Experimental Evidence for Chaotic Behaviour of a Parametrically Forced Pendulum // Phys. Lett.- 1983.-V. 96A, No 5.-PP.219-224.

548. *Kornev V. K and Semenov V. K.* Chaotic and Stochastic Phenomena in Superconducting Quantum Interferometers // IEEE Trans. Mag.- 1983.-V. 19, No 3.-PP.633-636.

549. *Kuramoto Y. and Tsuzuki T.* On the Formation of Dissipative Structures in Reaction-Diffusion Systems // Progr. Theor. Phys. Japan.- 1975.-V. 54, No 3.-PP.687-699.

550. *Kuramoto Y. and Yamada T.* Turbulent State in Chemical Reactions // Progr. Theor. Phys. Japan.- 1976.-V. 56, No 3.-PP.679-681.

551. *Lanford O. E.* An Introduction to the Lorenz System // Papers from the Duke Turb. Conf. (1976), Paper No 4.-PP.1-21.

552. *Lanford O. E.* Computer Pictures of the Lorenz Attractor // Lect. Notes in Math., No 615.- Berlin-Heidelberg-N. Y.: Springer, 1977.-PP.113-116.

553. *Lasota A. and Yorke J. A.* On the Existence of Invariant Measures for Piecewise Monotonic Transformations // Trans. Amer. Math. Soc.- 1973.-V. 186.-PP.481-488.

554. *Ledrappier F.* Some Relations Between Dimension and Lyapunov Exponents // Comm. Math. Phys.- 1981.-V. 81, No 2.-PP.229-238.

555. *Leipnik R. B.* Double Strange Attractors in Rigid Body Motion with Linear Feedback Control // Phys. Lett.- 1981.-V. 86A, No2.-PP.63-67.

556. *Leonov G. A., Abramovich S. M. and Bunin A. I.* Global Stability of the Lorenz System // Nonlinear and Turbulent Processes in Physics. V. 3.- N.-Y.: Gordon and Breach, Harwood Academic Publishers, 1984.

557. *Leray J.* Sur le mouvement d'un liquide visqueux emplissant l'espace // Acta Math.- 1934.-V. 63.-PP.193-248.

558. *Levandowsky M., Childress W. S., Spiegel E. A. and Hutner S. H.* A Mathematical Model of Pattern Formation by Swimming Microorganismos // J. Protozool.- 1975.-V. 22, No 2.-PP.296-306.

559. *Leven R. W. and Koch B. P.* Chaotic Behaviour of a Parametrically Excited Damped Pendulum // Phys. Lett.- 1981.-V.86A, No 2.-PP.71-74.

560. *Leven R. W., Pompe B., Wilke C. and Koch B. P.* Experiments on Periodic and

Chaotic Motions of a Parametrically Forced Pendulum // Physica D.- 1985.-V. 16, No 3.-PP.371-384.

561. *Libchaber A., Fauve S. and Laroche C.* Two-Parameter Study of the Routes to Chaos // Physica D.- 1983.-V. 7D, No 1-3.-PP.73-84.

562. *Linsay P. S.* Period Doubling and Chaotic Behaviour in a Driven Anharmonic Oscillator // Phys. Rev. Lett.- 1981.-V. 47, No 19.-PP.1349-1352.

563. *Lorenz E. N.* Deterministic Nonperiodic Flow // J. Atmos. Sci.- 1963.-V. 20, No 2.-PP.130-141.

564. *Lorenz E. N.* Noisy Periodicity and Reverse Bifurcation Sequences // Ann. N. Y. Acad. Sci.- 1980.-V. 357.-PP.282-291.

565. *Lotka A. J.* Undamped Oscillations Derived from the Law of Mass Action // J. Amer. Chem. Soc.- 1920.-V. 42, No 8.-PP.1595-1599.

566. *Lotka A. J.* Elements of Phisical Biology.- Baltimore, 1925.

567. *Lücke M.* Statistical Dynamics of the Lorenz Model // J. Stat. Phys.- 1976.-V. 13, No 6.-PP.455-475.

568. *Mackey M. C. and Glass L.* Oscillation and Chaos in Physiological Control Systems // Science.- 1977.-V. 197, No 4300.-PP.287-289.

569. *Malraison B., Atten P., Bergé P. and Dubois M.* Dimension of Strange Attractors: an Experimental Determination for the Chaotic Regime of Two Convective Systems // J. Physique Lettres.- 1983.- T. 44, No 22.-PP.897-902.

570. *Mandelbrot B. B.* Fractals: Form, Chance and Dimension.- San Francisco: Freeman Comp., 1977.

571. *Mañé R.* On the Dimension of the Compact Invariant Sets of Certain Non-Linear Maps // Lect. Notes in Math.- 1981.-V. 898.-PP.230-242.

572. *Manneville P. and Pomeau Y.* Intermittency and the Lorenz Model // Phys. Lett.- 1979.-V. 75A, No 1-2.-PP.1-2.

573. *Manneville P. and Pomeau Y.* Different Ways to Turbulence in Dissipative Dynamical Systems // Physica D.- 1980.-V. 1, No 2.-PP.219-226.

574. *Maschke E. K. and Saramito B.* On the Transition to Turbulence in Magneto-Hydrodynamic Models of Confined Plasmas // Physica Scripta.- 1982.-V. T2: 2.-PP.410-417.

575. *Matsumoto K. and Tsuda I.* Noise-induced Order // J. Stat. Phys.- 1983.-V. 31, No 1.-PP.87-108.

576. *May R. M.* Simple Mathematical Models with Very Complicated Dynamics // Nature.- 1976.-V. 261, No 6.-PP.459-467.

577. *Mayer-Kress G. and Haken H.* Intermittent Behaviour of the Logistic System // Phys. Lett.- 1981.-V. 82A, No 4.-PP.151-155.

578. *McGuinness M. J.* The Fractal Dimension of the Lorenz Attractor // Phys. Lett.- 1983.-V. 99A, No 1.-PP.5-9.

579. *McGuinness M. J.* A Computation of the Limit Capacity of the Lorenz Attractor // Physica D.- 1985.-V. 16D, No 2.-PP.265-275.

580. *McLaughlin J. B.* Stochastic Behaviour in Slightly Dissipative Systems // Phys. Rev. A.- 1979.-V. 20A, No 5.-PP.2114-2119.

581. *McLaughlin J. B.* Period-Doubling Bifurcations and Chaotic Motion for a Parametrically Forced Pendulum // J. Stat. Phys.- 1981.-V. 24, No 2.-PP.375-388.

582. *McLaughlin J. B. and Martin P. C.* Transition to Turbulence in a Statically Stressed Fluid System // Phys. Rev. A.- 1975.-V. 12A, No 1.-PP.186-203.

583. *Miles J.* Resonantly Forced Motion of Two Quadratically Coupled Oscillators // Physica D.- 1984.-V. 13D, No 1, 2.-PP.247-260.

584. *Milonni P. W., Ackerhalt J. R. and Galbraith H. W.* Chaos and Nonlinear Optics: a Chaotic Raman Attractor // Phys. Rev. A.- 1983.-V. 28A, No 2.-PP.887-891.

585. *Moon F. C. and Li G.-X.* The Fractal Dimension of the Two-Well Potential Strange Attractor // Physica D.- 1985.-V. 17D, No 1.-PP.99-108.

586. *Mori H.* Fractal Dimensions of Chaotic Flows of Autonomous Dissipative Systems // Progr. Theor. Phys. Japan.- 1980.- V. 63, No 3.-PP. 1044-1047.

587. *Mori H. and Fujisaka H.* Statistical Dynamics of Turbulence // Lect. Notes in Phys. V. 132.- Berlin: Springer, 1980.-PP. 181-197.

588. *Mors E. and Hedlund G. A.* Symbolic Dynamics. I, II // Amer. J. Math.-1938.-V. 60.-PP. 813-866; 1940.-V. 62.

589. *Myrberg P.J.* Iteration von Quadratwurzeloperationen // Ann. Acad. Sci. Fennica. Ser. A.- 1958.-V. 259.-PP. 1-10.

590. *Myrberg P.J.* Iteration der reellen Polynome Zweiten Grades // Ann. Acad. Sci. Fennica. Ser. A.- 1958.-V. 256.-PP. 1-10; 1959.-V. 268.-PP. 1-10; 1963.-V. 336.-PP. 1-10.

591. *Nakatsuka H., Asaka S., Itoh H., Ikeda K. and Matsuoka M.* Observation of Bifurcation to Chaos in an All-Optical Bistable System // Phys. Rev. Lett.- 1983.-V. 50, No 2.-PP. 109-112.

592. *Narducci L.M., Sadiky H., Lugiato L.A. and Abraham N.B.* Experimentally Accessible Periodic Pulsations of a Single-Mode Homogeneously Broadened Laser (the Lorenz Model) // Opt. Comm.- 1985.-V. 55, No 5.-PP. 370-376.

593. *Nauenberg M. and Rudnick J.* Universality and the Power Spectrum at the Onset of Chaos // Phys. Rev. B.- 1981.-V. 24 B, No 1.-PP. 493-495.

594. *Neyer A. and Voges E.* Dynamics of Electrooptic Bistable Devices with Delayed Feedback // IEEE J. Quant. Electr.- 1982.-V. QE-18, No 12.-PP. 2009-2015.

595. *Nicolis G.* Stability and Dissipative Structures in Open Systems Far from Equilibrium // Advances in Chemical Physics. V. 19./Ed. I. Prigogine, S. A. Rice.- N. Y.-L.-Sydney- Toronto: Wiley-Intersci., 1971.-PP. 209-324.

596. *Olsen L.F. and Degn H.* Chaos in an Enzyme Reaction // Nature.-1977.-V. 267, No 5607.-PP. 177-178.

597. *Ostlund S., Rand D., Sethna J. and Siggia E.* Universal Properties of the Transition from Quasi-Periodicity to Chaos in Dissipative Systems // Physica D.- 1983.-V. 8, No 3.-PP. 303-342.

598. *Ott E.* Strange Attractors and Chaotic Motions of Dynamical Systems // Rev. Mod. Phys.- 1981.-V. 53, No 4.-P. 1.-PP. 655-671.

599. *Ott E., Hanson J.D. and Antonsen T.M.* The Influence of Diffraction on Waves with Ergodic Ray Trajectories // Bull. Amer. Phys. Soc.- 1981.-V. 26, No 7.-PP. 1013.

600. *Packard N.H., Crutchfield J.P., Farmer J.D. and Shaw R.S.* Geometry from a Time Series // Phys. Rev. Lett.- 1980.-V. 45, No 9.-PP. 712-716.

601. *Percival I.C.* Regular and Irregular Spectra // J. Phys. B: Atom. and Molec. Phys.- 1973.-V. 6, No 9.-PP. L229-232.

602. *Percival I.C.* Semiclassical Theory of Bound States // Advances in Chemical

Phys. V. 36./Ed. I. Prigogine, S.A.Rice.-N.Y.London - Sydney -Toronto: Wiley-Interscience, 1977.-PP. 1-61.

603. *Perez J. and Jeffries C.* Direct Observation of a Tangent Bifurcation in a Nonlinear Oscillator // Phys. Lett.- 1982.-V. 92A, No 2.-PP. 82-84.

604. *Pesin Ya.B.* On the Notion of the Dimension with Respect to a Dynamical System // Ergod. Theory and Dyn. Systems.-1984.-V. 4, No 3.-PP. 405-420.

605. *Pikovski A.S.* A Dynamical Model for Periodic and Chaotic Oscillations in the Belousov-Zhabotinsky Reaction // Phys. Lett.- 1981.-V. 85A, No 1.-PP. 13-16.

606. *Pikovski A.S.* A New Type of Intermittent Transition to Chaos // J. Phys. A: Math. Gen.-1983.-V. 16.-PP. 109-112.

607. *Pikovski A.S. and Rabinovich M.I.* Stochastic Oscillations in Dissipative Systems // Physica D.- 1981.-V. 2, No 1.-PP. 8-24.

608. *Poincaré H.* Calcul des Probabilités.- Paris: Gautier-Villard, 1912.

609. *Pomeau Y. and Manneville P.* Intermittent Transition to Turbulence in Dissipative Dynamical Systems // Comm. Math. Phys.- 1980.-V. 74, No 2.-PP. 189-197.

610. *Procaccia I., Grassberger P. and Hentschel H.G.E.* On the Characterization of Chaotic Motions // Lect. Notes in Physics.- No 179.- Berlin: Springer, 1983.-PP. 212-221.

611. *Rand D., Ostlund S., Sethna J. and Siggia E.D.* Universal Transition from Quasiperiodicity to Chaos in Dissipative Systems // Phys. Rev. Lett.- 1982.-V. 49, No 2.-PP. 132-135.

612. *Renyi A.* Probability Theory.-Amsterdam: North-Holland, 1970.

613. *Reynolds O.* An Experimental Investigation on the Circumstances which Determine whether the Motion of Water shall be Direct or Sinuous, and of the Law of Resistance in Parallel Channels // Phil. Trans. Roy. Soc.-1883.-V.174.-PP. 935.

614. *Richardson L.F.* The Problem of Contiguity: an Appendix of Statistics of Deadly Quarrels // General Systems Yearbook.-1961.-V. 6.-PP. 139-187.

615. *Riela G.* Loss of Stability and Disappearance of Two-Dimensional Invariant Tori in a Dissipative Dynamical System // Phys. Lett.- 1982.-V. 91A, No 5.-PP. 203-204.

616. *Roux J.C.* Experimental Studies of Bifurcations Leading to Chaos in the

Belousoff-Zhabotinsky Reaction // Physica D.- 1983.-V. 7D, No 1-3.-PP. 57-68.

617. *Roux J.C., Simoyi R.H. and Swinney H.L.* Observation of a Strange Attractor // Physica D.- 1983.-V. 8, No 1-2.-PP. 257-266.

618. *Rössler O.E.* An Equation for Continuous Chaos // Phys. Lett.- 1976.-V. 57A, No 5.-PP. 397-398.

619. *Rössler O.E.* Chaotic Behavior in Simple Reaction Systems.-Z. Naturforsch.- 1976.-B. 31a, H. 3/4.-S. 259-264.

620. *Rössler O.E.* Chemical Turbulence: Chaos in a Simple Reaction-Diffusion System // Z. Naturforsch.-1976.-B. 31a, H. 10.-S. 1168-1172.

621. *Rössler O.E.* Chaos in Abstract Kinetics Two Prototypes.- Bull. of Math. Biol.- 1977.-V. 39, No 2.-PP. 275-289.

622. *Rössler O.E.* An Equation for Hyperchaos // Phys. Lett.- 1979.-V. 71A, No 2,3.-PP. 155-157.

623. *Rössler O.E. and Wegmann K.* Chaos in the Zhabotinsky Reaction // Nature.- 1978. -V. 271.-PP. 89-90.

624. *Ruelle D.* Some Comments on Chemical Oscillatioins // Trans. N. Y. Acad. Sci.- 1973.-V. 35, Ser. II, No 1.-PP. 66-71.

625. *Ruelle D.* The Lorenz Attractor and the Problem of Turbulence // Lect. Notes in Math. No 565.-Berlin: Springer, 1976.-PP. 146-158.

626. *Ruelle D.* Strange Attractors // Math. Intellengencer.-1980.-V. 2, No 3.-PP. 126-137.

627. *Ruelle D. and Takens F.* On the Nature of Turbulence // Comm. Math. Phys.- 1971.-V. 20, No 2.-PP. 167-192.

628. *Russel D.A., Hanson J.D. and Ott E.* Dimension of Strange Attractors // Phys. Rev. Lett.- 1980.-V. 45, No 14.-PP. 1175-1178.

629. *Sacker R.* Invariant Surfaces and Bifurcation of Periodic Solutions of Ordinary Differential Equations.-N.Y.: Courant Institute of Mathematical Sciences, 1964.

630. *Saltzman B..* Finite Amplitude Free Convection as an Initial Value Problem // J. Atmos. Sci.-1962.-V. 19, No 4.-PP. 329-341.

631. *Sano M. and Sawada Y.* Transition from Quasiperiodicity to Chaos in a System of

Coupled Nonlinear Oscillators // Phys. Lett.- 1983.-V. 97A, No 3.-PP. 73-76.

632. *Savage C.M. and Walls D.F.* Optical Chaos in Second-Harmonic Generation // Opt. Acta.-1983.-V. 30, No 5.-PP. 557-561.

633. *Scheffer V.* Géométrie Fractale de la Turbulence Équations de Navier-Stokes et Dimension de Hausdorff // C. R. Ac. Sci., Paris.-1976, PP. 121-122.

634. *Schmits R.A., Graziani K.R. and Hudson J.L.* Experimental Evidence of Chaotic States in the Belousov-Zhabotinsky Reaction // J. Chem. Phys.-1977.-V. 67, No 7.-PP. 3040-3044.

635. *Schmutz M. and Rueff M.* Bifurcation Schemes of the Lorenz Model // Physica D.-1984.-V. 11D, No 1,2.-PP. 167-178.

636. *Scholz H.J., Yamada T., Brand H. and Graham R.* Intermittency and Chaos in a Laser System with Modulated Inversion // Phys. Lett.- 1981.-V. 82A, No 7.-PP. 321-323.

637. *Schreiber I. and Marek M.* Transition to Chaos via Two-Torus in Coupled Reaction-Diffusion Cells // Phys. Lett.- 1982.-V. 91A, No 6.-PP. 263-267.

638. *Schreiber I. and Marek M.* Strange Attractors in Coupled Reaction-Diffusion Cells // Physica D.- 1982.-V. 5D, No 2+3.-PP. 258-272.

639. *Schulman J.N.* Chaos in Piecewise-Linear Systems // Phys. Rev. A.-1983.-V. 28, No 1.-PP. 477-479.

640. *Schulmeister Th. and Sel'kov E.E.* Folded Limit Cycles and Quasi-Stochastic Self-Oscillations in a Third-Order Model of an Open Biochemical System // Studia Biophysica.-1978.-B. 72, H.2.-S. 111-112.

641. *Seydel R.* Attractors of a Duffing Equation.-Dependence on the Exciting Frequency.-Physica D, 1985, V. 17D, No 3.-PP. 308-312.

642. *Shannon C.E.* A Mathematical Theory of Communication. P. I-III // Bell. System Techn. J.-1948.-V. 27, No 3.-P. 379-423; No 4.-PP. 623-656.

643. *Shaw R.* Strange Attractors, Chaotic Behavior, and Information Flow // Z. Naturforsch.-1981.-V. 36a, H.1-S. 80-112.

644. *Shenker S.J.* Scaling Behavior in a Map of a Circle onto Itself: Empirical Results // Physica D.- 1982.-V. 5, No 2+3.-PP. 405-411.

645. *Shepelyansky D.L.* Some Statistical Properties of Simple Classically Stochastic

Quantum Systems // Physica D.- 1983.-V. 8D, No 2.-PP. 208-222.

646. *Shibata A. and Saito N.* Time Delays and Chaos in Two Competing Species // Math. Biosciences.-1980.-V. 51, No 3/4.-PP. 199-211.

647. *Shimada I. and Nagascima T.* A Numerical Approach to Ergodic Problem of Dissipative Dynamical Systems // Progr. Theor. Phys. Japan.-1979.-V. 61, No 6.-PP. 1605-1616.

648. *Shimizu T.* Analytic Form of the Simplest Limit Cycle in the Lorenz Model // Physica A.-1979.-V. 97, No 2.-PP. 383-398.

649. *Shimizu T. and Morioka N.* Transient Behavior in Periodic Regions of the Lorenz Model // Phys. Lett.- 1978.-V. 69A, No 3.-PP. 148-150.

650. *Shimizu T. and Morioka N.* Period-Doubling Bifurcations in a Simple Model // Phys. Lett.- 1981.-V. 83A, No 6.-PP. 243-250.

651. *Shraiman B., Wayne C.E. and Martin P.C.* Scaling Theory for Noisy Period-Doubling Transition to Chaos // Phys. Rev. Lett.- 1981.-V. 46, No 14.-PP. 935-939.

652. *Simoyi R.H., Wolf A. and Swinney H.L.* One-Dimensional Dynamics in a Multicomponent Chemical Reaction // Phys. Rev. Lett.- 1982.-V. 49, No 4.-PP. 245-248.

653. *Smale S.* Dynamical Systems and Turbulence // Lect. Notes in Math. No 615.- Berlin: Springer, 1977.

654. *Smoes M.L.* Period of Homogeneous Oscillations in the Ferro-in-catalized Zhabotinskii System // J. Chem. Phys.-1979.-V. 71, No 11.-PP. 4669-4679.

655. *Sparrow C.* The Lorenz Equations: Bifurcations, Chaos and Strange Attractors // Appl. Math. Sci. V. 41.-Berlin: Springer, 1982.

656. *Swinney H.L.* Observations of Order and Chaos in Nonlinear Systems // Physica D.- 1983.-V. 7D, No 1-3.-PP. 3-15.

657. *Takens F.* Detecting Strange Attractors in Turbulence // Lect. Notes in Math., No 898.-Berlin-Heidelberg-N.Y.: Springer, 1981.-PP. 366-381.

658. *Testa J., Perez J. and Jeffries C.* Evidence for Uniersal Chaotic Behavior of a Driven Nonlinear Oscillator // Phys. Rev. Lett.- 1982.-V. 48, No 11.-PP. 714-717.

659. *Thompson J.M. and Graffari R.* Chaotic Dynamics of an Impact Oscillator //

Phys. Rev. A.-1983.-V. 27A, No 3.-PP. 1741-1743.

660. *Tomita K. and Tsuda I.* Chaos in the Belousov-Zhabotinsky Reaction in a Flow System // Phys. Lett.- 1979.-V. 71A, No 5,6.-PP. 489-492.

661. *Tomita K. and Tsuda I.* Towards the Interpretation of Hudson's Experiment on the Belousov-Zhabotinsky Reaction // Progr. Theor. Phys. Japan.-1980.-V. 64, No 4.-PP. 1138-1160.

662. *Tomita K. and Tsuda I.* Towards the Interpretation of the Global Bifurcation Structure of the Lorenz System // Progr. Theor. Phys. Suppl. (Japan).-1980, No 69.-PP. 185-199.

663. *Tresser C. and Coullet P.* Iterations d'endomorphismes et groupe de renormalisation // C. R. Acad. Sci. Paris.-1978.-T 287A, No 7.-PP. 577-584.

664. *Tresser C. and Coullet P.* Critical Transition to Stochasticity // Intrinsic Stochasticity in Plasmas/Ed. G. Laval, D. Grassilon.-Orsay: Les éditions de physique courtaboeuf, 1979.-PP. 365-372.

665. Turbulence and Chaotic Phenomena in Fluids / Ed. T. Tatsumi.-Amsterdam: North Holland, 1984.

666. *Turing A.M.* The Chemical Basis of Morphogenesis // Philos. Trans. Roy. Soc.-1952.-V. B237, No 641.-PP. 37-72.

667. *Turner J.S., Roux J.C., McCormick W.D. and Swinney H.L.* Alternating Periodic and Chaotic Regimes in a Chemical Reaction.-Experiment and Theory // Phys. Lett.- 1981.-V. 85A, No 1.-PP. 9-12.

668. *Tyson J.J.* On the Appearance of Chaos in a Model of the Belousov Reaction // J. Math. Biol.-1978.-V. 5, No 4.-PP. 351-362.

669. *Ueda Y. and Akamatsu N.* Chaotically Transitional Phenomena in the Forced Negative-Resistance Oscillator // IEEE Trans. on Circuits and Systems.-1981.-V. CAS-28, No 3.-PP. 217-223.

670. *Ulam S.M. and von Neumann J.* On Combination of Stochastic and Deterministic Processes // Bull. Amer. Math. Soc.-1947.-V. 53, No 11.-PP. 1120.

671. *Vlasova O.F. and Zaslavsky G.M.* Dissipation Effect on Chaos Onset in a Two-Resonance Overlap Case // Phys. Lett.- 1983.-V. 99A, No 9.-PP. 405-410.

672. *Walker G.H. and Ford J.* Amplitude Instability and Ergodic Behavior for Conservative Nonlinear Oscillator Systems // Phys. Rev.-1969.-V. 188, No 1.-PP. 416-432.

673. *Wegmann K. and Rössler O.E.* Different Kinds of Chaotic Oscillations in the Belousov-Zhabotinsky Reaction // Z. Naturforsch.-1978.-B. 33a, H.10.-S.1179-1183.

674. *Weiss C.O., Godone A. and Olafsson A.* Routes to Chaotic Emission in a CW He-Ne Laser // Phys. Rev. A.-1983.-V. 28A, No 2.-PP. 892-895.

675. *Weiss C.O. and King H.* Oscillation Period Doubling Chaos in a Laser // Opt. Commun.-1982.-V. 44, No 1.-PP. 59-61.

676. *Wersinger J.M., Finn J.M. and Ott E.* Bifurcation and «Strange» Behavior in Instability Saturation by Nonlinear Three-Wave Mode Coupling // Phys. Fluids.-1980.-V. 23, No 6.-PP. 1142-1154.

677. *Wersinger J.M., Finn J.M. and Ott E.* Bifurcations and Strange Behavior in Instability Saturation by Nonlinear Mode Coupling // Phys. Rev. Lett.- 1980.-V. 44, No 7.-PP. 453-456.

678. *Wigner E.P.* On the Quantum Correction for Thermodynamic Equilibrium // Phys. Rev.-1932.-V. 40, No 5.-PP. 749-759.

679. *Williams R.F.* The Structure of Lorenz Attractors // Lect. Notes in Math. No 615.-1977.-PP. 94-112.

680. *Wolf A. and Swift J.* Universal Power Spectra for the Reverse Bifurcations Sequence // Phys. Lett.- 1981.-V. 83A, No 5.-PP. 184-188.

681. *Wolf A., Swift J.B., Swinney H.L. and Vastano J.A.* Determining Lyapunov Exponents from a Time Series // Physica D.- 1985.-V. 16, No 3.-PP. 285-317.

682. *Yamada T.* Chaotic State in an Electronic Circuit Analysis of Gollub et al's Experiment // J. Phys. Soc. Japan.-1982.-V. 51, No 11.-PP. 3423-3430.

683. *Yamada T. and Graham R.* Chaos in a Laser System under a Modulated External Field // Phys. Rev. Lett.- 1980.-V. 45, No 16.-PP. 1322-1324.

684. *Yamada T. and Kuramoto Y.* A Reduced Model Showing Chemical Turbulence // Progr. Theor. Phys. Japan.-1976.-V. 56, No 2.-PP. 681-683.

685. *Yamazaki H., Oono Y. and Hirakawa K.* Experimental Study on Chemical

Turbulence // J. Phys. Soc. Japan.-1978.-V. 44, No 1.-P. 335-336; 1979.-V. 46, No 2.-PP. 721-728.

686. *Yorke J.A. and Yorke E.D.* Metastable Chaos: The Transition to Sustained Chaotic Behavior in the Lorenz Model // J. Stat. Phys.-1979.-V. 21, No 3.-PP. 263-277.

687. *Yoshimura K. and Watanabe S.* Chaotic Behavior of Nonlinear Evolution Equation with First Order Dispersion // J. Phys. Soc. Japan.-1982.-V. 51, No 9.-PP. 3028-3035.

688. *Young L.S.* Capacity of Attractors // Ergod. Theory and Dyn. Systems.-1981.-V. 1, P. 3.-PP. 381-388.

689. *Young L.S.* Dimension, Entropy and Lyapunov Exponents // Ergod. Theory and Dyn. Systems.-1982.-V. 2, P. 1.-PP. 109-124.

690. *Zaslavsky G.M.* The Simplest Case of a Strange Attractor // Phys. Lett.- 1978.-V. 69A, No 3.-PP. 145-147.

691. *Zaslavsky G.M.* Stochasticity in Quantum Systems // Phys. Rep.-1981.-V. 80, No 3.-PP. 159-250.

692. *Zeghlache H. and Mandel P.* Influence of Detuning on the Properties of Laser Equations // J. Opt. Soc. Amer.-1985.-V. B2, No 1.-PP. 18-22.

693. *Zisook A.B.* Universal Effects of Dissipation in Two-Dimensional Mappings // Phys. Rev. A.-1981.-V. 24, No 3.-PP. 1640-1642.

694. *Mira C.* Chaotic Dynamics // World Scientific.

695. *Schuster H.G.* Deterministic Chaos // Phys.-Verlag, Weinheim, 1984.

696. Chaos / Ed. A.V. Holden-Manchester University Press, 1986.

697. *Leonov N.N. and Neimark Yu.I.* On Steady-State Regimes of Operation of a Class of Control Systems // *Avtomatika i telemekhanika*.-1974.-No 5.-PP. 23-31 (*Russian*).

698. *Halsey T.C., Jensen M.H., Kadanoff L.P., Procaccia I. and Shraiman B.I.* Fractal Measures and their Singularities: The Characterization of Strange Sets // Phys. Rev. A.-1986.-V. 33, No 2.-PP. 1141-1151.

699. *Broomhead D.S. and King G.P.* Extracting Qualitative Dynamics from Experimental Data // Physica D.- 1986.-V. 20, Nos. 2,3.-PP. 217-236.

700. Dimensions and Entropies in Chaotic Systems.-Springer-Verlag. Berlin-

Heidelberg, 1986.

701. *Goldberger A.L. et al.* Nonlinear dynamics in sudden cardiac death syndrome-heart-rate oscillations and bifurcations // Experimentia, 1988.-V. 44, No 11.-PP. 983-987.

702. *Genkin M.D., Dobrovolsky V.A. and Perminov S.M.* Analysis of Large-Scale Flow Structures of a Viscous Incompressible Fluid in a Parallel-Plate Channel.- Preprint No 155 *IOF AN SSSR*, Moscow, 1986 (*Russian*).

703. Casati G. On the properties of quantum systems with the classical chaotic limit // Problems in Non-Linear and Turbulent Processes in Physics. P. 2. *Tr. 11 Mezhd. Nab. Gruppy* / Eds. A. S. Davydov and V. M. Chernousenko - Kiev: Naukova Dumka, 1985. - PP. 50-56 (*Russian*).

INDEX

adapted basis 36
amplitudinal chaos 395
anharmonic oscillator 5
aperiodicity 20
auxiliary mapping 136

Belousov-Zhabotinsky reaction 387
bifurcation point 103, 104, 106
bifurcation value 103
billiard 85
billiard problem 85
binary storage 43
Brusselator 384
Burgers equation 30
butterfly effect 75

capacity 255
catastrophes 177
chaotic attractor 175
chaotic behaviour 84
chaotic motion 20
conservative oscillator 5
conservative rotor 6
contractive mapping 136
correlation dimension 255
critical power index 266

dangerous 181
diffusion equation 26
dispersion 28
dispersion equation 27
dual representation (of a system) 53
Duffing equation 11

embedding dimension 263
entropy 85
envelope velocity 28
excitatory neuron 373

Feigenbaum universality 189
Fibonacci numbers 290
figure eight 166
flicker noise 303
flip-flop 42, 43
fractal dimension 255

general contraction 44
generalized dimension 257
generators with inertial non-linearity 341
gyrostat 323

Hamiltonian system 18
hard 181
harmonic oscillator 2
Hausdorff dimension 256
Hausdorff measure 255
Heisenberg representation 434
Hill equation 13
homoclinic structure 44, 89
hyperbolicity 87
hyperchaos 399
hysteresis effect 12

inertial intercoupling 322
information dimension 254, 255
inhibitory neuron 373
intermittency 24, 276

KAM theory 18
Korteweg-de Vries equation 30

linear oscillator 3
Liouville operator 433
local instability 44
Lorenz attractor 154
Lorenz equations 15
Lyapunov dimension 258
Lyapunov exponents 252

macrodescription 79
manifolds S- and S+ 149
Markov partition 246
Mathieu equation 13
matryoshka 190
metric entropy 254, 265
microdescription 79
Morse-Smale system 87

Navier-Stokes equation 39
negative friction 4
negative resistance 4
nested structure 178
noise generator 401
non-autonomous oscillator 11
non-autonomous rotator 11
non-equilibrium phase transition 266
non-isochronal oscillator 5
normal form 99

omega-explosion 107
one-parametric bifurcation 104

Oregonator   384, 387
oscillator   2, 3, 4, 5, 10, 11, 17, 181

partial model   37
Poincaré loop   145, 150
Poincaré rotation number   184
pointwise dimension   255
Poiseulle flow   381
Poisson stability   20
predator-prey pair   400

quasi-attractor   282
quasi-random sequence   79
quasi-strange attractor   282

rank of a synchronism   222
rapidly rotating phase
resonant term   103
Reynolds number   94
Riemann waves 29
Riemann wave equation 29
rotator   10, 11

Schrödinger representation   434
screw-type chaos   397
shumotron   401
Smale horseshoe   88, 153
soft oscillation 181
spiral chaos   395
stadium billiard 86
standard mapping   439
stationary wave 28
stochastic synchronism   169, 171, 220
stochasticity amplifier   61, 69
strange attractor   20, 95
structurally stable system   86
switch 43
symbolic dynamics   144
symbolic notation   234
synchronism   169
synchronization 50-52
synchronization threshold   264, 353
systemes grossiers   63, 86

tangential bifurcation   276
topological Markov chain   85
transition from the negative to the positive   143

van der Pol oscillator   10

weakly interacting harmonic oscillators   17
well-stirredness 85
Wigner density function   434

MIX
Papier aus verantwortungsvollen Quellen
Paper from responsible sources
FSC® C105338

If you have any concerns about our products,
you can contact us on
ProductSafety@springernature.com

In case Publisher is established outside the EU,
the EU authorized representative is:
**Springer Nature Customer Service Center GmbH
Europaplatz 3, 69115 Heidelberg, Germany**

Printed by Libri Plureos GmbH
in Hamburg, Germany